Technician's Guide to Programmable Controllers

6th Edition

Terry R. Borden

Richard A. Cox

Australia • Brazil • Japan • Korea • Mexico • Singapore • Spain • United Kingdom • United States

Technician's Guide to Programmable Controllers, Sixth Edition
Terry R. Borden, Richard A. Cox

Vice President, Editorial: Dave Garza
Director of Learning Solutions: Sandy Clark
Acquisitions Editor: Stacy Masucci
Managing Editor: Larry Main
Senior Product Manager: John Fisher
Vice President, Marketing: Jennifer Baker
Marketing Director: Deborah Yarnell
Senior Marketing Manager: Erin Brennan
Marketing Coordinator: Jillian Borden
Senior Production Director: Wendy Troeger
Production Manager: Mark Bernard
Content Project Manager: David Barnes
Senior Art Director: David Arsenault
Technology Project Manager: Joe Pliss
Cover Image: Dainis Derics/www.shutterstock.com

Library of Congress Control Number: 2011937980

ISBN-13: 978-1-111-54409-6

ISBN-10: 1-111-54409-3

Delmar
Executive Woods
5 Maxwell Drive
Clifton Park, NY 12065
USA

Cengage Learning is a leading provider of customized learning solutions with office locations around the globe, including Singapore, the United Kingdom, Australia, Mexico, Brazil, and Japan. Locate your local office at **www.cengage.com/global**

Cengage Learning products are represented in Canada by Nelson Education, Ltd.

To learn more about Delmar, visit **www.cengage.com/delmar**

Purchase any of our products at your local bookstore or at our preferred online store **www.cengagebrain.com**

Printed in the United States of America
2 3 4 5 6 7 17 16 15 14

TABLE OF CONTENTS

Chapter 1

Chapter 2

Chapter 3

Chapter 4

Chapter 5

Chapter 6

Chapter 7

Chapter 8

Chapter 9

Chapter 10

Chapter 11

Chapter 12

Chapter 13

Chapter 14

Chapter 15

Chapter 16

Chapter 17

Chapter 18

Chapter 19

Chapter 20

Chapter 21

Chapter 22

PREFACE

The programmable logic controller, first introduced in 1969, has become an unqualified success. This computer-based device has become the industry standard, replacing the hard-wired electromechanical devices and circuits that had controlled the process machines and driven equipment of industry in the past.

Programmable logic controllers, or **PLCs** as they are referred to, vary in size and sophistication. When PLCs were first introduced, they typically used a dedicated programming device for entering and monitoring the PLC program. The programming device could only be used for programming a specific brand of PLC. These dedicated programmers, while user friendly, were very expensive and could not be used for anything except programming a PLC. With the increased use of the personal computer, software was designed that allowed a personal computer to be used for PLC programming. While dedicated programming devices are still available, the most common programming device used today is a personal computer running Windows®-based programming software.

Many electricians and/or technicians seem apprehensive about PLCs and their application in industry. One of the purposes of this text is to explain PLC basics using a plain, easy-to-understand approach, so that electricians and technicians with no PLC experience will be more comfortable with their first exposure to programmable logic controllers.

Half the battle of understanding any programmable logic controller is to first understand the terminology of the PLC field. This text covers terminology, as well as explaining the input/output section, processor unit, programming devices, memory organization, and much more.

A chapter has been included to explain not only ladder diagrams but also Relay Ladder Logic, which is the programming language used in the majority of programmable controllers today.

Examples of basic programming techniques used with typical PLCs are discussed and illustrated, as are the commonly used commands and functions. There are a variety of PLCs on the market today, and it would be impossible to write a book that explains how they all work and are programmed. Instead, this book is intended to discuss PLCs in a somewhat general, or generic, sense and to cover the basic concepts of operation that are common to all. Many of the examples used in this text are based on the Allen-Bradley PLC-5, SLC 500, MicroLogix, and Logix5000 family of PLCs.

New to This Edition: Programming instructions along with examples of the Allen-Bradley Logic5000 controllers has have been included throughout this edition. This will help the reader gain an understanding of tag-based memory as well as the additional flexibility that comes with Logix5000 controllers. Two new chapters have been added to this edition, "Function Block Diagram and Structured Text Programming" and "Sequential Function Chart Programming." These programming languages are gaining in popularity with many programmers and manufacturers. Function block diagram programming is typically found in process control applications in which there is more data handling and calculations. Structured text programming is best used for complex mathematical operations or specialized array/table loop processing. Sequential function chart programming is most often used with an application that is sequential in nature and has definable steps that would typically be executed in sequential order.

Although this text only scratches the surface of available commands and other advanced programming capabilities of the PLCs on the market today, the reader will gain a basic understanding of the most commonly used commands and how they work. A chapter, "PLC Programming Examples," will provide the reader with examples of PLC programming code/logic utilizing many of the instructions covered in previous chapters. The examples are intended to help the reader gain a better understanding of the various PLC instructions and how they can be combined to provide simple control logic solutions.

As with any new skill, a firm base of understanding is required before an electrician or technician can become proficient. After completing the text, the reader will possess a good foundation upon which additional PLC skills and understanding can be built.

The best teacher, of course, is experience, and the only way to really understand any given PLC is to work with that PLC. If a PLC is not available, the next best thing is a workshop or seminar sponsored by a local PLC distributor. If a workshop or seminar is not available, obtain as much literature and other information as possible from a local electrical distributor or PLC representative.

The PLC manufacturers are reducing prices as well as adding new features and program capabilities every day. With the rapid advancements in PLCs, the electrician without an electronics background need not feel intimidated. The manufacturers are doing everything possible to make the PLC easy to install, program, troubleshoot, and maintain.

STUDENT SITE

A new chapter, "23 Alarm and Event Programming," is located on a Student Companion site.

Accessing a Student Companion site from CengageBrain:

1. GO TO http://www.cengagebrain.com
2. TYPE Technician's Guide to Programmable Controllers or 9781111544096 in the search window.
3. Locate the desired product and CLICK on the **Access** button for free study tools.
4. CLICK on **Student Downloads** under Book Resources to gain access to the new chapter.

INSTRUCTOR SITE

An Instruction Companion Website containing supplementary material is available. This site contains an Instructor Manual with answers to the chapter review questions, testbank, image gallery of text figures, and chapter presentations done in PowerPoint. Contact Delmar Cengage Learning or your local sales representative to obtain an instructor account.

Accessing an Instructor Companion Website from SSO Front Door

1. GO TO http://login.cengage.com and log in using the Instructor email address and password.
2. ENTER author, title, or ISBN in **the Add a title to your bookshelf** search box, and CLICK on **Search** button.
3. CLICK **Add to My Bookshelf** to add Instructor Resources.
4. At the Product page, click on the **Instructor Companion site** link.

NEW USERS

If you're new to Cengage.com and do not have a password, contact your sales representative.

About the Authors

Richard A. Cox is the executive director of COXCO Training and Consulting in Spokane, Washington, and is a retired member of the Electrical/Robotics Department at Spokane Community College. He holds a Bachelor of Science degree from the University of New York and a Master of Science degree from Eastern Washington University, and is also a retired member of the International Brotherhood of Electrical Workers, Local 73.

Terry R. Borden is the Manager of Hydro Production for Pend Oreille County PUD and is a former member of the Electrical Maintenance and Automation Department at Spokane Community College in Spokane, Washington. He holds a degree in Industrial Automation and Robotics. He has worked as a control engineer, and was a partner in Applied Solutions, LLC.

The authors wish to acknowledge the cooperation and assistance of the different manufacturers whose product information and photographs are used throughout the text to illustrate PLC concepts and components. All these manufacturers are leaders in the PLC field, and each offers a full line of programmable logic controllers. It is not a question of which PLC is best, but rather which PLC best fits your needs and individual applications.

CHAPTER 1 What is a Programmable Logic Controller (PLC)?

Objectives

After completing this chapter, you should have the knowledge to:

- Describe several advantages of a programmable logic controller (PLC) over hardwired relay systems.
- Identify the four major components of a typical programmable logic controller and describe the function of each.
- Define the term *discrete*.
- Define the term *analog*.
- Identify different types of programming devices.

A programmable logic controller is a solid-state system designed to perform the logic functions previously accomplished by components such as electromechanical relays, drum switches, mechanical timers/counters, etc., for the control and operation of manufacturing process equipment and machinery.

Even though the electromechanical relay (control relays, pneumatic timer relays, etc.) has served well for many generations, often, under adverse conditions, the ever-increasing sophistication and complexity of modern processing equipment requires faster acting, more reliable control functions than electromechanical relays and/or timing devices can offer. Relays have to be hardwired to perform a specific function, and when the system requirements change, the relay wiring has to be changed or modified. In extreme cases, such as in the auto industry, complete control panels had to be replaced since it was not economically feasible to rewire the old panels with each model changeover.

It was, in fact, the requirements of the auto industry and other highly specialized, high-speed manufacturing processes that created a demand for smaller, faster acting, and more reliable control devices. The electrical/electronics industry responded with modular-designed, solid-state electronic devices. These early devices, while offering solid-state reliability, lower power consumption, expandability, and elimination of much of the hardwiring also brought with them a new language. The language consisted of AND gates, OR gates, NOT gates, OFF RETURN MEMORY, J-K flip flops, and so on.

What happened to simple relay logic and ladder diagrams? That is the question the plant engineers and maintenance electricians/technicians asked the solid-state device manufacturers. The reluctance of the end user to learn a new language and the advent of the microprocessor gave the industry what is now known as the **programmable logic controller (PLC).** The first programmable logic controller was invented in 1969 by Richard (Dick) E. Morley, who was the founder of the Modicon Corporation.

Internally there are still AND gates, OR gates, and so forth in the processor, but the design engineers have preprogrammed the PLC so that programs can be entered using RELAY LADDER LOGIC. While RELAY LADDER LOGIC may not have the mystique of other computer languages such as PASCAL, FORTRAN, and C++, it is a high-level, real-world, graphic language that is understood by most electricians and technicians. RELAY LADDER LOGIC programming is the most common programming language used today but other programming languages such as Sequential Function Chart, Structured Text, and Function Block languages can also be found.

The National Electrical Manufacturing Association (NEMA) defines a programmable controller as follows:

> A programmable controller is a digital electronic apparatus with a programmable memory for storing instructions to implement specific functions, such as logic, sequencing, timing, counting, and arithmetic to control machines and processes.

What does a PLC consist of, and how is it different from a computer system? The PLC consists of a programming device (computer or handheld programmer), processor unit, power supply, and an input/output (I/O) **interface** such as the computer system illustrated in Figure 1–1. And while there are similarities, there are also some major differences.

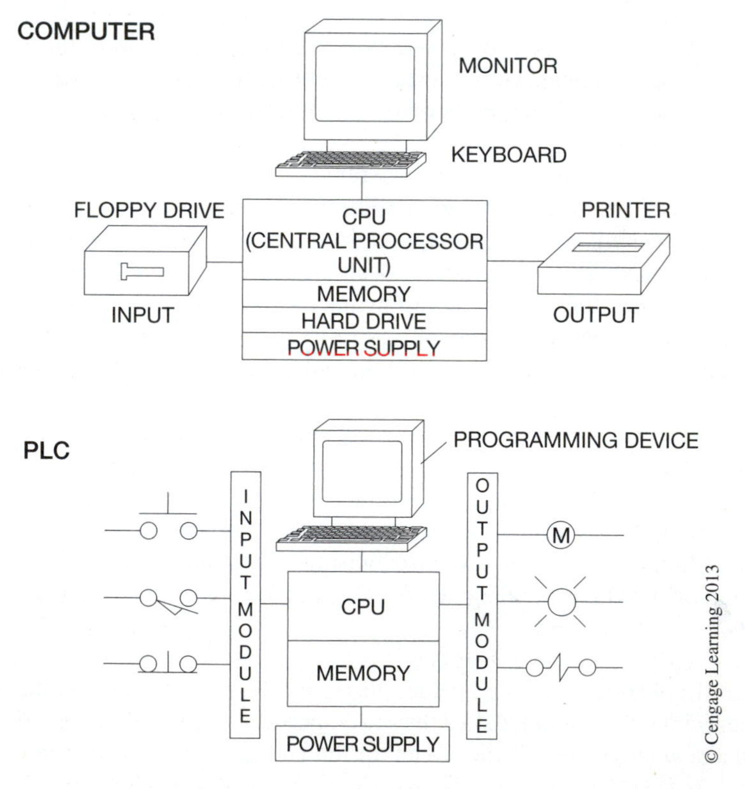

Figure 1–1 Comparison of a Computer System and a PLC

Note: *An interface occurs when two systems come together and interact, or communicate. In the case of the PLC, the communication or interaction is between the inputs (limit switches, push buttons, sensors, and the like), outputs (coils, solenoids, lights, and so forth), and the processor. This interface happens when any input or output voltage (AC or DC) or current signal is changed to or from a low-voltage DC signal that the processor uses internally for the decision-making process.*

PLCs are designed to be operated by plant engineers and maintenance personnel with limited knowledge of computers. Like the computer, which has an internal memory for its operation and storage of a program, the PLC also has a memory for storing the user program, or LOGIC, as well as a memory for controlling the operation of a process machine or driven equipment. But unlike the computer, the PLC is typically programmed in RELAY LADDER LOGIC, *not* one of the computer languages. It should be stated, however, that some PLCs will use other forms of PLC language, such as Structured Text, Sequential Function Chart, and Function Block to program the PLC. A brief description of Structured Text, Sequential Function Chart, and Function Block programming will be provided in later chapters.

The PLC is also designed to operate in the industrial environment with wide ranges of ambient temperature, vibration, and humidity, and is not usually affected by the electrical noise that is inherent in most industrial locations.

Note: *Electrical noise is discussed in Chapter 2.*

Maybe one of the biggest, or at least most significant, differences between the PLC and a computer is that PLCs have been designed for installation and maintenance by plant electricians who are not required to be highly skilled electronics technicians. Troubleshooting is simplified in most PLCs because they include fault indicators, blown-fuse indicators, input and output status indicators, and written fault information that can be displayed on the programmer.

Although the PLC and the personal computer are different in many ways, the personal computer is often used for programming and monitoring the PLC. Using personal computers in conjunction with PLCs will be discussed in later chapters.

A typical PLC can be divided into four components. These components consist of the **processor unit,** the **power supply,** the **input/output section** (interface), and the **programming device.**

The processor unit houses the processor, which is the decision-maker, or "brain" of the system. The brain is a microprocessor-based system that replaces control relays, counters, timers, sequencers, and so forth, and is designed so that the user can enter the desired program in RELAY LADDER LOGIC. The processor then makes all the decisions necessary to carry out the user program, based on the status of the inputs and outputs for control of a machine or process. It can also perform arithmetic functions, data manipulation, and communications between the local I/O section, remotely located I/O sections, and/or other networked PLC systems. Figure 1–2 shows Allen-Bradley SLC-5/05 and PLC-5/20C, and LOGIX 5550 processor units.

Note: *Some manufacturers refer to the processor as a* **CPU** *or central processing unit.*

The power supply is necessary to convert 120 or 240 volts AC voltages to the low voltage DC required for the logic circuits of the processor, and for the internal power required for the I/O modules.

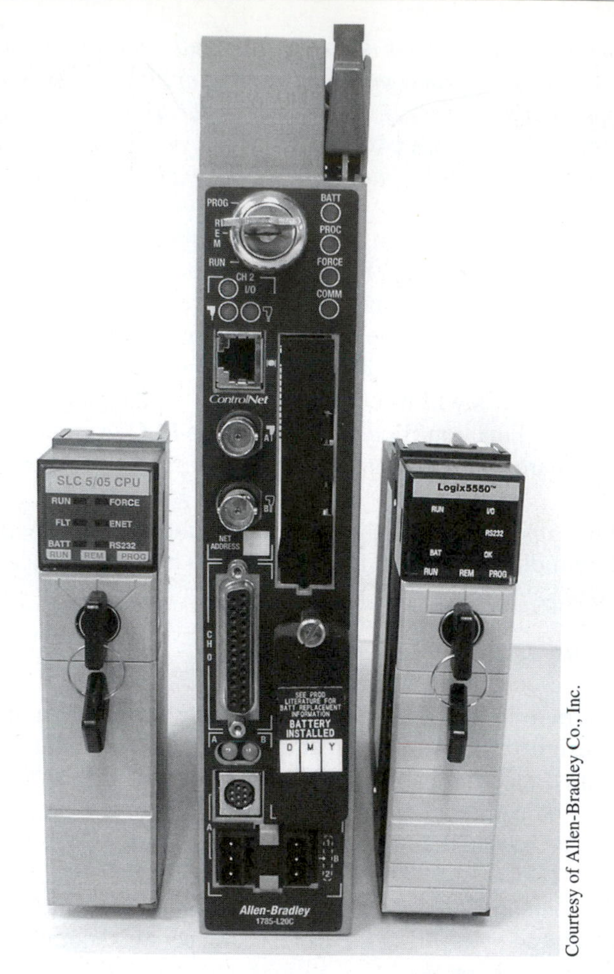

Courtesy of Allen-Bradley Co., Inc.

Figure 1–2 Allen-Bradley SLC-5/05 and PLC-5/20C, and LOGIX 5550 Processor Units

The power supply can be a separate unit as, shown in Figure 1–3, one of modular design that plugs into the processor rack, as shown in Figure 1–4, or, depending on the manufacturer, one that is an integral part of the processor.

Note: *The power supply does not supply power for the actual input or output devices themselves; it only provides the power needed for the internal circuitry of the input and output modules. DC power for the input and output devices, if required, must be provided from a separate source.*

The power supply can be broken down into four basic parts as shown in Figure 1–5. The first block, or section, of the power supply consists of a step-down transformer. The step-down transformer

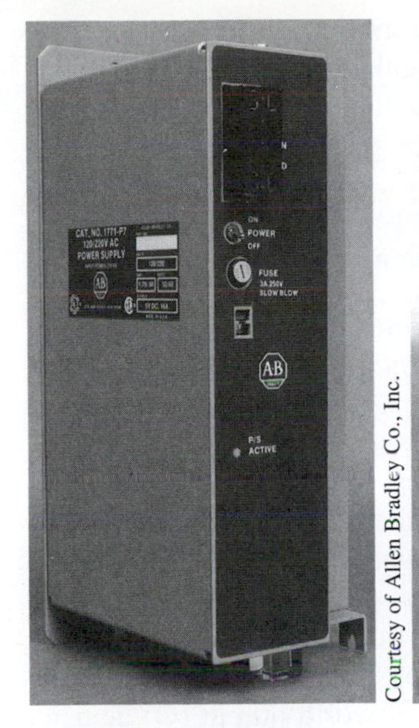

Courtesy of Allen Bradley Co., Inc.

Figure 1–3 Allen-Bradley Separate Power Supply

Courtesy of GE Fanuc Automation

Figure 1–4 Modular (Plug-in) Power Supply

reduces the voltage level of the incoming AC power. Many power supplies use a step-down transformer that is also a constant voltage transformer. A constant voltage transformer maintains a constant output voltage, even if the incoming power is fluctuating. The second portion of the power supply is the rectifier section, and contains the full wave bridge rectifier(s) to convert the AC sine wave from the secondary of the transformer to a pulsating DC voltage (shown by the wave form in Figure 1–5). The pulsating DC voltage must be further conditioned before it can be used by the processor and I/O modules. The third section of the power supply, the filter section, uses filter

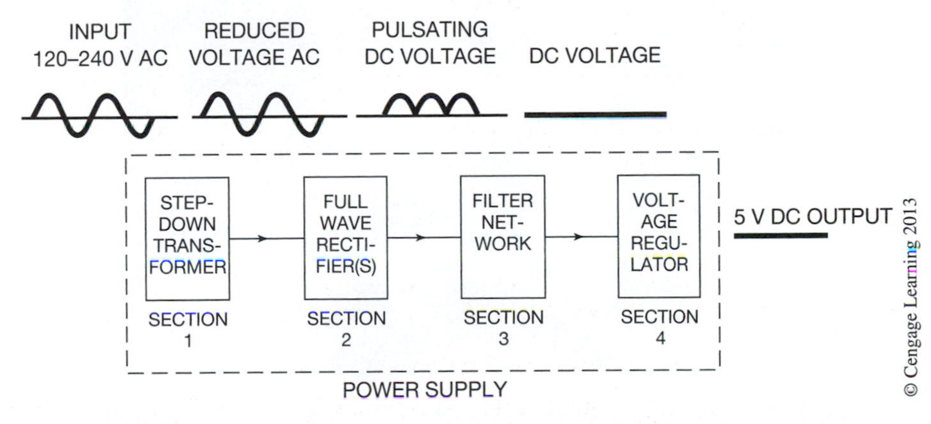

Figure 1–5 Block Diagram of a Typical Power Supply

devices and/or networks to filter and smooth the DC voltage coming from the rectifier section. The final section of the power supply consists of a voltage regulator. The regulator's function is to maintain a constant DC output voltage, even if the incoming AC voltage fluctuates or varies due to load changes or line disturbances.

The size or amperage rating of the power supply is based on the size, number, and type of I/O modules that are to be used. Power supplies are normally available with output current ratings of 3–20 amps.

Note: *Consider future needs and the possibility of expansion when initially sizing the power supply. It is cheaper in the long run to install a larger power supply initially than to try to add additional capacity at a later date.*

The input/output section consists of input modules and output modules. The number of input and output modules necessary is dictated by the requirements of the equipment that is to be controlled by a PLC. Figure 1–6 shows an input/output section. Modules are "plugged in" or added as required.

Input and output modules, referred to as the I/O (I for input and O for output), are where the real-world devices are connected. The real-world input (I) devices can be push buttons, limit switches, analog sensors, pressure switches, selector switches, etc., while the real-world output (O) devices can be hardwired motor starter coils, solenoid valves, indicator lights, positioning valves, and the like. The term *real world* is used to distinguish actual devices that exist and must be physically wired from the internal functions of the PLC system that duplicate the function of relays, timers, counters, and so on, even though none physically exists. This may seem a bit strange and hard to

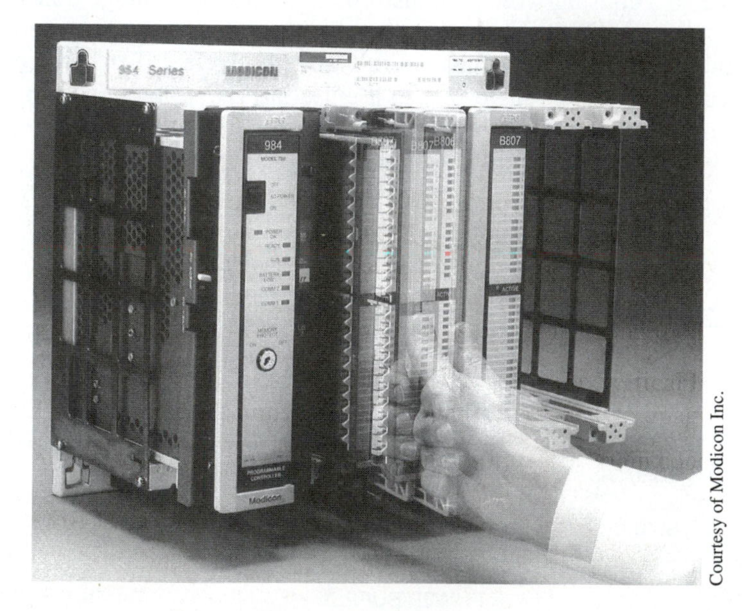

Courtesy of Modicon Inc.

Figure 1–6 Inserting a 32-Point Input Module into a Modicon I/O Rack

understand at this point, but the distinction between what the processor can do internally—which eliminates the need for all the previously used control relays, timers, counters, and so forth—will be graphically shown and readily understandable later in the text.

Real-world input and output devices are of two types: discrete and analog. Discrete I/O devices are either ON or OFF, open or closed, while analog devices have a range of possible values. Examples of discrete devices are limit switches, push buttons, motor starter coils, and indicator lamps. Examples of analog devices are pressure sensors, temperature probes, panel meters, variable speed drive signals, and modulating valves. When reference is made to an I/O device, the terms *discrete input device, discrete output device, analog input device,* and *analog output device* are commonly used to describe the type of device.

A reference was made earlier in this chapter to the I/O section as an interface. Although not a common reference, it is an accurate one. The I/O section contains the circuitry necessary to convert input voltages of 120–240 V AC or 0–24 V DC, etc., from *discrete input devices* to low-level DC voltages (typically 5 V) that the processor uses internally to represent the status or condition (*ON* or *OFF*). Similarly, the I/O section changes low-level DC signals from the processor to 120–240 V AC or DC voltages required to operate the *discrete output devices*. The I/O section also converts varying voltage or current signals from *analog input devices* into corresponding decimal values by way of an Analog-to-Digital converter (ADC). This same process, but reversed via a Digital-to-Analog converter (DAC), is used by the I/O section to convert decimal values into corresponding voltage or current signals necessary to operate *analog output devices*. The field signals from both digital and analog devices are normally isolated from the low-level logic circuitry of the processor by means of optical coupling. This is a brief overview of the I/O section and its function. How input and output devices are wired to I/O modules, optical coupling, and more information about the module circuitry itself is covered in Chapter 2.

The programming device is used to enter the desired program or sequence of operation into the PLC memory. The program is entered using RELAY LADDER LOGIC, or one of the other PLC programming languages, and it is this program that determines the sequence of operation and ultimate control of the process equipment or driven machinery. The programming device can be one of two types: hand-held or personal computer. The personal computer, or PC, is the most common programming device used today. The dedicated hand-held programmer (Figure 1–7) was once very popular but has been largely replaced with portable personal computers.

A personal computer (PC) is used to program most of the PLCs on the market today. The PLC programming software that is installed on the PC and a communications cable is sometimes all that is required to program the PLCs. At other times special hardware keys and/or communication cards are required to be installed on the PC for it to work successfully as a programming device. The PC provides the benefit of a large viewing screen that allows more of the program to be viewed at one time and makes troubleshooting and memory access much easier. It also provides program storage, as well as runs all the various software packages we have come to depend on today, such as spreadsheets, word processing, and graphics. Figure 1–8 shows a laptop personal computer that, with the appropriate software, is used to program and monitor a programmable logic controller.

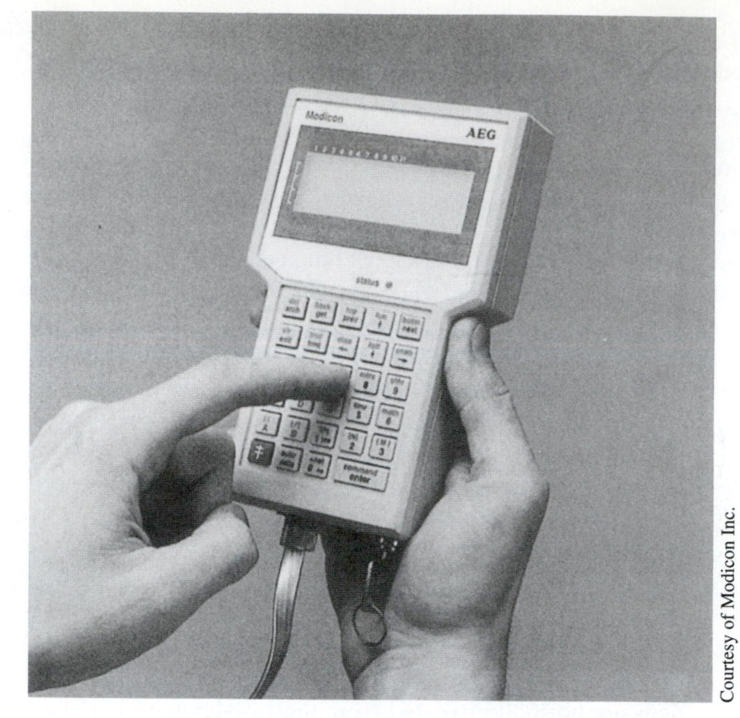

Courtesy of Modicon Inc.

Figure 1–7 Hand-Held Programmer

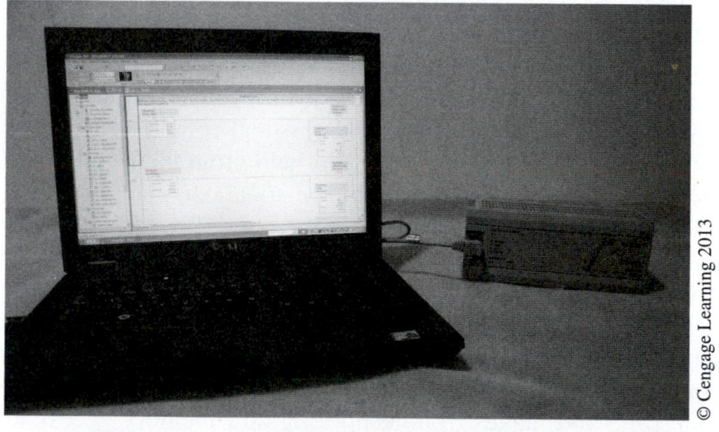

Figure 1–8 Laptop Computer Connected to a AB MicroLogix 1000

Chapter Summary

Programmable logic controllers (PLCs) have made it possible to precisely control large process machines and driven equipment with less physical wiring and lower installation costs than is required with standard electromechanical relays, pneumatic timers, drum switches, and so on. The programmability allows for fast and easy changes in the RELAY LADDER LOGIC to meet the changing needs of the process or driven equipment without the need for expensive and time-consuming rewiring. Designed to be "technician friendly," the modern PLC is easier to program and can be used by plant engineers and maintenance technicians who have little or no electronic background.

Review Questions

1. List the four main components of a programmable logic controller.
2. Define the term *interface*.
3. Define the term *real world*.
4. Define the term *discrete*.
5. Explain the following initials or acronyms:

DC	ADC	AC
CPU	DAC	PC
PLC	NEMA	I/O

6. Define the term *analog*.
7. List the two types or styles of programming devices.
8. RELAY LADDER LOGIC is a high-level graphic computer language.
   T F
9. What is the major advantage of a PLC system over the traditional hardwired control system?
10. Draw a block diagram and label the main components of a typical DC power supply.

CHAPTER 2 Understanding the Input/Output (I/O) Section

Objectives

After completing this chapter, you should have the knowledge to:

- Describe the I/O section of a programmable controller.
- Identify DIP switches.
- Describe how basic AC and DC input and output modules work.
- Define *optical isolation* and describe why it is used.
- Describe the proper wiring connections for input and output devices and their corresponding modules.
- Explain why a hardwired emergency-stop function is desirable.
- Define the term *interposing*.
- Describe what I/O shielding does.
- List environmental concerns when installing PLCs.

I/O SECTION

The input/output section, or I/O section, is the major reason that PLCs are so versatile when used with process machines or driven equipment. The I/O section has the ability to change virtually any type of voltage or current signal into a logic-level signal (typically 5 V DC) that is compatible with the processor. The I/O section automatically makes the conversions necessary for the processor to interpret input signals and to activate output devices, even when the input and output devices are of various voltage and current levels.

A DC input module, for example, can be used with a 24 V DC proximity switch to turn on a 240 V AC motor starter coil that is connected to an AC output module. The conversion and interfacing is all accomplished automatically in the I/O section of the PLC, and it is the ease with which the interfacing is accomplished that has made the PLC such a viable tool in industrial and process control.

The input modules of the I/O section provide the status (*ON* or *OFF*) of push buttons, limit switches, proximity switches, and the like, to the processor so decisions can be made to control the machine or process in the proper sequence. Outputs, such as motor starter coils, indicator lights, and solenoids are interfaced to the processor through the output section of the I/O. Once a decision has been made by the processor, a signal is sent to the output section to control the flow of current to the output device. In general, the status of the inputs are relayed to the processor and, based on the logic of the program that has been written, a decision is made to turn the outputs to *ON* or *OFF*. All of the different types and levels of signals (voltages and currents) used in the control process are interfaced in the I/O section.

The I/O section generally can be divided into two categories: fixed I/O and modular I/O.

Fixed I/O

PLCs with fixed I/O typically come in a complete unit that contains the processor, I/O section, and power supply. The I/O section contains a fixed number of inputs and outputs. For example, the Allen-Bradley MicroLogix PLCs shown in Figure 2–1a have a combination of digital and/or analog inputs and outputs in a self-contained base unit. Like most small PLCs the Allen-Bradley MicroLogic can be DIN-rail or panel mounted. Figure 2–1b shows the DIN–rail mounting instructions for the GE Fanuc Micro PLC.

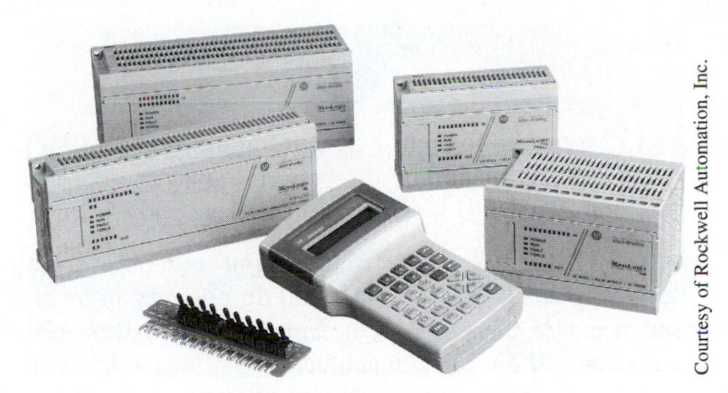

Courtesy of Rockwell Automation, Inc.

Figure 2–1a MicroLogix PLCs

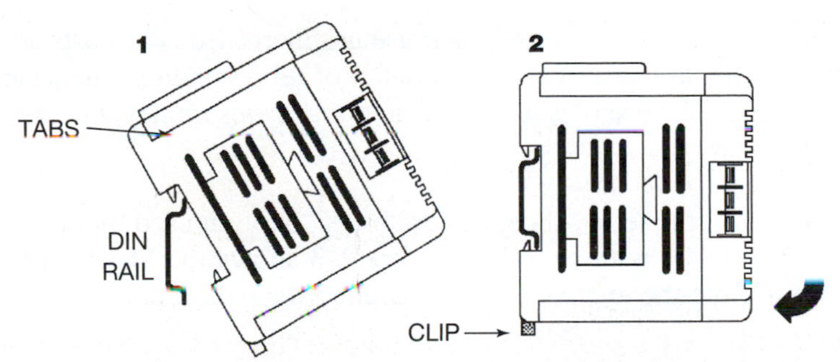

Position the upper edge of the unit over the DIN rail, so that the rail is behind the tabs as shown above.

Pivot the unit downward (for a unit being mounted right-side up) until the spring-loaded clip in the bottom of the unit clicks firmly into place.

Courtesy of GE Fanuc Automation

Figure 2–1b DIN Rail Mounting

If more I/O capability is required or different voltages are needed, expansion units with various I/O configuration can be added. Figure 2–2 shows an Allen-Bradley SLC 500 fixed I/O controller with an optional two-slot expansion chassis.

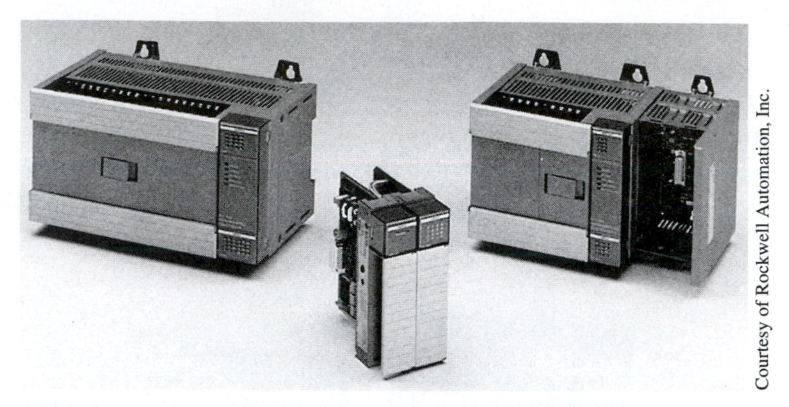

Figure 2–2 SLC 500 Fixed I/O Chassis with Optional Two-Slot Expansion

Small PLCs with fixed I/O typically have a discrete input and output section. As discussed in Chapter 1, discrete-type I/O signals are *ON* or *OFF* and do not vary in level. For example, when a 120 V limit switch closes or is *ON,* the signal to the input section will be 120 V, and the signal will be 0 V when the switch is open (*OFF*). Many manufacturers offer models with analog inputs and/or outputs. In addition, expansion I/O units can be added to their fixed I/O PLCs.

While these PLCs are small in size, they are big on features. Most include full-feature instruction sets that include timers, counters, sequencers, shift registers, word moves, data compare, and much more. One should consult the specific dealer for a full list of features.

As the cost of these compact units has decreased, their use has increased. The costs are so competitive that any control processes that use only a small number of relays and/or timers can now be accomplished using a small PLC. The use of a small PLC not only saves money, but also gives added reliability and flexibility.

Because of their shape and size, the term "shoebox" or "brick" is often used by manufacturers and users alike when referring to PLCs with fixed I/O. Figure 2–3 shows the Modicon Micro and the expandable Modicon A120. Note the relative size compared to the ballpoint pen.

Modular I/O

Modular I/O, as the name implies, is modular in nature, more flexible than fixed I/O units, and provides added versatility when it comes to the type and number of input and output devices that can be

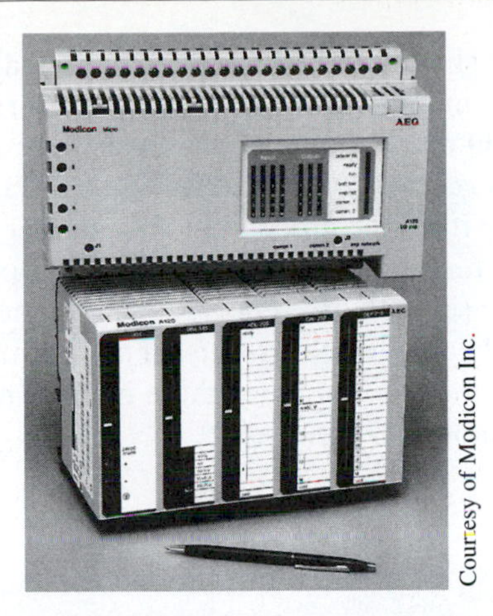

Courtesy of Modicon Inc.

Figure 2–3 Modicon Micro and A120 PLCs

connected to the system. The various types of input and output modules that make up the I/O section are housed, or installed, in an I/O rack or chassis.

The I/O rack or chassis is a framework or housing into which modules are inserted. Figure 2–4a shows three different sizes of racks. Figure 2–4b shows the rack with the I/O modules installed and the processor ready for installation.

Courtesy of Modicon Inc.

Figure 2–4a I/O Racks

Courtesy of Modicon Inc.

Figure 2–4b I/O Rack with Processor and I/O Modules

Racks or chassis come in many shapes and sizes, and typically allow 4, 8, 12, or 16 modules to be inserted. Racks that contain I/O modules and the processor are referred to as *local* I/O. Racks that contain I/O modules, remote I/O communication cards, and are mounted separately or away from the processor are referred to as *remote* I/O. An advantage of remote I/O racks is that they can be mounted up to 10,000 feet away from the processor. The number of remote I/O racks that a processor can control varies with each manufacturer. The communication between the remote rack and the processor is accomplished using several different types of communication methods. These methods include coaxial cable, twin axial cable, shielded-twisted pair, or fiber optics. If distance or electrical noise are considerations, the fiber optic communication method may be the best option. Figure 2–5 shows a local rack and three remote I/O racks.

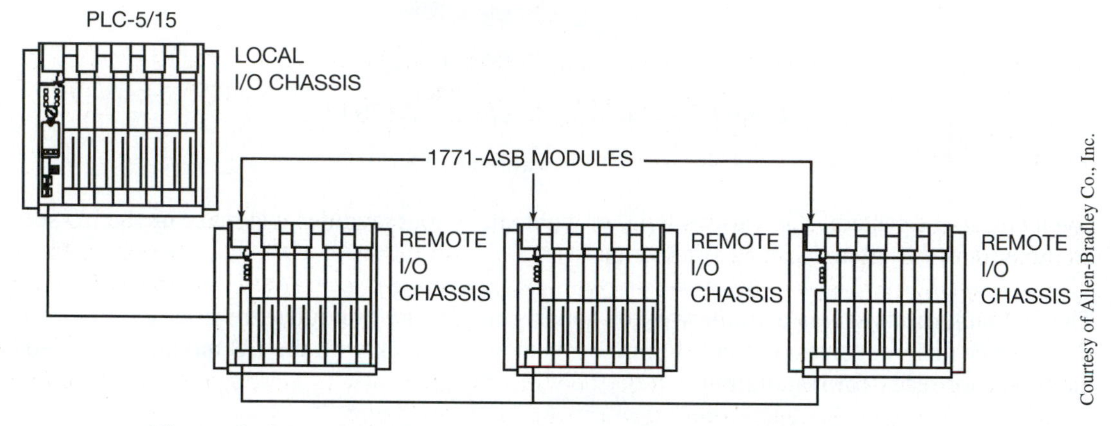

Figure 2–5 Local Rack with Processor and Three Remote I/O Racks

Whether local or remote, racks normally have jumpers or switches that have to be set or configured in order for the racks to communicate with the processor. A common switch used for rack configuration is referred to as a DIP switch. DIP is short for dual-in-line package, a common electronic package design, or style, for use on printed circuit boards. These DIP switches are either *ON* or *OFF,* and when set in the proper sequence, are used to assign an address to a rack, such as Rack 1, Rack 2, Rack 3, etc. DIP switches are also used to set fault parameters as well as other processor functions. DIP-switch settings will be specified by the PLC manufacturers and are found in the installation manual.

Note: *Under no circumstances should a pencil be used to change a DIP switch position. The graphite in the pencil tip can break off, causing the switch to short. Instead, use the tip of a ballpoint pen or other nonconducting pointed object to change switch positions.*

DIP switches are generally mounted on a printed circuit board located in the back of the I/O rack or chassis. This printed circuit board is often referred to as the *backplane*. Figure 2–6 shows a backplane printed circuit board and a DIP switch group assembly.

Within each rack, individual input and output device connections must have a distinct **address** so the processor knows where the device is located, and in return, can send and receive signals,

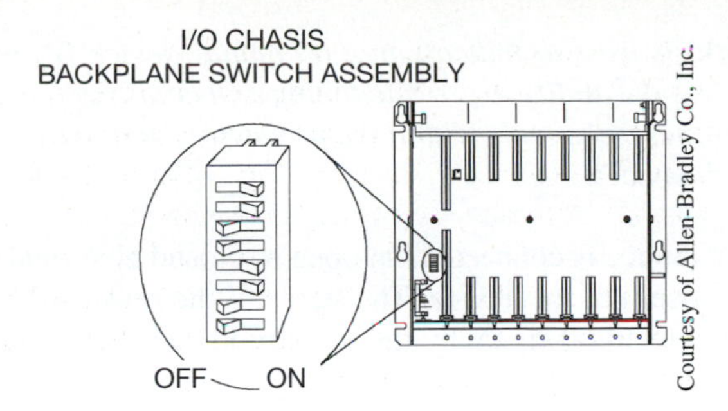

Figure 2–6 Backplane and I/O DIP Switches

enabling the processor to monitor and/or control the device. Allen-Bradley, for example, uses the rack number, location of a module within a rack, and the terminal number of a module to which an input or output device is connected to determine the device's address. Addresses and addressing of input and output devices will be covered in Chapter 4.

Also mounted on the backplane of the I/O rack are prewired slots or connectors into which the individual I/O modules are inserted. When inserting the modules, proper alignment is assured by card guides (also referred to as printed circuit board guides) that are mounted on the top and bottom of the I/O rack, as seen in Figures 2–4a and 2–4b.

Input and output modules can be separated into three basic groups: discrete or digital input/output modules, analog input/output modules, and specialized modules.

DISCRETE I/O MODULES

Discrete I/O modules are types of modules that only accept digital or *ON*- and *OFF*-type signals. These modules only recognize these two states or conditions which, again, are *ON* or *OFF*. If a discrete device, such as a limit switch, is connected to this type of module, the module determines the state, or position, of the limit switch, and communicates the state, or status, to the processor. If the limit switch is open (*OFF*), the module indicates to the processor that the limit switch is *OFF*. This *OFF* condition is stored in the processor memory as a zero (0). Had the limit switch been in a closed position, the module would have sent a signal to the processor indicating that the limit switch was *ON,* or closed. The *ON* condition would have been stored in the processor memory as a one (1). All information stored in the processor memory about the status or condition of discrete I/O devices is always in ones and zeros.

Discrete modules are the most common type used in a majority of PLC applications and can be divided into two groups: input and output.

Discrete Input Module

The discrete input module communicates the status of the various real-world input devices connected to the module (*ON* or *OFF*) to the processor.

Note: *The term* real world *is used to indicate that an actual device is involved. As you will learn later in the text, the PLC has the ability to provide timing and counting functions for a machine, even though the timers and counters exist only within the processor, and are not wired into the circuit as with real-world, or actual devices.*

Once the real-world input device is connected, an open or closed electrical circuit exists, depending on the position (open or closed) of the device. The status of the real-world input device is then converted to a logic-level DC electrical signal by the input module and sent to the processor.

Discrete input modules come in a wide range of voltages for various applications. Some of the more common voltage modules are 120 V AC, 240 V AC, 24 V DC, and 12–24 V DC. Some manufacturers give their modules an AC/DC rating to increase their flexibility and reduce required inventory. It is important to note, however, that while the module may be used with either AC or DC input voltages, the voltages *cannot* be intermixed on the same module.

Input modules can be purchased with a wide range of input terminals or points, which determine the number of individual field devices that can be connected to the module. Common sizes, depending on the manufacturer, are 8, 16, and 32 points. Sixteen- and 32-point modules are often referred to as high-density modules since they are physically the same size as an 8-point module. High-density modules usually provide lower cost per point, or input device, but are also more difficult to wire. The increased difficulty in wiring is caused by the closer proximity of the wiring terminals and the increased number of wires in the wiring harness.

AC Discrete Input Module

Figure 2–7 shows a simplified diagram of one of the input circuits of a typical AC discrete input module. Resistors are used to drop the incoming voltage; then a bridge rectifier is used to convert the AC input voltage to DC. Next, a filtering circuit is used to condition the DC and guard against electrical noise. Electrical noise can cause a short-duration DC pulse that is sometimes interpreted by the processor as a closed signal. This false, or erroneous, signal could be interpreted as a valid signal, and a 1 would be placed in memory to indicate the device was *ON,* even though it was not. To eliminate the possibility of faulty operation due to electrical noise, the filter section of the module delays an actual input signal from being sent to the processor for 15 to 25 milliseconds (msec).

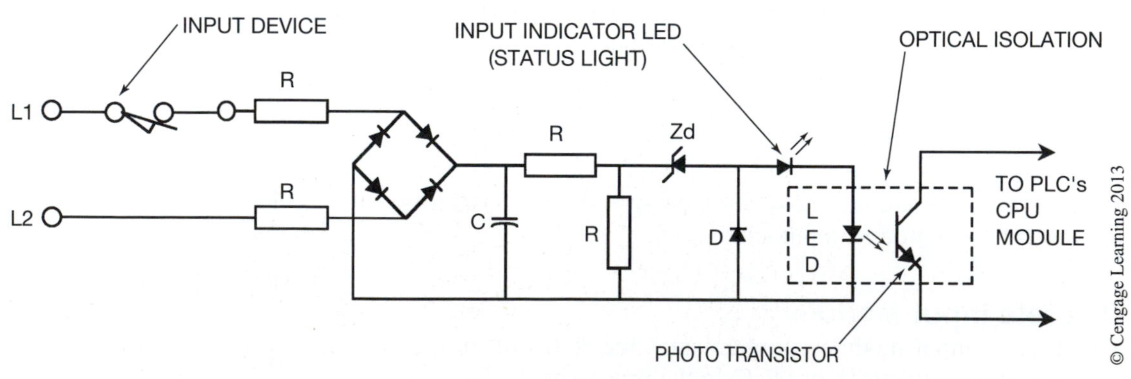

Figure 2–7 Simplified AC Input Module Circuit with Indicator Light

The filter requires that the AC signal be not only of a specific value but also be present for a specific amount of time before the module views it as a real signal and communicates the results to the processor. A valid signal is relayed through an **optically coupled** circuit, across the backplane of the I/O rack to the processor.

The optically coupled circuit uses a light emitting diode (LED) to turn *ON*, or forward bias, a photo transistor to complete the electrical circuit to the processor. When the LED is turned *ON* to indicate that the actual input device has closed, the light from the LED is picked up by the photo transistor, which makes the transistor conduct or switch on, completing a 5 V DC logic circuit, and the status of the input is communicated to the processor. This form of optical coupling is also referred to as **optical isolation.** By employing optical coupling, or isolation, there is no actual electrical connection between the input device and the processor. This eliminates any possibility of the input line voltage, i.e. 120 or 240 V AC, from coming in contact with and damaging the low-voltage DC section of the processor. Optical isolation also protects the processor from electrical noise, voltage transients, or spikes. In summary, optical isolation prevents any unwanted voltage from the I/O section from reaching the logic section of the processor.

Individual status lights are provided for each device that is connected to an input terminal (Figure 2–7). The status light is ON when the input device is closed and is OFF when the input device is open. With the status lights showing the actual position of the various input devices connected to the input module, they make a valuable troubleshooting aid. The electrician or technician need only look at the status lights on the input module to determine the position, or status, of any input device.

A typical I/O module consists of two parts: a printed circuit board and a terminal assembly. The printed circuit board plugs into a slot, or connector, in the I/O rack and contains the solid-state electronic circuits that interface the I/O devices with the processor. The terminal assembly then attaches to the front edge of the printed circuit board, which may or may not have a protective cover, depending on the manufacturer. Figure 2–8, left to right, shows a typical 16-point AC input module, a 16-point AC output module, and an analog input module.

Figure 2–8 AC Input and Output Modules and Analog Input Module

Figure 2–9 shows how the input module (Figure 2–8) is installed in the I/O rack.

Figure 2–9 Installing a Module in an I/O Rack

After the input modules have been installed in the I/O rack, they are ready to have one side of each input device connected to their terminals or wiring arms (Figure 2–10).

Figure 2–10 I/O Module Field Wiring Arm

While each input device has two wires connected, only one wire is connected directly to the input module. The other wire of each input device is connected to Line 1 (Figure 2–11a). The same connection scheme is used for 8-, 16-, or 32-point input modules. Figure 2–11b shows the wiring connections for an Allen-Bradley 16-point input module.

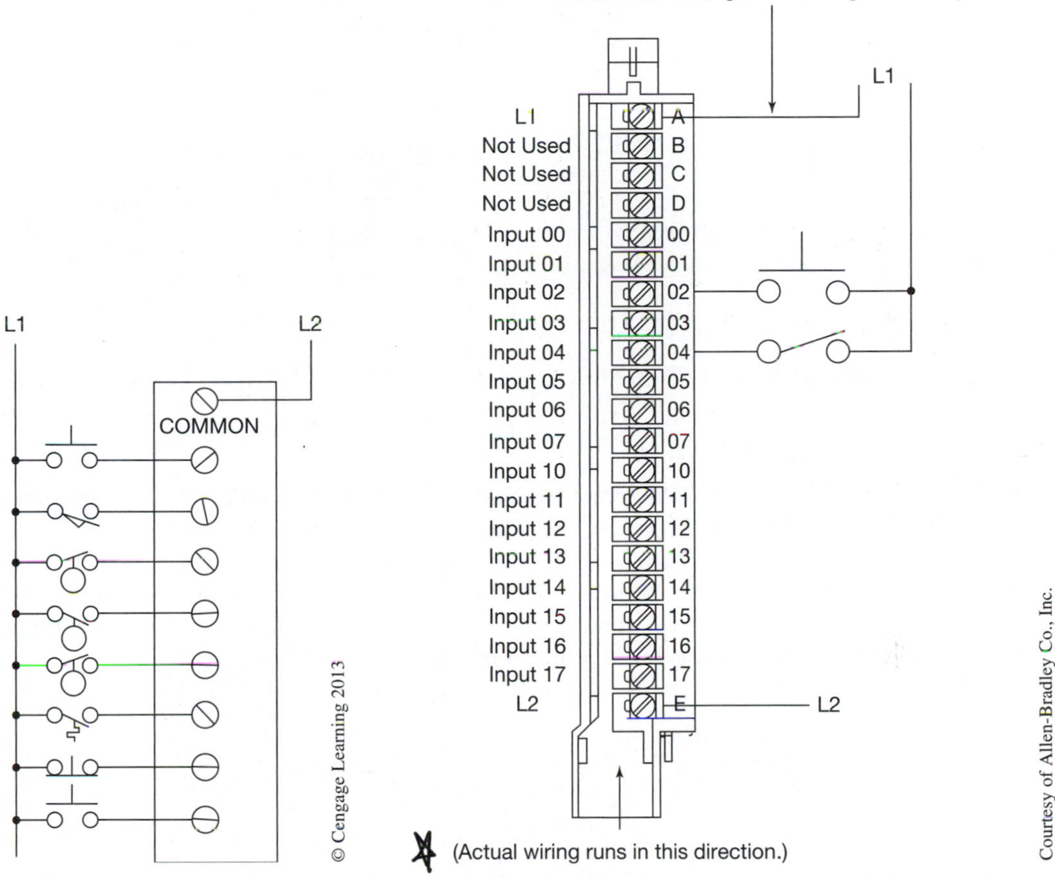

Figure 2–11a Field Wiring for AC Module

Figure 2–11b Connections for an Allen-Bradley 16-Point Input Module

The wires from the individual devices are referred to as field wiring, since the wires are external to the PLC and are connected in the field. On larger process machines, the field wiring that is brought into the I/O rack consists of hundreds of wires. The basic rule is that one side of each input device is wired to a hot conductor (L1 for AC or + for DC), and the other side of the device is wired to an input terminal on the input module. The input module has a common connection for the neutral, or grounded potential (L2), for AC modules, and for the negative (–) for DC modules. Consult the literature that comes with each input module to ensure that the correct wiring connections are made.

Figure 2–12 shows two simplified circuits. In the first, or traditional circuit, the input device (single-pole switch) is connected to, and controls, the light. In the PLC circuit, the input device is connected to the input module instead of the light. The module converts the 120 V AC input signal to 5 V DC, and communicates the status of the single-pole switch to the processor which, in turn, controls a light that is connected to an output module.

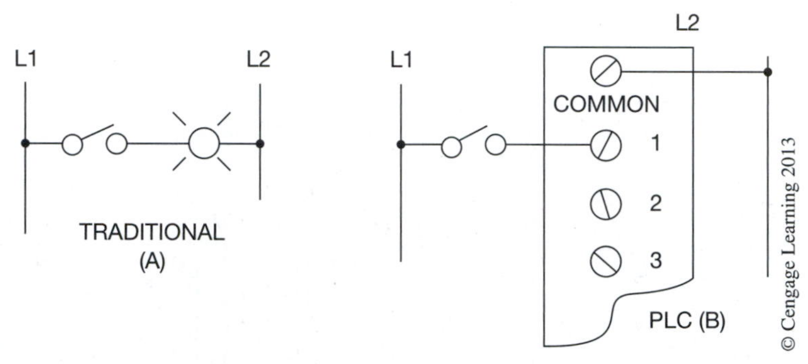

Figure 2–12 Traditional Wiring for Single Pole Switch (A) Compared to PLC Wiring (B)

DC Discrete Input Module

Figure 2–13 shows a simplified diagram for a typical DC input module. With the exception of the bridge rectifier used in the AC module, the principles are the same. Resistors are used to lower, or drop, the incoming voltage. Filtering circuits condition the low-voltage signal and add an additional delay in the response time. This period of delay is slightly less than the delay used in the AC input module, but it is also used to verify that the signal received is a valid signal of a proper duration, and not a signal caused by electrical noise, voltage transients, or the like.

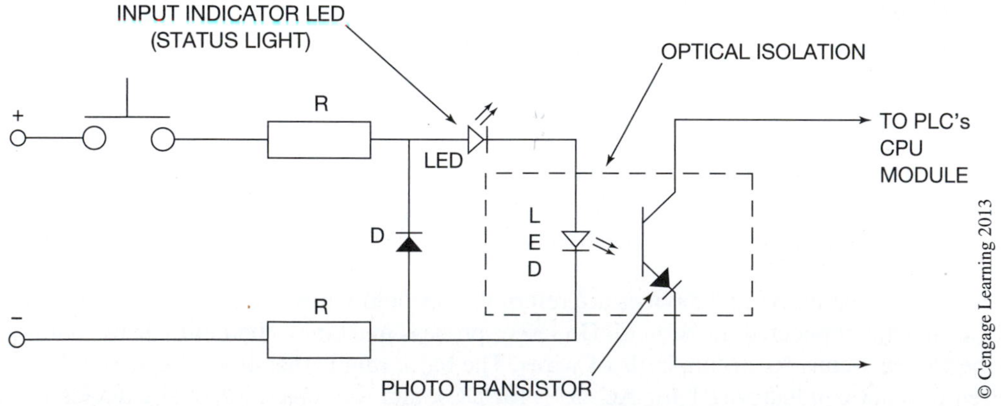

Figure 2–13 Simplified Circuit for DC Input Module with Indicator Light

Optical isolation is also used on the DC input module to isolate the processor from the higher voltage of the input devices. When the LED is turned *ON* by the closing of the input device, the photo transistor conducts, and the status of the input device is communicated across the backplane of the I/O rack to the processor. Status lights, shown in the diagram, are also provided for each input device to indicate whether the input device is open or closed. Figure 2–14 shows how a typical DC input module is wired.

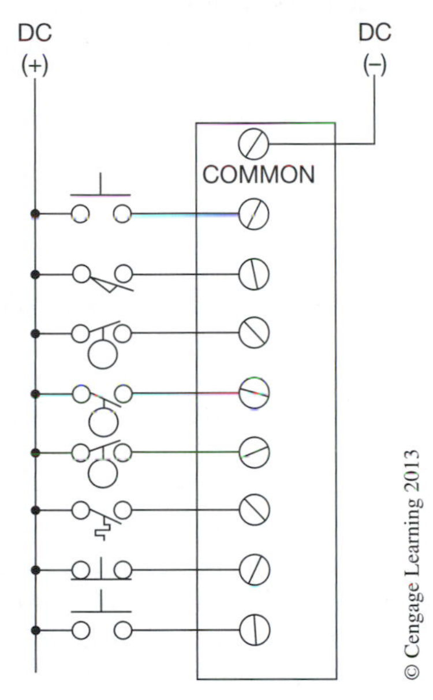

Figure 2–14 Field Wiring for DC Input Module

Fast-Responding DC Input Modules

Fast-responding DC input modules are used when the process requires fast-acting sensors to respond to high-speed and/or high-volume applications. Encoders, other types of sensors that respond with many pulses for each rotation of a shaft, or proximity switches that are counting parts or products produced by high-speed machines are all examples of the benefits of a fast-responding module. The internal operation of the module is the same as the discrete DC input module, with the exception of the delay created in the filtering circuit. The fast-responding module has only a 1-msec or less delay. With this short time, normal mechanical contact devices may not work correctly due to the contact bounce that is inherent in some mechanical switches and contacts. The contact bounce is counted as an additional input signal and processed by the processor. This extra count has an obvious adverse affect on machine operation and is not a proper application of this type of input module.

Discrete input modules come in a variety of types and styles that fit most applications requiring an *ON*- or *OFF*-type of signal. Allen-Bradley, for example, manufactures nearly twenty different types of discrete input modules. Selecting the proper module for any given application is relatively easy when the voltage level, voltage type, current, and response time are known.

Discrete Output Modules

The purpose of a discrete output module is to control the current flow to real-world devices such as motor starter coils, pilot lights, control relays, and solenoid valves. As was true for the discrete input module, the discrete output module also works on a digital, or *ON* and *OFF,* basis. When the processor has made a decision to turn a specific device *ON,* the processor places a 1 in the memory location assigned to that output device, and later in the process, the information is communicated by way of the backplane of the I/O rack to the output module, and the required real-world device will be turned *ON*. Similarly, when the processor determines that a device needs to be turned *OFF,* a 0 is placed in the device's memory location and the device will be turned *OFF*. The output module acts like a remote control switch that is controlled by the processor for turning output devices *ON* and *OFF*.

Output modules are generally classified as AC and DC. Both types offer a wide range of voltages, typical for the input modules discussed earlier. Read the information sheets carefully to determine that the module selected is appropriate for the load (output device) that is to be controlled.

Output modules are sized by the number of output devices that can be connected to them. The numbers of terminals, or points, for connecting output devices are typically 8, 16, and 32. Output modules have a current rating for each terminal or connection point, as well as an overall rating for the module. The individual terminal rating indicates the maximum current, or load, that can be controlled. This rating is a continuous duty rating. A surge rating is usually provided, indicating the maximum current that can be controlled and the length of time. The time rating may be given in milliseconds or in cycles. A typical current rating for a 120 V AC output module would be: 1.5 amperes maximum continuous duty, with a surge current rating of 4 amperes for 8.3 msec (1/2 cycle).

The surge current rating is necessary for the inrush, or "pull-in," current that is present when motor starter coils, solenoids, and other inductive loads are initially energized. Once the load has been energized, the "hold-in," or "seal-in," current is much less, and the continuous duty rating of the module is sufficient.

Each module also has a total current rating. The total current rating must be determined from the manufacturer's literature, not by simply adding the ampere rating per point. To further clarify this point, consider the rating of one manufacturer's 16-point 120 V AC output module. Each point is rated for 2 amperes continuous duty, yet the maximum current rating for the module is only 8 amperes. Why is the total rating not 2 amperes times 16 points, or 32 amperes? The answer has to do with the way the module dissipates the heat generated by the current flow in the module. Normally, no fans or other external methods of cooling are used, and the heat that is generated within the module is dissipated using heat sinks. Heat sinks work on convection alone, and this limits the amount of heat that they can effectively dissipate. The total current rating that a module is given, therefore, is determined by the ability of the module to dissipate heat. Thus, the lower current rating of the 16-point 120 V AC module shows that the module can only satisfactorily dissipate the heat from 8 amperes of continuous current flow, not the full 32 amperes that would be possible if all loads were operating at the maximum 2 amperes rating. In reality, however, how likely is it that all 16 loads connected to the module would be on and operating at their full 2-amperes capacity?

Subjecting the module to higher-than-rated current loads creates excess heat in the module. This excess heat has a detrimental affect on the electronic components and leads to shortened operating life and/or component failure. The ambient temperature that the I/O operates in is another factor that must be considered. PLCs are normally designed to operate in a temperature range of 32°–140°F or 0°–60°C. Operating in temperatures above these limits affects the module's ability to dissipate heat and can lead to early component failure.

Another PLC manufacturer rates their 16-point, 120 V AC output module at .5 amperes per point, but starts to derate the current level when more than 50% of the points are *ON,* and/or when the temperature exceeds 40°C. With each manufacturer rating their output modules differently, it is necessary to carefully review the literature when specifying modules.

AC Output Module

The internal circuitry for one point of a typical AC output module is shown in Figure 2–15. The AC output module usually consists of a **Triac** (shown in the figure). However, some manufacturers use a **Silicon-Controlled Rectifier (SCR)** instead of a Triac. When the processor determines that the output is to be turned *ON,* a signal is sent across the backplane of the I/O rack and the LED (light emitting diode) is turned *ON.* The light from the LED causes the photo transistor to conduct, which provides current for the gate of the Triac. This portion of the output module optically isolates the logic section of the processor from the line voltage of the output devices.

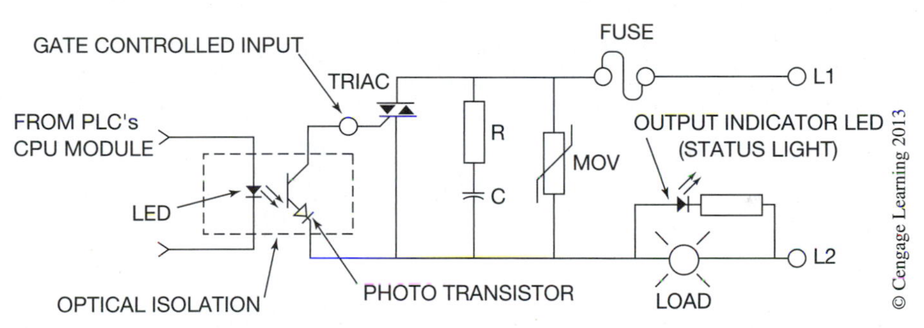

Figure 2–15 Typical AC Output Module Circuit

The Triac is used as an electronic switch to turn output devices *ON* and *OFF*. The Triac itself is the equivalent of two silicon-controlled rectifiers in inverse parallel connection with a common gate. The gate controls the switching state (*ON* or *OFF*) of the device. Once a signal is applied and the break-over voltage point is reached on the gate (normally 1 to 3 V), the Triac freely conducts in either direction, completing the path for current flow to the output device.

A Triac is a solid block of crystalline material and is more sensitive to applied voltages and currents than standard relay contacts. Triacs are also limited to a maximum peak applied voltage, and, if this value is exceeded, a "dielectric-type" breakdown can result, causing a permanent short-circuit condition. A snubber circuit that consists of a resistor in series with a capacitor and a metal oxide varistor (MOV) is used to protect the Triac from damage from electrical noise and voltage spikes. The resistor and capacitor form an RC circuit that is used to control the rate at which voltage builds up in the circuit. If the voltage rises too fast, the capacitor absorbs, and the resistor dissipates, excess voltage. The metal oxide varistor is designed to break down, or conduct, at certain voltage levels. In a 120 V AC circuit, the peak voltage is approximately 170 V. The MOV would typically be set to break down at 190 V. When the MOV conducts, it clips the excess voltage and thus prevents damage to the module.

Triacs constructed of a solid "block of material" have some characteristics that are not found with standard relay contacts. The Triac, rather than having *ON* and *OFF* states, actually has low and high resistance levels, respectively. In its *OFF* state (high resistance), a small leakage current still flows through the Triac. This leakage current, which is usually only a few milliamperes or less, normally causes no problem. When low-resistive pilot lights are connected to AC output modules, a faint glow of the filament may be detectable, even when the module is *OFF*. Similarly, the coils of some small control relays or solenoids may produce a detectable hum due to the Triac leakage current, even though the Triac is technically *OFF*. This small leakage current also causes false readings in some digital and analog meters. Troubleshooting techniques for triacs are covered in Chapter 21.

While Triacs are capable of carrying surge currents higher than their continuous current ratings, such surges must be of short duration (1/2 to 1 cycle) and not repetitive. Exceeding the manufacturer's listed surge values or the maximum continuous current rating, usually referred to as maximum RMS on-state current, results in a permanent short circuit.

After an output module is installed in the I/O rack, the actual real-world output devices are connected. Figure 2–16 shows the proper termination for an 8-circuit AC output module. Each output device has two wires; one wire from each output device is wired to the L2 potential. The other wire from each device is wired to one terminal of the output module, as shown in the figure. L1 is then connected to a common terminal on the module to supply the other potential for the output devices.

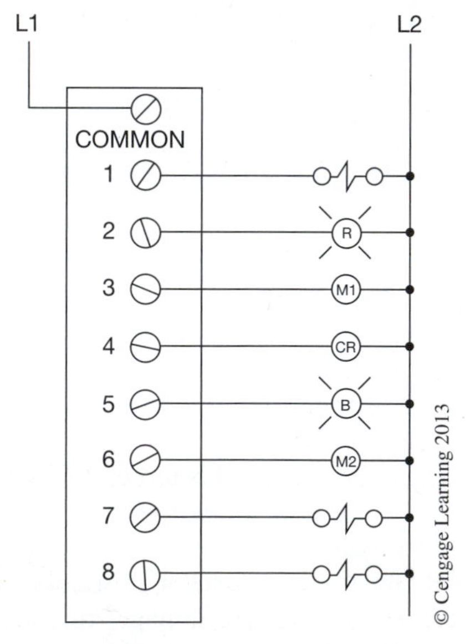

Figure 2–16 Field Wiring for AC Output Module

Figure 2–17 shows two simplified circuits, each including an output device. In the first circuit, the output is wired to the L2 potential, and the single-pole switch is used to control the L1 potential to complete the circuit. In the PLC circuit, the output is again wired to L2, but the switch is replaced by the output module, which is wired to the L1 potential. Simply stated, the output module may be viewed as an electronic switch that is controlled by the PLC's processor.

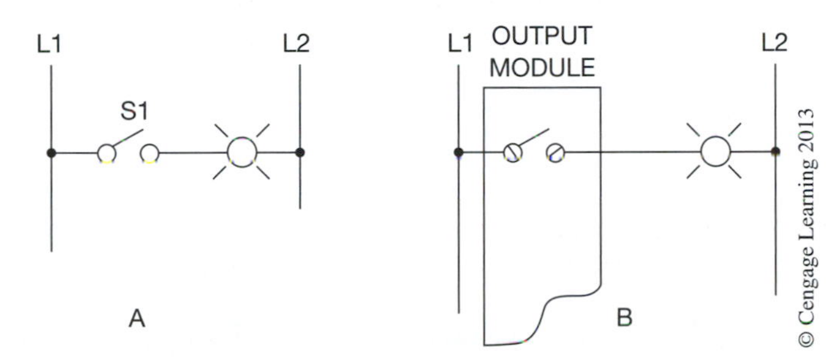

Figure 2–17 Traditional Wiring for an Output Device (A), Compared to PLC Wiring (B)

Output Fuses

In order to prevent damage to output modules, it is important not to exceed the current rating or to exceed its inrush capability. It is also important that output modules be protected from short circuits and ground faults. To provide protection for overcurrent, short circuits, and ground faults, output modules are always fused. The number of fuses used varies with each manufacturer. Some modules have an individual fuse for each output terminal, or point, others have one fuse for each 8 outputs, while still others use one fuse to protect all 16 points on their output module. Some PLCs, like the Allen-Bradley SLC 500, do not come with internal fuses, and fuses must be added to protect the outputs. Some newer output modules have internal electronic fusing to prevent too much current from flowing though the module.

Most PLCs come with "blown-fuse" indicators to show that a fuse has blown. Some modules have a "blown-fuse" indicator for each output terminal. If a fuse is blown on output terminal 6, an indicator lamp at terminal 6 will be lit to indicate the location of the blown fuse. Other modules have only one "blown-fuse" indicator for the whole module. This one indicator lamp only indicates that a fuse has blown; it does not indicate its location, and it is up to the electrician or technician to determine the blown fuse(s). When individual indicators are used for each output terminal, troubleshooting is greatly simplified. Figure 2–18 shows an Allen-Bradley AC output module with the blown-fuse indicator light identified.

As you may expect, access to the fuses varies with the manufacturer. Some modules must be removed from the I/O rack and a cover removed before the fuses can be changed, while other modules provide direct access to the fuses on the front edge of the module.

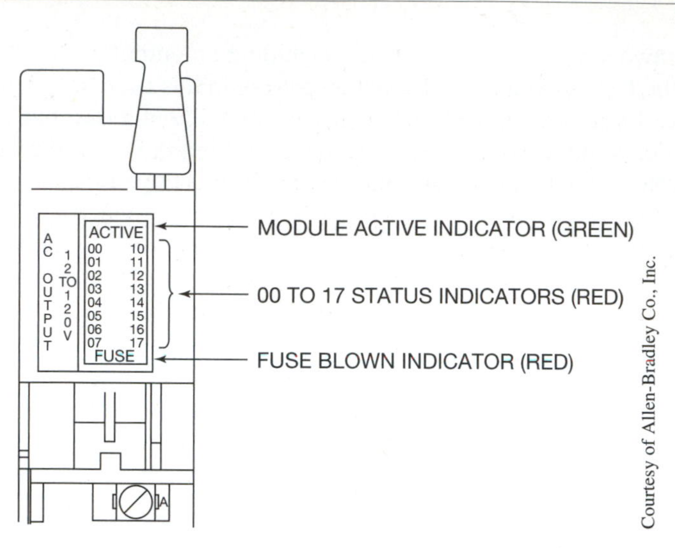

Figure 2–18 AC Output Module with Blown-Fuse Indicator Light

It is important that when a blown fuse is removed, its replacement fuse be of the same voltage rating, amperage rating, interrupting rating, response time, and physical size as recommended by the manufacturer.

To speed troubleshooting and fuse replacement, many plants now add additional fuses to each output circuit. Terminal blocks are available that have built-in fuse holders and individual blown-fuse indicators. A separate fuse can then be wired in series with each output device, external to the output module. The fused terminal blocks are then mounted in a convenient location for easy access and troubleshooting.

Status Lights

Status lights are provided for each output point to indicate when that point has been turned *ON*. For troubleshooting purposes, it is important to understand how these status lights are wired. If the power for the lights comes from the processor side of the module, an illuminated light indicates that the processor has sent a signal to turn *ON* the output attached to *that* particular point. It is *not* an indication that current is flowing to the output device. If the status light is powered from the actual output power, then it is an indication that the output is turned on.

Some modules have two status lights. One is referred to as the logic light, and the other as the output light. When the logic lamp is lit, it indicates that the logic to turn on the output device has been sent from the processor. When the output light is lit, it indicates that there is a path for current flow to the output device.

The status lights are a powerful troubleshooting tool, but it is important to understand exactly what the status lights indicate, and not to read more information into them than they are able to provide.

Module Keying

Figure 2–19 shows an Allen-Bradley 120 V AC Input module. This AC Input module looks just like the AC or DC input module discussed earlier (Figure 2–8). This is true not only for Allen-Bradley modules, but also for other manufacturer's products as well. In all instances, the manufacturers label and/or color code the fronts of all modules to distinguish between the different types (AC input, DC output, Analog, and so on). Most manufacturers have designed each module so it can be **keyed.** In Figure 2–19, item 6, the module has been notched in two places. Installing keying bands on the Allen-Bradley I/O rack backplane connector (Figure 2–20) where a specific module is to be installed prevents any module, other than the type for which the connector is keyed, from being installed in that connector or slot. Figure 2–21 shows a close-up view of what the keying band looks like and how it is installed. Each type of module has a unique combination of notches. This feature prevents inadvertent or accidental replacement of the wrong type of module, say an input module, into a slot that is already wired to output devices. To prevent damage and downtime, it is important that the keying system be used.

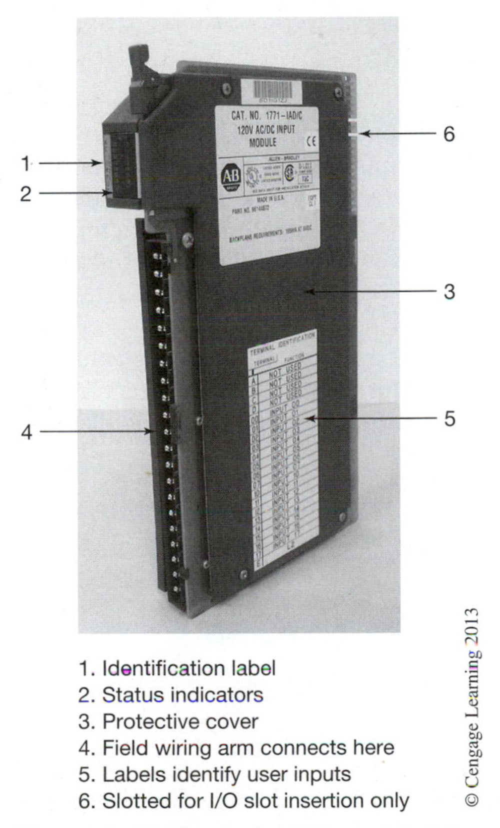

Figure 2–19 Notched AC Input Module

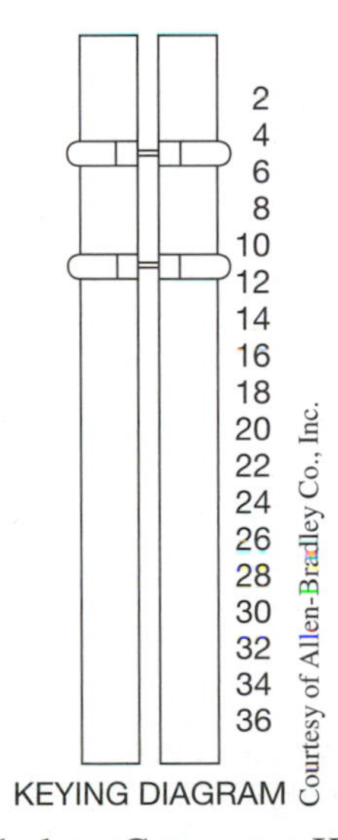

Figure 2–20 Backplane Connector Keying Diagram

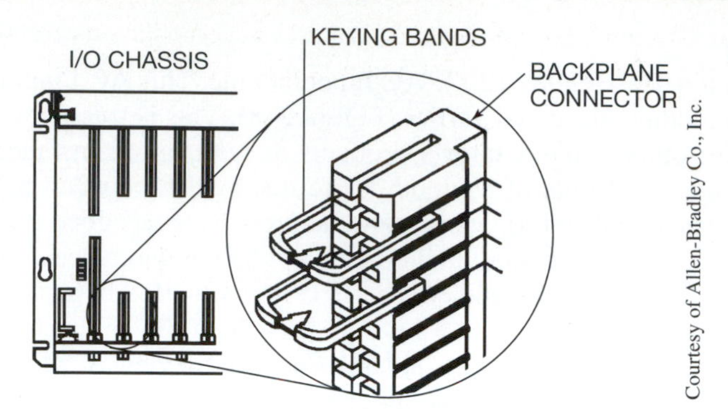

Figure 2–21 Connector Keying Pin

Some manufacturers have an electronic keying feature that automatically compares the expected module to the physical module before I/O communication begins. If enabled, electronic keying helps to prevent communication to a module that does not match the type and revision expected.

DC Output Modules

DC output modules are basically the same in operation as AC output modules. The difference is the use of a power transistor instead of a Triac for the control of output current. The power transistor has a quicker switching capability than the Triac; therefore, the response time for DC modules is faster than for AC modules. The RC circuit and MOV used in the AC output module are replaced with a diode to provide protection from electrical noise and spikes. A typical DC output circuit is shown in Figure 2–22.

DC output modules are available in ranges from 12 to 240 V DC depending on the manufacturer. It is important to check to make sure that the module is appropriate for the voltage and current level of the output device that is intended for connection to the module. DC modules will be fused like their

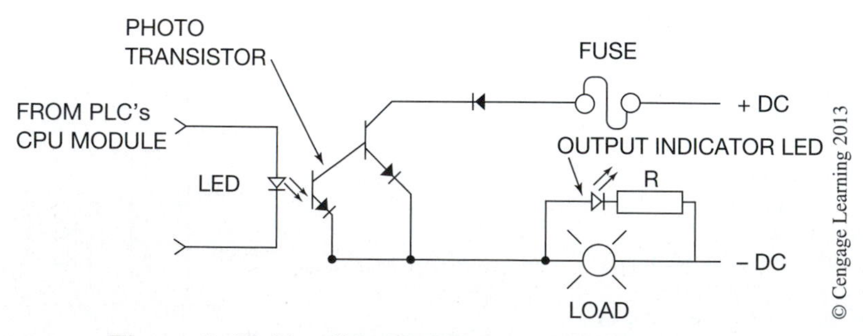

Figure 2–22 Simplified DC Output Module Circuit

AC counterparts and also have blown-fuse indicators, as well as status lights. Figure 2–23 shows how the output devices are wired to a DC output module.

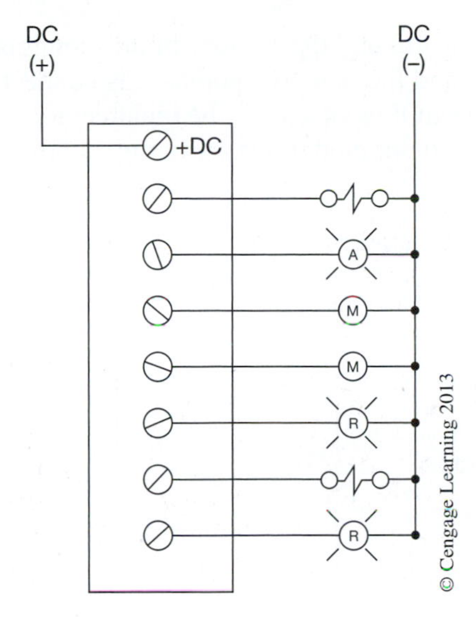

Figure 2–23 Field Wiring for a DC Output Module

Sourcing and Sinking

The terms *sourcing* and *sinking* refer to the manner in which DC devices are wired. To properly interface DC devices with the real world, the difference between sourcing and sinking must be understood.

Figure 2–24 is an example of a sourcing application. The positive potential is connected to the input module and the negative potential is connected to the input device. Using conventional current flow (+ to −), it is said that the input module is the source of supply for the real-world input device.

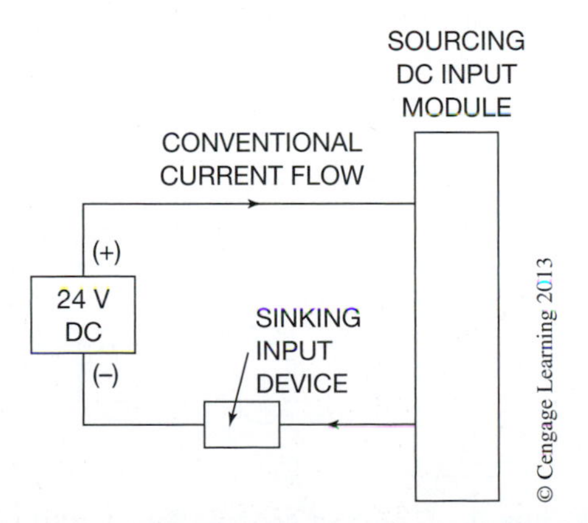

Figure 2–24 Input Module Sourcing Application

Stated simply, the input device receives current from the input module. In electronics, if a device (input module) provides current, or is the source of current, it is said to be sourcing.

Figure 2–25 is an example of a sinking application. In this configuration, the positive potential is connected to the input device and the negative potential is connected to the input module. In this case, using a conventional current flow of + to –, the input device is said to be providing current to the input module. If the device (input module) is receiving current, it is said to be sinking.

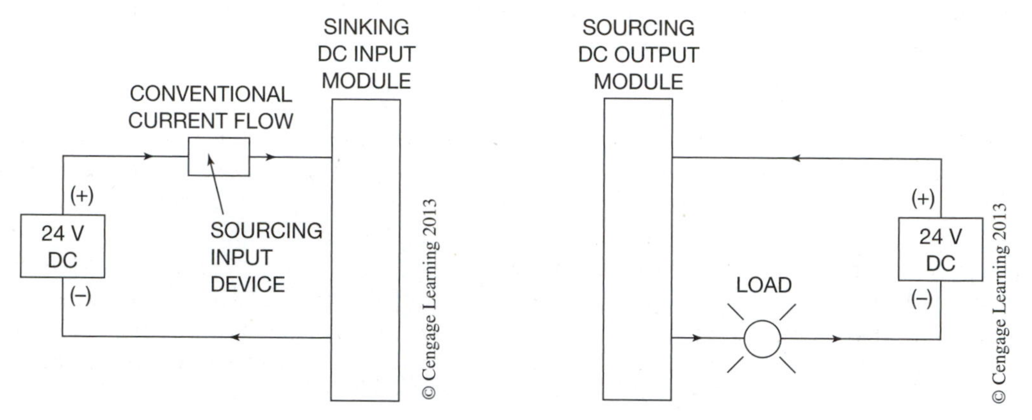

Figure 2–25 Input Module Sinking Application **Figure 2–26** Output Module Source Capability

When an output module is connected positive, as shown in Figure 2–26, and provides current to the real world (output device), the module is referred to as sourcing.

Figure 2–27 shows an output module wired negative, and the output device wired positive. When the output device (real world) provides the current to the module, the module is referred to as sinking.

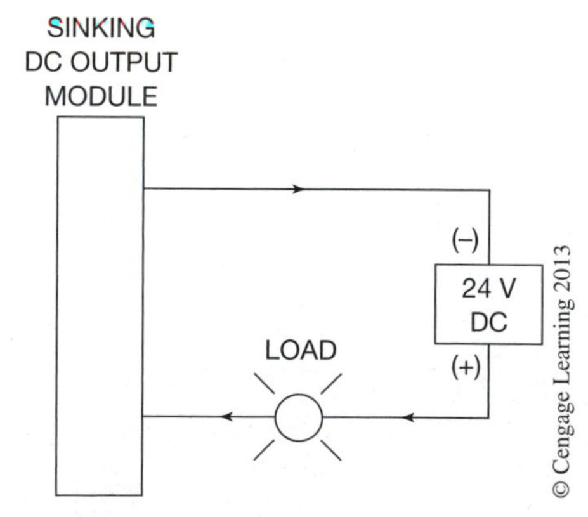

Figure 2–27 Output Module Sink Capability

There is always a great deal of confusion surrounding sourcing and sinking as it applies to input and output modules and devices. As a general rule, sinking modules are used with output modules when interfacing with electronic equipment (TTL or CMOS compatible), while sourcing modules are used for such DC loads as solenoids. The safest way to make sure that connections to DC devices are correct is to select a module that works for the application, based on the manufacturer's specifications and wiring diagram.

Contact Output Modules

Contact modules have electromechanical relays mounted on a printed circuit board that is inserted, or plugged into, the I/O rack. A signal from the processor energizes the coil of the electromechanical relay which, in turn, opens or closes a set of contacts. Each set of contacts is isolated and can be ordered normally open (N.O.) or normally closed (N.C.). This type of module is used when extra current ratings are required or when it is desirable to isolate loads of different voltages, or phases, from the same source. A contact output module is also used when the leakage current of a standard AC output module would affect the control process.

For example, a contact module can be used with a Variable Speed Drive application. Assume that the drive has been programmed for multiple speed settings. To select a given speed, a circuit must be completed at two points on a terminal strip mounted on the drive. One set of contacts is used for each speed setting. As the drive supplies its own power, all that is needed is to switch the control power through the contacts on the contact module. As the contacts are isolated, there is no possibility that the power from the PLC system can damage the drive.

Interposing Relay

When it is necessary to control loads larger than the rating of an individual output circuit, a standard control relay, which has a small inrush and sealed current value, is connected to the output module. The contacts of the control relay, which are generally rated at 10 amps, can then be used to control a larger load. This method of control is a common practice for NEMA size 4 and large motor starters, depending on the rating of the output module. When a control relay is used in this manner, it is called an interposing relay (Figure 2–28).

Reed Relay Output Module

The reed relay type output module is used when dry reed relays are desirable. They may be used for low-level switching (small current–low voltage), multiplexing analog signals, or for interfacing controls with different voltage levels. The voltage range of the reed relay contacts is normally in the range of 0–24 V AC or DC with a current rating of .1 ampere. Reed relay output modules are cheaper than normal solid-state AC/DC output modules. Reed relay modules are available with normally open (N.O.) contacts, normally closed (N.C.) contacts, or a combination of both N.O. and N.C. contacts, again depending on the manufacturer.

Transistor-Transistor Logic (TTL) I/O Modules

TTL input modules are designed to be compatible with other solid-state controls, sensing instruments, many types of photoelectric sensors, and some 5 V DC level control devices. TTL output modules are used for interfacing with discrete or integrated circuit (IC) TTL devices, LED displays, and various other 5 V DC devices.

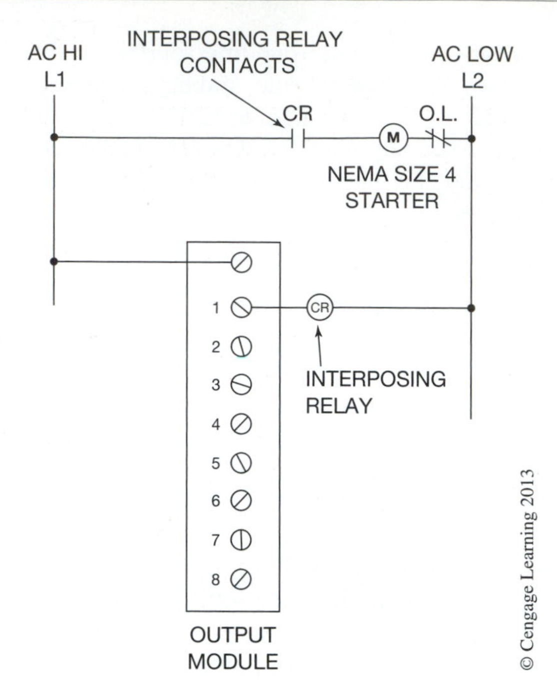

Figure 2–28 Interposing Relay

ANALOG I/O MODULES

Analog input modules are used to convert analog signals (i.e., 4–20 mA, 0–10 V DC) that sense such process variables as temperature, pressure, speed, and position to Binary or Binary-Coded Decimal (BCD) values, depending on the manufacturer, for use by the PLC logic as required. The conversion from analog to digital is accomplished with an analog-to-digital converter (ADC). The analog output module changes Binary or BCD values generated in the PLC logic into analog signals using a digital-to-analog converter (DAC). These analog output signals can be used for controlling variable speed controllers, valve positioners, displays, etc. Binary and BCD are covered in Chapter 5.

Figure 2–29 shows an analog input module and how a two-wire transmitter is wired.

Another type of analog input module is the Resistance Temperature Detector (RTD) input module. The module senses RTD signals and converts them to a corresponding temperature. Figure 2–30 shows typical wiring for an RTD device.

Note: *PLC manufacturers are introducing new and special application modules almost daily. A few modules have been discussed in this chapter for a basic understanding only. The local PLC representative(s) should be contacted for full and complete list(s) of modules that is available.*

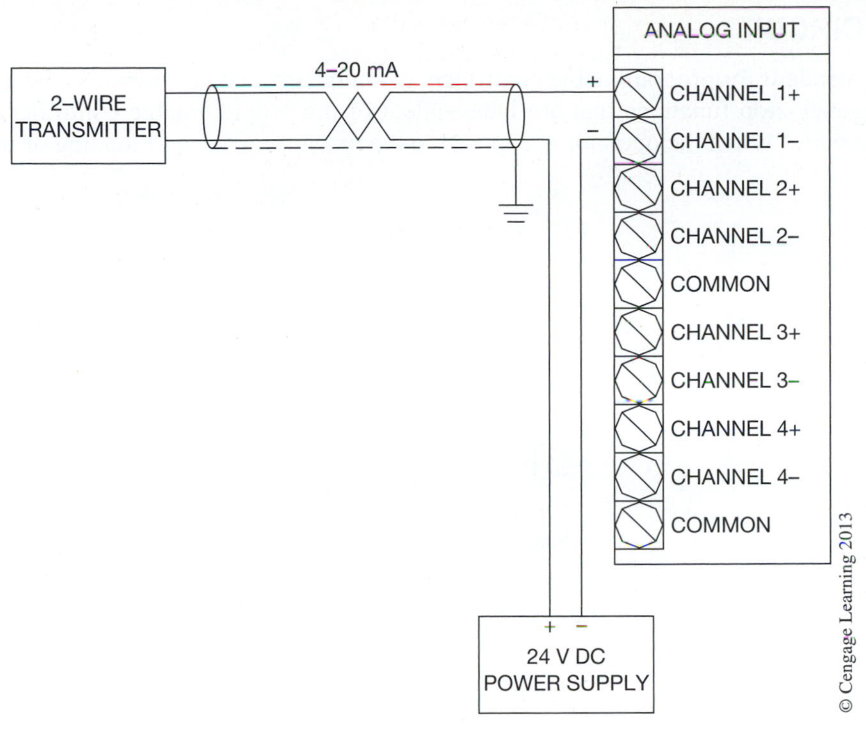

Figure 2–29 Analog Input Module

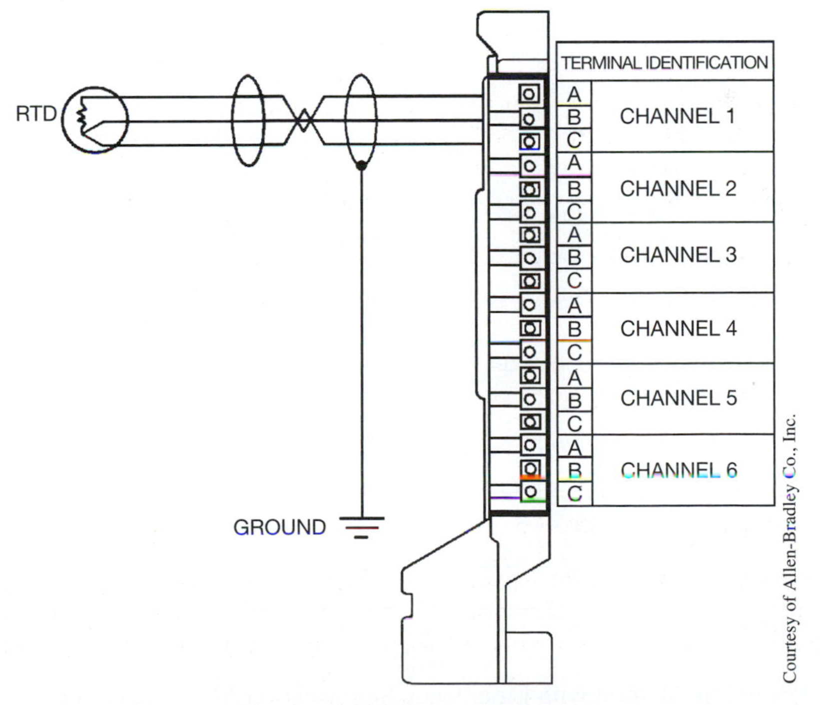

Figure 2–30 RTD Input Module

SAFETY CIRCUIT

The NEMA standard for programmable controllers recommends that consideration be given to the use of emergency-stop functions that are independent of the programmable controller. The standard reads in part: "When the operator is exposed to the machinery, such as loading or unloading a

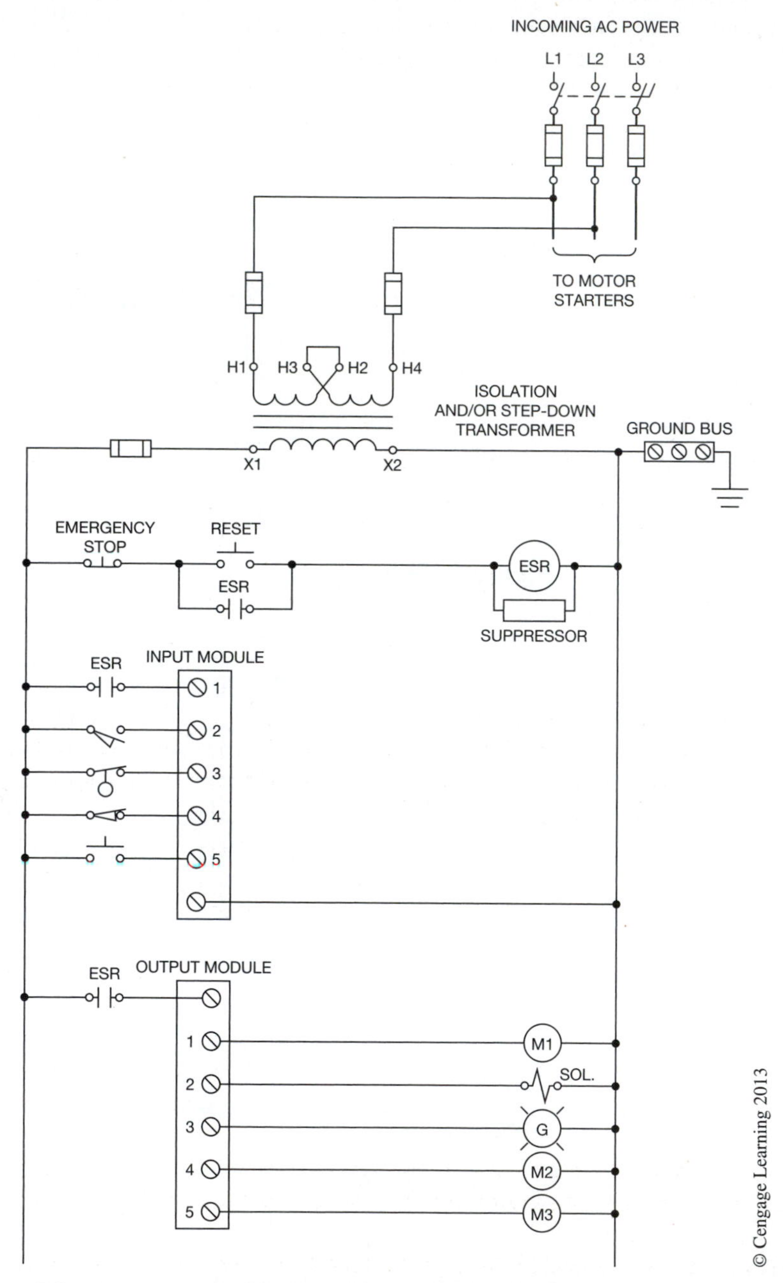

Figure 2–31 Power Distribution with Emergency Stop Relay (ESR) for a Grounded AC System

machine tool, or where the machine cycles automatically, consideration should be given to the use of an electromechanical override or other redundant means, independent of the controller, for starting or interrupting the cycle."

While programmable logic controllers of today are rugged and dependable, where safety is concerned *do not* depend on solid-state devices and circuitry of the PLC, or PLC program. The NEMA recommendation recognizes the importance of a hardwired emergency-stop, or E-Stop (shown in Figure 2–31), to remove power to the output devices. A second Emergency Stop Relay (ESR) contact is typically wired to a PLC input module so that the program logic can take appropriate action for a safe restart of the system.

It is also worth noting that solid-state output devices usually (though not always) fail shorted, rather than in an *open* condition. Because they fail in a shorted or *ON* condition, an added safety hazard is possible if a hardwired emergency stop (E-Stop) is not included as part of the PLC installation.

RACK INSTALLATION

Before installing a rack or chassis, consideration must be given to the following:

- temperature
- dust
- vibration
- humidity
- field wiring distances
- troubleshooting accessibility

The ambient temperature of the proposed location should not be lower than 32°F or higher than 140°F (0°C and 60°C). Fans are normally not used with I/O racks, and all cooling of the electrical or electronic components is accomplished by convection. Convection cooling is accomplished when warm air caused by heat in the components rises and creates a movement of air. This movement of air draws cool air in through the bottom of the rack and expels warm air out through the top. To maintain efficient convection cooling, it is important that the rack be installed correctly and not used as a shelf for notebooks or other material that would impede or block the natural flow of heat up through the rack.

During initial installation it is common practice to cover the top of the rack to prevent any scrap wire, stripped insulation, screws, and nuts from falling into the I/O modules, power supply, or processor, which could cause a short circuit or other electrical failure. The protective cover must be removed after installation to assure that proper cooling can take place.

Under adverse conditions, when the ambient temperatures exceed the manufacturers' recommended maximums, enclosure fans, ventilation louvers, or air condition units can be used.

When the temperature is expected to go below 32°F, a thermostatically controlled heater is used inside the enclosure to prevent condensation.

Dust can also cause a problem in the I/O rack when it accumulates on the electronic components of the modules, power supply, or processor. Accumulated dust prevents the components from dissipating heat effectively. A dust-tight enclosure with a cover and gasket can be used to prevent problems that dust can create. It is important to remember that any enclosure used to house PLC components must be large enough to allow for proper air circulation and heat dissipation. If the enclosure is too small, the heat will build up inside the enclosure and have a detrimental effect on the electrical or electronic components. The installation manual usually specifies the minimum size of enclosure that can be used.

Excessive vibration can also lead to early component failure. It is important to mount PLC equipment on solid, nonvibrating surfaces. Vibration effects from equipment must be minimized to assure proper longevity for the equipment.

Humidity, while normally not a problem, must be considered when installing a PLC. Allen-Bradley, for example, rates their PLCs for operation in a humidity range of 5%–95% (without condensation). Some manufacturing processes, however, create high humidity (high moisture content) conditions. Exposing electronic equipment to extremely high humidity environments over an extended period of time can reduce component life and affect operation. Evaluate the environment carefully and mount equipment in an area that minimizes the exposure to high humidity and moisture.

While it is important that the controller and programming unit be installed or mounted in a control center or other central location, the use of remote I/O racks allows the input and output modules to be installed close to the actual operating equipment (Figure 2–32).

By mounting the I/O rack close to the actual equipment, the amount of conduit, cable, and other associated wiring and labor costs will be decreased. The only wiring needed for communication back to the processor will be a shielded-twisted pair, twin axial cable, fiber optic cable, or the like. By having the input and output modules located close to the process or driven equipment, troubleshooting is also easier and more efficient. As discussed earlier in this chapter, each input and output module has status lights that indicate whether an input or output device is *ON* or *OFF*. Having this capability close to the actual equipment shortens troubleshooting time and increases production.

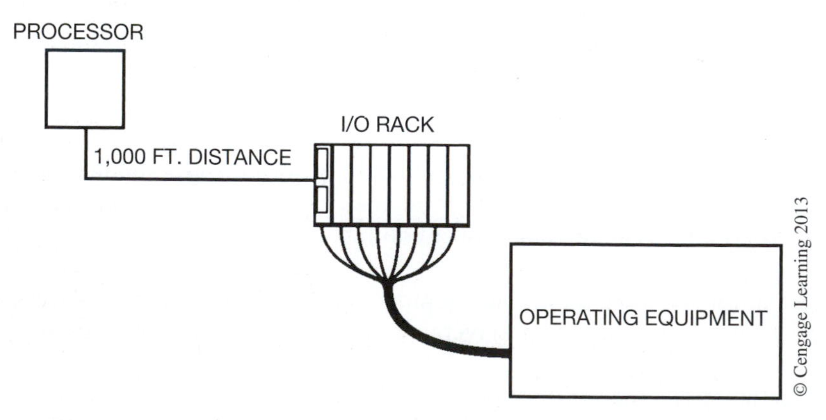

Figure 2–32 Remote I/O Rack Close to Operating Equipment

Before mounting the racks and other associated equipment, give careful consideration to location and accessibility. If access is restricted or the equipment is difficult to reach, troubleshooting, repair, and maintenance will be more difficult and time consuming.

ELECTRICAL NOISE (SURGE SUPPRESSION)

Electrical noise is generated whenever inductive loads such as relays, solenoids, motor starters, and motors are operated by "hard contacts" such as push buttons, selector switches, and relay contacts. The noise, or high transient voltages (spikes), is caused by the collapsing magnetic field when the inductive device is switched *OFF*. The level of the voltage spike can be very high and is capable of causing erratic operation of the processor and/or output module, or can cause permanent damage to the module. The interference caused by these voltage spikes and the accompanying electrical noise is often called Electromagnetic Interference (EMI). There are several steps that can be taken to reduce or eliminate the effects of EMI. Two of the most common are isolation and suppression.

Isolation of the electrical noise is accomplished by installing an isolation transformer for the PLC system (Figure 2–33) to supply the power for the controller and the input circuits. The figure

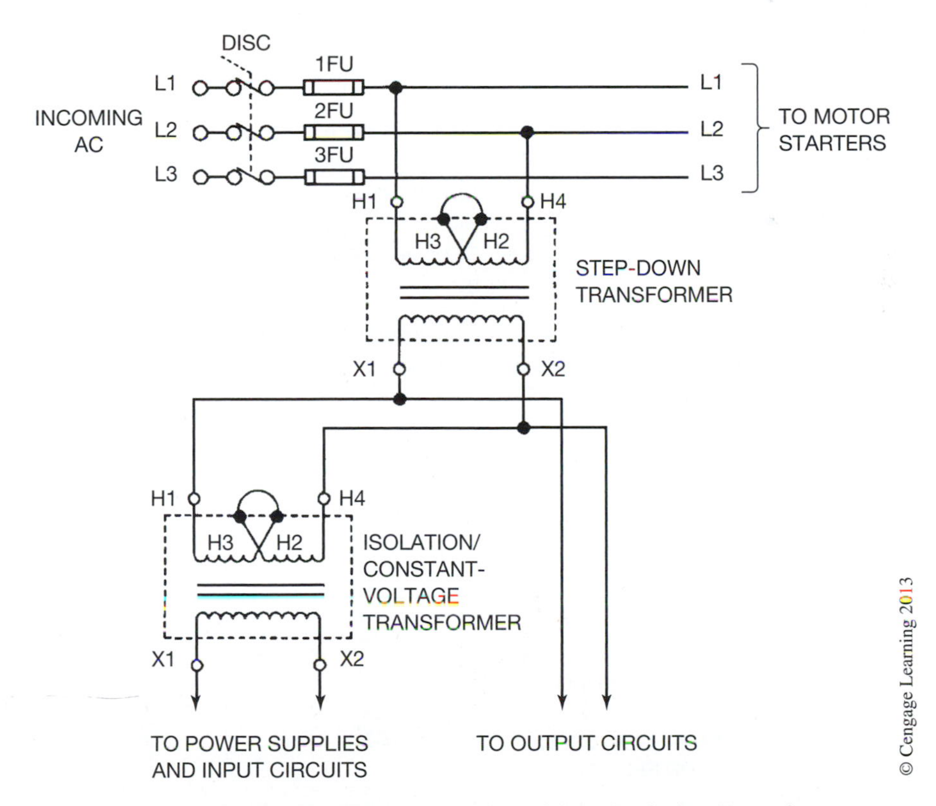

Figure 2–33 Reducing Electrical Noise with an Isolation Transformer

shows a constant voltage transformer, but a standard step-down transformer of the proper size also effectively isolates electrical noise.

A second method in reducing EMI is to install surge suppression networks or devices on the individual motor starters, motors, and solenoids. These suppression devices can consist of an RC circuit (resistor/capacitor), an MOV, or an RC combination for AC loads and a diode for DC coils. The collapsing magnetic field of the inductive device is, in a sense, dissipated by the suppression network and reduces the effects of EMI. Figure 2–34 shows typical installations of the various types of suppression devices.

The type of surge suppressor to use depends on the size and type of load. An equipment representative or local electrical distributor should be consulted for help with selection and application.

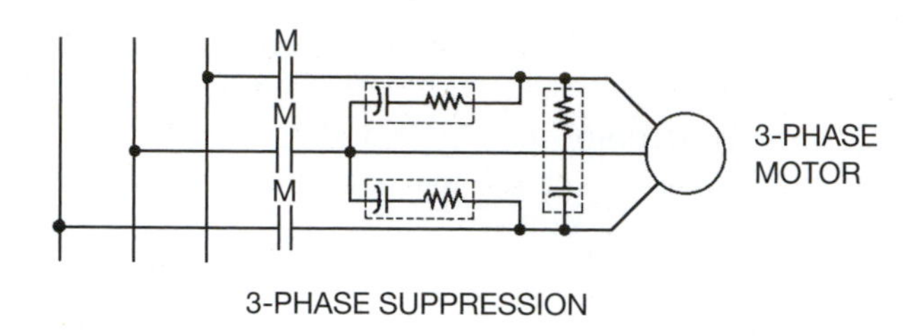

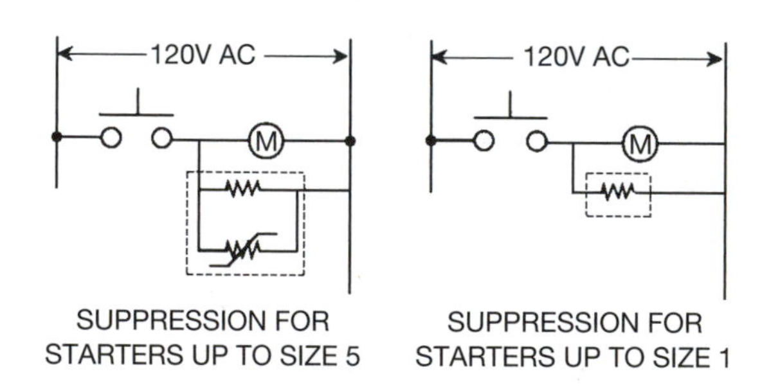

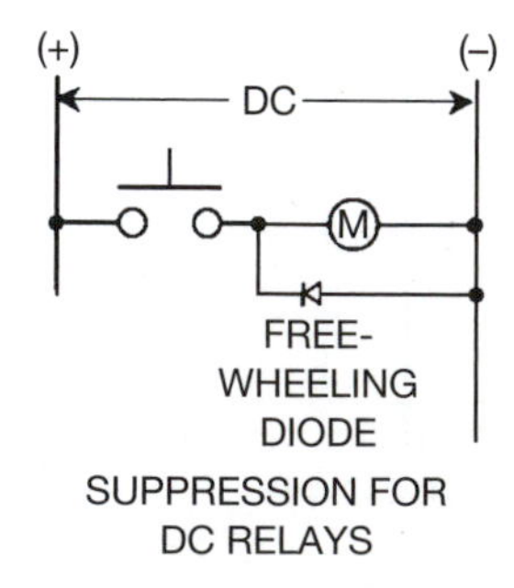

Figure 2–34 Typical Suppression Devices

GROUNDING

With solid-state control systems, proper grounding helps eliminate the effects of electromagnetic induction. Figure 2–35 shows a typical installation using an equipment grounding conductor to connect several PLCs and/or I/O racks together. The equipment grounding conductor is attached to the metal frame of the PLC and/or I/O rack with a ground lug. A detail of the connection is shown in Figure 2–36.

Note: *Check local codes and manufacturers' specifications to ensure proper installation.*

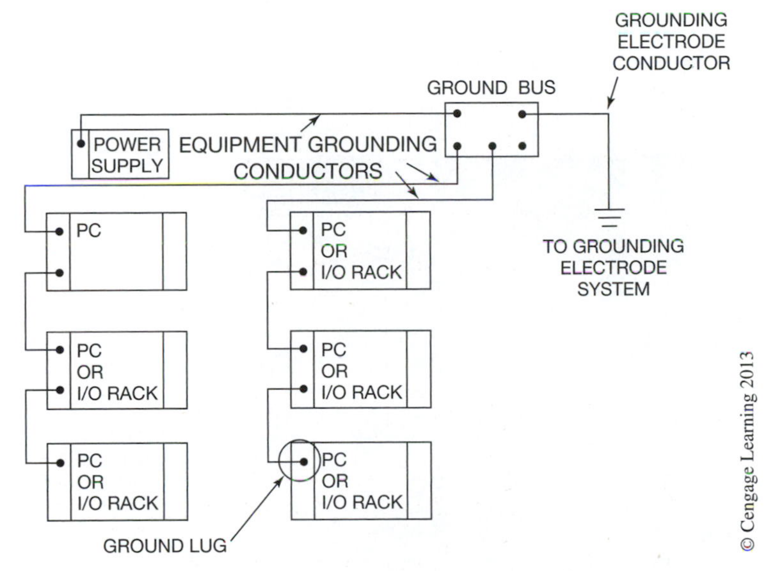

Figure 2–35 Typical Equipment Grounding Configuration

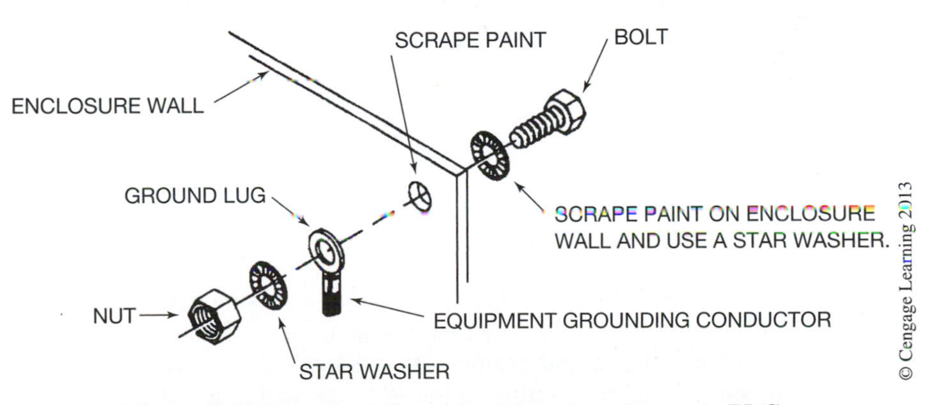

Figure 2–36 Detail of Grounding Lug Attachment to PLC

I/O SHIELDING

Certain I/O modules such as TTL, analog, and thermocouples require shielded cable to reduce the effects of electrical noise. The cable shield, which surrounds the cable conductors, shields the conductors from electrical noise.

When installing shielded cable, it is important that the shield only be grounded at one end. If the shield is grounded at both ends, a ground loop is created, which can introduce ground currents that may result in faulty input signals and/or operation of the processor.

As a properly grounded I/O rack is already connected to earth ground through an equipment grounding conductor, the shield should be terminated at the I/O rack, *not* at the device end.

Figure 2–37 shows the shield of a shielded cable connected to the I/O rack frame.

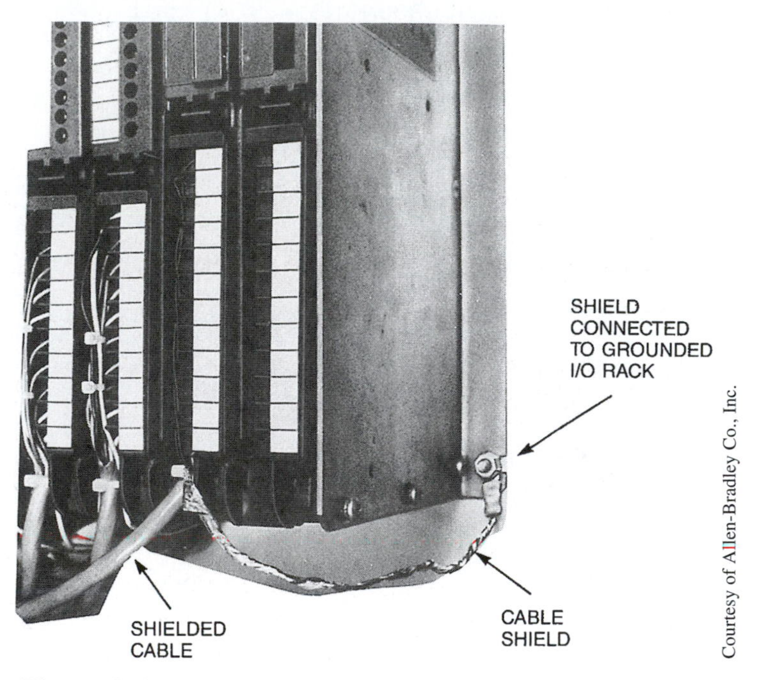

Figure 2–37 Cable Shield Connected to Grounded I/O Rack

Figure 2–38 shows a shielded-twisted pair cable connected to a sensing device and I/O Rack. In the first example the shield is connected to the I/O rack frame as some manufacturers recommend. An alternative method is shown in the second example. In this method terminal blocks are used within the PLC enclosure to connect the field wiring to the I/O rack wiring and a third terminal block is

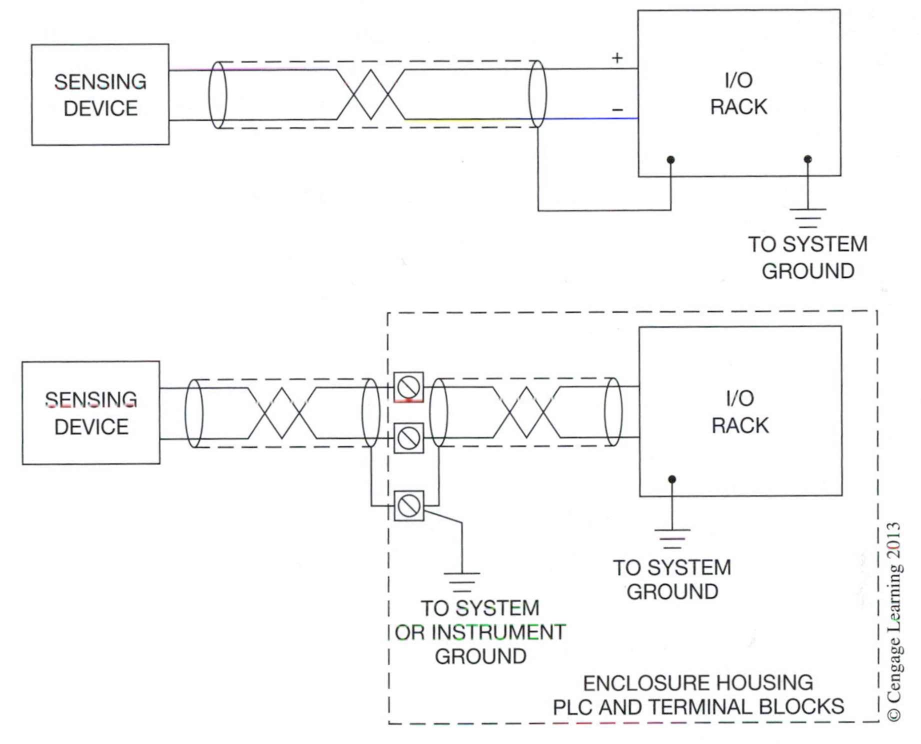

Figure 2–38 Shields Connected to Ground at One End Only

used to connect the shields to the system or instrument ground. In either case the shields are to be connected at one end only, preferably the I/O rack end.

Additional methods of noise reduction are as follows:

- Mount equipment in metal enclosures, when possible, because metal helps protect against EMI.
- Separate I/O and PLC wiring from the motor and other large loads to reduce the possibility of induction in the control circuits. This is usually accomplished by installing the control wiring in one raceway, or cable-tray, and the power circuits in another raceway or cable-tray with physical distance or separation between the two.

Chapter Summary

The I/O rack houses the individual input and output modules that are connected to real-world devices. The input modules act as an interface between the actual input devices and the processor, while the output modules act as an interface between the actual output devices and the processor. The status (*ON* or *OFF*) of the input devices is communicated to the processor; the processor makes a decision, and in turn communicates to the output modules to turn *ON* or *OFF* the output devices that are connected to the output module. The processor may be connected to the I/O rack by way of interconnecting cable(s), through a bus duct, or it may be mounted in the same rack as the I/O.

The I/O section is divided into two categories: fixed and modular. Discrete I/O modules operate on digital, or *ON* and *OFF* signals, whereas analog I/O operates on a variety of signal levels and types. Input and output modules are available in a variety of voltages and normally can control 8, 16, or 32 individual input or output devices. Optical coupling, or isolation, is used to protect the low-voltage (5 V DC) side of the processor from the line-voltage input and output signals that can be as high as 240 V.

AC output modules typically use Triacs for switching *ON* and *OFF* the actual output devices. When in the high-resistive, or *OFF* state, Triacs have a small leakage current that flows through the output device. When Triacs fail, they normally fail in the *ON* condition. Fuses used for protection of output modules are carefully selected by the manufacturer for current and time characteristics, and only fuses recommended by the manufacturer should be used to prevent possible damage to the equipment.

PLC troubleshooting is simplified by the addition of status lights on the I/O modules. The lights indicate which inputs are *ON* or *OFF* and which outputs are *ON* or *OFF*. Indicator lights also indicate if an output module has a blown fuse. To prevent an incorrect module from being installed in a given rack slot, the modules are often keyed.

A wide variety of input and output modules are available that fit almost any application. Care must be taken to ensure that the module has the correct voltage, current, and time characteristics. The various PLC manufacturers continue to introduce innovative new modules to meet the changing requirements of automated equipment and process control.

Proper installation of PLC equipment requires that the environment (dust, heat, humidity, and vibration) be considered, as well as the physical location for access and troubleshooting. Reduction and/or elimination of electrical noise, voltage spikes, voltage variation, and the like is necessary to ensure proper operation of the system.

Review Questions

1. Describe briefly the purpose of the I/O section.
2. State two reasons for employing optical isolation.
3. Draw an AC input module with four input devices, show all necessary electrical connections, and identify potentials L1 and L2.
4. Draw an AC output module with four output devices, show all necessary electrical connections, and identify potentials L1 and L2.

5. Triacs are susceptible to "dielectric-type" breakdown if the maximum peak voltage level is exceeded.
   T F
6. Briefly describe why a hardwired emergency-stop circuit is recommended for PLC installations.
7. Briefly describe the function of an interposing relay.
8. I/O modules are keyed to prevent unauthorized personnel from removing them from the I/O rack.
   T F
9. Which of the following are *not* normally sources of electrical noise?
   a. solenoid
   b. relay
   c. indicator lamp
   d. motor starter
   e. motor
   f. overload heaters
10. To ensure maximum benefit of shielding, the shield of a shielded cable must be terminated and grounded at both ends.
    T F
11. E-Stop refers to
    a. extra stop
    b. emergency-stop
    c. every stop
    d. elevator stop
    e. energy stop
12. Electromagnetic interference (EMI) can be reduced with the proper grounding of equipment.
    T F
13. Solid-state output devices tend to
    a. never fail
    b. fail in the *open* or *OFF* condition
    c. fail in the *shorted* or *ON* condition
    d. not be affected by overload
14. List three environmental considerations when installing PLC equipment.
15. What type of tool or object should be used to change the position of DIP switches?

CHAPTER 3 Processor Unit

Objectives

After completing this chapter, you should have the knowledge to:

- Describe the function of the processor.
- Describe a typical program scan.
- Identify the two distinct types of memory.
- Describe the function of the *watchdog timer.*
- Identify various memory designs.
- Describe the two types of programming devices.
- Explain the terms *on-line* and *off-line* programming.

The processor unit houses the microprocessor, memory module(s), and the communications circuitry necessary for the processor to operate and communicate with the I/O and other peripheral equipment. The DC power required for the processor is provided either by a power supply that is an integral part of the processor unit, or by a separate power supply unit, depending on the manufacturer. The processor, or "brain," of the programmable logic controller is the decision-maker that controls the operation of the equipment to which it is connected. The processor controls the operation of the output devices that are connected to the output modules based on the status of the input devices and the program that has been entered into memory (Figure 3–1). The processor is often referred to as the central processing unit or CPU.

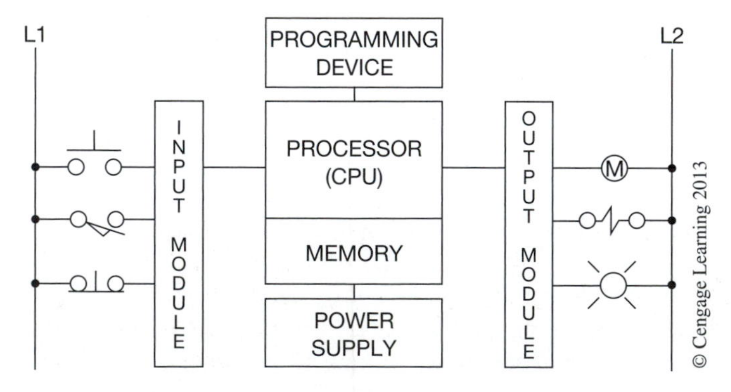

Figure 3–1 Basic PLC Configuration

Processors are available that control as few as 8, or as many as 40,000, real-world inputs and/or outputs. The size of the processor unit to be used is dependent on the size of the process(es) or driven equipment to be controlled. The larger the number of input and output devices that are required for the process, the more powerful the processor must be to properly control the number of I/O that will be connected. One PLC can control more than one machine or process line and is limited only by the I/O required, physical distance, and memory capacity of the PLC used.

Note: *It is difficult to discuss processor unit configuration due to the differences in PLC hardware from the various manufacturers. The discussion that follows is general and is not intended to cover all the PLCs on the market today. It should also be noted that when pictures are used to illustrate a given configuration or concept, one of a particular manufacturer's models is illustrated; however, the manufacturer also has other models larger and/or smaller, and in different configurations.*

THE PROCESSOR

The processor may be a self-contained unit, or may be modular in design, and plug directly into the I/O rack as shown in Figure 3–2. Whatever the configuration, the processor consists of a microprocessor, memory chips, circuits necessary to store and retrieve information from the memory, and communication circuits required for the processor to interface with the programmer and other peripheral devices. The memory and communication circuits can be modules separate from the microprocessor module. The actual hardware configuration will depend on the PLC.

Courtesy of Rockwell Automation, Inc.

Figure 3–2 ControlLogix XT with a Logix 5563 Processor Installed

The microprocessor is the device that

1. Monitors the state or status (*ON* or *OFF*) of the input devices.
2. Systematically solves the logic of the user program.
3. Controls the state of the output devices (*ON* or *OFF*).
4. Communicates with other devices (operator interface terminals, personal computers, etc.).
5. Manages memory and updates timers, counters, and internal registers.

The execution or completion of these tasks is referred to as the processor scan.

When the PLC is powered up or turned *ON,* the processor runs an internal self-diagnostic, or self-check, prior to initiating its first scan. If any part of the processor system is not functioning, such as a faulty memory, improper communication with the I/O section, or failure in a remote rack, the processor fault light or other indicator light comes *ON*. With some systems, if a monitor is connected, a written explanation or fault code will appear on the screen. Some systems use status words to indicate the hardware or software that has malfunctioned. Status words can be included in the program so that when a malfunction is detected, an alarm will sound to alert the operator that there is a problem.

Once the processor has passed the self-diagnostic check, it is ready to go to work. Figure 3–3 illustrates a typical four-step PLC scan. In the first step of the scan, the processor determines the status of the input devices. It does so by looking at the memory locations that have been designated for all the input devices. Remember, as stated earlier in the text, the actual status (*ON* or *OFF*) of any input device is stored in a memory location as either a 1 or a 0. A 1 indicates that a device is *ON* or closed, while a 0 indicates that the input device is *OFF* or open. Based on the 1s and 0s, the processor determines the actual condition of all the input devices.

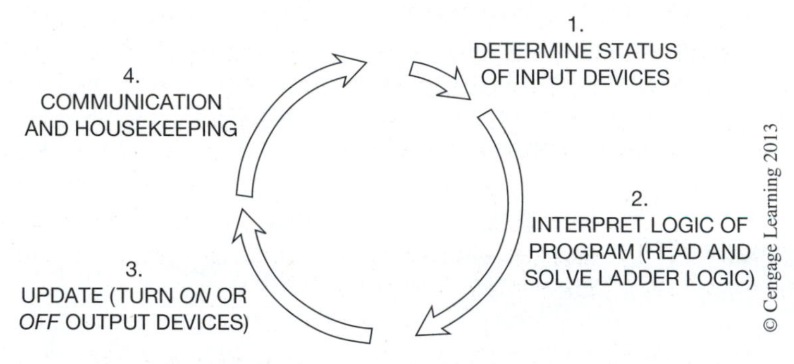

Figure 3–3 Typical Processor Scan

The second step in the processor scan is to interpret the logic of the program that has been written and stored in the processor memory. Based on the program requirement, the processor will turn the required output devices *ON* or *OFF,* which is the third step in the processor scan. This third step is referred to as **updating** the outputs. This updating process occurs once during each scan. The fourth step of the scan is often referred to as housekeeping. During this part of the scan the processor will perform any necessary memory management (housekeeping). Memory management will include updating timers and counters as well as internal registers, etc. In some processors during this part of the scan, the processor will communicate with any connected devices (communications). Some PLCs today have two processors, one that interprets the logic and a second that handles the I/O updates.

The scan is continuous and the four-step process is repeated over and over every few milliseconds. To summarize, the four steps of the scan are:

1. Determine the status of the input devices (*ON* or *OFF*).
2. Read and solve the logic of the program (ladder logic).
3. Update the output devices (turn *ON* or *OFF*).
4. Evaluate communications and housekeeping procedures.

The time it takes to complete one scan will vary from a fraction of a millisecond to 50+ milliseconds. Scan time is dependent on many factors such as size and number of program tasks, program instruction types, and memory used. Some manufacturers allow you to specify the percentage of processors' time that is devoted to communication and background functions, which also impacts overall scan times.

Note: *As part of the processor's internal self-diagnostic system, a* **watchdog timer** *is used. The watchdog timer is preset to an amount of time that is slightly longer than the scan time would be under normal conditions. At the start of each scan, the watchdog timer is turned* ON *and starts to accumulate time. If the program is correct, the program scan will be completed prior to the time set on the watchdog timer, and at the end of each scan, the watchdog timer is reset to 0. If for some reason the program scan is not completed in the allotted time, indicating that there is a problem with the program, the watchdog timer will time out, which puts the processor into a faulted condition. The range of the timer is software selectable (adjustable) on many PLCs.*

Normally, before any output devices can be turned *ON* or *OFF,* the processor has to scan the entire program that is in user memory. The program may be only a few rungs long or it may be hundreds of rungs in length, depending on the equipment that is being controlled. Some input devices operate so fast that by the time the user program can be read and solved and outputs updated, the input device may have changed positions more than once since the processor originally determined its status at the start of the scan. The same may be true for an output device that needs to be updated sooner than a regular scan will allow. To solve this problem, many PLCs have special program instructions that allow critical or high-speed input and output devices to be updated sooner than would be possible under normal scan conditions. The special instructions actually interrupt the scan when it is reading the program and allow I/O devices to be updated immediately.

As mentioned earlier, some PLCs have two processors, one that interprets the logic and a second that handles the I/O updates. With this type of processor the I/O is being updated during the execution of the program. This means the I/O updates **asynchronous** to the execution of the logic, which improves I/O update times.

Note: *Additional information, and a more detailed discussion on how the processor scans the user program, are provided in Chapter 9.*

The memory section of the processor consists of hundreds or thousands of locations where information is stored. In the broadest sense, the memory is divided into two classifications: user and storage. The **user memory** is for the storage of the user program that contains the relay logic, or instructions that control the driven equipment or the process. The **storage memory** is used to store information such as input/output status, timer or counter preset and accumulated values, and internal control relays, etc., which is necessary for the processor to control the equipment or process. The actual memory structure of various PLC manufacturers will be covered in Chapter 4, while the purpose or use of each memory is covered later in this chapter.

Memory chips used in the processor can be separated into two distinct groups: **volatile** and **non-volatile.** A volatile memory is one that loses its stored information when power is removed. Even momentary loss of power erases any information stored or programmed on a volatile memory chip. A nonvolatile memory has the ability to retain stored information when power is removed, accidentally or intentionally. To protect a volatile memory, backup batteries are included in the processor or power supply. The batteries may be rechargeable nickel cadmium, lead acid, or nonrechargeable alkaline or lithium types.

Caution: Extra care must be exercised when disposing off batteries, since they are classified as hazardous waste. Special care must be taken with lithium batteries because they may explode when exposed to, or dropped into, water.

When batteries are included, they may be located in the processor, or in the power supply, depending on the PLC. Wherever they are located, there is a battery indicator light(s) to indicate the condition, or state of charge, of the batteries. Common indicator lights are *BAT OK* and *BAT LOW*. A simpler system uses one light to indicate that the battery condition is normal. When the light goes out, it is a warning that the batteries need to be replaced.

When the battery indicator light comes on (or goes out), indicating that the batteries need to be replaced, the memory is still protected for a minimum of two weeks. Depending on the size of the memory and the type of batteries used, in many cases the memory remains protected for one year or more with fully charged batteries. In reality, rarely is the power interrupted or off for more than a few hours.

The type of batteries used and the number required will vary with each manufacturer. Because alkaline and lithium batteries are not rechargeable and must be replaced periodically, care must be taken to always replace the batteries with the type specified, paying special attention to the orientation of each battery in the battery holder to ensure that proper polarity is maintained. Some batteries are available with leads that simply plug into a connector on the PLC. Figure 3–4 shows a lithium battery with leads that is mounted on the back of the faceplate cover, and is used to backup the memory of the GE Fanuc PLC.

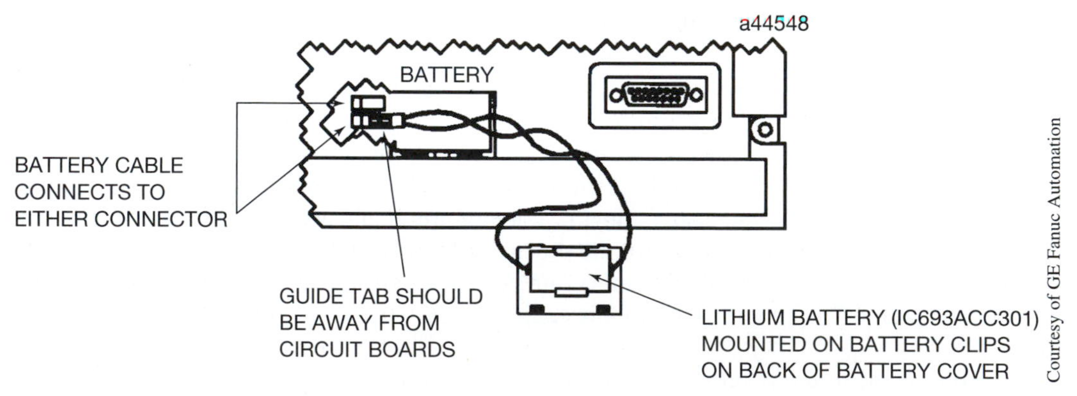

Figure 3–4 Lithium Battery with Leads

Caution: As a general rule, a copy is made of the current program prior to changing the batteries. This copy is referred to as a "backup" copy, and is used to replace the original program if for some reason the program in memory is lost. The batteries in several PLCs can be changed without turning off the main power. The processor unit of the GE Fanuc PLC, shown in Figure 3–5, uses a single lithium battery that protects the volatile memory. Note that a second battery connector has been added and wired in parallel. The new battery is attached to the second connector in order to protect the memory before the old battery is removed. Changing batteries is one of the few maintenance requirements of a PLC. Failure to change the batteries in a timely manner may have serious consequences if a backup copy of the program is not made. Common sense dictates that a backup copy be made of every PLC program.

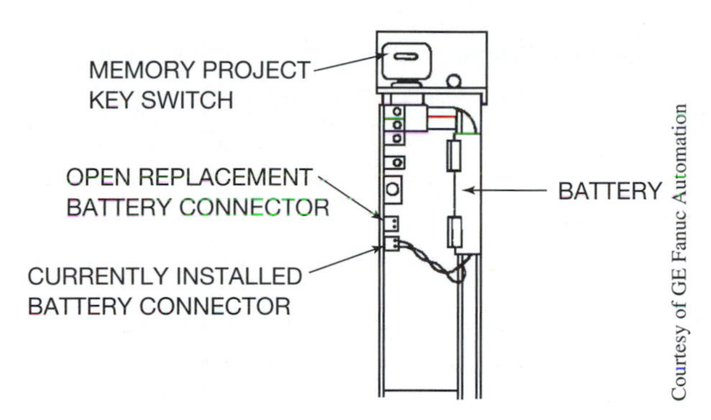

Figure 3–5 Parallel Battery Connection Used with GE Fanuc 90-70 CPU

MEMORY TYPES

No attempt will be made to explain solid-state memory types in more than a generalized way for basic understanding. Detailed explanations of solid-state memory types are available in the electronics section of most libraries.

The most common type of volatile memory is **Random Access Memory (RAM).** Information can be written into, or read from, a RAM chip, and it is often referred to as read/write memory. Information stored in memory can be retrieved or read, while "write" indicates that the user can program or write information into the memory. Random access refers to the ability of any location (address) in the memory to be accessed or used. RAM is used for both the user memory and storage memory in many PLCs. Since RAM is volatile, it must have battery backup to retain or protect the stored program. Various forms of RAM include MOS, HMOS, and CMOS-RAM (Complimentary Metal Oxide Semiconductor), one of the most popular, to name just a few.

CMOS-RAM is popular because it has a very low current drain when not being accessed (15 μ amperes), and the information stored in memory can be retained by as little as 2 V DC. A typical fully charged lithium battery is rated 2.95 V at 1.75 amperes/hour and normally holds or protects a program for 60 days or longer.

Nonvolatile memories are memories that retain their information or program when power is lost, and do not require battery backup. A common type of nonvolatile memory is **Read Only Memory (ROM).** "Read only" indicates that the information stored in memory can only be read, and cannot be changed. Information in ROM is placed there by the manufacturer for the internal use and operation of the PLC, and the manufacturer does not want the information changed or altered. PLCs, like other computer-based systems, undergo constant change. When changes are made in the way a system operates, or when new features are added, ROM chips can be replaced to upgrade the PLC.

Other types of nonvolatile memory are PROM, UVPROM, EPROM, EEPROM, and FLASH.

PROM Programmable Read Only Memory allows initial and/or additional information to be written into the chip. PROM may be written into only once after being received from the PLC manufacturer, and programming is accomplished by pulses of current. The current melts fusible links in the device, preventing it from being reprogrammed. This type of memory is used to prevent unauthorized program changes.

Note: *Regardless of the memory type, the memory can also be protected by a key switch located on the front of the processor, or on the programming device. With the programmer "locked-out," the program in the processor can be run but not changed. The key switch can also be used to lock the processor out completely and prevent it from running the program.*

Another popular method of restricting access to the program is to use passwords. Passwords restrict access to the program to only those personnel who know the correct password and how to enter it using the programming device. Passwords are often referred to as "software" locks, whereas key switches are referred to as "hardware" locks.

UVPROM–EPROM Ultra Violet Programmable Read Only Memory is ideally suited when program storage is to be semipermanent, or additional security is needed to prevent unauthorized

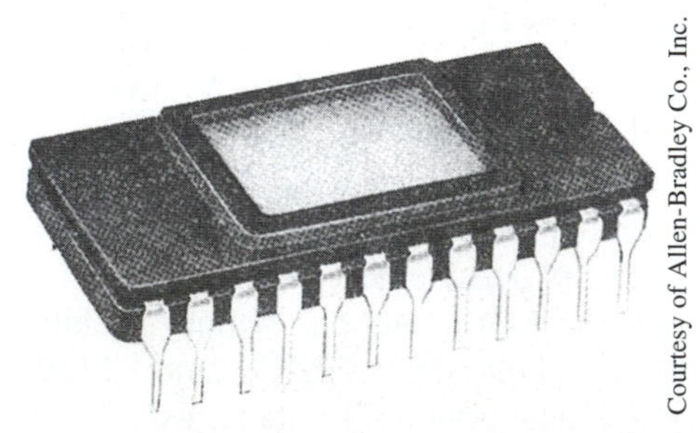

Courtesy of Allen-Bradley Co., Inc.

Figure 3–6 Typical UVPROM or EPROM Memory Chip

program changes. The UVPROM chip is also referred to as **EPROM** (**E**rasable **P**rogrammable **R**ead **O**nly **M**emory) (Figure 3–6). The EPROM chip has a quartz window over a silicon material that contains the electronic integrated circuits. This window is normally covered by an opaque material, but when the opaque material is removed and the circuitry exposed to ultraviolet light, the memory content can be erased. Once erased, the EPROM chip can be reprogrammed, using a special programmer. After programming, the chip window must once again be covered with an opaque material, such as electrician's tape, to avoid undesirable alteration of the memory.

Caution: Special care and handling of the UVPROM, or for that matter any integrated circuit (IC) chip, must be exercised to ensure that the pins do not become dirty, bent, or subjected to any static electric charges.

EEPROM Electrically Erasable Programmable Read Only Memory is also referred to as Double EPROM and E2PROM. EEPROM is a chip that can be programmed using a standard programming device and can be erased by the proper signal being applied to the erase pin. EEPROM is used primarily as a nonvolatile backup for the user program in RAM. If the user program in RAM is lost or erased, a copy of the program stored on an EEPROM chip can be downloaded into RAM. It is common on some PLCs for the processor to load the program from the E2PROM chip into RAM memory each time the processor is powered up or after a power failure. Figure 3–7 shows an EEPROM memory card used with the Modicon 984-120 Compact PLC to store the user program. This credit card size device offers a convenient method for copying and/or loading user programs.

FLASH Flash memory is a nonvolatile memory chip that can be electrically erased and reprogrammed. It is a specific type of EEPROM that is erased and programed in large blocks. The CompactFlash memory card is being used with some PLCs on the market today.

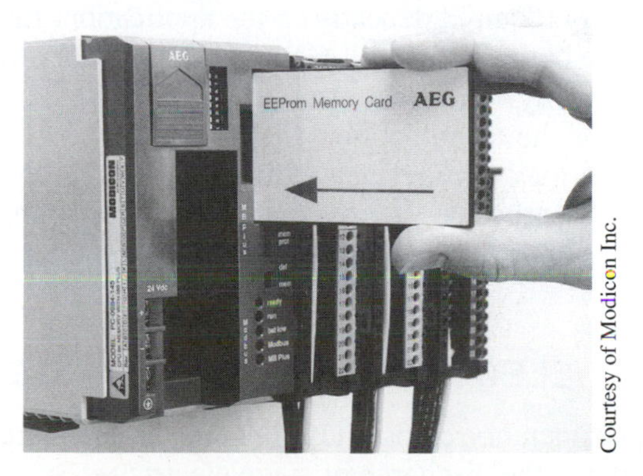

Courtesy of Modicon Inc.

Figure 3–7 EEPROM Memory Card

MEMORY SIZE

PLCs are available with memory sizes ranging from as little as 256 words for small systems up to 32 meg (million) for the larger systems. Memory size is usually expressed in K values: 2K, 4K, 16K, and so on. K, or kilo, which usually stands for 1,000, actually represents 1,024 in computerese. The difference between a standard K (1,000) and the 1,024K value used with processors and computers is due to the way the words were counted. One of the counting or numbering systems used with PLCs is the **binary system.** The binary system has a base 2, as contrasted to the decimal system we use every day that is base 10. Base 10 represents the numbers 0 through 9, which is 10 digits. The binary numbering system with a base 2 only has 2 digits. The digits are 1 and 0. As with the decimal system that has place values (tens, hundreds, thousands), the binary system also has place values. These are 1, 2, 4, 8, 16, and so on, and each place value is equal to twice the value of the previous number. Base 2^0 represents the number 1; base 2^1 represents 2; base 2^2 represents the number 4 ($2 \times 2 = 4$); base 2^3 represents the number 8 ($2 \times 2 \times 2 = 8$); and so on. Counting in this fashion, 2^{10} would equal 1,024. While 1,024 is actually larger than the 1,000 that K actually represents, K (with a value of 1,024) is used in PLCs, and personal computers as well. This also explains the reason for the odd memory sizes of individual memory chips: 256 (1×2^8); and 512 (1×2^9). A memory chip of 256 words would be ¼K and 512 words would be ½K. A PLC with a total memory of 64K would actually have 65,536 words of memory (64 times 1,024). Words, word structure, and numbering systems are covered in Chapter 6.

While it is common for PLCs to measure their memory capacity in words, it is important to know the number of bits in each word. A PLC that uses 8-bit words would have half the memory capacity of a PLC that uses 16-bit words. For example, the PLC that uses 8-bit words has 65,536 bits of storage with an 8K word capacity ($8 \times 8 \times 1024 = 65{,}536$), whereas a PLC using 16-bit words has 131,072 bits of storage with the same 8K memory ($16 \times 8 \times 1024 = 131{,}072$). It is important to know the word size of any given PLC before memory size can be accurately compared.

Note: *Personal computers and some PLC manufacturers, such as Simatic T.I., size memory in bytes, not words. A byte is 8 bits, or half of a 16-bit word. A 32-bit word would have 4 bytes.*

The actual size of the memory required depends on the application. In the event that future expansion is planned, there are two options: buy a PLC with more memory than is presently necessary to allow for future expansion, or buy a PLC that meets present needs and add memory (upgrade) when the need arises. Depending on the manufacturer, adding memory may be as simple as replacement of the memory module, or it may require that additional memory chips be added to the existing memory module. Some processors have no provisions for memory expansion and must be replaced if the memory needs to be increased.

GUARDING AGAINST ELECTROSTATIC DISCHARGE (ESD)

A major cause of failure of memory chips and other sensitive electronic components is electrostatic discharge. ESD is simply the discharge of static electricity. Static electricity can build up on the surface of a workstation, on clothing, the carpet on the floor of the workplace, on plastic cups, styrofoam, cellophane, and other such materials. To reduce the possibility of damage from ESD, the following precautions should be taken:

1. Use nonstatic floor coverings.
2. Handle chips correctly.
3. Ground the work surface.
4. Wear a wrist strap.

Grounding the work surface and wearing a wrist strap are the two most important precautions to take. Be sure to put on the wrist strap before starting to handle memory chips. Make sure it fits snugly, that the metal pad of the strap is touching your skin, and that the wrist strap ground wire is securely fastened to ground.

Although some manufacturers indicate that ESD is not a problem with their products, this is misleading; always follow the ESD precautions when handling memory chips and other sensitive electronic components.

For PLCs in which the memory can be expanded by adding volatile RAM chip(s) to the memory module, the following procedures should be used:

1. Record a copy of the current user program on disk.
2. Remove main power from the PLC.
3. Remove the memory module and take to a clean area.
4. Carefully remove any screws necessary to gain access to the printed circuit board where the extra RAM sockets are located. If the backup battery is located on the module, disconnect the battery before removing or installing memory chip(s).

Note: *The RAM chip(s) will come packaged in a conductive plastic bag (often referred to as a "static" bag). Within the bag, each RAM chip will be inserted into a conductive sponge-like material. The conductive, yet highly resistive, material is used to keep all the pins of the chip at the same electrical potential.*

Caution: When working with RAM chips, do not handle cellophane covered articles such as cigarette packages or candy wrappers, plastic, styrofoam, or other materials that can cause a static charge. Do not install the chip in carpeted or contaminated areas where pins may become fouled. And never slide the RAM chip across any surface, store a RAM chip in a non-conductive plastic bag, or insert the chip into non-conductive material.

The volatile RAM chips used today are not as susceptible to damage from static charges as they were a few years ago. But rather than just removing the chip from the conductive material and installing it into the proper socket, the following precautions still should be used:

1. Ground all tools before contacting the RAM chip, including yourself.
2. Wear a conductive wrist strap that has a minimum 200K ohm resistance and is connected to earth ground as shown in Figure 3–8.
3. Control relative humidity at 40 percent to 60 percent, if possible.

Remove the chip from the conductive foam. Be careful to touch only the chip base. *Do not* touch the pins. Inspect the pins for proper alignment. If any pins have been bent, gently straighten them

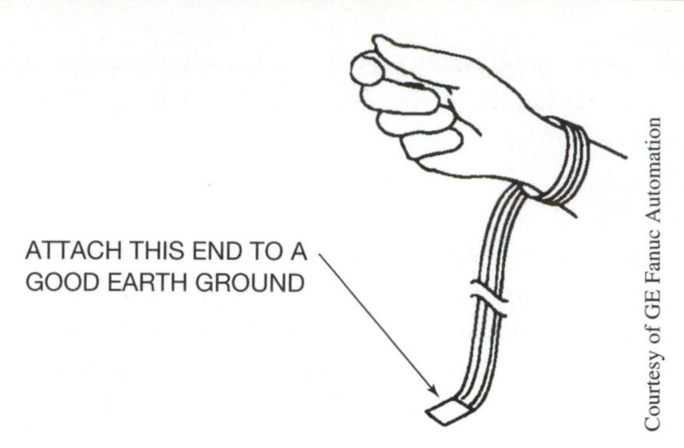

Figure 3–8 Wrist Strap Grounding Device

using needle-nose pliers that have been grounded. A dot or notch on the case of the chip is used for proper orientation of the chip into the socket. Grasp the chip by both ends and gently set it in the socket. *Do not insert.* Be sure the chip is positioned so the dot or notch of the chip matches the dot or notch on the socket.

Before attempting to insert the chip into the socket, check each pin to make sure it lines up properly with the corresponding socket point. Make any necessary pin adjustments as outlined above.

When pin alignment is ensured, insert the chip into the socket. Insertion is accomplished by pressing *gently* on the case of the chip until the chip is fully seated into the socket.

Carefully reassemble the memory module, remembering to reconnect the backup battery if one was mounted on the module. Reinstall the module in the processor and reapply power to the system. The user program can now be reentered into the processor and any additional user program can be added using the new memory chip(s).

MEMORY STRUCTURE

As indicated earlier, the processor memory is divided into two general classifications: user memory and storage memory.

User memory contains the instructions programmed by the user. The instructions are entered by a programming device.

Storage memory is where the status (*ON* or *OFF*) of all input and output devices is stored. Numeric values of timers and counters (preset and accumulated), numeric values for arithmetic instructions, and the status of internal relays also are stored in this memory.

While the information presented in this section applies generally to all PLCs, more specific information and memory structure can only be obtained by reviewing the specifications and literature of individual manufacturers. In subsequent chapters, the memory structure of specific PLCs will be discussed and illustrated, but the text does not cover all the PLCs on the market today.

PROGRAMMING DEVICES

A programming device is needed to enter, modify, and troubleshoot the PLC program, or to check the condition of the processor. Once the program has been entered and the PLC is running, the programming device may be disconnected. It is not necessary for the programming device to be connected for the PLC to operate, but it can be used to monitor the PLC program while the program is running.

Programming devices, or programmers as they are most often called, come in two types: hand-held and computer (Figures 3–9, 3–10).

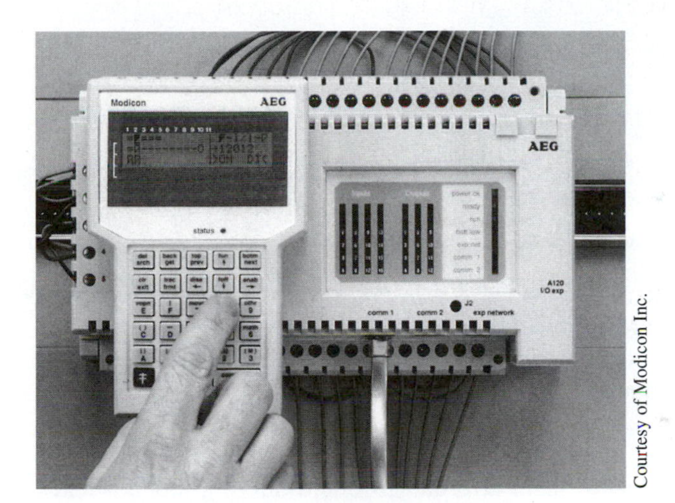

Courtesy of Modicon Inc.

Figure 3–9 Hand–Held Programmer

Figure 3–10 Laptop Computer

HAND-HELD PROGRAMMERS

Hand-held programmers are smaller, cheaper, and more portable than personal computer programmers. While the portability is a real plus, the hand-held programmer has some limitations.

Unlike the personal computer that can display a complete circuit network, hand-held programmers have limited display capabilities. Some hand-held programmers display a rung of logic with up to four horizontal lines, while others only display one line or one element at a time. The display is either LED or liquid crystal. Figure 3–11 shows an Allen-Bradley hand-held programmer with liquid crystal display.

The hand-held programmer may not have the full programming features of the personal computer programmer, and requires more "keystrokes" to actually enter a program. Hand-helds typically have restricted access to the processor memory.

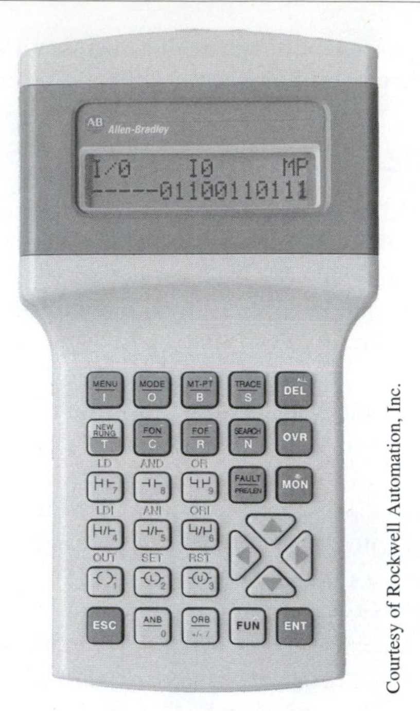

Courtesy of Rockwell Automation, Inc.

Figure 3–11 Allen-Bradley Hand-Held Programmer

On the plus side, hand-held programmers are well suited for installations that require constant changes in circuit requirements because they are lightweight (normally less than 2 pounds), portable, and ruggedly constructed. It is much easier to connect the hand-held programmer to the processor for changing program parameters or for troubleshooting than it is to bring out the large, heavier personal computer programmer.

While the relatively low cost of hand-helds makes them affordable troubleshooting tools, it takes more time to go through the program one contact or rung at a time. The extra time is the trade-off for the lower initial cost of the programming device.

COMPUTER PROGRAMMERS

With software available for all major brands of PLCs, the personal computer is the most common programming device used today. Personal computers are available in the small laptop variety (shown in Figure 3–10) or the even smaller notebook style. These small computers make for excellent programming devices because of their portability.

The personal computer usually has a color monitor, and the monitor shows multiple rungs of program logic, as well as highlighting the circuit elements to indicate status. The computer has the added ability to interface the PLC software program with other software programs for "cut and paste" program development and editing. The PLC software provides for documentation capabilities of the PLC program. The documentation may be in the form of labeling each element, or writing rung comments. Added graphic capabilities are also normally a part of a PLC software program.

When a program is first developed and edited, it is done in the program mode or the **off-line** mode. Off-line indicates that the program has not yet been loaded into the processor memory. The program is not operational until it is loaded into the processor memory, and the processor is placed in the RUN mode. Once the program has been checked, the program is loaded into the processor for testing and further verification and/or modification. Changes that are made after the program has been loaded into the processor are called **on-line** programming.

Caution: Making changes to the program while the program is running and the driven equipment is operational (on-line programming) must only be done by trained personnel who not only understand the PLC program, but also thoroughly understand the driven equipment and/or process.

As mentioned earlier, the monitor of a personal computer shows multiple rungs of program logic, as well as highlighting the circuit elements to indicate status. When the PLC program has been downloaded into the user memory of the PLC processor, the processor placed in the *RUN* mode, and the circuit activated, the computer monitor gives a visual display of the circuit condition.

Actual circuit condition is shown on the computer display in basically two ways: some PLCs intensify, or make brighter, all contacts, interconnecting lines, and coils that are passing current or have power flow; others intensify or use reverse video to indicate which contacts and coils have power flow. Figure 3–12a illustrates how a circuit appears before the *START* button is pushed for a system that intensifies contacts, interconnecting lines, and coils. Figure 3–12b shows the computer display after the *START* button is pushed and the holding contacts close.

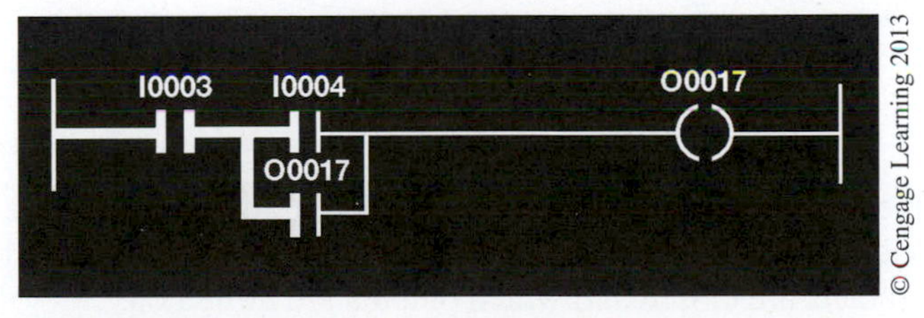

Figure 3–12a Display Prior to START Button Being Depressed

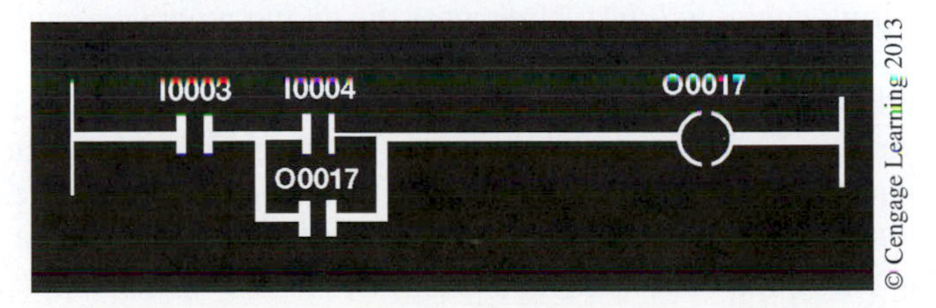

Figure 3–12b Display After START Button Is Depressed and Holding Contacts Close

The terms "passing current" or "power flow" are holdovers from hardwired circuits. In fact, there is no power flow, or current flow, as we normally think of it; rather, it is logic continuity or logically true statements.

Figure 3–13a illustrates how a display using reverse video looks before the *START* button is pushed, and Figure 3–13b shows the display after the *START* button is depressed and the holding contacts close.

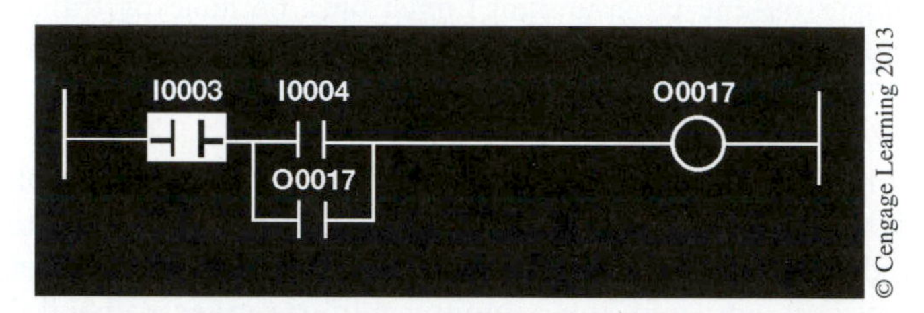

Figure 3–13a Reverse Video Display Prior to START Button Being Depressed

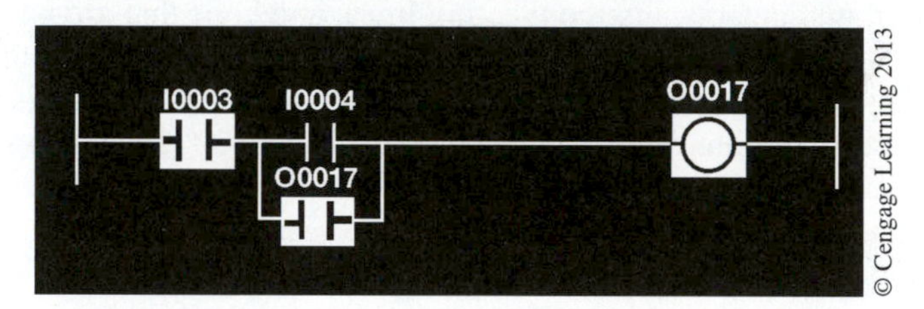

Figure 3–13b Reverse Video Display After START Button Is Depressed and Holding Contacts Close

No matter which method is used, this feature of the programming software is a powerful troubleshooting aid. By viewing the display on the computer monitor, the electrician or technician can determine which contacts are closed and which outputs are turned on.

Note: *An output coil that is intensified only indicates that the output module circuit is* ON. *It does not guarantee that the actual output device is* ON. *However, if the output device and associated wiring are complete, the output device will be* ON *anytime the output module circuit is* ON.

To provide a dependable backup of the program in case the memory fails or is inadvertently cleared or altered, a flash drive, CD disk, or hard drive is used to record and/or load the user program. In fact, a major advantage of using a personal computer for programming is the ability of the personal computer to store the program on disk and/or on the hard drive. If for some reason the program is lost, the restoration of the program is simple. Merely copy the program from the disk to the processor memory.

Chapter Summary

The processor contains the circuitry necessary to monitor the status (*ON* or *OFF*) of all inputs and control the condition (*ON* or *OFF*) of all outputs. It also has the ability to solve and execute the individual program steps in the user program, has a memory for storing the user program and other numeric information, and has the ability to retrieve and use any and all information stored in memory. The memory used in PLCs is of two distinct groups: volatile and nonvolatile. Volatile memory requires a battery backup to prevent the program from being lost due to a power failure; nonvolatile memory holds the program when power is lost or turned off. Programs can be stored on various types of memory chips, as well as on disks or on the hard drive of a computer programmer.

The programming device, or programmer, is used to enter, modify, and monitor the user program. Which type of programming device to use will vary with each application. Contacts and coils are either intensified or displayed in reverse video to indicate power flow or logic continuity. Programming the PLC is not difficult, but time must be spent to become familiar with the specific PLC and its programming techniques.

Review Questions

1. The processor is often referred to as the ______________ of the programmable controller.
2. Briefly describe *volatile memory.*
3. Briefly describe *nonvolatile memory.*
4. 1K of memory is actually
   a. 1,000 words
   b. 1,010 words
   c. 1,024 words
   d. 1,042 words
5. Calculate the actual number of words in an 8K memory.
6. The most common type of volatile memory is
   a. PROM
   b. EAROM
   c. EEPROM
   d. RAM
7. Which of the following are types of nonvolatile memory?
   a. EEPROM
   b. PROM
   c. RAM
   d. EAROM
   e. FLASH
8. List the two broad categories of memory (not volatile and nonvolatile).
9. List the two types of programming devices.
10. What is meant by the term *Reverse Video*?

11. What is a *watchdog timer*?
12. What special precautions should be taken with lithium batteries?
13. When a PLC is first turned *ON,* it will run a self-diagnostic or self-check test.
    T F
14. Describe the four steps of a typical PLC processor scan.
15. The actual scan time, or time it takes the PLC to complete a four-step scan, decreases as the number of program words increases.
    T F

CHAPTER

4 Memory Organization

Objectives

After completing this chapter, you should have the knowledge to:

- Identify the two broad categories of memory and describe the function of each.
- Identify the types of information stored in each category of memory.
- Define the term *byte*.
- Define the acronym *bits*.
- Define *holding registers*.
- Define the term *tag*.
- Identify the different data types.
- Understand the difference between tasks, programs, and routines.

MEMORY WORDS AND WORD LOCATIONS

For the programmable controller to function properly and control a process or driven equipment, it must be able to perform the user program repeatedly and accurately. The system must also be able to perform its control function with great speed, which is achieved by processing all information in binary signals. The key to the speed with which binary information can be processed is that there are only two states, each of which is distinctly different. Binary signals fall into one of two states, which are 1 and 0. The 1 and 0 can represent *ON* or *OFF,* true or false, voltage or no voltage, high or low, or any other two conditions depending on the system. There is no in-between state or condition, and when information is processed, the decision is either *yes* or *no.* There is no *maybe, almost,* or any other alternative.

As indicated in Chapter 3, the processor memory consists of hundreds or thousands of locations that are referred to as words. Each word is capable of storing binary data in the form of binary digits, or **bits** (**BI**nary digi**TS**). A binary digit, like a binary signal, can only be a 1 or a 0. The number of bits that a word can store will depend on the system or PLC. Words can be made up of 32 bits, 16 bits, or 8 bits. Figure 4–1 shows a 16-bit word.

If a memory size is 256 words, then it can actually store 4,096 bits of information using 16-bit words (256 words × 16 bits per word) or 2,048 bits using an 8-bit word (256 words × 8 bits per word).

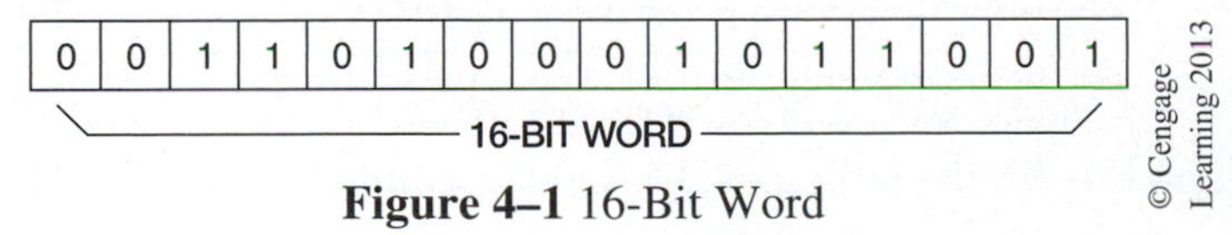

Figure 4–1 16-Bit Word

When comparing memory sizes of different PLC systems, it is important to know the number of bits per word of memory. Bits can also be grouped within a word into **bytes.** A byte is a group of 8 bits.

So that information stored in each word can be located, each word is numbered or given an address. Allen-Bradley Logix family of PLCs use tag-based memory, which will be covered later in this chapter. Addressing words in the memory serves the same function as the addresses used for homes or apartments. Word 100, for example, represents a specific word location in memory, just like 100 N. Lincoln represents the address of an apartment building. The bits in word 100 are found by referencing a given bit number, just like the occupant of the apartment complex is found by a given apartment number.

Since a bit of information can only be a 1 or a 0 (*ON* or *OFF*), how is the status of bits within a word determined? Words that store the status of individual bits for input devices are set to 1 (*ON*) or 0 (*OFF*), depending on the status (*ON* or *OFF*) of the input devices that the bit locations represent. Other bits are set to 1 or cleared to 0 by the processor in response to the logic of the user program, Relay ladder logic, or special instructions, which, in turn, control the status (*ON* or *OFF*) of other bits that represent output devices.

A simple example of how this works is illustrated in Figure 4–2.

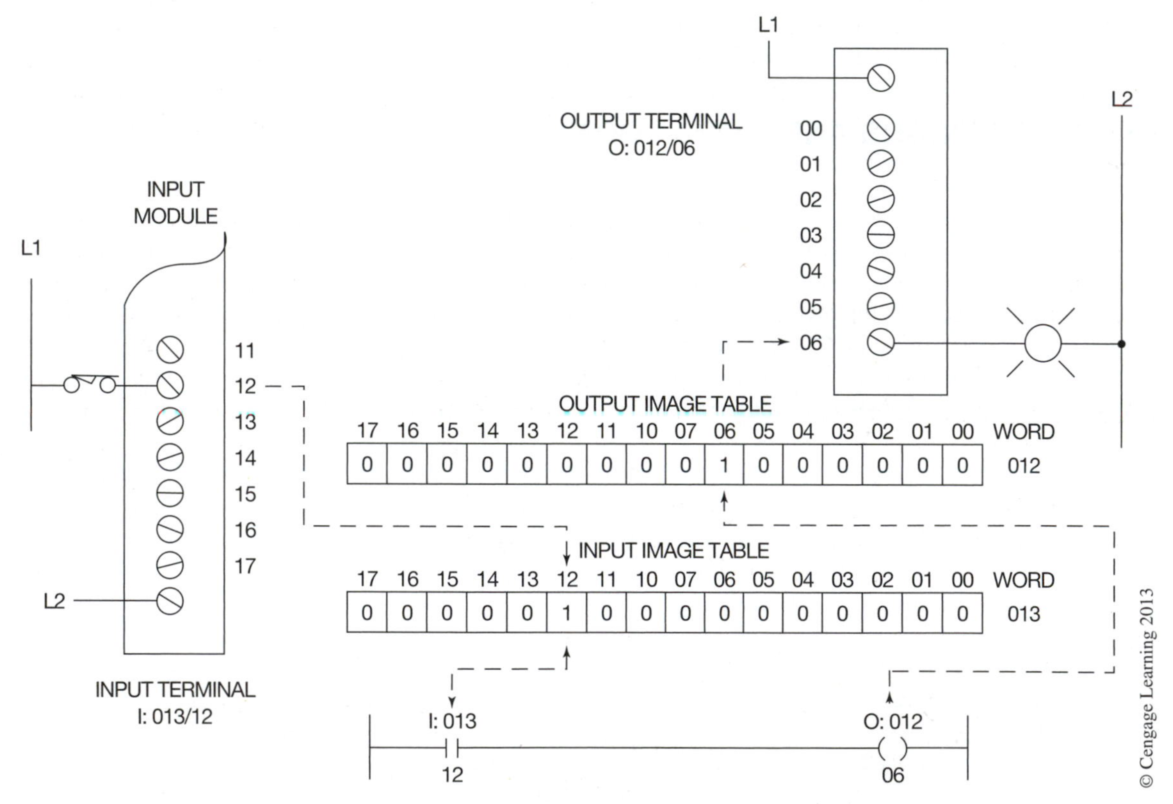

Figure 4–2 Relationship of Bit Address to Input and Output Devices

Note: *The example uses memory organization and addressing utilized by the Allen-Bradley PLC-5 family. While the example is specific to Allen-Bradley, the concepts illustrated are common to all PLCs. Allen-Bradley uses an octal numbering system to address bit locations. Notice that the 16 bits are numbered 00 through 07 and 10 through 17. In the octal numbering system, the numbers 8 and 9 are never used. The octal numbering system will be covered in detail in Chapter 5.*

Assume that when a given limit switch is closed, the closure will turn an indicator lamp *ON*. The limit switch is connected to an input module in the I/O rack, while the indicator lamp is connected to an output module. Chapter 2 discussed DIP switches that were set in a prescribed sequence to identify the I/O rack number for the processor, and that the location of each terminal point of each I/O module within the rack determined the address of a given device. In Figure 4–2, the limit switch is connected to terminal 12 on an input module, and is given an address of I:013/12. This indicates that bit 12 of input image table word 013 stores the status (*ON*-[1] or *OFF*-[0]) of the limit switch. The indicator lamp is connected to terminal 06 of an output module and is given an address of O:012/06. This address indicates that bit 06 of output image table word 012 controls the status (*ON*-1 or *OFF*-0) of the lamp.

By programming a simple circuit into the user memory of the processor as shown at the bottom of Figure 4–2, the processor controls the indicator lamp using the logic of the user program. The logic states that if contact I:013/12 closes, lamp O:012/06 should light, or go *ON*. When power is applied to the processor, the processor starts its scan and looks at bit 12 of input image word 013 to see if the bit is set to 1 or 0. If the limit switch is open, the bit will be set to 0, or *OFF*. If the limit switch is closed, as indicated in Figure 4–2, the input module sends a signal to the processor, and bit 12 of input image word 013 will be set to 1, or *ON*.

The next part of the scan solves the user program. The logic of the ladder diagram, or user program, indicates that when contact I:013/12 (bit 12 of input word 013) is closed, or *ON,* the indicator lamp O:012/06 should be turned *ON*. The processor reads the logic, and during the third step of the scan, sets bit 06 of output word 012 to 1, which turns the lamp connected to the output module terminal 06 *ON*.

The address I:013/12 also tells us that the limit switch is an input device, and is wired to terminal 12 of module group 3 of rack 1.

Figure 4–3 illustrates the significance of each letter/digit or group of digits used for addressing the Allen-Bradley PLC-5 family of programmable logic controllers.

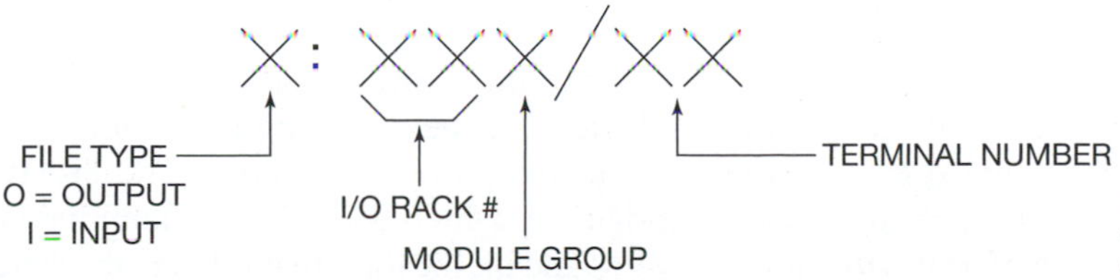

Figure 4–3 Allen-Bradley PLC-5 Address Format

The first letter is used to indicate the type of file (input, output, timer, counter, etc.). The letter I represents an input device, and the letter O represents an output.

The next two digits identify the rack number. One rack can control 128 I/O points. Rack numbers start at 00. A rack is different than a chassis. The chassis is the physical frame that actually holds the input and output modules that make up a rack. Depending on the density of the I/O modules used, a rack may require a 16-slot chassis or only 4 slots. If 8-point I/O are being used, it will take 16 slots of 8-point I/O modules to make 128 I/O points ($8 \times 16 = 128$), whereas if 32-point I/O are being used, it will only take a 4-slot chassis to make a rack ($32 \times 4 = 128$).

The next number identifies the module group within the rack. This is always a number from 0 through 7. The last two digits identify the actual terminal number to which the device is wired.

Figure 4–4 reviews the concept using the address of the limit switch I:013/12.

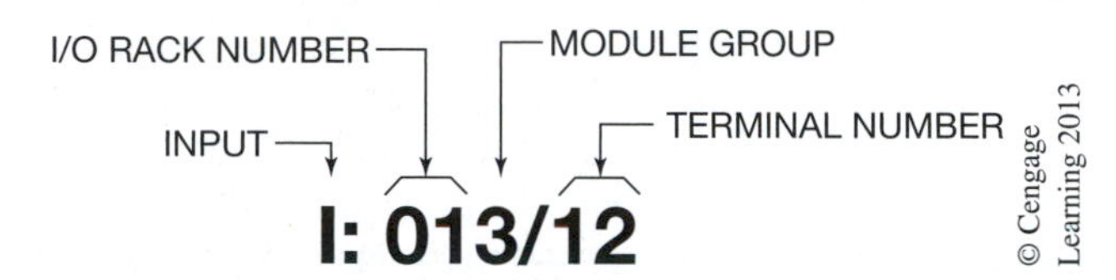

Figure 4–4 Limit Switch Address I:013/12

The letter I tells us that the address represents an input device. The next two digits, 01, tell us that the device is located in I/O rack number 01. The next digit, which is a 3, further identifies the location as module group number 3. The last two digits, 1 and 2, identify the actual terminal (12) on the input module to which the limit switch is connected.

Another example of this concept is shown in Figure 4–5. The limit switch address I:013/12 gives us a hardware location for an input device in rack 01, module group 3, terminal 12. This same address, I:013/12, tells us that the status (*ON* or *OFF*) or state of the limit switch is reflected by bit 12 of word 013 in the input image table.

This same addressing scheme gives us a hardware location for the indicator lamp addressed O:012/06. The letter O indicates an output device. The next two digits, 01, tell us that the I/O rack location is 01. The next digit identifies the module group as group 2. The last two digits locate terminal 06 as the terminal on the output module to which the indicator lamp is wired. Again, the address O:012/06 also locates the memory word and bit location that reflects the status (*ON* or *OFF*) of the indicator lamp, as shown in Figure 4–6.

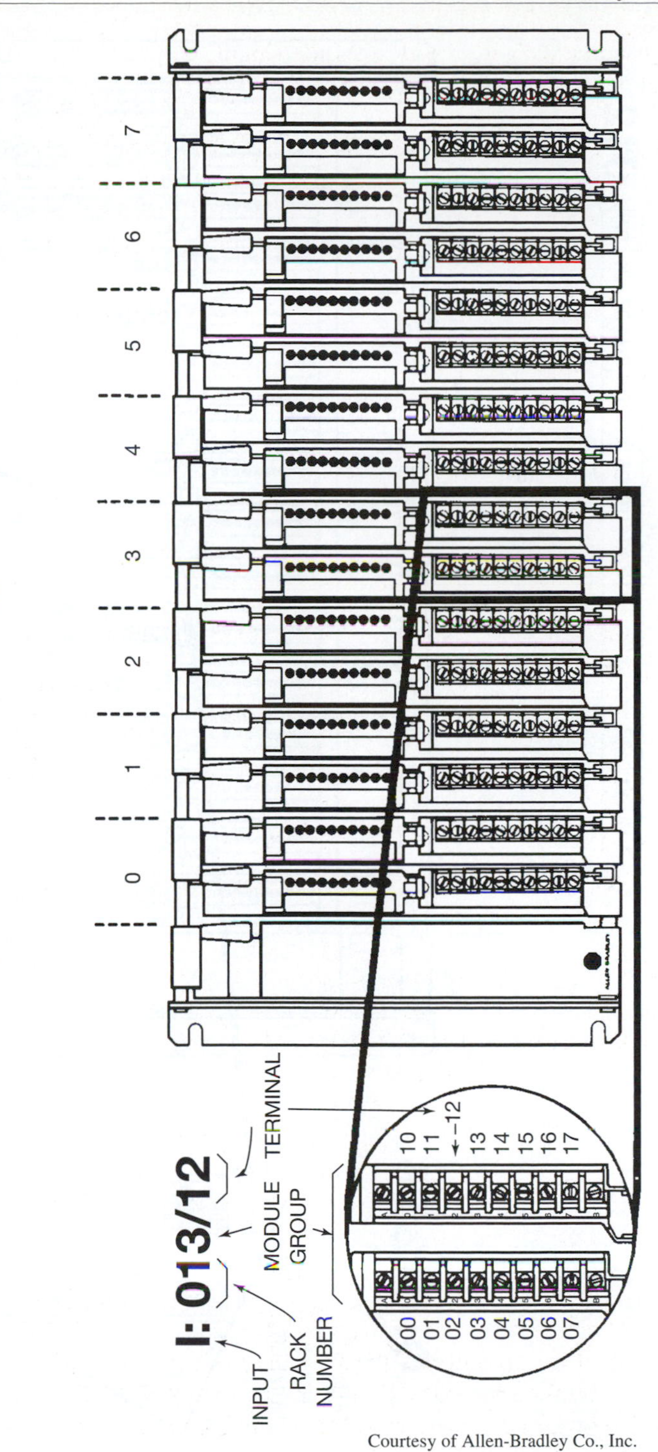

Figure 4–5 Relating Input Address I:013/12 to Actual Hardware Location

Courtesy of Allen-Bradley Co., Inc.

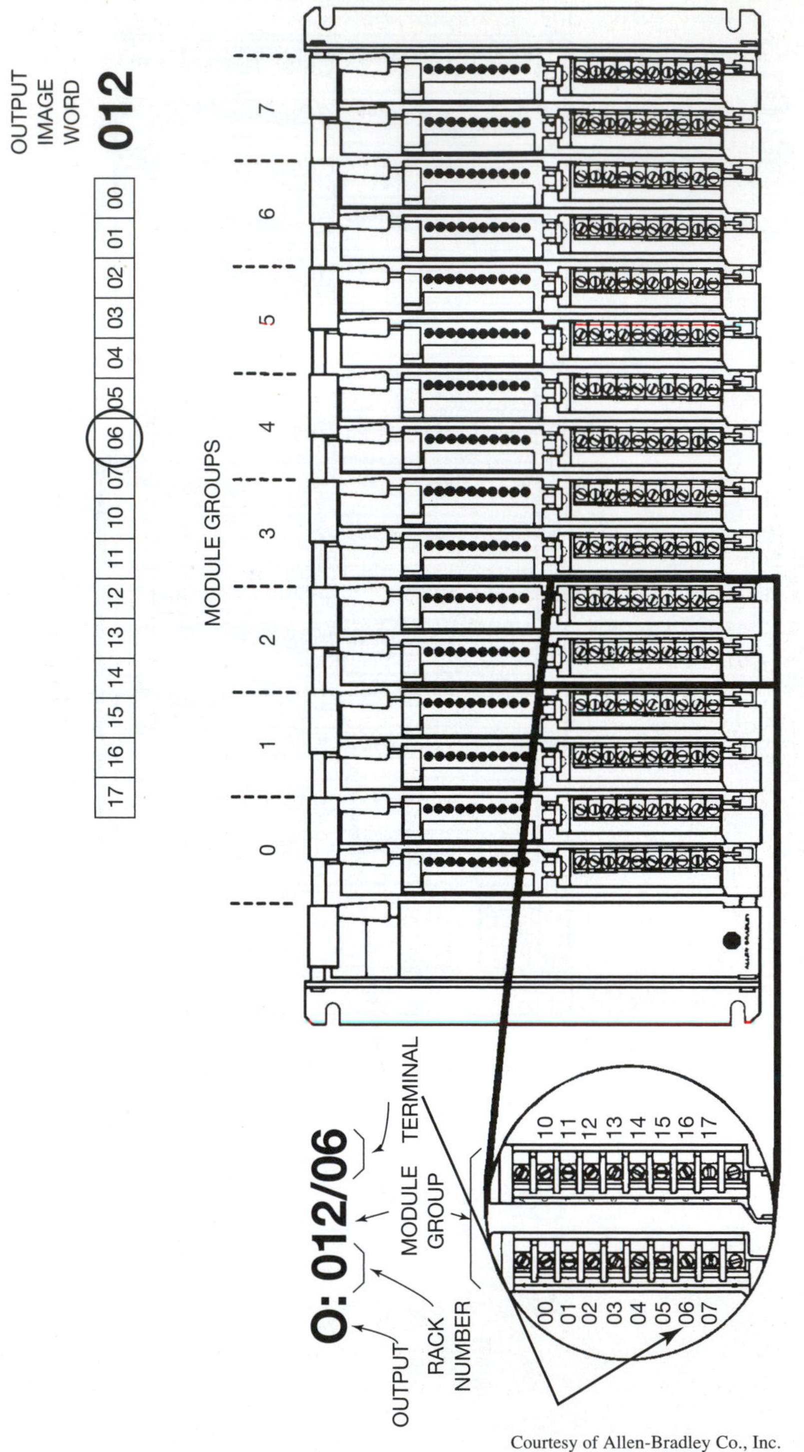

Courtesy of Allen-Bradley Co., Inc.

Figure 4–6 Relating Output Address O:012/06 to Actual Hardware Location

While the address system discussed is specific to the Allen-Bradley PLC-5 family, most PLC manufacturers use an addressing scheme that identifies memory word locations, and may also give hardware locations.

SLC 500 AND MICROLOGIX ADDRESSING SCHEME

Like the PLC-5 family, the SLC 500 and MicroLogix family use the letter I for an input address and the letter O for an output address.

I = external input device
O = external output device

A typical address for an input device would be I:0.1/6. The I, of course, indicates that this is the address for an input device. The colon (:) is called an element delimiter. This means that the colon separates the input designator, I, from the rest of the address. The next character, the number 0, indicates the slot number that holds the actual input module. The slot number can range from slot 0, adjacent to the power supply in the first chassis, to a maximum of 30.

If the number of inputs for that slot is more than 16, a period (.), which is called a word delimiter, would be used after the slot number. The forward slash (/) is called the bit delimiter. The number that follows the forward slash is the bit number of the word as well as the terminal number of the I/O module. The digit 6 indicates the terminal number where the input device is wired. Figure 4–7 illustrates the addressing scheme for the SLC 500 and the MicroLogix family.

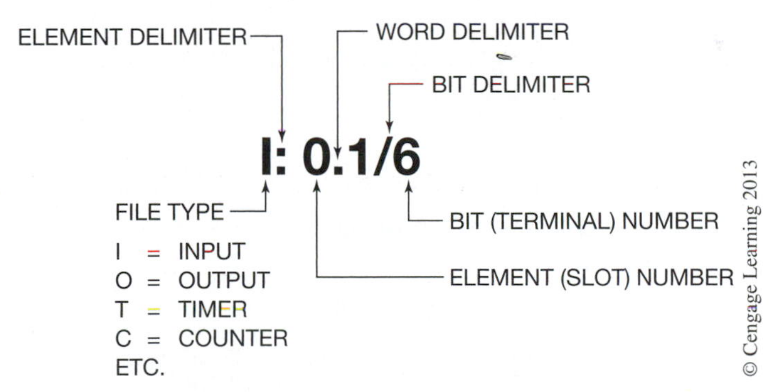

Figure 4–7 Allen-Bradley SLC 500 and MicroLogix Addressing Scheme

In the address shown in Figure 4–7, the I indicates that the file type is an Input file and also indicates this is the address of an input device. The 0 after the colon(:), which is the element delimiter, indicates that the input device is connected to slot 0. The period (.) after the slot number indicates that the inputs exceed 16 and require two words in the input image table. The number 1 indicates that this is word 1 in slot 0. The number 6 after the forward slash is the bit number.

Note: *The SLC 500 and MicroLogix do not use the octal numbering system like the PLC-5 family, but instead use the decimal numbering system.*

The SLC 500 controller is available in either fixed or modular I/O. The fixed I/O units have fixed I/O of 20 (12 inputs and 8 outputs), 30 (18 inputs and 12 outputs), and 40 (24 inputs and 16 outputs).

For fixed I/O controllers, all of the I/O are in slot 0. Figure 4–8 shows a fixed I/O controller that has 24 inputs and 16 outputs. Also shown are the input and output image tables for Data File 0 (output) and

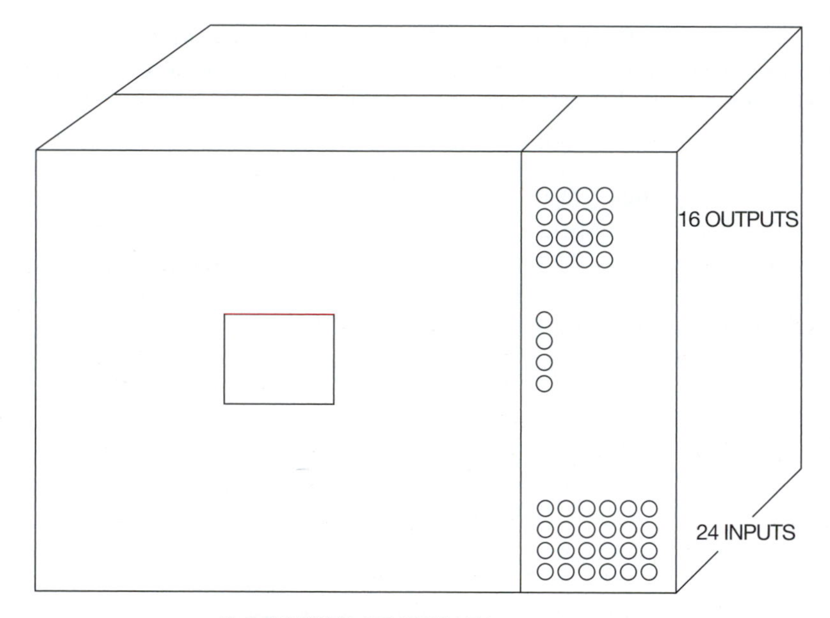

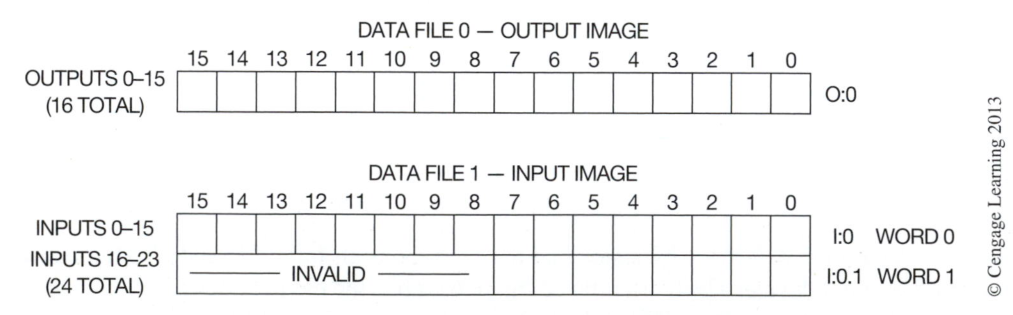

Figure 4–8 SLC 500 with 40 Fixed I/O

Data File 1 (input). Note that for 24 inputs, the Input Image table uses two words for slot 0. All 16 bits of word 0 (I:0) are used, whereas only the first 8 bits of word 1 (I:0.1) are used (bits 0 through 7). The unused bits of word 1—bits 8 through 15—are marked invalid and are not available for use.

Figure 4–9 shows an SLC 500 modular controller that consists of a seven-slot chassis interconnected to a ten-slot chassis.

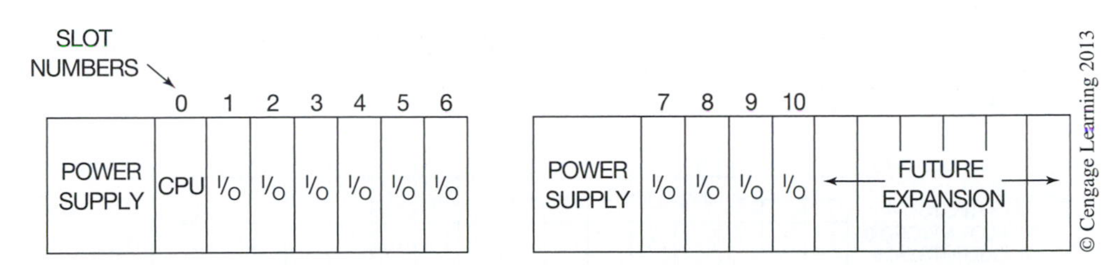

Figure 4–9 SLC 500 Modular Controller (Seven-Slot Chassis Connected to a Ten-Slot Chassis)

In this configuration, slot 0 contains the CPU, so slot 0 becomes an invalid I/O slot number. The controller is configured as follows:

SLOT	INPUTS	OUTPUTS
1	6	6
2	32	None
3	None	16
4	8	8
5	None	32
6	16	None
7	16	None
8	8	None
9	None	16
10	None	16

Figure 4–10 shows Data File 0 (the output image table) and Data File 1 (the input image table) as they would appear for the mix of I/O described above. Note that wherever 32-point I/O modules are used, the data table will require two 16-bit words. Slot 2 has a 32-point input module, and the input image table shows word I:2 (word 0 for inputs 0 through 15) and word I:2.1 (word 1 for inputs 0–15, which are actually inputs 17 through 32). Likewise, where 32 outputs are used in slot 5, the output image table for slot 5 shows that two words are used: O:5 and O:5.1.

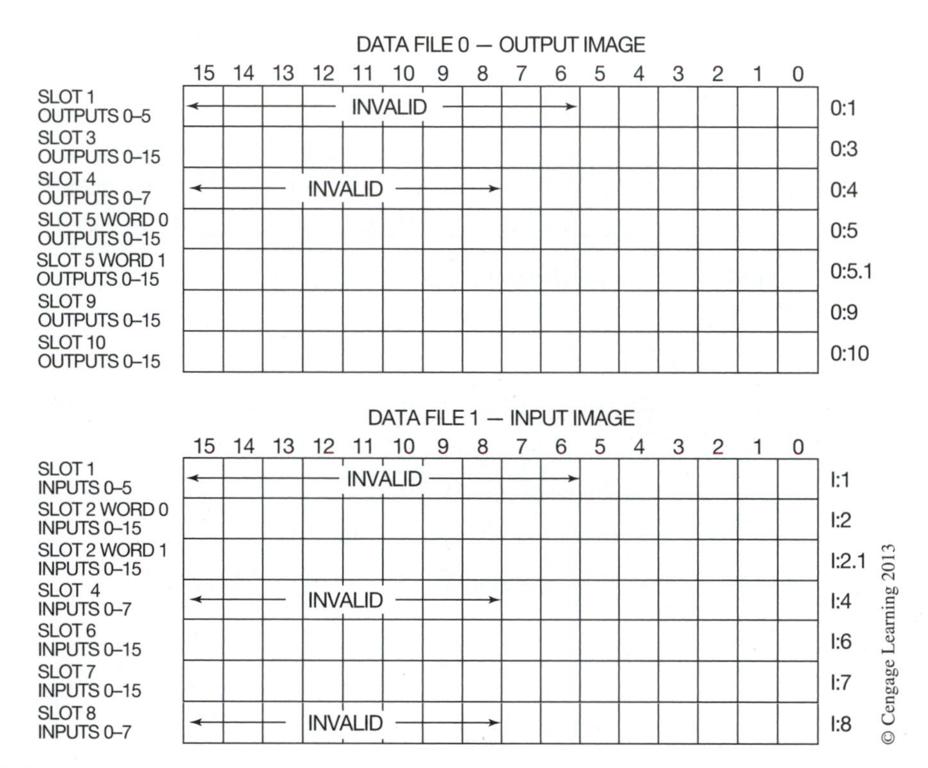

Figure 4–10 Output and Input Image Tables for the SLC 500 Configuration Shown in Figure 4–9

Note that slot 4 of both the input and output image tables has 8 inputs and 8 outputs. While only 8 bits of each word in the input and output image table are needed, the unused bits cannot be used for any other programming, as indicated by the word *invalid* on the image tables.

Figure 4–11 shows an SLC 500 seven-slot chassis connected to a four-slot chassis using an Allen-Bradley 1746-C9 communications cable. The processor is installed in slot 0 of the first chassis, which is adjacent to the power supply. Figure 4–11 shows an exploded view of the output/input module installed in slot 3. The output/input module is shown with the door that covers

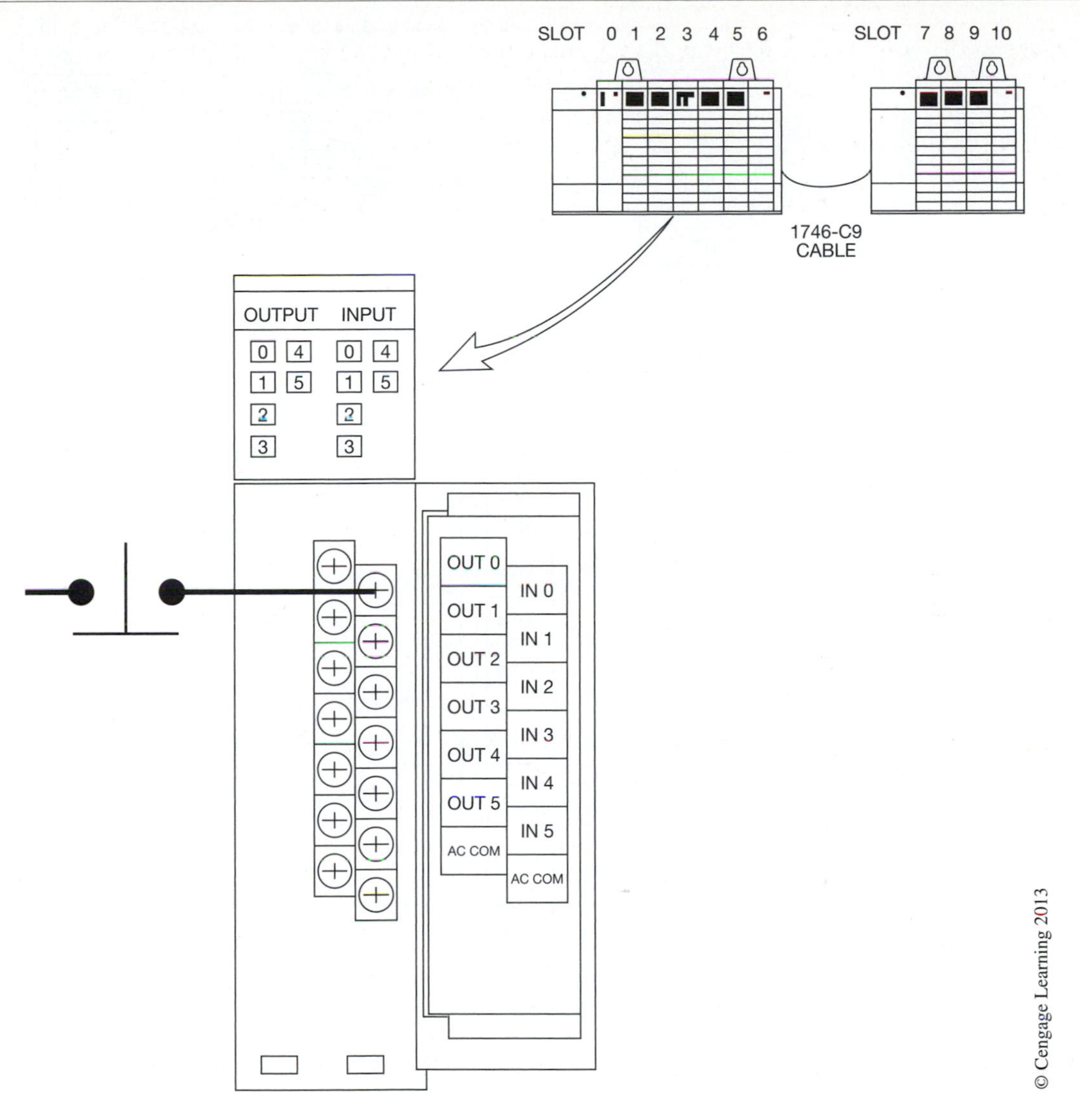

Figure 4–11 SLC 500 Modular Controller with Input Device Connected

the terminals open. The inside of the door shows a pictorial view of the terminal layout. The first vertical row of terminal screws is for the six outputs plus an AC common connection. The second vertical row of terminals is for the six input devices and an AC common connection. The normally open push button that is shown would have an address of I:3/0. I is for input, the 3 indicates that the input device is installed in slot 3, and the 0 indicates the input device is connected to input terminal 0.

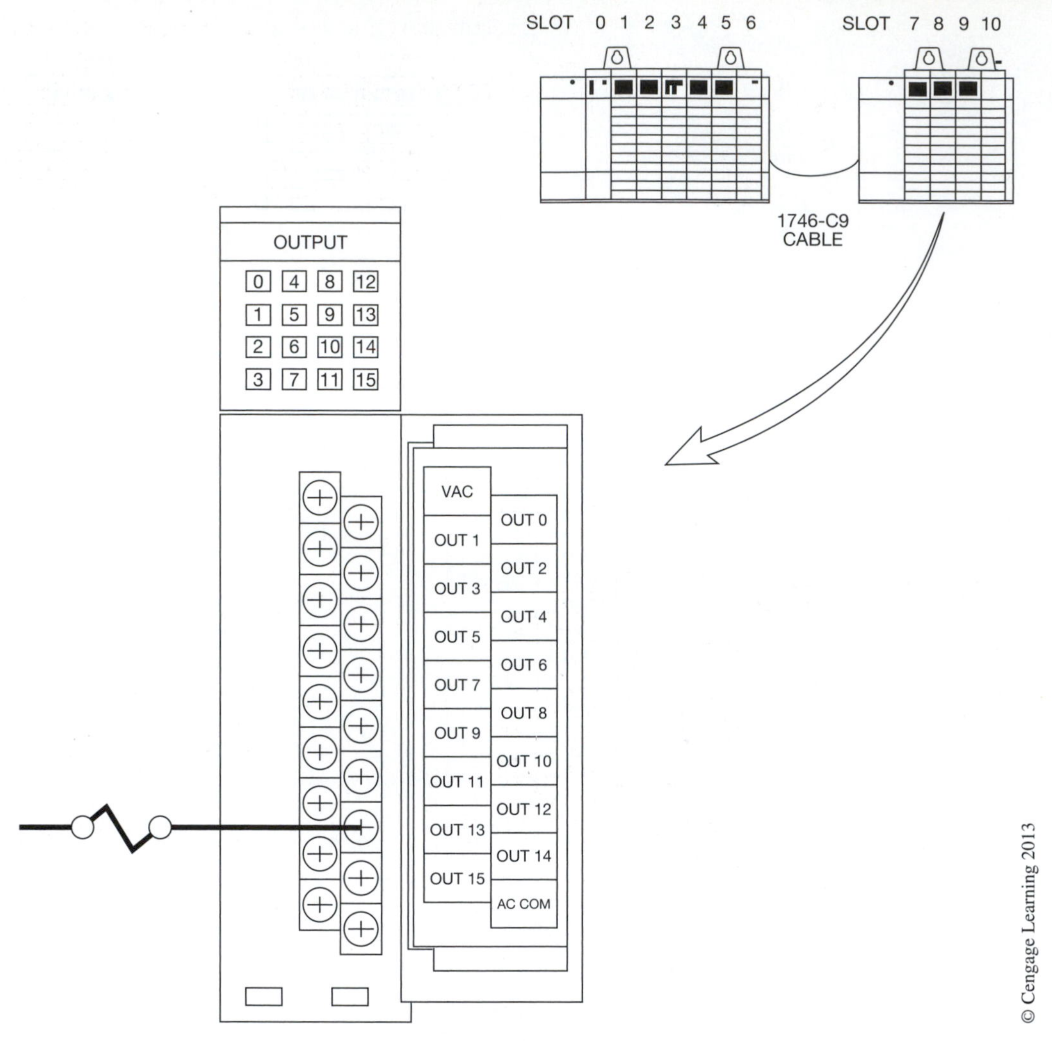

Figure 4–12 SLC 500 Modular Controller with Output Device Connected

Figure 4–12 shows the same seven-slot chassis connected to a four-slot chassis. In this figure, an output device (solenoid) is connected to one terminal of a 16-point output module that is installed in slot 8. As before, the terminal door is open and shows the layout of the terminals. The first terminal is for connecting the AC power. Notice that the output numbers alternate from Out 0, then Out 1, Out 2, and so on. The very last terminal is for the AC common, or neutral connection. The address for the solenoid connected as shown would be O:8/12. The O indicates an output device, the 8 indicates that the module is installed in slot 8, and the 12 indicates that the solenoid is connected to

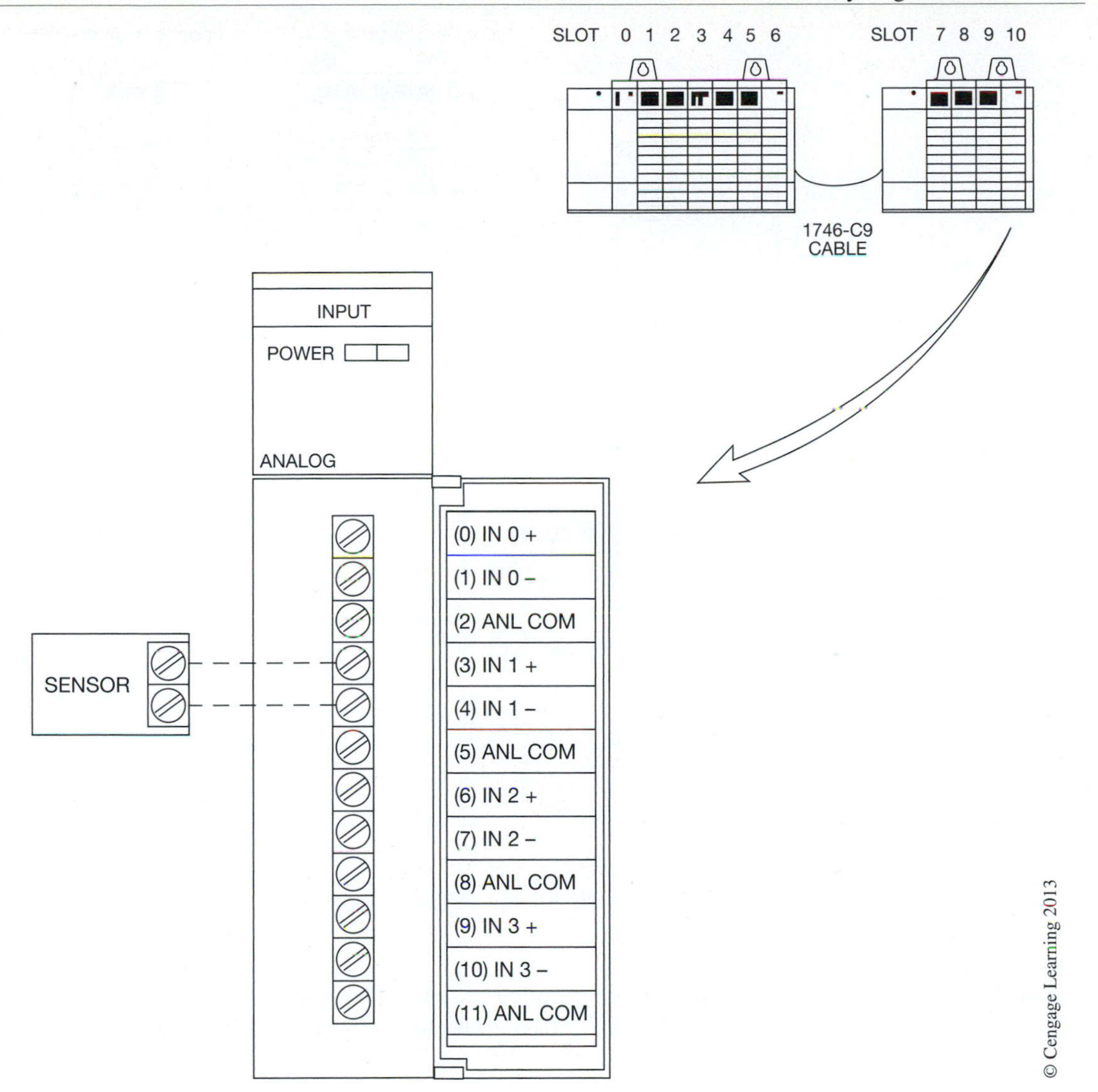

Figure 4–13 SLC 500 with Analog Input Device Connected

output terminal 12. This address also indicates that the status of the output device is found in bit 12 of Output Image Table word 8.

Figure 4–13 shows an analog input module installed in slot 10. The exploded view of the module shows a sensor connected to the module at terminals 3 and 4. As connected, the address of the analog sensor is I:10.1. From the layout on the terminal door, we can see that terminals 3 and 4 are identified as Input 1+ and Input 1–.

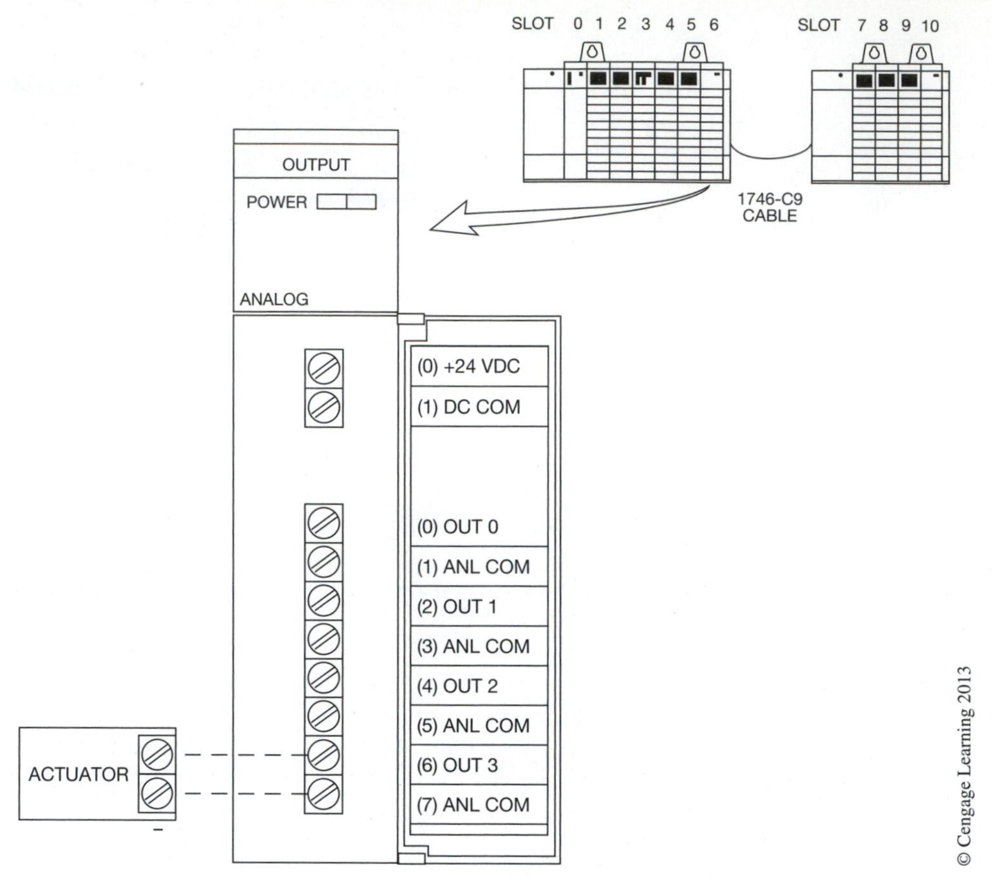

Figure 4–14 SLC 500 with Analog Output Device Connected

Figure 4–14 shows an analog output module installed in slot 6. The analog output in this illustration is an actuator connected to terminals 6 and 7. The address for the actuator is O:6.3.

MEMORY ORGANIZATION

As discussed in Chapter 3, there are two general classifications of memory: storage memory and user memory (Figure 4–15).

Storage Memory

Storage memory is that portion of memory that will store information on the status of input and output devices, preset and accumulated values of timers and counters, internal relay equivalents, numerical values for arithmetic functions, and so on. The entire storage memory is called a data table, a register table, or other names, depending on the PLC manufacturer. A register is defined as an area for storing information (logic or numeric). Although the names or titles that are given to sections or subsections of the storage memory vary, the principles involved do not.

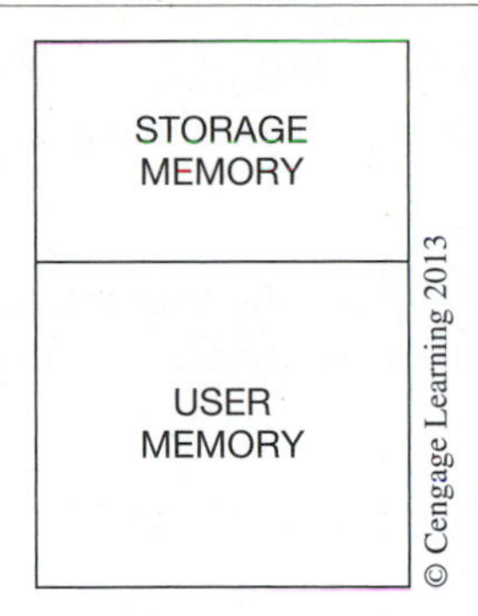

Figure 4–15 Two Broad Categories of Memory

For example, the section of the memory that stores the status of the real-world input devices may be referred to as an input image table, input register, input status table, or external input section. No matter what name is used, the information is stored in the same way. The status (*ON* or *OFF*) of each input device is stored as either a 1 or a 0 (*ON* or *OFF*) in one bit of a memory word. When the processor is executing the user program (ladder diagram), it scans the input device status stored in the storage memory to determine which inputs are *ON* or *OFF*.

The section of storage memory set aside for output status may be referred to as the output image table, output register, output status table or external output section. Again, the name does not change the function of this section of the storage memory, or the method by which information is placed in memory for control of the actual output devices. As the processor executes the user program, it sends binary data (1s or 0s) to the output section of memory to control the output devices. Each output device is represented by one bit of a memory word.

Numeric information for timer or counter preset and accumulated values, arithmetic functions, sequencer functions, data manipulation, etc., uses a part of the storage memory that is called data registers or internal storage. Information is entered and stored in this part of memory using the binary, BCD, or hexadecimal numbering systems (the various numbering systems are covered in Chapter 5). The numbering system(s) used depends on the PLC hardware and system requirements. The storage of numeric information requires that several bits of one word be used to represent numbers. In a practical sense, any word used to store numerical information is not available for additional storage, even if all the bits of the word are not used.

Internal relays will replace the numerous control relays used in most hardwired control circuits. Many PLCs have a portion of memory set aside just for internal relays. The concept and use of internal, or dummy, relays is covered later in the text.

User Memory

The user memory, or logic memory as it is sometimes called, is where the programmed ladder logic is stored. Within the user memory, words are set aside as **holding registers.** Holding registers typically store information generated and used by the processor when it is solving the user program. Holding registers that are set aside to store intermediate values or other short-term bits of information are sometimes referred to as *scratch areas* or *scratch pads*.

The user memory accounts for most of the total memory of a given PLC system. A system with an 8K memory (8192 words) typically has a storage memory of 2K or less, and the balance of memory (6K) is available for user memory.

Once the user program has been entered into the user memory by a programming device, the programmable controller is ready to control the process or driven equipment in accordance with the user's logic.

ALLEN-BRADLEY PLC-5 FILE STRUCTURE

The Allen-Bradley PLC-5 processors are usually programmed with a personal computer and software specific to the PLC-5 family, and the areas of memory are often referred to as files. Although there are still two memory sections (storage [data] and user [program]), the PLC-5 memory map, or structure, is very flexible in the way that the memory can be allocated. Figure 4–16 shows the PLC-5 **default** memory structure. Default refers to the initial value, setting, or configuration prior to any user changes.

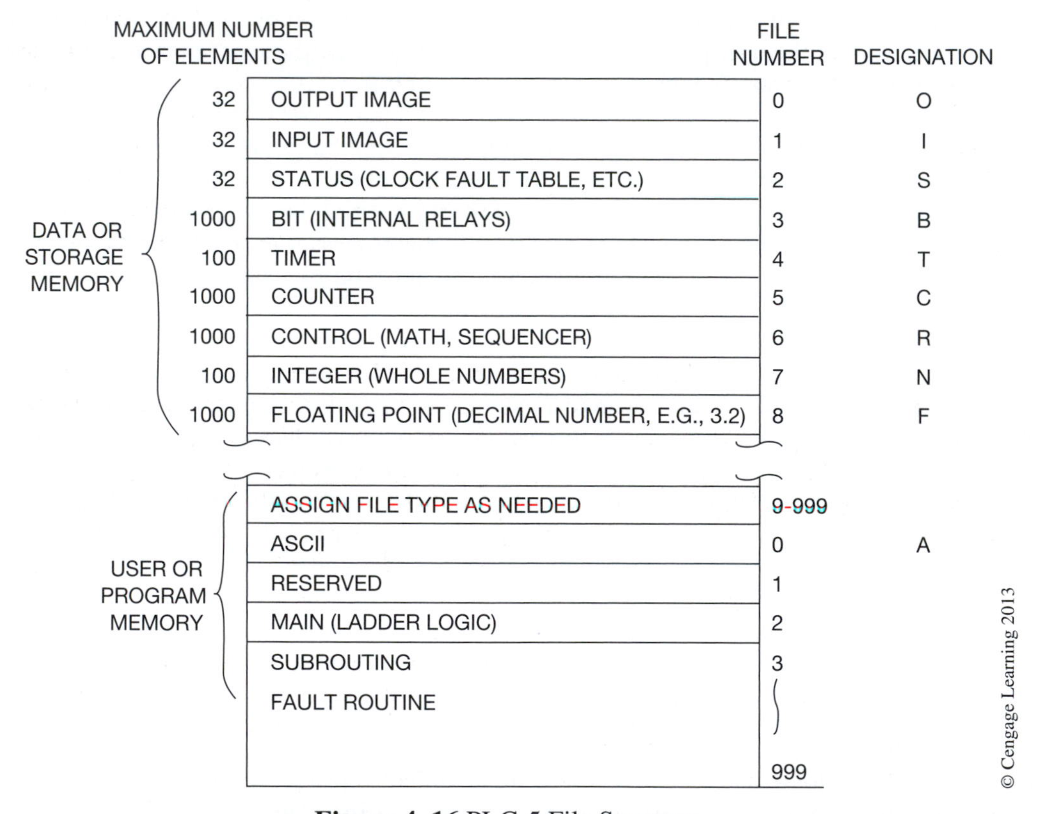

Figure 4–16 PLC-5 File Structure

In the data or storage memory section file 0 is the output image file. This file has 32 words of 16 bits each, and can hold the status of 512 real-world output devices (32 × 16). The status of the outputs

(*ON* or *OFF*) is updated once each scan. On some PLC-5 models, like the PLC-5/25, the size of the file can be increased to accommodate more output devices.

File 1 is the input image file. This file, like file 0, has 32 words of memory and can store the status of 512 input devices. The status (*ON* or *OFF*) of the input devices, like the output image file, is updated once each scan and can be increased in size on many PLC-5 models.

Both files 0 and 1 use the octal numbering system, and the memory locations (bits) are also numbered using the octal numbering system (there are no 8s or 9s in the octal numbering system). The digits are 0–7, 10–17, 20–27, and so forth.

File 2 is the status, or S file. This file is used to store information on general processor status, fault codes, real-time clock and calendar, major and minor fault bits, and program scan times in msec. Information from this file is used or incorporated into the user program. The size of this file changes depending on the processor that is being used.

File 3 is the B, or bit file, and is used primarily for internal or dummy relays. The default size of this file is one word, but can be expanded to 1000 words if needed. All addresses from this file must start with B3. Another B or bit file may be created using the other areas of the memory. A B10 file could be created that would also have internal or dummy relays. The addressing B10 versus B3 is used for organization and ease of identification. The B3 file may be associated with one piece of equipment, while the B10 file could be associated with another piece of equipment and/or operation. Up to 999 files can be created in a PLC 5/15 processor memory, as long as you do not try to allocate more memory than is physically available in the processor. The B3 file is typical of the remaining files in flexibility, as well as being addressed using the decimal numbering system.

File 4 is the T, or timer file. All timer addresses must start with T4 unless new timer files have been created (e.g., T9, T10, and T11). When files are being created, the same number cannot be used twice. If file 10 is used as a B file (B10), then file 10 cannot be used as a timer file. Each timer that is programmed uses three words of memory from its timer file.

File 5 is the C, or counter file. All counters that are programmed have C5 as the start of their addresses. Each counter, as with timers, uses three words of the counter file memory.

File 6 is the R, or control file. The words in this file are used with special functions like sequencer, file moves, word to file moves, and math functions. File 7 stores whole numbers (integers) and is called the N, or integer file. The integer file is used to store numeric values for data compare, arithmetic functions, and the like. For storing numbers with a decimal point, or floating point, file 8, the F file is used. Files 9–999 can be used or assigned as needed. They may be used to expand the size of binary, timer, and counter files, etc.

The program portion, or user portion, of memory (Figure 4–16) is used to store information that relates to the user program, or for information that is needed for the processor to operate. File 0 is used to store ASCII information, while File 1 is reserved for internal use by the processor. File 2 is where the user program is stored in relay ladder logic. Files 3–999 are for storing subroutines, fault routines, and selectable timed interrupt (STI) as they are needed.

Figure 4–17 shows the data file structure used by the various PLC-5 models. Note that the I/O section varies from 32 words for the input image table and 32 words for the output image table for the PLC-5/10, 5/12, 5/15, 5/11, 5/20, and 5/20E, while there are 192 words for both the input and output image tables for the PLC-5/60, 5/60L, and 5/80.

*DATA TABLE FILES

PLC-5 MEMORY

data table

program

FILE DESCRIPTION		NUMBER (Default file)	MAXIMUM SIZE OF FILE (16-BIT WORDS)						MEMORY USED	MEMORY USED
			PLC -5/10, -5/12, -5/15	PLC -5/11, -5/20, -5/20E	PLC -5/25	PLC -5/30	PLC -5/40, -5/40E, -5/40L	PLC -5/60, -5/60L, -5/80	CLASSIC PLC-5 PROCESSORS	ENHANCED PLC-5 PROCESSORS
OUTPUT	O	0	32	32	64	64	128	192	2/file + 1/word	6/file + 1/word
INPUT	I	1	32	32	64	64	128	192	2/file + 1/word	6/file + 1/word
STATUS	S	2	32	128	32	128	128	128	2/file + 1/Word	6/file + 1/word
BIT (BINARY)	B	3-999 (3)	1000						2/file + 1/word	6/file + 1/word
TIMER	T	3-999 (4)	1000 structures of 3						2/file + 3/structure	6/file + 3/structure
COUNTER	C	3-999 (5)	1000 structures of 3						2/file + 3/structure	6/file + 3/structure
CONTROL	R	3-999 (6)	1000 structures of 3						2/file + 3/structure	6/file + 3/structure
INTEGER	N	3-999 (7)	1000						2/file + 3/word	6/file + 3/word
FLOATING POINT	F	3-999 (8)	1000						2/file + 2/float word	6/file + 2/float word
ASCII	A	3-999	1000						2/file + 1/2 per character	6/file + 1/2 per character
BCD	D	3-999	1000						2/file + 1/word	6/file + 1/word
BLOCK TRANSFER[1]	BT	3-999	1000 structures of 6							6/file + 6/structure
MESSAGE[1]	MG	3-999	585 structures of 56							6/file + 56/structure
PID[1]	PD	3-999	399 structures of 82							6/file + 82/structure
SFC STATUS[1]	SC	3-999	1000 structures of 3							6/file + 3/structure
ASCI STRING[1]	ST	3-999	780 structures of 42							6/file + 42/structure
EXTRA STORAGE		3-999								

[1]enhanced PLC-5 processors only.

Courtesy of Allen-Bradley Co., Inc.

Figure 4–17 Data Table Map File Structure for the PLC-5 Family

The second column of the chart shows the file numbers for the default files. Files 0, 1, and 2 are fixed and cannot be changed. Files 3–8, however, can be changed from the default settings and used as required. For example, if one wanted to use file 3—the binary file—for a timer file, it would be necessary to delete the binary file. The binary file is deleted from the data table map screen and then used as a timer file. Because files 3–8 can be changed, files 3–999 can then be used for timer files, counter files, and the like. However, it is much easier to use files 9–999 when additional files are needed.

SLC 500 AND MICROLOGIX FILE STRUCTURE

When a PLC program for either the SLC 500 or the MicroLogix is being developed, the information needed for the program to function properly is created and stored in processor files. These files are classified into two general types: Program Files (User) and Data Files (Storage).

Program files typically contain controller information (type of processor, I/O configuration, etc.), the ladder logic program, subroutine programs, and interrupt subroutines. The specific program files for the SLC 500 are shown in the tables on the next page.

SLC 500 Program Files	
System program file (file 0)	Used to store information about the processor and the I/O configuration.
Reserved file (file 1)	Reserved for internal use of the processor and is not user-accessible.
Main ladder program file (file 2)	Stores the instructions entered by the user that determine controller operation.
Subroutine ladder program file (file 3–255)	Stores any subroutines not created in the main ladder diagram.

SLC 500 Data Files	
Output-O (file 0)	Stores the status, *ON* or *OFF,* of output devices wired to the controller.
Input-I (file 1)	Stores the status, open or closed, of the input devices wired to the controller.
Status-S (file 2)	Stores controller operation information. This file is useful for troubleshooting controller and program operation.
Bit-B (file 3)	Stores the logic for internal or dummy relays.
Timer-T (file 4)	Stores the preset values, accumulated values, and status bits for timers.
Counter-C (file 5)	Stores the preset values, accumulated values, and status bits for counters.
Control-R (file 6)	Stores information on sequencers and shift registers.
Integer-N (file 7)	Stores numeric values.
Floating Point-F (file 8)*	Stores numbers with a decimal point or floating point.
String-ST (user-defined file)*	
ASCII-A (user-defined file)*	

*This file applies to selected SLC 500 processors.

MicroLogix processors have the same program files as the SLC 500, and three additional files.

MicroLogix Program Files	
System program file (file 0)	Used to store information about the processor and the I/O configuration.
Reserved file (file 1)	Reserved for internal use of the processor and is not user-accessible.
Main ladder program file (file 2)	Stores the instructions entered by the user that determine controller operation.
User error fault routine (file 3)	File is executed when a recoverable fault occurs.
High-speed counter interrupt (file 4)	File is executed when a high-speed counter interrupt occurs. Can also be used for a subroutine ladder program.
Selectable timed interrupt (file 5)	Executed when a selectable timed interrupt occurs. Can also be used for a subroutine ladder program.
Subroutine ladder program file (file 6–15)	Stores any subroutines that have been created in the main ladder diagram.

MicroLogix Data Files	
Output-O (file 0)	Stores the status, *ON* or *OFF,* of output devices wired to the controller.
Input-I (file 1)	Stores the status, open or closed, of the input devices wired to the controller.
Status-S (file 2)	Stores controller operation information. This file is useful for troubleshooting controller and program operation.
Bit-B (file 3)	Stores the logic for internal or dummy relays.
Timer-T (file 4)	Stores the preset values, accumulated values, and status bits for timers.
Counter-C (file 5)	Stores the preset values, accumulated values, and status bits for counters.
Control-R (file 6)	Stores information on sequencers and shift registers.
Integer-N (file 7)	Stores numeric values.

The MicroLogix programmable controllers use both volatile RAM and nonvolatile EEPROM memory. Program data downloaded into the processor from a programming device are stored first in the RAM memory and then copied to the EEPROM memory where they are stored as both backup data and retentive data. It is important to remember that all data in the RAM memory are lost in the event of a power failure, whereas information stored on the EEPROM memory is not affected by power loss.

LOGIX MEMORY

The Allen-Bradley Logix family of PLCs have a processor memory that is separated into two isolated sections (Figure 4–18). The logic and data memory area stores the program code (ladder logic) and tag data. The I/O memory area stores the I/O data, force tables, message buffers, and produced/consumed tags. Tags will be covered later in this section. As you can see this is somewhat different from the "Storage" and "User" memory discussed earlier.

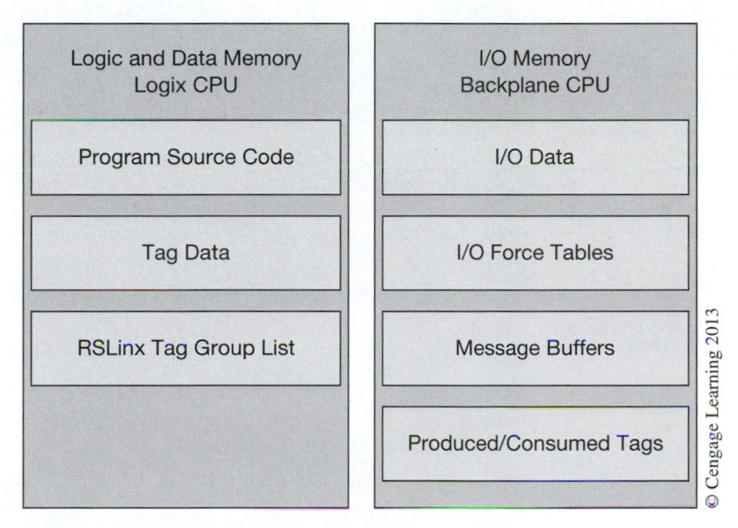

Figure 4–18 Allen-Bradley Logic Memory

The Logix CPU in Figure 4–18 executes the user's program code and messages. The backplane CPU on the other hand communicates with the I/O modules and sends/receives data from the backplane. The backplane CPU also operates independently from the Logix processor, which means the I/O information is being updated asynchronous to the program execution. As you recall from Chapter 3, the I/O is typically updated at the end of the program scan in a single CPU controller. In the Logix PLC, the I/O is being continually updated by the backplane CPU since it operates independently from the processor executing the program.

Most PLCs typically have register-based memory as discussed earlier in this chapter. Logix memory on the other hand is a tag-based memory that is common to users that have a computer background. In the Logix processor, the user determines the memory structures and can adjust these to match the application. A user can create tag names and define the data type as needed. There is no predefined memory layout as is typical with other PLCs.

For example: When the user configures an I/O module (see Figure 4–19) in the Logix PLC, the Logix processor automatically creates and configures the necessary I/O memory (tags) and data types. Each I/O tag name created follows this format:

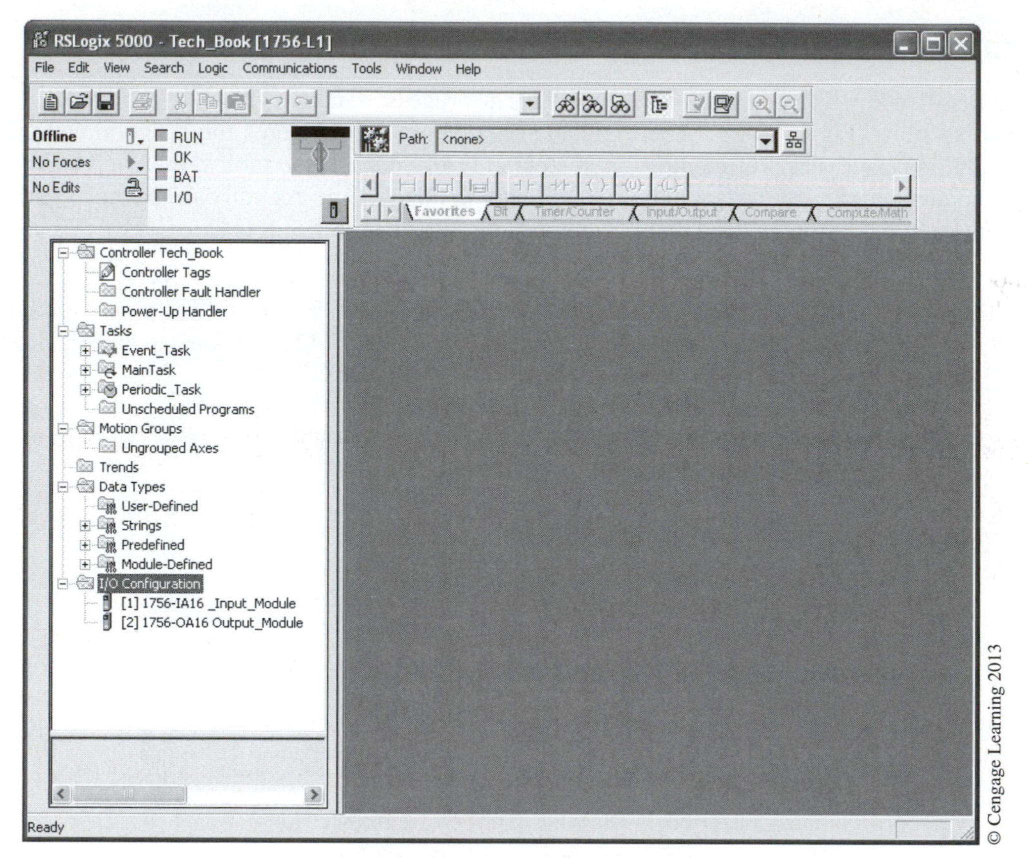

Figure 4–19 Logix I/O Configuration

Location:SlotNumber:Type.MemberName.SubMemberName.Bit

- Location—Identifies the network location of the I/O module, Local or the Adapter Name. If the module is located in the local chassis with the processor, then "Local" is placed in the location field of the tag. If the I/O module is located remotely from the processor chassis, than the remote adapter or bridge name is placed in the location field.
- SlotNumber—Identifies the slot number of the I/O module in its chassis.
- Type—Type of data being addressed. This can be I (input), O (output), C (configuration), or S (status).

- MemberName—Is the specific data from the I/O module; depends on what type of data the module can store. For a digital module, a "Data" member usually stores the input or output bit values. For an analog module, a Channel member (CH#) usually stores the data for a channel.
- SubMemberName (optional)—Specific data related to a MemberName.
- Bit (optional)—Specific point on the I/O module; depends on the size of the I/O module.

The good news is that the Logix processor automatically creates the correct I/O tags (controller scope tags) for the modules that are installed.

If you notice in Figure 4–19 there are two I/O modules that have been configured, a 1756-IA16 input module and a 1756-OA16 output module. The number in [ ] is the slot number location of the module. When the I/O tags are created for each module, they are created in the controller tag area of memory. Figure 4–20 shows the tags that were created for the two modules. Notice at the top of the tag edit screen (right pane) under "Tag Name" the tags created for each module. Just as previously described, the tag name begins with "Local" because each module is located in the local chassis

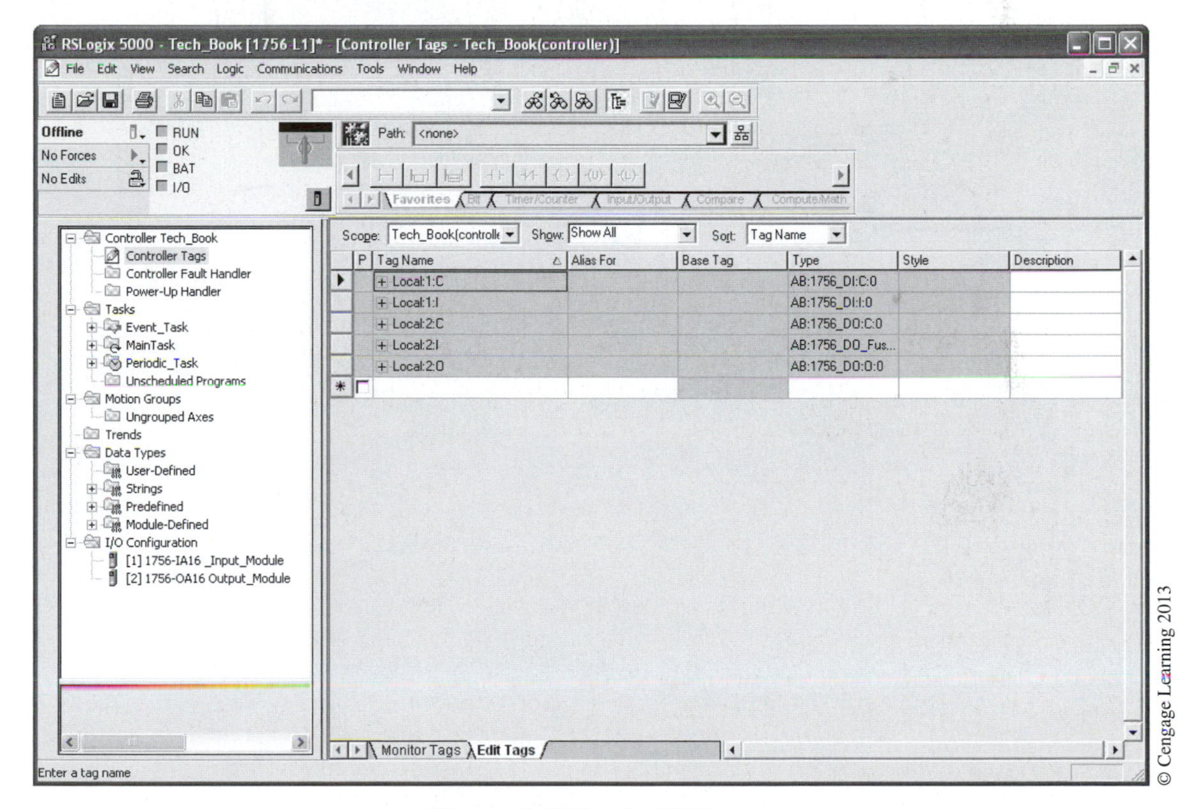

Figure 4–20 Logix I/O Tags

with the processor. Next is the slot number of the module and then the data type. Two data types are shown for the 1756-IA16 input module and three are shown for the 1756-OA16 output module. Can you guess the data types for each module? If you recall, the "C" indicates a configuration type tag and each module has one. The "I" indicates an input data type tag and the "O" indicates an output data type tag.

Note: *It is worth mentioning that the Logix family of controllers and their associated I/O modules have many features that most traditional PLCs do not. Some of those features include diagnostic information, time stamping, fuse blown indication, etc. This book makes no attempt to cover all of the features and capabilities. It is only intended to give the reader a basic understanding of the different memory and addressing methods used today.*

If you expand the "Local:1:I" tag in Figure 4–20 by clicking on the "+" sign to the left of the tag, you will see all of the input data tags associated with the 1756-IA input module located in slot 1 of the local chassis (see Figure 4–21). Notice that there are two input tags. One is a fault input tag and the other is a data input tag. The fault tag would contain fault information associated

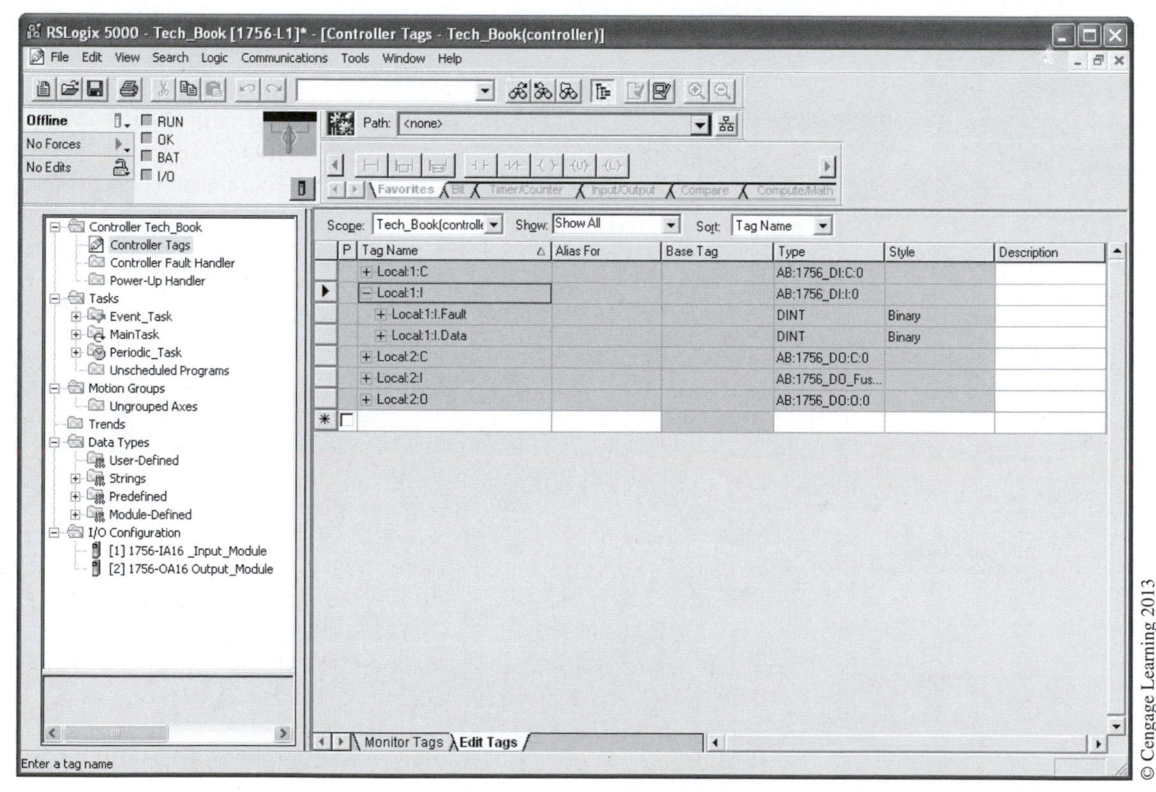

Figure 4–21 Input Data Tags For Input Module in Slot 1

with that module and the data tag would contain the actual input status of the inputs wired to that module.

If we again expand the "Local:1:I.Data" tag we will see all of the individual input data tags for the 1756-IA input module (Figure 4–22). It is worth noting that the Logix family of controllers are based on 32-bit operations and store all data in a minimum of 4 bytes or 32 bits of data. That's why there are more than sixteen data tags shown. Only the first 16 are actually used by this input module.

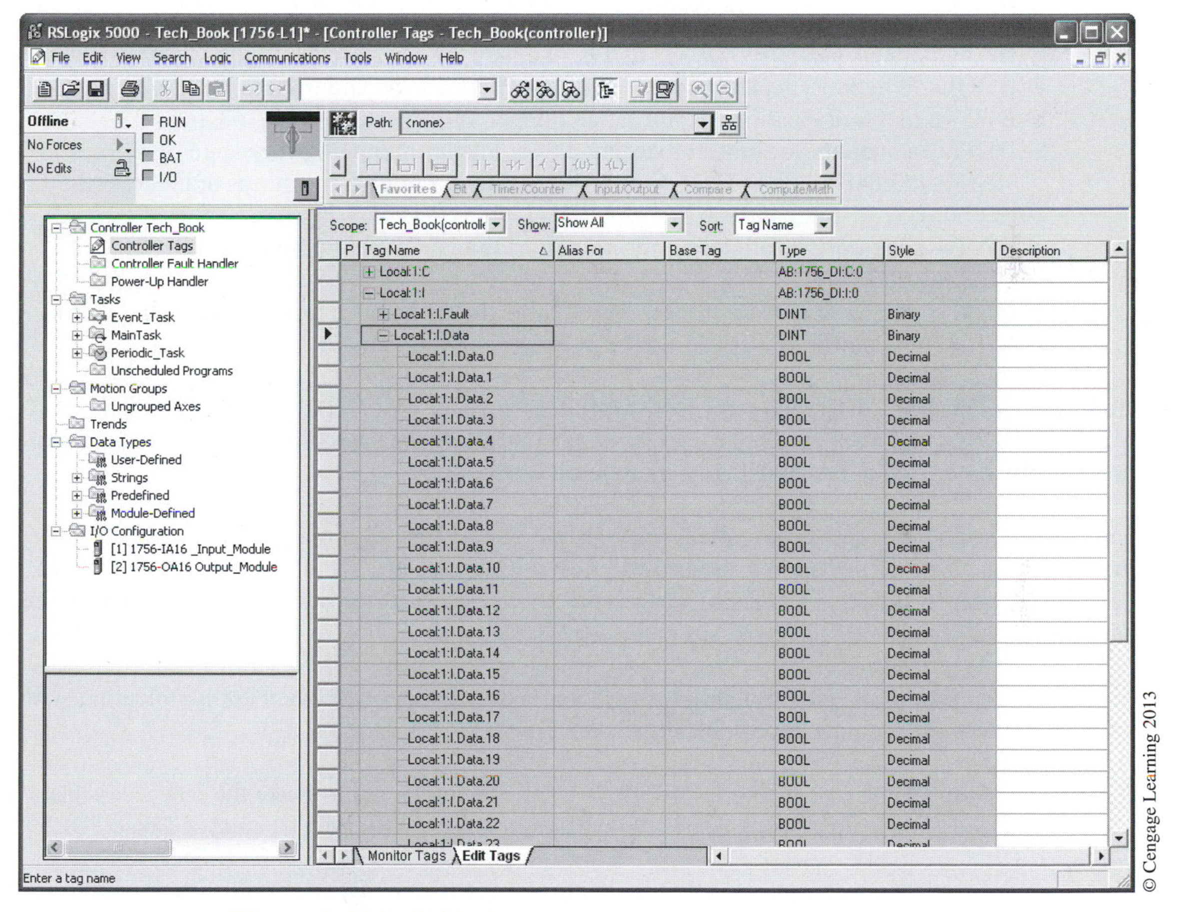

Figure 4–22 Individual Input Data Tags For Each Input Point

If you were to address the first input wired to your 1756-IA16 input module, the address tag would be "Local:1:I.Data.0".

Data Types

The Logix family of controllers supports IEC 61131-3 atomic data types, such as REAL, BOOL, SINT, INT, DINT, and LINT. The controllers also support compound data types, such as predefined structures, arrays, and user-defined structures. As previously mentioned, the Logix controllers read and manipulate 32-bit data values versus the 16-bit data values found with traditional PLC controllers. Since the CPU manipulates 32-bit data values, the minimum memory allocation for data in a tag is 4 bytes or 32 bits. When creating a tag that stores data that are less than 4 bytes, the controller uses what is needed and the remainder becomes unused memory, as was the case with 1756-IA16 input module data tag discussed in Figure 4–22.

The data type of a tag defines the amount and function of bits and bytes (words) of memory assigned to the tag. The predefined data types (DINT, BOOL, etc.) are used to store the following types of data:

- BOOL—a memory location for a single bit where 1 = on and 0 = off.
- INT—a memory location for storing an integer value between –32,768 and +32,78.
- DINT—a memory location for storing a base integer number in the range of –2,147,483,648 to +2,147,483,647. DINT stands for double integer or double word (32 bits).
- SINT—a memory location for storing a short integer (8 bits) number in the range of –128 to +127.
- REAL—a memory location for a 32-bit value that contains a mantissa or an exponent (raised by a power of 10) that can be very large or very small.

Additional predefined data types are also available such as CONTROL, COUNTER, TIMER, MESSAGE, PID, etc. They are used for storing and controlling data associated with a specific function such as timers, counters, and data manipulation instructions.

You should use DINT data types whenever possible as they use less memory and execute faster than other data types. They should be used for most numeric values and array indexes. Arrays will be covered later in this chapter. REAL data types should be used for manipulating floating-point analog values.

SINT and INT should be used primarily in user-defined structures and when communicating with external devices that do not support the DINT data type.

It is best to group BOOL values into DINT arrays to save memory and to make the bits accessible to some specialized program instructions such as File Bit Comparison (FBC) and Diagnostic Detect (DDT).

Keep in mind that the minimum memory allocation for any tag is 32 bits (DINT). When assigning data types such as BOOL, INT, and SINT to a tag, the Logix controller still allocates a full 4 bytes (DINT or 32 bits) but only uses part of it, as shown in Figure 4–23.

Arrays

An array allocates a contiguous block of memory in the controller of the same data type. Each data type in the array is a single tag and each tag is considered to be one element in the array. The elements in the array occupy memory in order, meaning the array starts at 0 and extends to the numbers of elements in the array.

Minimum Memory Allocation of One DINT	Data Type
31 30 29 28 27 26 25 24 23 22 21 20 19 18 17 16 15 14 13 12 11 10 9 8 7 6 5 4 3 2 1 0	
Unused Memory	BOOL
Unused Memory	SINT
Unused Memory	INT
	DINT
	REAL

Figure 4–23 DINT Memory Allocation

For example, the array in Figure 4–24 is a one-dimensional, ten-element array of the data type DINT.

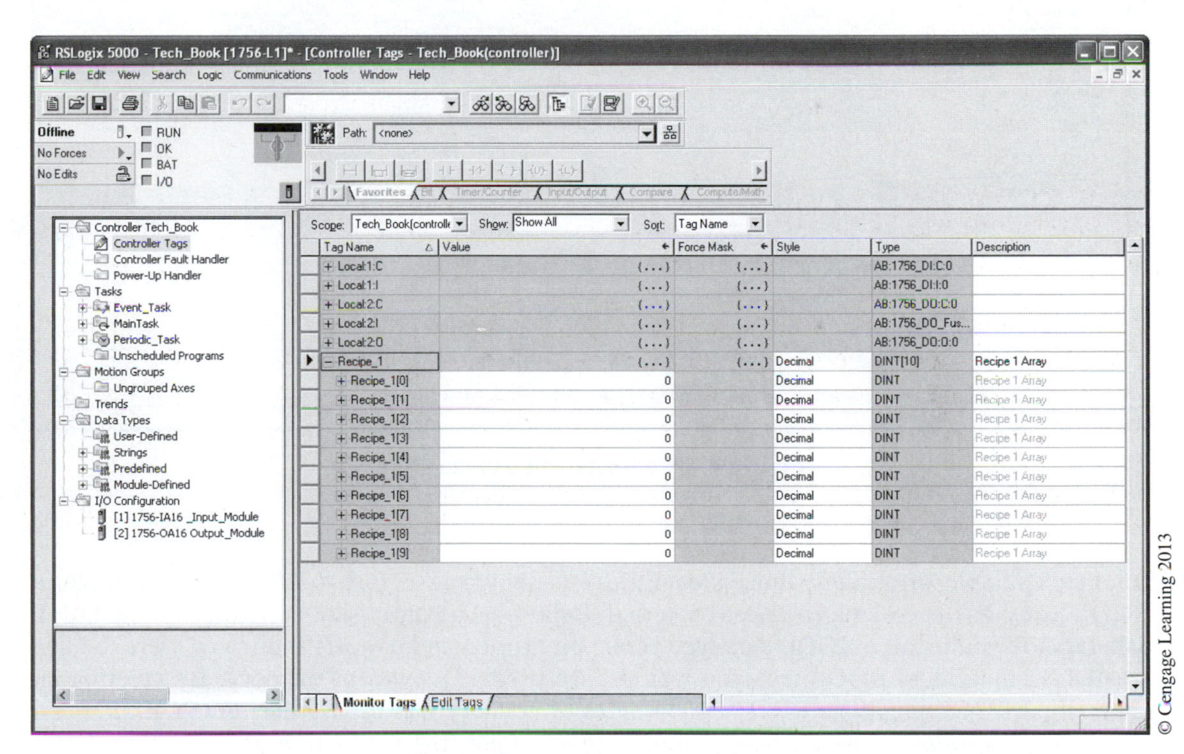

Figure 4–24 Ten-Element Array

In this example, the array is named "Recipe_1" and has ten elements or DINI tags assigned to the array. Since the elements in the array occupy memory in consecutive order, the array starts with tag Recipe_1[0] and ends with tag Recipe_1[9].

Since the Logix controller does not automatically group data of the same type in memory, you can use arrays to help group data of the same type. For example, in Figure 4–25 there is a 15-element timer array that was created called "Packing_Station_Timers." Now all of the timers associated with the Packing Station logic are grouped together. Single-dimensional arrays like this, can help to organize data and conserve memory.

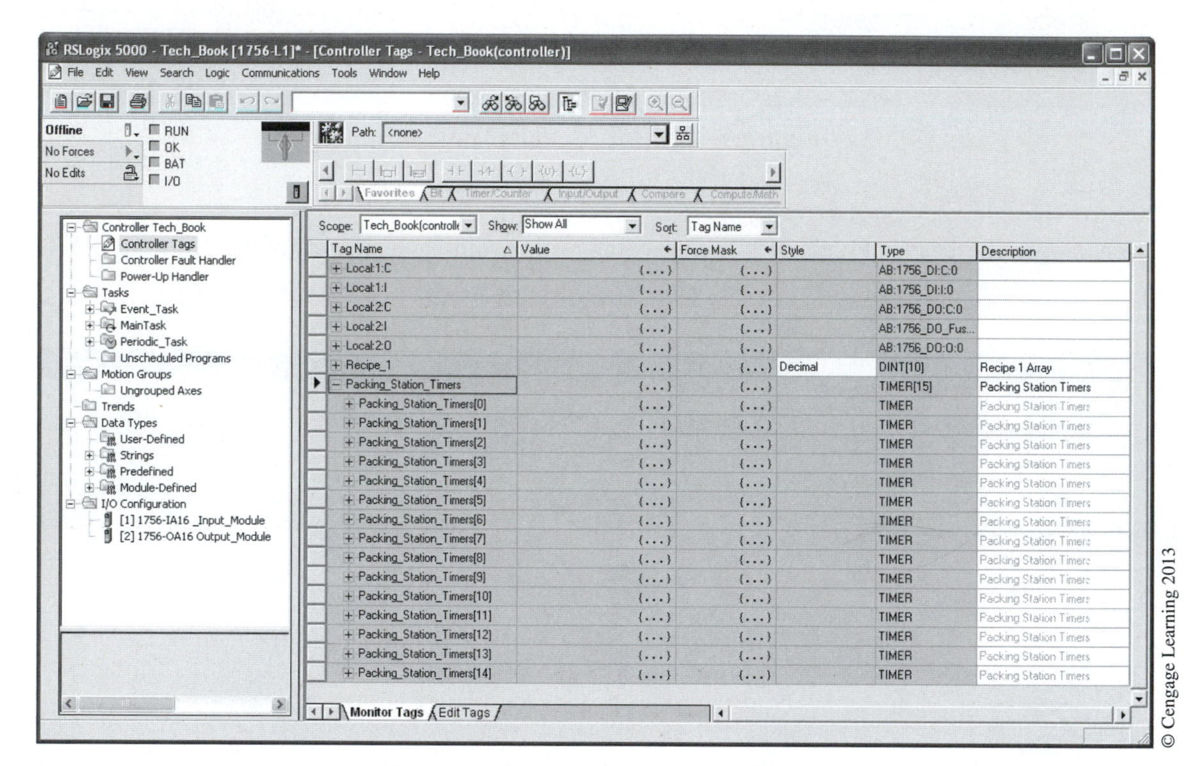

Figure 4–25 15-Element Timer Array

Another example of a single-dimensional array is shown in Figure 4–26. In this example, a BOOL array has been created called "Internal_Relays." The array size is 32 bits or one DINT data type. If you recall a BOOL data type is one bit (1 = on and 0 = off) and if you were to have created 32 individual BOOL tags you would have used 128 bytes of memory. By creating an array of 32 BOOL tags (one DINT) only 4 bytes of memory was used. *Note: BOOL arrays can only be used with bit instructions; if you need to use a file type instruction, then create a DINT array.*

Arrays can be one-, two-, or three-dimensional as illustrated in Figure 4–27. A three-dimensional array might store information such as, part number, color, and size. As you can see, arrays open up all sorts of possibilities for storing specific data as a table of values.

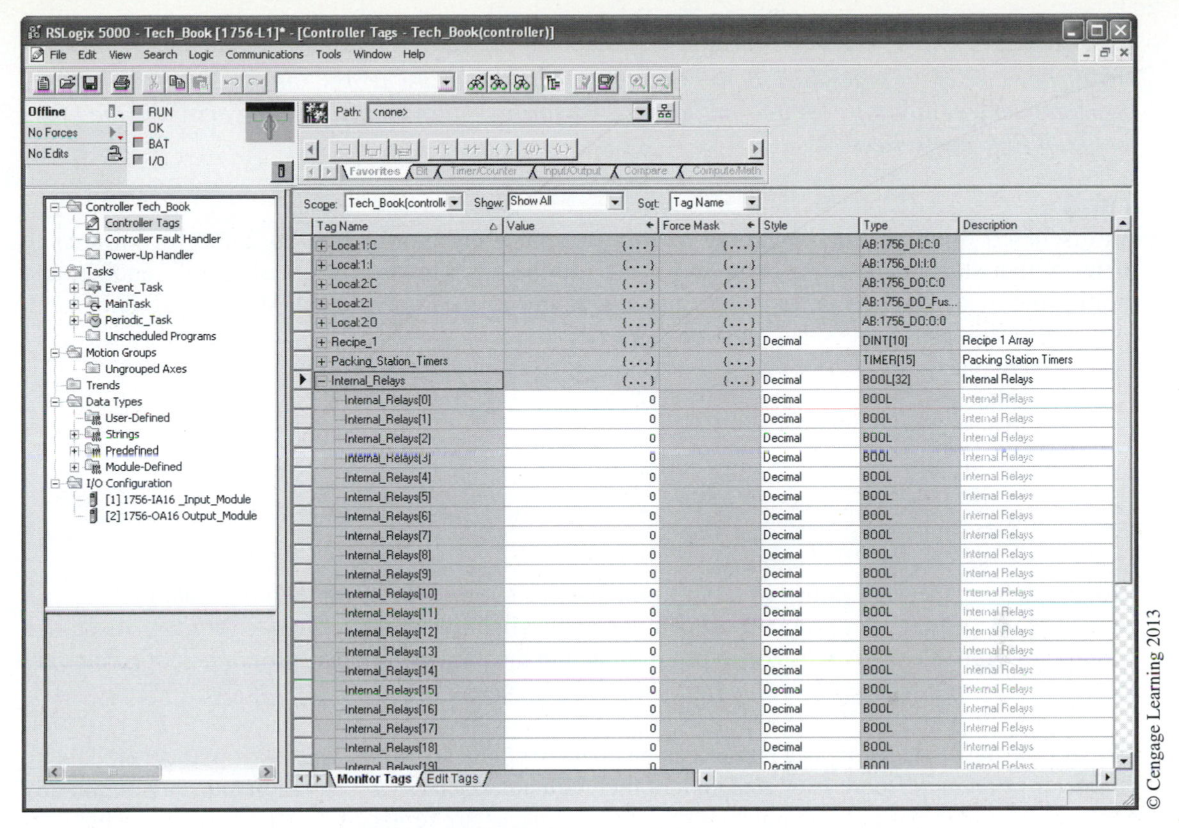

Figure 4–26 32-Element BOOL Array

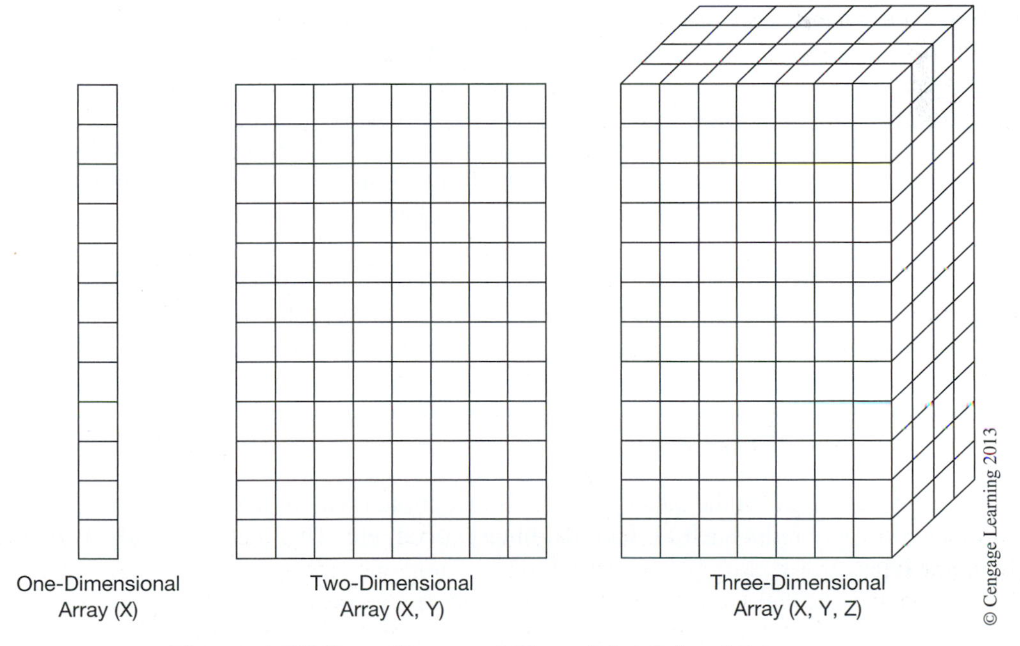

Figure 4–27 One-, Two-, and Three-Dimensional Arrays

Note: *Because of the many variations of creating and addressing arrays, the reader should consult the manufactures literature for information on configuring, addressing, and using arrays in the Logix family of PLC controllers.*

Tags

As discussed briefly at the beginning of this section, a tag is a text-based name for an area of memory that stores data in the controller. Tags are the basic means for creating, referencing, and monitoring data. The controller stores tags in memory as they are created. When a tag is created, there are certain parameters that must be defined:

- Scope
- Name
- Tag Type
- Data Type
- Style
- Description

Scope—defines the availability of a tag to the user programs. A tag can be designated as either a controller-scoped tag or a program-scoped tag. Controller-scoped tags, such as I/O tags discussed earlier, are available to every task and program within the project, whereas program-scoped tags are available only to the program with which they are associated.

Name—the tag name itself. Tag names can be up to 40 characters long, must start with an alphabetic character or an underscore [_], and cannot end with an underscore. The tag name can contain any combination of alphabetic and numeric characters. Spaces are replaced with underscores when you enter in a tag name.

Tag Type—the type of tag: Base, Alias, Produced, or Consumed. Base tags are tags that store a value for use by the logic within the project and are the actual named area of memory. Alias tags use a different or second name for an existing tag's data area of memory. Alias tags are commonly used to simplify long or complex naming structures such as I/O tags or to allow programming of the logic before the actual I/O drawings are complete. When the value of a base tag changes, so do all alias tags that reference the base tag. Figure 4–28 shows an alias tag (Start_Push_Button) for the first input point of the 1756-IA16 module discussed earlier.

A produced (broadcast) tag is any tag that is shared with other controllers over the backplane, ControlNet™ network, or Ethernet/IP network. A consumed (receive) tag is a tag that holds the value of a produced tag.

Data Type—the data type of the tag, which can be either a predefined data type (DINT, REAL, TIMER, etc.) or a user-defined data type.

Style—the display radix for the data type. This is an optional feature that allows the user to change to a different display radix such as decimal, binary, octal, etc. This feature is only available with certain data types.

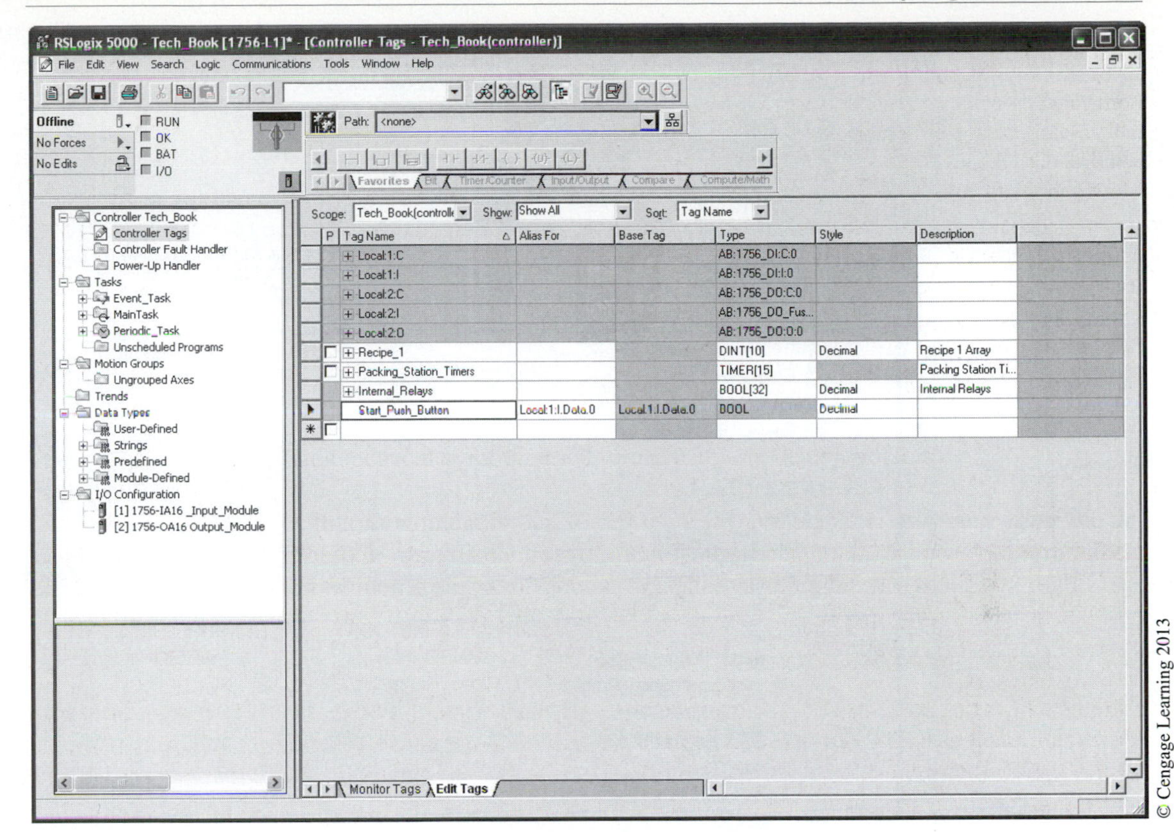

Figure 4–28 Alias Tag

Tasks

The Logix controller is a preemptive multitasking PLC system, meaning that only one task (a set of one or more programs) can be active at a time but has the ability to interrupt the active task and switch to a different task, then return. A task contains programs, each with its own routines. The routines contain the executable code. Figure 4–29 shows the project organization and how tasks, programs, and routines fit together.

A task triggers the execution of its scheduled programs. Tasks can either be continuous, periodic, or event.

Continues Task—a task that runs in the background anytime other operations such as periodic tasks are not executing. A continuous task runs all the time and automatically restarts after each completion. A project does not require a continuous task, but one is created by default with each new project and can be deleted or modified as desired. It is worth noting that there can only be one

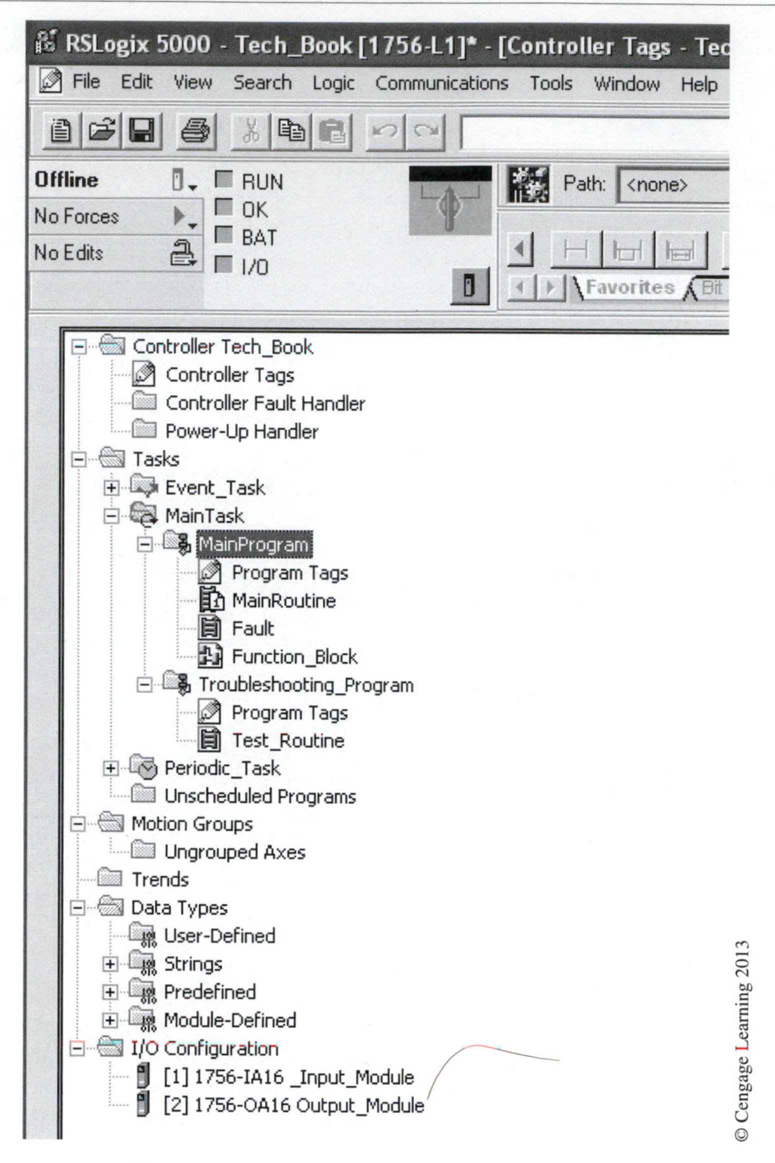

Figure 4–29 Project Organization Window

continuous task in a project. A continuous task is always interrupted by a periodic or event task and, by default, has the lowest priority.

Periodic Task—a task that is triggered at a repeated time interval. Whenever the periodic task is triggered by the controller, it interrupts any lower-priority tasks, executes one time, and returns to where the previous task left off. Periodic tasks have a time range of 1 millisecond to 2000 seconds and are used for applications requiring accurate and deterministic execution.

Event Task—a task that is triggered only when a specific event (trigger) occurs. Whenever an event task is triggered, the event task interrupts any lower-priority tasks, executes one time, and returns to where the previous task left off.

Programs

A program is the second level below the task. As you recall, each task can have up to 32 programs. When the task is triggered, the programs defined for that task are executed in sequence from the first scheduled to the last scheduled. The ladder logic with each program can modify controller-scoped and local program-scoped data. Developing multiple programs can be useful in helping to organize major equipment pieces or work areas. They can also be useful to isolate machine operations or during development by multiple programmers. When a Logix project is developed, a default program is created and scheduled in the default MainTask (Figure 4–30).

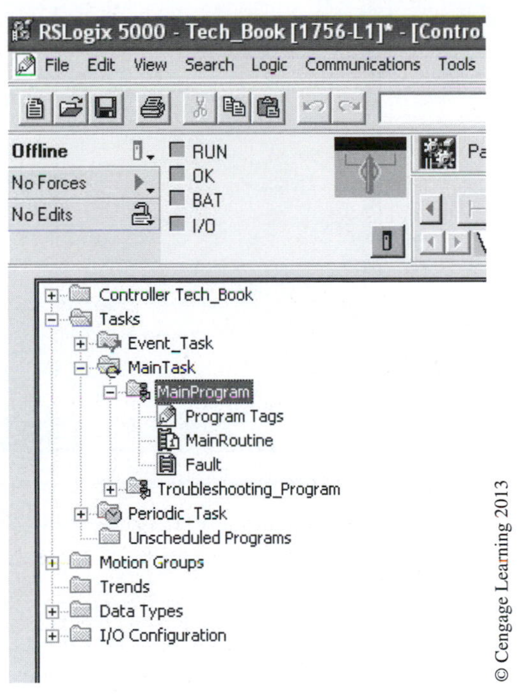

Figure 4–30 Main Task Program File

Programs are scheduled in a specific task or left unscheduled. For example, a technician may create a troubleshooting or test program and only schedule the program when needed. Another example might be that an original equipment manufacturer develops one project and then schedules

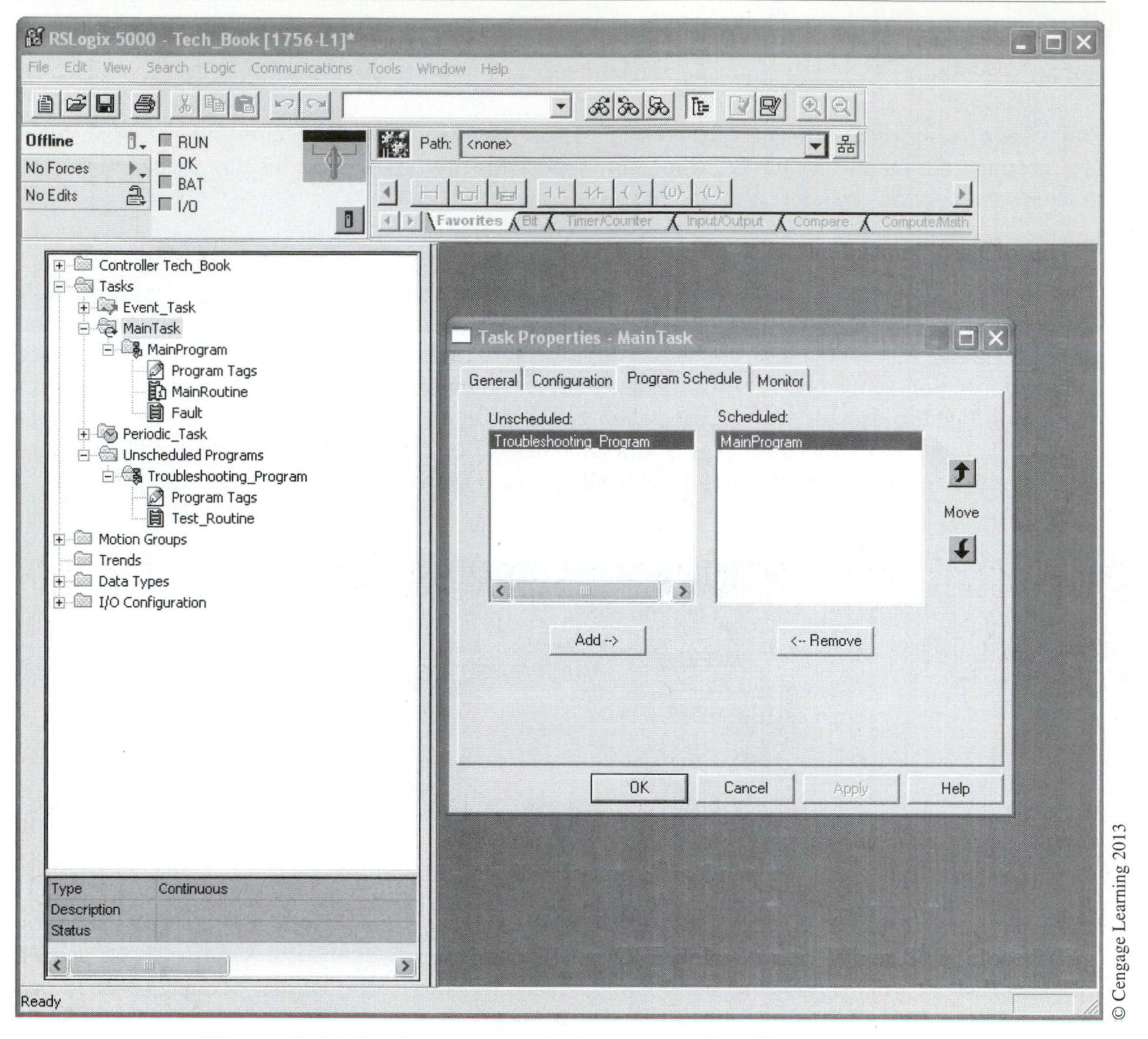

Figure 4–31 Configuring Unscheduled and Scheduled Programs

or unschedules programs based on the needs of that particular machine. In Figure 4–31 the author has created a troubleshooting program that is configured as unscheduled. Notice how a folder titled "Unscheduled Programs" has been created in the controller organizer window on the left and the program placed under that folder. The configuration screen on the right clearly shows how programs can be scheduled and unscheduled. If there are multiple scheduled programs they can be arranged in the sequence desired, from the first scheduled to the last scheduled.

Routines

A routine contains the actual ladder logic within a program. The routine contains a set of logic instructions that provides the executable code to the controller. The routine is programmed in a single programming language, such as ladder logic. The Logix PLC is capable of four programming languages; relay ladder logic, function block diagram, sequential function chart, and structured text. Each of these programming languages will be covered in later chapters.

There is no limit to the number of routines that can be created under a program. Use routines to isolate machine functions or modularize code into subroutines. For example, you could develop a ladder logic routine that controlled the wash cycle and another for the rinse cycle. In this way, the logic is organized by function and can be executed only when needed.

A routine can be assigned as the main routine that executes automatically when the controller triggers the associated task and program, or a fault routine that executes if the controller finds a fault within any routines in the associated program. Routines can be created that are only executed when instructed by logic in the main routine; these types of routines are called subroutines.

In Figure 4–32 the author has created two routines; one is our rinse cycle routine and the other is our wash cycle routine. Since both routines are triggered from the main routine, as seen in Figure 4–32, they are called subroutines and are only executed when the buttons are pushed.

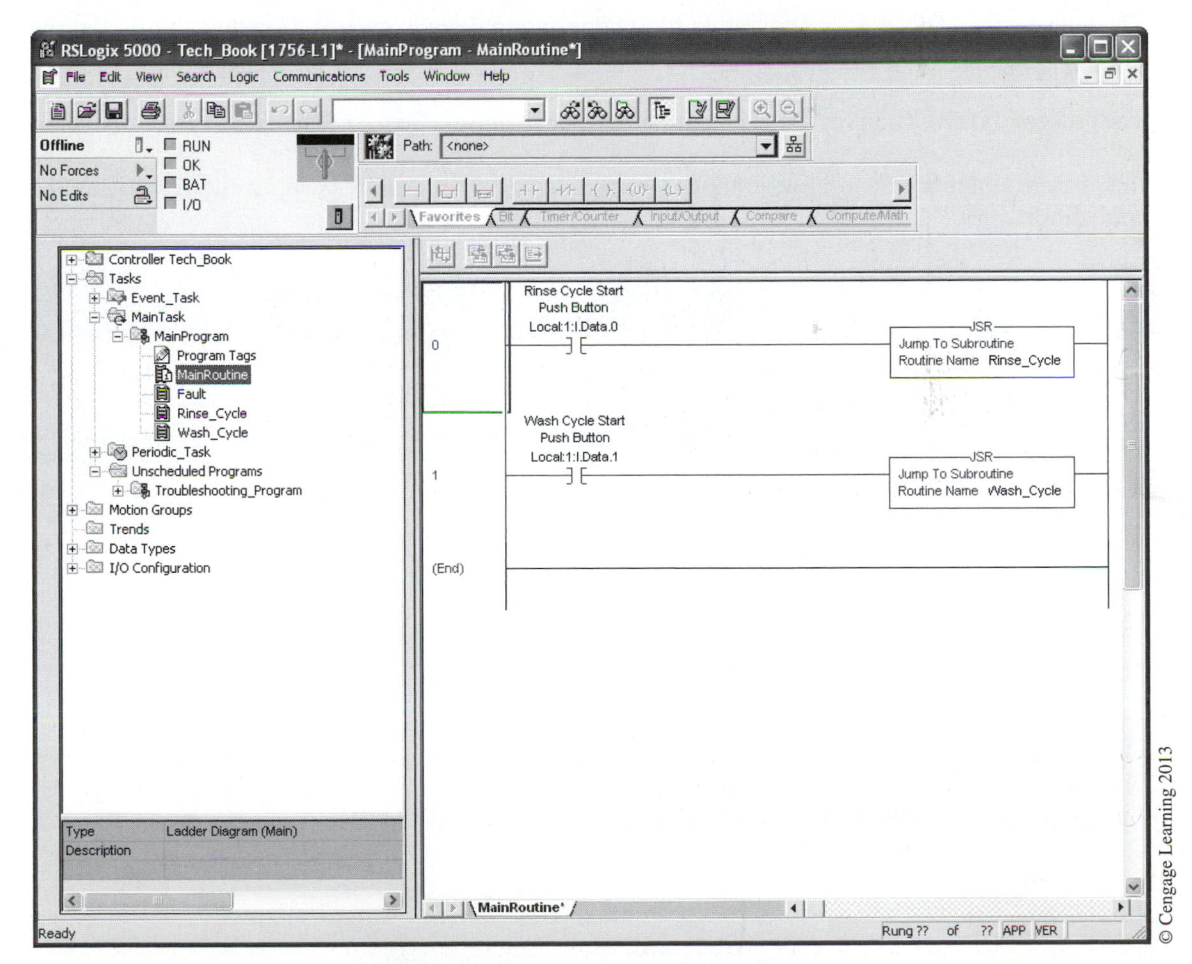

Figure 4–32 Routines and Subroutines

Keep in mind that only one task can be assigned as the continuous task. Programs assigned to the continuous task will execute according to their assigned order, and only one routine in each program can be assigned as the main or continuous routine.

As the names and structure vary between PLC families, the only way to really understand the memory structure is to obtain the literature for the specific PLC that you are dealing with. Salespeople and technical representatives are all invaluable resources when you are trying to gather information or clarification about a particular PLC.

Chapter Summary

All data, logic, and numerics are stored with binary digits that are represented as either a 1 or a 0. By storing binary data, the processor can rapidly scan and execute the user program and update the I/O section. I/O addresses in many cases not only identify the word and bit that is associated with the I/O, but also indicate hardware location (rack, module group, and terminal).

The names given to memory sections or subsections are unique to each PLC manufacturer, but the memories all work in basically the same manner.

The processor memory stores the I/O status, the user program, and numeric data used by the processor.

In tag-based memory the user determines the memory structures and can adjust these to match the application.

Review Questions

1. The following types of information are normally found and/or stored in one of the PLC's two memory categories (user and storage). Place an S (for storage memory) or a U (for user memory) before the information type to indicate in which category it is normally found and/or stored.
   a. status of discrete input devices
   b. preset values of timers and counters
   c. numeric values of arithmetic
   d. holding registers
2. Identify the following PLC-5 files:
   a. I b. O c. N
   d. S e. B f. T
   g. R h. F i. C
3. Define the term *byte*.
4. In a PLC-5, data file 5 is what type of file?
5. What word and bit number are represented by PLC-5 address O:010/01?

6. Using SLC 500 addressing, what do the following addresses indicate?
   a. O:3/15
   b. I:2.1/3
   c. O:5/0
   d. I:7/8
7. Using Allen-Bradley PLC-5 address format, what would address I:013/12 indicate?
8. Using SLC 500 or MicroLogix address format, what would address I:1.0/4 indicate?
9. Referring to Figure 4–14, what would address O:6.2 indicate?
10. Define the term *scope tags*.
11. Describe the following data types:
    a. REAL b. BOOL
    c. SINT d. DINT
12. What is an *array*?

CHAPTER 5 Numbering Systems

Objectives

After completing this chapter, you should have the knowledge to:

- Understand decimal, binary, octal, hexadecimal, and binary coded decimal (BCD) numbering systems.
- Convert from one numbering system to another.
- Express negative numbers in 2s complement.
- Add signed numbers.
- Convert a negative binary display to its decimal equivalent.
- Complete a subtraction problem using 2s complement and addition.

Electricians, technicians, or other personnel who are required to program, modify, or maintain a PLC must have a "working" knowledge of the different numbering systems that are used. For example, the input/output addresses may use the octal numbering system; the timer and counter addresses may use the decimal numbering system; accumulated and preset values of the timers and counters may use the binary numbering system; operator interfaces, such as thumbwheels and seven-segment displays, may require information to be sent and received using the BCD format; and the hexadecimal system may be used for loading information into sequencers. The numbering system used in each area discussed varies with the different PLC manufacturers, but it is obvious that to fully understand and program a PLC, an understanding of the various numbering systems is necessary.

DECIMAL SYSTEM

The decimal numbering system is used every day by electricians and technicians, and it is a system they are comfortable with. This system uses ten unique numbers, or digits, which are 0 through 9. A numbering system that uses 10 digits is said to have a base of 10. The value of the decimal number depends on the digit(s) used, and each digit's place value. Each position can be represented as a power of 10, starting with 10^0 as shown in Figure 5–1. In the decimal system, the first position to the left of the decimal point is called the units place, and any digit from 0–9 can be used. The next position to the left of the units place is the tens place; next is the hundreds place, the thousands place, and so on, with each place extending the capability of the decimal system by ten, or a power of ten.

Note: *Any number that uses an exponent of 0, such as 10^0, has a place value of 1. Exponent 10^0 equals 1.*

A specific decimal number can be expressed by adding the place values, as shown in Figure 5–2.

Mathematically, each place value is expressed as a digit number times a power of the base, or 10, in the decimal numbering system.

Another example is shown in Figure 5–3 using the decimal number 239.

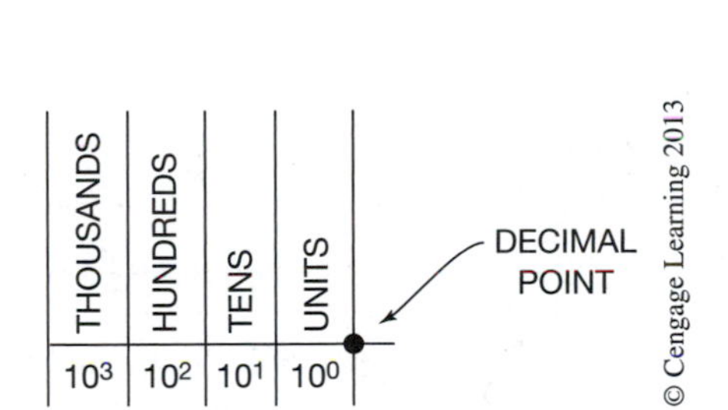

Figure 5–1 Place Value and Corresponding Power of Ten

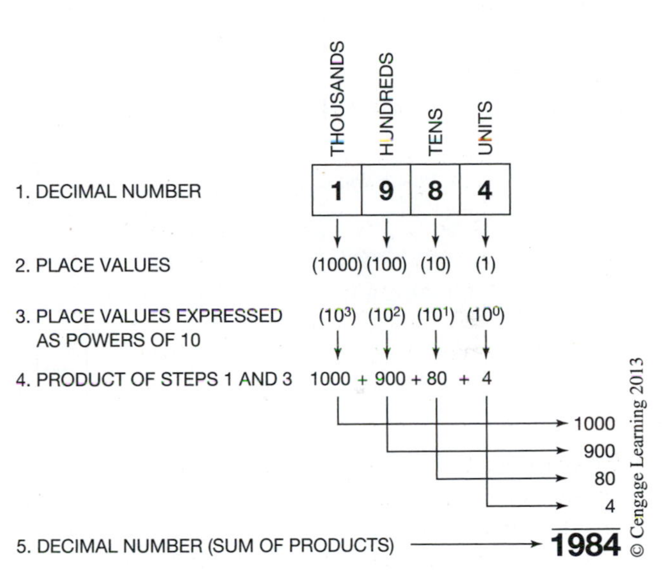

Figure 5–2 Decimal Numbering System

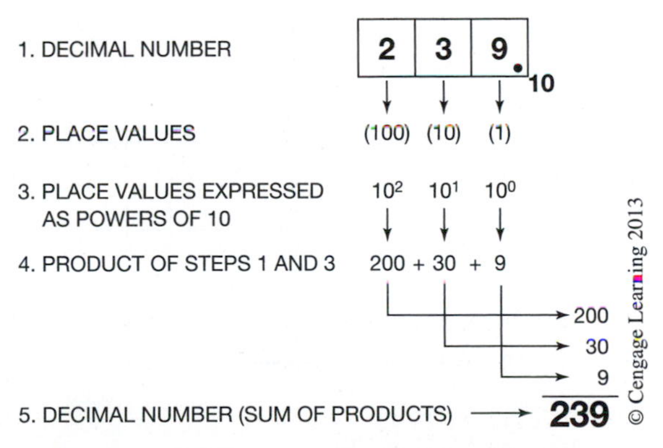

Figure 5–3 Decimal Numbering System

BINARY SYSTEM

The binary system uses only two digits: 1 and 0. Since only two digits are used, this system has a base of 2. Like the decimal system—and all numbering systems for that matter—each digit has a certain place value. The first place to the left of the starting point, or binary point, is the units or 1s location (base 2^0). The next place, to the left of the units place, is the 2s place, or base 2^1, as shown in Figure 5–4. The next place value is the 4s place, or base 2^2, then the 8s place, or base 2^3, and so forth. A binary number is always indicated by placing a 2 in subscript to the right of the units digit. Figure 5–4 illustrates how a binary number is converted to a decimal equivalent number. Note the subscripted 2 at the lower right-hand corner of the binary number line that indicates a base 2, or binary number.

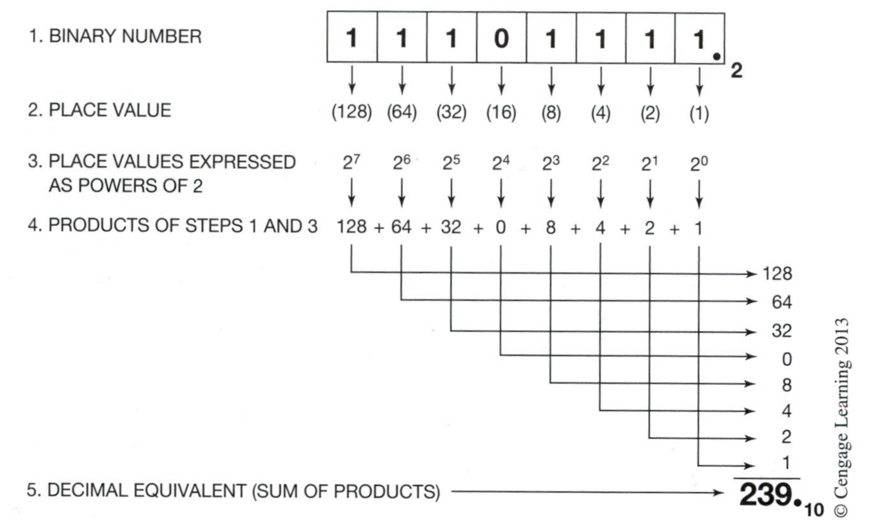

Figure 5–4 Converting a Binary Number to a Decimal Number

To convert a decimal number into a binary number, or to any numbering system for that matter, use the following procedure, as shown in Figure 5–5. Divide the decimal number by the base you wish to

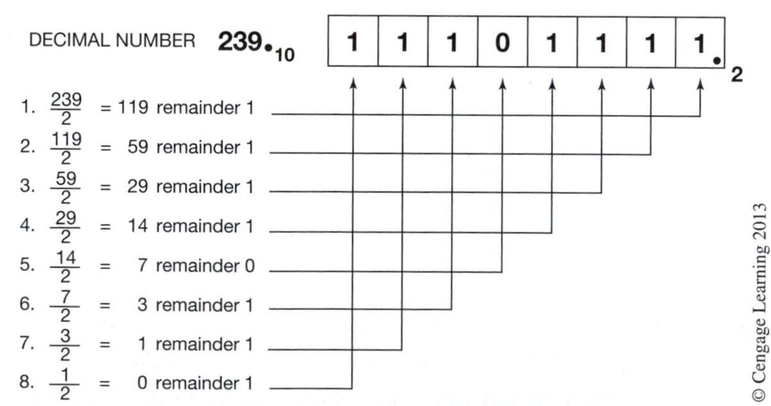

Figure 5–5 Converting a Decimal Number to a Binary Number

convert to, in this case 2. The remainder is the 1s value (see Step 1 in the figure). Now divide the quotient from the first division again; the remainder becomes the value that is placed in the 2s location (see Step 2). The quotient of each preceding division is then divided by the base 2 until the base can no longer be divided (see Step 8), and the remainder (1) becomes the last digit in the binary number.

It is important to arrange the remainders correctly when making the decimal-to-binary conversion. The first digit placed in the 1s position is called the *least* significant digit, whereas the last digit is called the *most* significant digit. The last digit placed has the highest place value (128s) which is why it is called the most significant digit. This reference to least and most significant digits is common, and refers to the relative position of any given digit within a number.

The following steps summarize this decimal-to-binary conversion.

Step 1. The decimal number is divided by 2 (base of the binary numbering system). The quotient is listed (119) as well as the remainder (1).

Step 2. Divide the quotient of Step 1 (119) by base 2, and list the new quotient (59) and the remainder (1).

Step 3. Divide the quotient of Step 2 (59) by base 2, and list the new quotient (29) and remainder (1).

Step 4. Divide the quotient of Step 3 (29) by 2, and list the new quotient (14) and the remainder (1).

Step 5. Divide the quotient of Step 4 (14) by 2, and list the new quotient (7) and remainder (0).

Step 6. Divide the quotient of Step 5 (7) by 2, and list the new quotient (3) and remainder (1).

Step 7. Divide the quotient of Step 6 (3) by 2, and list the new quotient (1) and remainder (1).

Step 8. Divide the quotient of Step 7 (1) by 2, and list the new quotient (0) and remainder (1).

Note: *When using a calculator to do the division, the value to the* right *of the decimal must be multiplied by the base to get the actual remainder. For example, when 239 is divided by 2 (Step 1) on a calculator, the answer is 119.5. To find the actual remainder, the 0.5 is multiplied by 2, the base, to find the remainder 1. This procedure is true for any numbering system. The base times the value to the right of the decimal point equals the actual remainder.*

The binary numbering system is used to store information in the processor memory in the form of *bits*.

2S COMPLEMENT

Virtually all programmable controllers, computers, and other electronic calculating equipment perform counting functions using the binary system. For those PLCs that are programmed to perform arithmetic functions, a method of representing both positive (+) and negative (–) numbers must be used. The most common method is 2s complement. The 2s complement is simply a convention for binary representation of negative decimal numbers.

Before going any further with a discussion of 2s complement, a review of adding binary numbers may be helpful. In decimal addition, numbers are added according to an addition table. A partial addition table is shown in Figure 5–6.

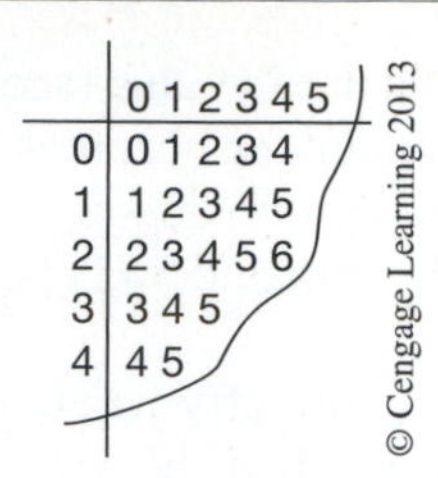

Figure 5–6 Decimal Addition System

To use the table, the first number to be added is located on the vertical line, and the second number on the horizontal line. The sum, or total, is found where the two imaginary lines intersect. For example, 3 + 2 = 5 (as shown in Figure 5–7).

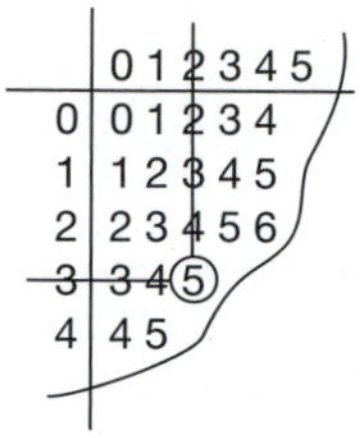

	0	1	2	3	4	5
0	0	1	2	3	4	
1	1	2	3	4	5	
2	2	3	4	5	6	
3	3	4	(5)			
4	4	5				

Figure 5–7 Adding 2 and 3

For binary addition, a similar addition table is constructed. The table is small because the binary system only has two digits (1 and 0) (Figure 5–8).

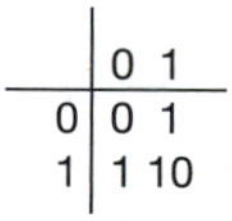

	0	1
0	0	1
1	1	10

Figure 5–8 Binary Addition Table

To use the table, the first number (digit) to be added is located on the vertical line, the second digit is located on the horizontal line. The sum, or total, is found where the two imaginary lines intersect. Figure 5–9 shows an example of adding 1 + 0 = 1.

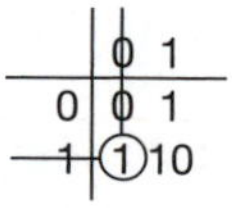

	0	1
0	0	1
1	(1)	10

Figure 5–9 Adding Binary 1 and 0

Notice that if 1 and 1 are added, the table shows 1 0, not 2, as might be expected. 1 0 is the binary representation of 2 (Figure 5–10).

8	4	2	1
0	0	1	0

$_2$

Figure 5–10 Binary Representation of 2

Figure 5–11 shows how binary numbers 1011_2 and 110_2 are added.

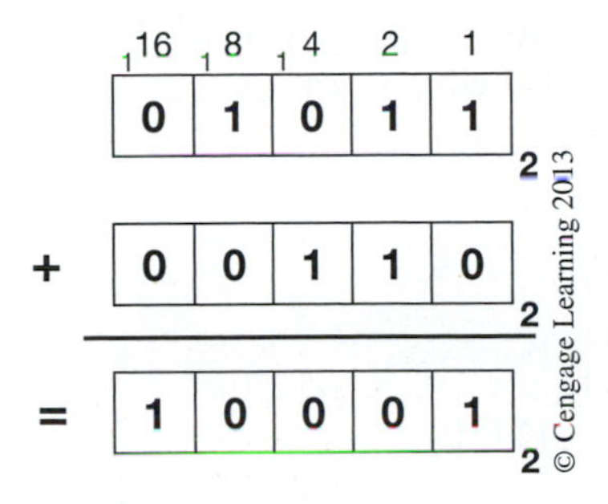

Figure 5–11 Adding Binary Numbers

In the 1s column 1 + 0 = 1.
In the 2s column 1 + 1 = 0, with a carryover of 1.
In the 4s column 1 + 0 + 1 = 0, with a carryover of 1.
In the 8s column 1 + 1 + 0 = 0, with a carryover of 1.
In the 16s column 1 + 0 + 0 = 1.
The sum (total) of 1011_2 and 110_2 is, therefore, 10001_2.

To verify our results we can convert the binary numbers to decimal equivalent numbers and add them, as shown in Figure 5–12.

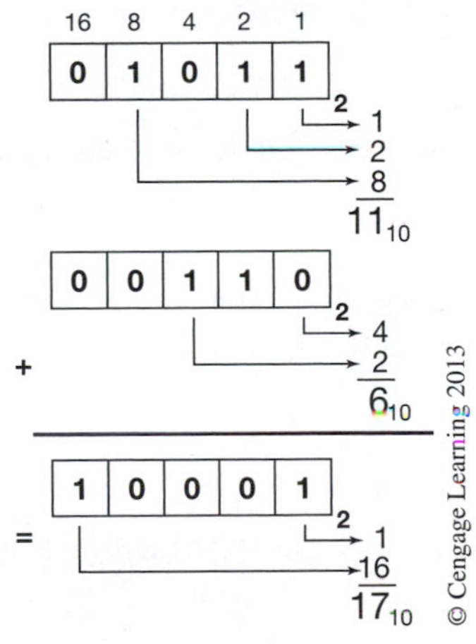

Figure 5–12 Converting Binary Numbers to Decimal Equivalents

Another example of adding binary numbers is shown in Figure 5–13, where 11011_2 and 11_2 are added.

$$\begin{array}{r} {}^{1\,1} \\ 11011 \\ +\quad 11 \\ \hline 11110_2 \end{array}$$

Figure 5–13 Addition of Binary Numbers

In the 1s column 1 + 1 = 0, with a carryover of 1.
In the 2s column 1 + 1 + 1 = 1, with a carryover of 1.

Note: *1 + 1 + 1 = 3. The binary equivalent of* 3_{10} *is* 11_2.

In the 4s column 1 + 0 = 1.
In the 8s column 1 + 0 = 1.
In the 16s column 1 + 0 = 1.
The sum of 11011_2 and 11_2 is 11110_2.

To verify this method, convert the binary numbers to decimal numbers, and add them, as shown in Figure 5–14.

$$\begin{array}{rcr} 11011_2 & = & 27_{10} \\ +\quad 11_2 & = & 3_{10} \\ \hline 11110_2 & = & 30_{10} \end{array}$$

Figure 5–14 Comparing Binary and Decimal Addition

To represent negative numbers using the binary numbering system, one bit is designated as a signed bit. If the designated bit is a 0 (zero), the number is positive, and if the bit is a 1, the number is negative.

Using a 4-bit word length, and using bit 4 as the designated signed bit, 0001_2 represents +1 decimal (see Figure 5–15).

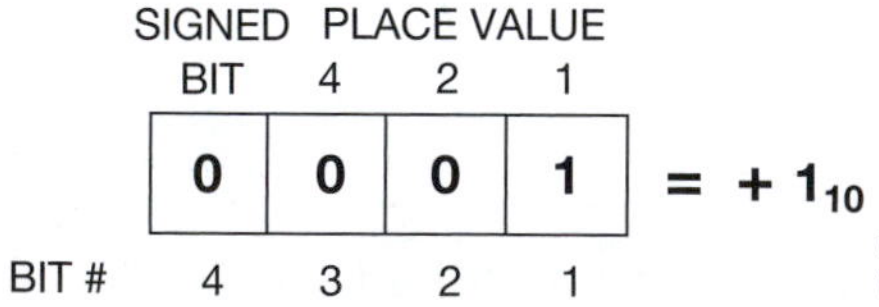

	SIGNED BIT	PLACE VALUE 4	2	1	
	0	0	0	1	$= +1_{10}$
BIT #	4	3	2	1	

Figure 5–15 4-Bit Word with a Signed Bit

The table in Figure 5–16 shows all of the possible numbers for a 4-bit word using 2s complement.

BINARY NUMBER	DECIMAL
0111	+7
0110	+6
0101	+5
0100	+4
0011	+3
0010	+2
0001	+1
0000	0
1111	–1
1110	–2
1101	–3
1100	–4
1011	–5
1010	–6
1001	–7
1000	–8

Figure 5–16 2s Complement Numbers for a 4-Bit Word

Notice that the negative numbers go to –8 while the positive numbers only go to +7. In this case, the signed bit is used for its place value, which is 8. The same holds true for 8- and 16-bit words. The maximum negative number is always one number *higher* than the maximum positive number.

To display a negative binary number requires that the same value positive number be complemented (all 1s changed to 0s and all 0s changed to 1s) and a value of 1 added. The result is the 2s complement of the number. Figure 5–17 shows the steps to express –5 in 2s complement using a 4-bit word.

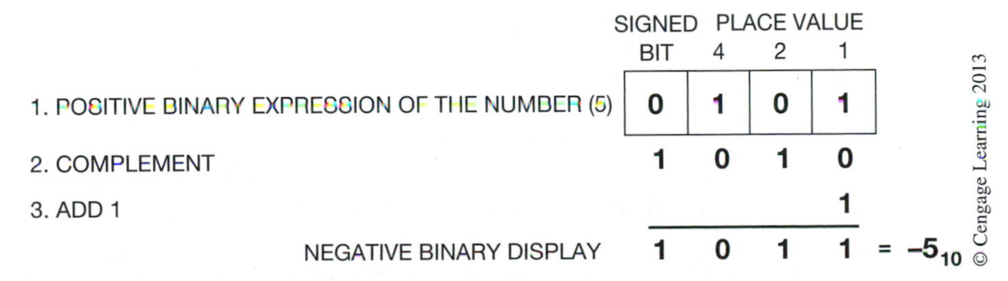

Figure 5–17 Expressing –5 in 2s Complement

Another example of 2s complement is shown in Figure 5–18 with the steps required to express –7 in 2s complement.

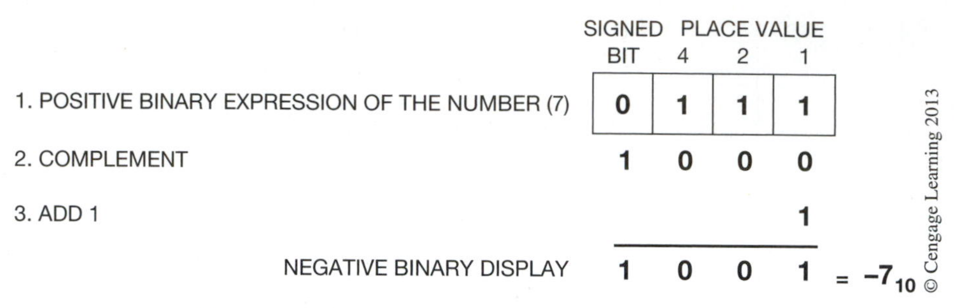

Figure 5–18 Expressing –7 in 2s Complement

To convert a negative binary number to the decimal equivalent, the negative binary display is complemented, 1 is added, the binary sum is converted to decimal, and the negative sign (–) is added. Figure 5–19 shows what steps are necessary to determine the negative value of 1110_2.

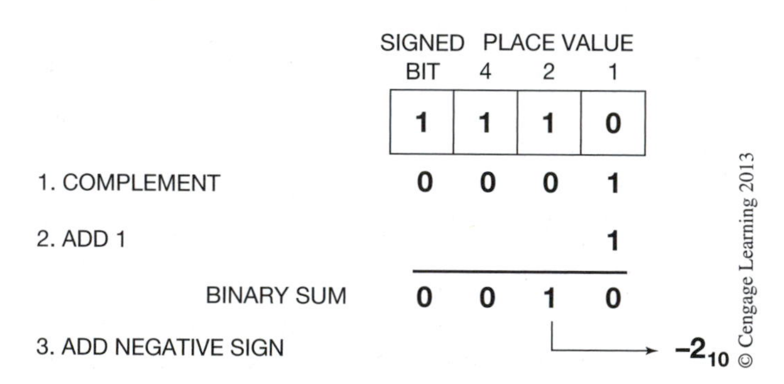

Figure 5–19 2s Complement to Decimal Equivalent

An easy way to convert a negative binary number to its equivalent negative decimal number is to subtract the place value of the signed bit from the value of the binary digits. In Figure 5–19, the binary sum 1110 is equal to -2_{10}, which is the decimal number –2. In this example, a four-bit word is used with bit 4 being the signed bit. The fourth bit normally would have a place value of 8. In this example, the value 8 is subtracted from the value of the binary number 110_2 which is equal to 6. 6 – 8 = –2.

Another example using this method of converting negative binary numbers to their decimal equivalent is shown in Figure 5–20, using an 8-bit word with bit 8 being the signed bit.

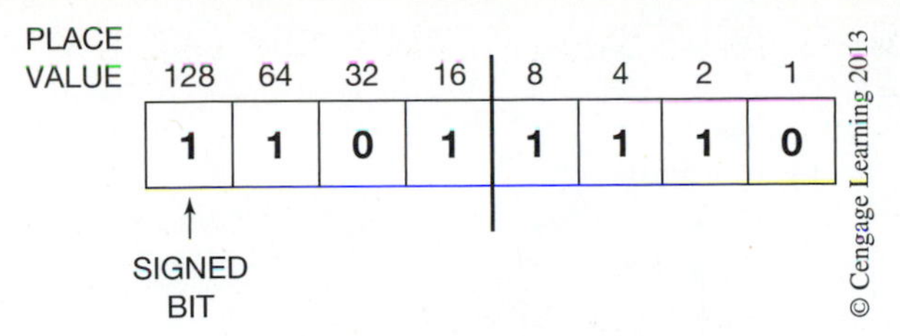

Figure 5–20 8-Bit Word 2s Complement

In this example, the signed bit would have a value of 128 (2^7), whereas the numeric value of the other bits would be 94, as shown in Figure 5–21.

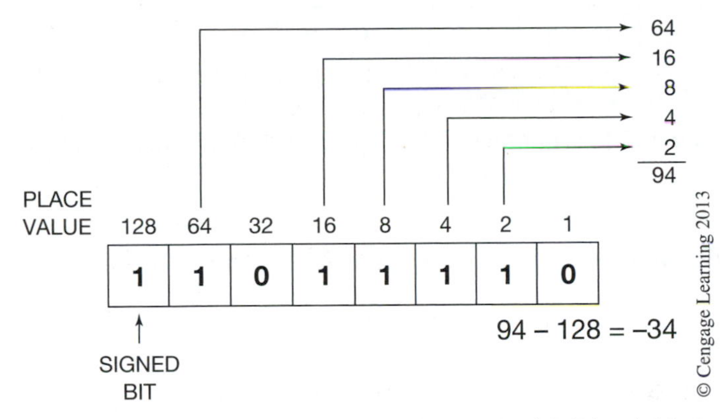

Figure 5–21 8-Bit Word (2s Complement) with Bits Added

Subtracting the place value of the signed bit (128) from the numeric value of the other bits (94) gives us the decimal number: 94 – 128 = –34. To verify this answer, complement the original binary number 1101 1110 to get 0010 0001. Then add 1.

$$\begin{array}{r} 0010\ 0001 \\ 1 \\ \hline 0010\ 0010 \end{array}$$

The answer is $2^5 + 2^1$ or $32 + 2 = 34$; adding the negative sign gives us a final answer of –34, the same answer we got when we subtracted the place value of the signed bit (128) from the numeric value of the binary number 94.

If the PLC system uses 2s complement for arithmetic, the highest positive number that a 16-bit word can represent is +32,767 as shown in Figure 5–22.

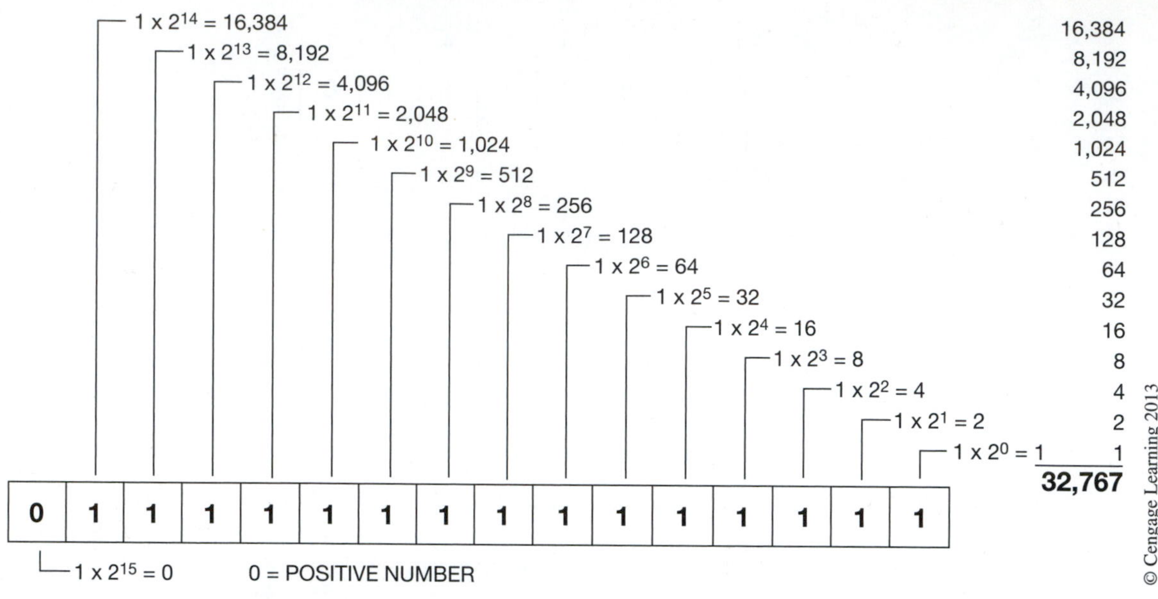

Figure 5–22 Maximum Positive Value of 2s Complement 16-Bit Word

The largest negative number that can be represented by a 16-bit word is –32,768, as shown in Figure 5–23.

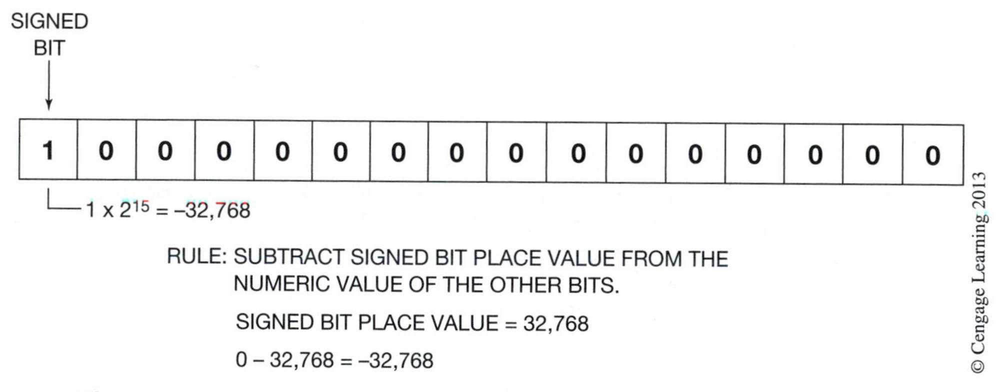

Figure 5–23 Maximum Negative Value of 2s Complement 16-Bit Word

Another method of converting positive numbers to 2s complemented negative numbers is as follows: starting at the least significant bit and working to the left, copy each bit up to and including the first 1 bit, and then complement or change each remaining bit. Figure 5–24 shows this alternate method of expressing –2 in 2s complement using a 4-bit word.

1. ORIGINAL POSITIVE NUMBER (+2)	0	0	1	0
2. COPY UP TO FIRST 1 BIT			1	0
3. COMPLEMENT THE REMAINING BITS	1	1	1	0
2S COMPLEMENT –2	1	1	1	0

Figure 5–24 Alternate Method of 2s Complement

A further example is shown in Figure 5–25 for 2s complementing the value 24 using an 8-bit word.

1. ORIGINAL POSITIVE NUMBER (+24)	0	0	0	1	1	0	0	0
2. COPY UP TO FIRST 1 BIT					1	0	0	0
3. COMPLEMENT REMAINING BITS	1	1	1	0	1	0	0	0
2S COMPLEMENT –24	1	1	1	0	1	0	0	0

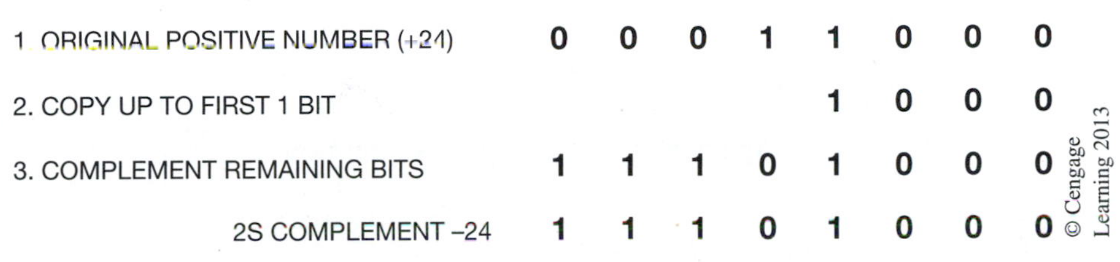

Figure 5–25 2s Complement of –24 Decimal

By using 2s complement, negative and positive values can now be added. The two steps for adding -7_{10} and $+5_{10}$ using 2s complement with a 4-bit word are shown in Figure 5–26.

	SIGNED BIT	PLACE VALUE 4	2	1
1. EXPRESS –7 IN 2S COMPLEMENT				
a. POSITIVE EXPRESSION OF NUMBER (+7)	0	1	1	1
b. COMPLEMENT	1	0	0	0
c. ADD 1				1
NEGATIVE BINARY DISPLAY	1	0	0	1
2. ADD –7 AND +5				
	1	0	0	1
+	0	1	0	1
BINARY SUM	1	1	1	0 = -2_{10}

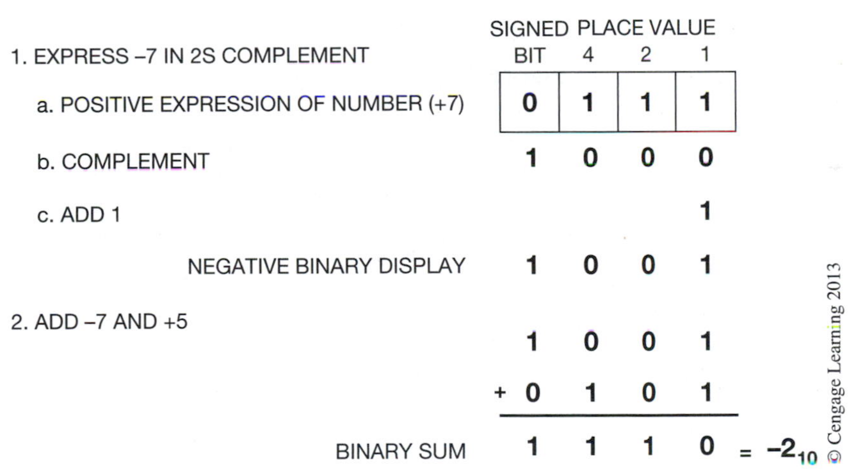

Figure 5–26 Adding Positive and Negative Numbers

Note: *When adding signed binary numbers, any carryover from the signed bit column is discarded.*

Once addition of signed numbers is possible, the other arithmetic functions (subtraction, multiplication, and division) are also possible, because they are achieved by successive addition on a PLC.

EXAMPLE: Subtracting the number 20 from 26 is accomplished by complementing 20 to obtain –20, and then performing addition.

Subtracting 20 from 26 by complementing 20 and performing addition using an 8-bit word is shown in Figure 5–27.

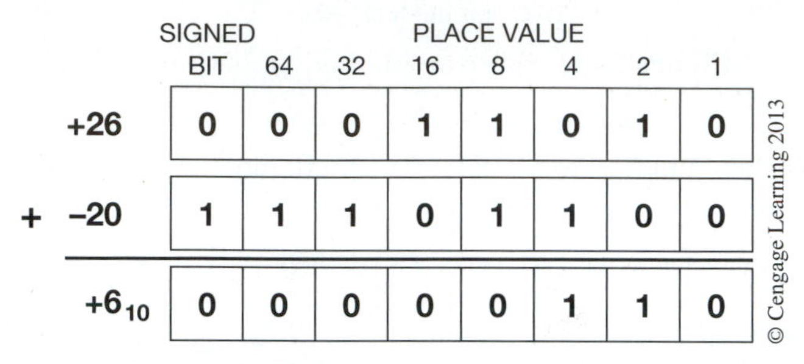

Figure 5–27 Subtraction by Addition

OCTAL SYSTEM

The octal system, or base 8, is made up of eight digits: numbers 0 through 7. The first digit to the *left* of the octal point is the units place, or 1s, and has a base or power of 8^0. The next place is eights (8s) or base 8^1. The next place is sixty-fours (64s) or base 8^2, followed by five hundred twelves (512s) or base 8^3, and four thousand ninety-sixes or base 8^4, and so on. An octal number will always be expressed by placing an eight in subscript to the right of the units digit, as shown in Figure 5–28.

$$\mathbf{357.}_8$$

Figure 5–28 Octal Number

The method of converting an octal number to a decimal equivalent number is illustrated in Figure 5–29.

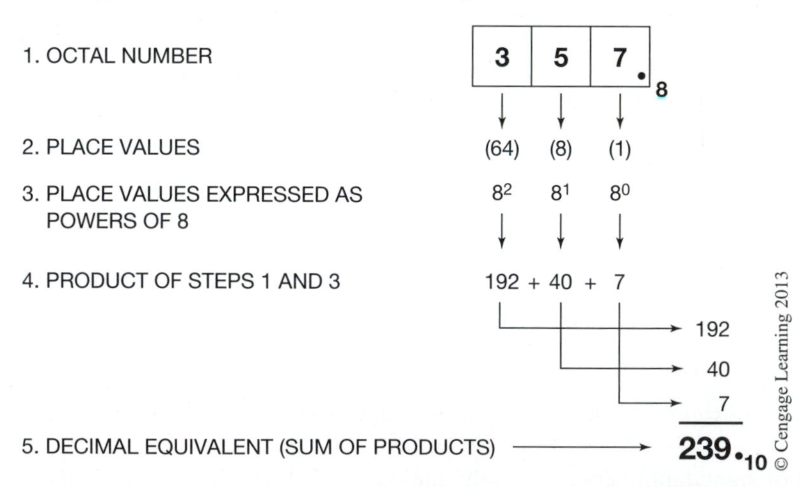

Figure 5–29 Converting an Octal Number to a Decimal Number

The decimal number 239 is converted to an octal number in Figure 5–30.

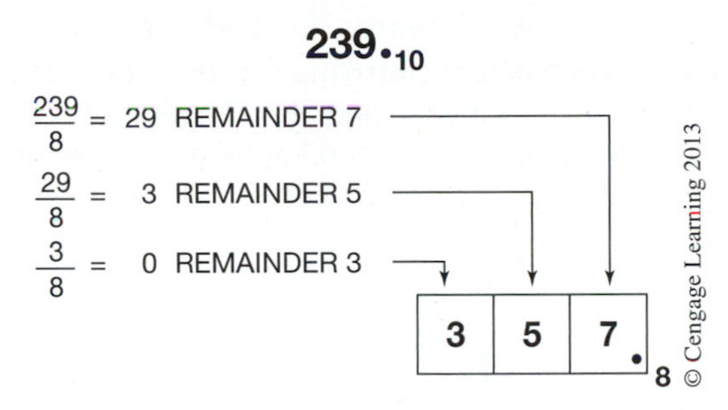

Figure 5–30 Converting a Decimal Number to an Octal Number

Step 1. The decimal number 239 is divided by 8 (base for the octal numbering system). The quotient is listed (29) as well as the remainder (7). A calculator shows the answer as 29.875. The quotient is 29, and the remainder is 0.875 × 8, or 7.

Step 2. Divide the quotient of Step 1 (29) by 8, and list the new quotient (3) and the remainder (5). A calculator gives the answer 3.625. The quotient is 3, and the remainder is 0.625 × 8, or 5.

Step 3. Divide the quotient from Step 2 (3) by 8, and list the new quotient (0) and remainder (3). The quotient 3 divided by 8 equals 0.375. The new quotient is 0, and the remainder is 0.375 × 8, or 3.

The decimal number 239 is the same as the octal number 357.

Since the largest single number that can be expressed using the octal numbering system is seven (7), each octal digit can be represented by using only three (3) binary bits (base 2). Figure 5–31 illustrates how to convert an octal number to a binary number. The figure shows three sets of binary bits and the place value of each bit. For the *least* significant digit (7), a one (1) must be placed in the 1s place, the 2s place, and the 4s place to equal 7. For the middle digit (5), a 1 is placed in the 1s place and the 4s place, while a 0 is placed in the 2s place. This combination equals 5. For the *most* significant digit (3), a 1 is placed in the 1s place and the 2s place, and a 0 is placed in the 4s place. This combination adds up to 3.

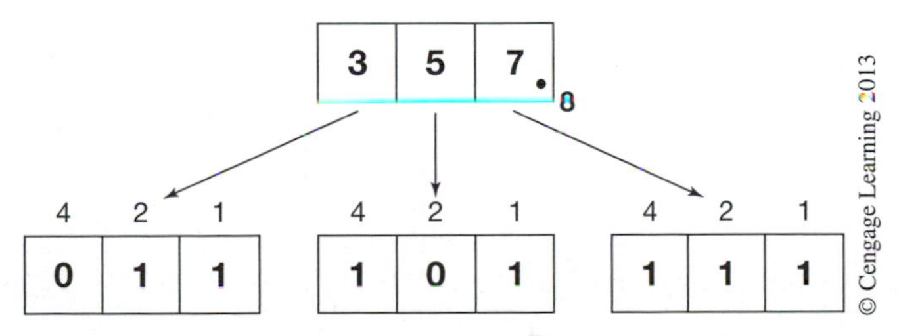

Figure 5–31 Conversion of Octal Number to Binary

Allen-Bradley uses the octal numbering system for I/O addressing for the PLC-5 family. The terminals of the input and output modules are labeled 00 through 07 and 10 through 17, rather than 0 through 15 as would be the case with decimal numbering (which is used for the SLC 500 and MicroLogix PLCs). When using the octal numbering system, words are labeled 000–007, 010–017, 020–027, and so forth, whereas the bits are labeled 00–07 and 10–17. Figure 5–32 shows a memory word with the internal bits addressed using the octal numbering system.

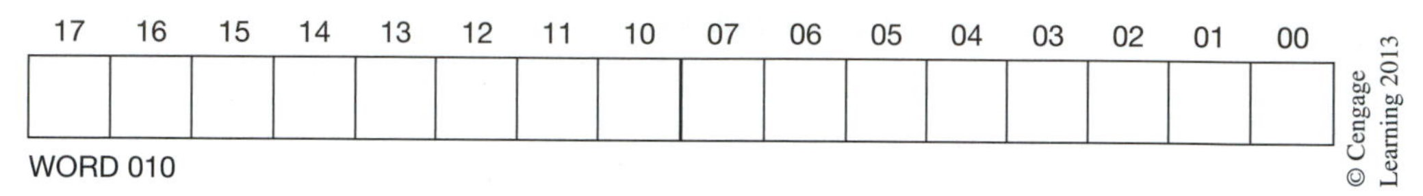

Figure 5–32 Word and Bit Labeling Using the Octal Numbering System

HEXADECIMAL SYSTEM

The hexadecimal system, often referred to as HEX, consists of a number system with base 16.

It seems logical that the numbers used in base 16 would be 0 through 15. However, only numbers 0 through 9 are used, and the letters A through F represent numbers 10–15, respectively. The place values from the hexadecimal point are 1s—16^0, 16s—16^1, 256s—16^2, 4096s—16^3, and so on.

Each hexadecimal digit is represented by four (4) binary digits. The binary equivalents are shown in the table in Figure 5–33.

HEXADECIMAL	BINARY	DECIMAL
0	0000	0
1	0001	1
2	0010	2
3	0011	3
4	0100	4
5	0101	5
6	0110	6
7	0111	7
8	1000	8
9	1001	9
A	1010	10
B	1011	11
C	1100	12
D	1101	13
E	1110	14
F	1111	15

Figure 5–33 Hexadecimal Equivalents for Binary and Decimal

The decimal number 4,780 is converted to hexadecimal as illustrated in Figure 5–34.

DECIMAL NUMBER 4780_{10}

1. $\frac{4780}{16}$ = 298 REMAINDER 12
2. $\frac{298}{16}$ = 18 REMAINDER 10
3. $\frac{18}{16}$ = 1 REMAINDER 2
4. $\frac{1}{16}$ = 0 REMAINDER 1

1	2	10	12
1	2	A	C

$_{16}$

Figure 5–34 Converting a Decimal Number to a Hexadecimal Number

Step 1. The decimal number is divided by 16 (base for the hexadecimal numbering system). The quotient is listed (298) as well as the remainder (12). A calculator provides the answer 298.75. The quotient is 298, and the remainder is 0.75 × 16, or 12.

Step 2. Divide the quotient of Step 1 (298) by 16, and list the new quotient (18) and the remainder (10). The answer is 18.625. The quotient is 18, and the remainder is 0.625 × 16, or 10.

Step 3. Divide the quotient from Step 2 (18) by 16, and list the new quotient (1) and the remainder (2). Eighteen divided by 16 equals 1.125. The quotient is 1, and the remainder is 0.125 × 16, or 2.

Step 4. Divide the quotient from Step 3 (1) by 16, and list the new quotient (0) and the remainder (1). One divided by 16 equals 0.0625. The quotient is 0, and the remainder is 0.0625 × 16, or 1.

Converting a hexadecimal number to a decimal number is illustrated in Figure 5–35.

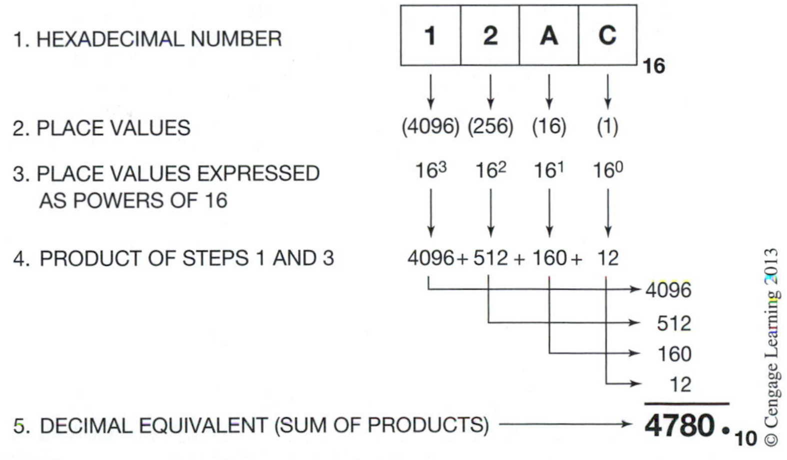

Figure 5–35 Converting a Hexadecimal Number to a Decimal Number

Note: *Remember that A is equivalent to 10, and C is equivalent to 12.*

The binary equivalent of the hexadecimal number 12AC is shown in Figure 5–36.

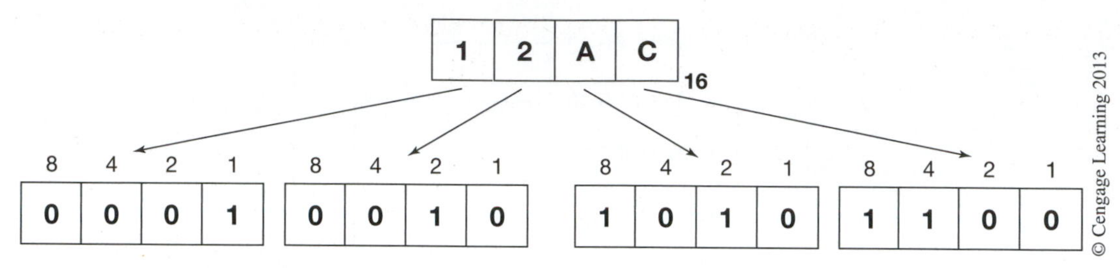

Figure 5–36 Binary Equivalent of a Hexadecimal Number

Since the largest number that can be displayed using the hexadecimal numbering system is 15, or F (as shown in the table in Figure 5–33), only four binary bits are needed to display each hexadecimal digit. The conversion to binary simply places the 1s in the correct binary locations to duplicate the hexadecimal digit (1 through F), as illustrated. The C has a value of 12, so 1s are placed in the 8s and 4s locations, while zeros (0) are placed in the 2s and 1s locations for a total binary value of 12. The same procedure is followed for the remaining digits A (10), 2, and 1.

The HEX system is used when large numbers need to be processed. The hexadecimal system is also used by some PLCs for entering output instructions into a sequencer.

Figure 5–37 shows the conversion of a 16-bit binary number to its hexadecimal equivalent.

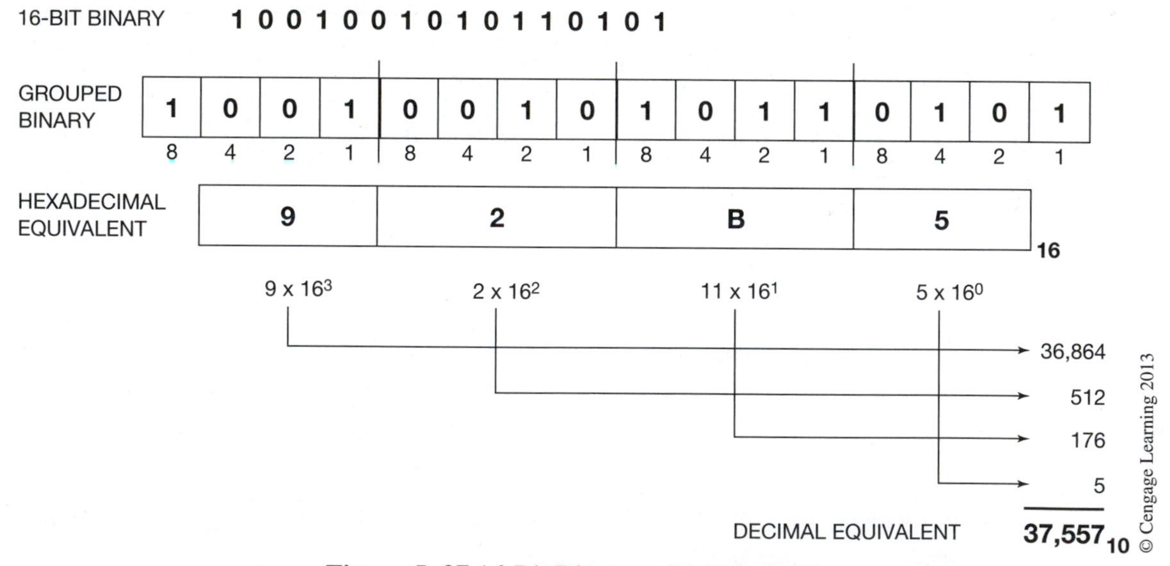

Figure 5–37 16-Bit Binary to Hexadecimal

The first step in converting 16-bit binary to hexadecimal is to group the 16-bit binary word into groups of four (conversion to BCD). Each group of four digits is converted to its hexadecimal equivalent. In Figure 5–37, the hexadecimal number is $92B5_{16}$.

The conversion of $92B5_{16}$ to decimal is $37{,}557_{10}$. Figure 5–38 shows the conversion of the original 16-bit binary number to its decimal equivalent.

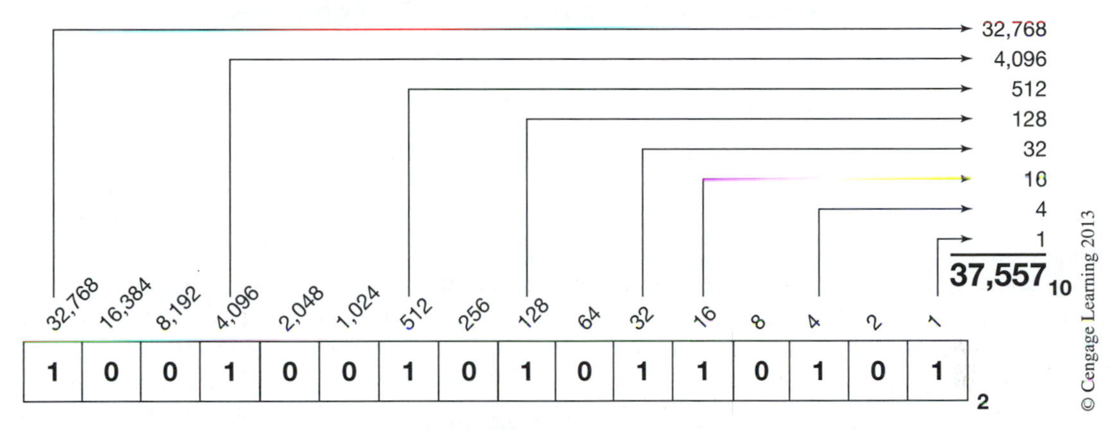

Figure 5–38 Converting a 16-Bit Digital Number to a Decimal Number

BCD SYSTEM

When large decimal numbers are to be converted to binary for memory storage, the process becomes somewhat cumbersome. To solve this problem and speed conversion, the BCD system was devised. In the BCD system, four binary digits (base 2) are used to represent each decimal digit. To distinguish the BCD numbering system from a binary system, the designation BCD is subscripted and placed to the lower right of the units place. Converting a BCD number to a decimal equivalent is shown in Figure 5–39.

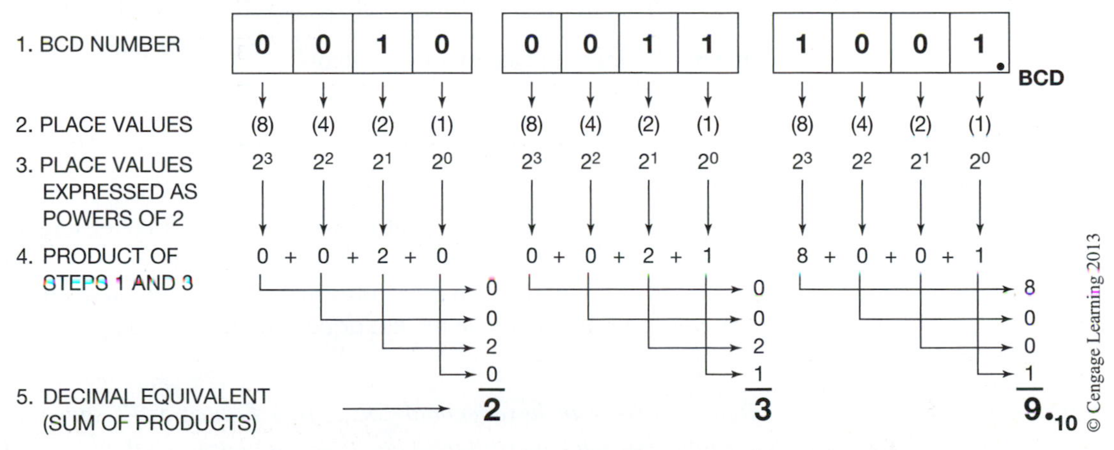

Figure 5–39 Converting a BCD Number to a Decimal Number

When using a BCD numbering system, three decimal numbers may be displayed using 12 bits (3 groups of 4), or 16 bits (4 groups of 4) may be used to represent four decimal numbers or digits.

When only three decimal digits are to be represented, using 12 bits, they are further identified as *most significant digit* (MSD), *middle digit* (MD), and *least significant digit* (LSD) (Figure 5–40).

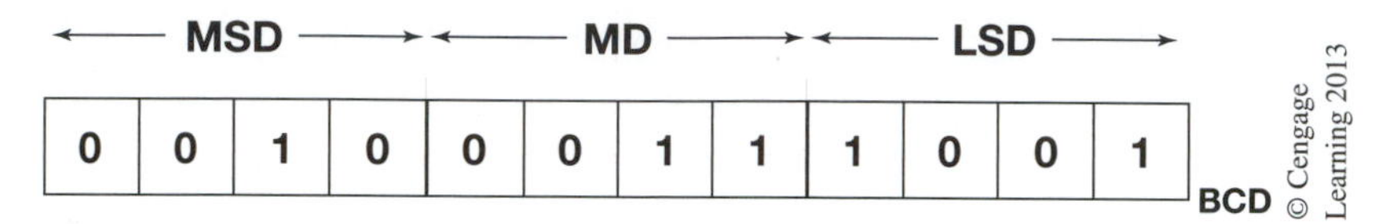

Figure 5–40 Significant Digits

When using the BCD system, the largest decimal number that can be displayed by any four binary digits is 9. The table in Figure 5–41 shows the four binary digit equivalents for each decimal number 0 through 9.

PLACE VALUE				DECIMAL EQUIVALENT
2^3 (8)	2^2 (4)	2^1 (2)	2^0 (1)	
0	0	0	0	0
0	0	0	1	1
0	0	1	0	2
0	0	1	1	3
0	1	0	0	4
0	1	0	1	5
0	1	1	0	6
0	1	1	1	7
1	0	0	0	8
1	0	0	1	9

Figure 5–41 Binary to Decimal Equivalents

USING NUMBERING SYSTEMS

The alphanumeric keys of many programming terminals generate standard ASCII characters and control codes. **ASCII** is an acronym for American Standard Code for Information Interchange. The ASCII code uses different combinations of 7 bit binary (base 2) information for communication of data. The data may be communicated to a printer, barcode reader, or be shown on the display of the programmer and/or computer.

Note: *ASCII information is often expressed in hexadecimal (base 16.) Figure 5–42 shows the 128 standard ASCII control codes and character set with both the binary and hexadecimal numbering systems.*

MSB
MOST SIGNIFICANT BIT

LSB
LEAST SIGNIFICANT BIT

BINARY →		000	001	010	011	100	101	110	111
	HEX →	0	1	2	3	4	5	6	7
000	0	NUL	DLE	SP	Ø	@	P	\	p
0001	1	SOH	DC1	!	1	A	Q	a	q
0010	2	STX	DC2	"	2	B	R	b	r
0011	3	ETX	DC3	#	3	C	S	c	s
0100	4	EOT	DC4	$	4	D	T	d	t
0101	5	ENQ	NAK	%	5	E	U	e	u
0110	6	ACK	SYN	&	6	F	V	f	v
0111	7	BEL	ETB	'	7	G	W	g	w
1000	8	BS	CAN	(	8	H	X	h	x
1001	9	HT	EM	)	9	I	Y	i	y
1010	A	LF	SUB	*	:	J	Z	j	z
1011	B	VT	ESC	+	;	K	[	k	{
1100	C	FF	FS	,	<	L	\	l	¦
1101	D	CR	GS	-	=	M	]	m	}
1110	E	SO	RS	.	>	N	^	n	~
1111	F	SI	US	/	?	O	–	o	DEL

Figure 5–42 Standard ASCII Control Code and Character Set

The digital or hexadecimal number is determined by first locating the vertical column where the code or character is located, and then the horizontal row.

EXAMPLE: The letter A is in column 4, row 1. The binary number that transmits the letter A is 100 0001. The hexadecimal number is 41. The symbol # is 010 0011 in binary and 23 in HEX.

An eighth bit is often used by programmers to provide error-checking of information that is transmitted. This eighth bit is called the **parity bit.**

For *even* parity, the parity bit (the eighth bit) is added to the seven bits that represent the ASCII codes and characters so that the number of 1s will always add up to an even number.

EXAMPLE: The binary number for the # symbol is 010 0011. The 1s add up to three, an odd number. By adding an eighth bit and making it a 1, the total of 1s is now 4, or even, as shown in Figure 5–43.

	PARITY BIT	ASCII CODE BITS
EVEN PARITY	1	0 1 0 0 0 1 1

Figure 5–43 Parity Bit Set to 1 for Even Parity

The letter A, which is the binary number 100 0001, has two 1s and is already even. In this case, the parity bit would be a 0, as shown in Figure 5–44.

	PARITY BIT	ASCII CODE BITS
EVEN PARITY	0	1 0 0 0 0 0 1

Figure 5–44 Parity Bit Set to 0 for Even Parity

The ASCII control code BS (backspace) is binary number 000 1000. For even parity, a 1 is added for the parity bit, as shown in Figure 5–45.

	PARITY BIT	ASCII CODE BITS
EVEN PARITY	1	0 0 0 1 0 0 0

Figure 5–45 Even Parity

By checking each character or control code that is sent for an even number of 1s, transmission errors can be detected when an odd number of 1s is found.

For systems that operate on *odd* parity, the parity bit is used to make the total of 1s add up to an odd number.

EXAMPLE: The number 5 has a binary number of 011 0101. The 1s add up to 4. The parity bit is set to 1, making the 1s total 5, or an odd number. Figure 5–46 illustrates this concept.

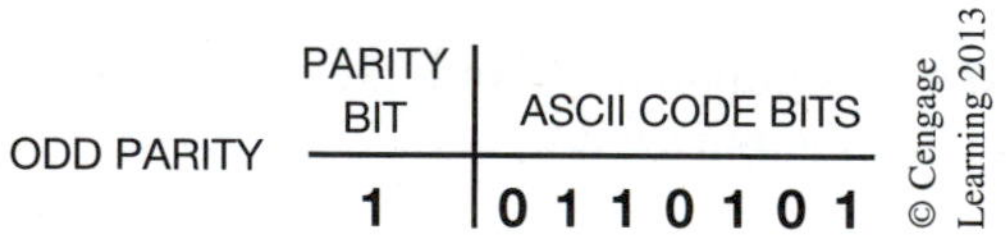

	PARITY BIT	ASCII CODE BITS
ODD PARITY	1	0 1 1 0 1 0 1

Figure 5–46 Parity Bit Set to 1 for Odd Parity

For systems that do not use a parity bit for error-checking, the eighth bit is always a zero (0).

Chapter Summary

There are several numbering systems that are used to store information in the form of binary digits (bits) into the memory system of a processor. The specific numbering system or the combination of numbering systems used depends on the hardware requirements of the specific PLC manufacturer. The important thing to remember, however, is that no matter which numbering system or systems are used, the information is still stored as 1s and 0s.

For programmable controllers to perform arithmetic functions, a way must be found to represent both positive and negative numbers. One of the most common methods used is called 2s complement. Using 2s complement, negative and positive numbers can be added, subtracted, divided, and multiplied. In reality, however, all arithmetic functions are accomplished by successive addition.

Review Questions

1. When information is stored using only 1s and 0s, it is called a ________________ system.
2. A *bit* is an acronym for ____.
3. The decimal numbering system uses 10 digits, or a base of 10. List the base for each of the following numbering systems.
   a. binary base ________________
   b. hexadecimal base ________________
   c. octal base ________________
4. Convert *binary* number 11011011 to a *decimal* number.
5. Convert *decimal* number 359 to a *binary* number.
6. Convert *hexadecimal* number 14CD to a *decimal* number.
7. Convert *decimal* number 3247 to a *hexadecimal* number.
8. Convert *decimal* number 232 to an *octal* number.
9. How do we prevent binary numbers 10 and 11 from being confused as decimal numbers?
10. Convert the following *binary* values to *decimal*.
    a. 10011000
    b. 01100101
    c. 10011001
    d. 00010101
11. Convert the following *BCD* values to *decimal*.
    a. 1001 1000
    b. 0110 0101
    c. 1001 1001
    d. 0001 0101
12. The BCD value 1001 0011 0101 is *not*
    a. 935 decimal
    b. 0011 1010 0111 binary
    c. 647 octal
    d. 3A7 hexadecimal
13. The hexadecimal value 2CB is *not*
    a. 715 decimal
    b. 1313 octal
    c. 0010 1100 1011 binary
    d. 0111 0001 0011 BCD

14. Express the following signed decimal numbers in 2s complement. Use 8-bit words. Show all work.
    a. (–)7
    b. (–)4
    c. (–)3
15. Convert the following decimal numbers to 2s complement and add. Use 8-bit words. Show all work.
    a. (+)4
       (–)7
    b. (–)10
       (+)22
    c. (+)22
       (+)33

CHAPTER

6 Understanding and Using Ladder Diagrams

Objectives

After completing this chapter, you should have the knowledge to:

- Identify a wiring diagram.
- Identify the parts of a wiring diagram.
- Convert a wiring diagram to a ladder diagram.
- List the rules that govern a ladder diagram.

There are basically two types of electrical diagrams: wiring diagrams and ladder diagrams.

WIRING DIAGRAMS

The wiring diagram shows the circuit wiring and its associated devices (relays, timers, motor starters, switches, and the like) in their relative physical locations (Figure 6–1). While this type of diagram assists in locating components and shows how a circuit is actually wired, it does not show the circuit in its simplest form. To simplify understanding of how a circuit works, and to show the electrical relationship of the components (not the physical relationship), a ladder diagram is used.

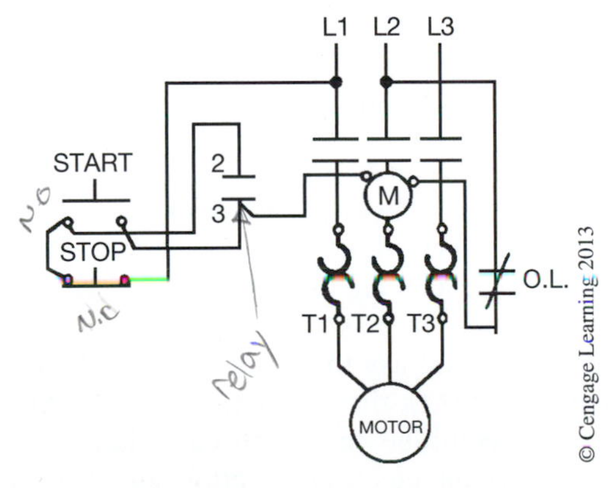

Figure 6–1 Wiring Diagram

LADDER DIAGRAMS

The ladder diagram, also referred to as a schematic or elementary diagram, is used by the electrician or technician to speed their understanding of how a circuit works. Figure 6–2 shows the same circuit as Figure 6–1, but in ladder diagram form.

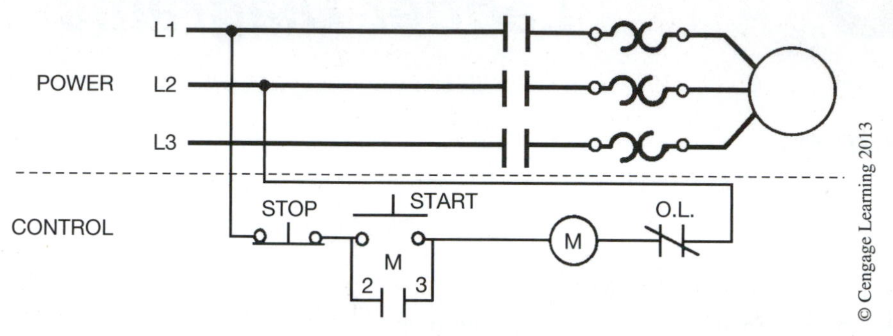

Figure 6–2 Ladder Diagram

To simplify the circuit and help to understand its configuration, the power portion of the circuit is shown separate from the control portion. No attempt is made to show the actual physical location of the components. Since the motor connections (power portion) are the same for any three-phase motor, it is common practice not to show the motor starter or the motor. By not showing the power portion of the circuit, a simplified ladder diagram is created, showing only the control portion of the diagram (Figure 6–3).

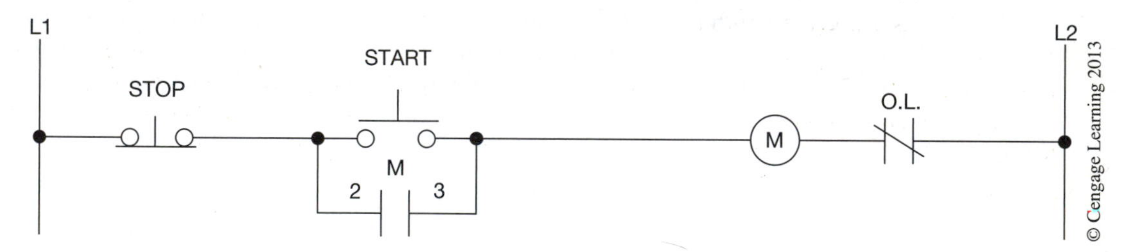

Figure 6–3 Simplified Ladder Diagram

The power required for the control circuit is always shown as two vertical lines, while the actual line(s) of logic are drawn as horizontal lines. The power lines, or rails as they are often called, are like vertical sides of a ladder, whereas the horizontal logic lines are like the rungs of a ladder and are referred to as rungs.

When referring back to Figure 6–1, it is easy to see the physical relationships between the *STOP/START* station, the motor starter coil (M), the overload contacts (O.L.), and the holding contacts (2 and 3), but it is difficult to determine the electrical relationships. The ladder diagram in Figure 6–3, however, clearly shows the electrical relationships between all of the control circuit components.

LADDER DIAGRAM RULES

Some basic rules for ladder diagrams are as follows:

1. A ladder diagram is read like a book; from left to right and from top to bottom.
2. The vertical power lines (rails) of the ladder diagram represent the *voltage potential* of the circuit. The potential could be AC or DC, and varies in voltage from 6 V to 480 V. Standard labeling for the rails is L1 and L2. L1 is AC high or hot for AC circuits, and positive or plus (+) for DC circuits. L2 is AC low or neutral for grounded AC circuits, and negative or minus (−) for DC circuits. The rails may also be marked X1 and X2 when the voltage potential is derived from a step-down transformer.
3. Devices or components are shown in order of importance whenever possible. In Figure 6–3 the *STOP* button is shown ahead of the *START* button. For safety reasons, the *STOP* button has a higher order of importance than the *START* button.
4. Electrical devices or components are shown in their normal condition. The normal condition of electrical diagrams is the circuit deenergized (*OFF*) and with no external forces such as pressure or flow, etc., acting on the device. The *STOP* button is shown closed because that is the normal position for the *STOP* button. The holding contacts (2 and 3) of coil M are shown open. This is the normal position for these contacts when coil M is deenergized. The normally open (N.O.) M holding contacts 2 and 3 do not close until there is a complete path for current flow to coil M. When coil M energizes, M contacts 2 and 3 close, providing a parallel path for current flow with the *START* button.
5. Contacts associated with relays, timers, motor starters, and the like always have the same number or letter designation as the device that controls them. This labeling method holds true no matter where the contacts(s) appear in the circuit. For example, in Figure 6–3 the N.O. holding contacts 2 and 3 are controlled (activated) by motor starter coil M. Therefore, the contacts are identified with the letter M.
6. All contacts associated with a device change position when the device is energized. Figure 6–4 shows a control relay (CR) controlled by a switch (S-1) on rung 1 of the ladder diagram. Rung 2 shows a normally closed (N.C.) control relay contact in series with a green indicator lamp. Rung 3 shows an N.O. control relay contact in series with a red indicator light.

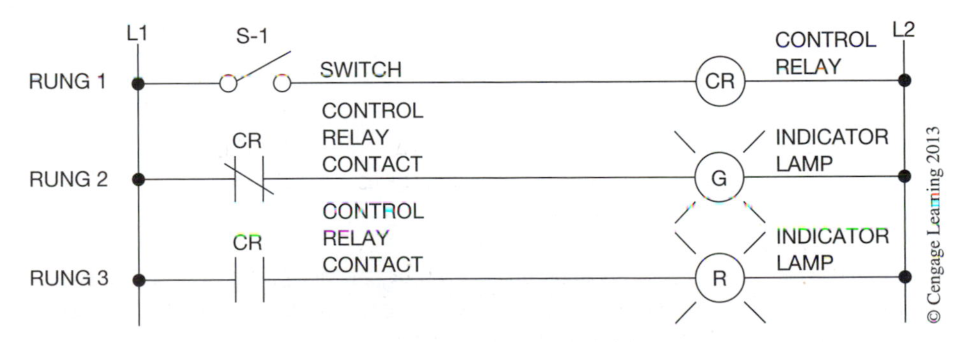

Figure 6–4 Three-Rung Ladder Diagram

When power is applied to the rails of the ladder diagram, the only device in the circuit that operates is the green indicator lamp. The green indicator lamp lights due to a complete path for current flow through the N.C. control relay contacts. These contacts are normally closed and only change position and open when the control relay in rung 1 is energized. When switch S-1 is closed, completing the path for current flow and energizing the CR in rung 1, the N.C. CR contacts in rung 2 open, while the N.O. CR contacts in rung 3 close. The action of the contacts will turn *OFF* the green lamp in rung 2 and turn *ON* the red lamp in rung 3. As long as the control relay remains energized through S-1, the normally closed contact in rung 2 remains open, and the normally open contact in rung 3 remains closed. When S-1 is opened and CR deenergizes, the contacts controlled by CR will return to their normal state (N.C. in rung 2 and N.O. in rung 3).

7. In a ladder diagram, devices that perform a *STOP* function are normally wired in series. Figure 6–5 shows two switches wired N.C. that control a green indicator lamp.

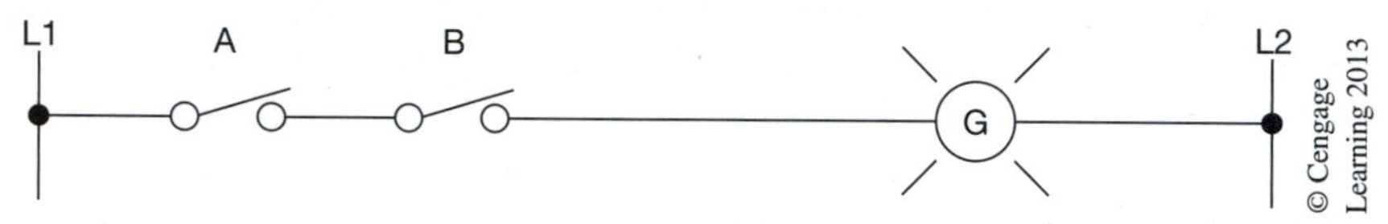

Figure 6–5 Two Switches Wired in Series

With the two switches wired in series, both A and B must remain closed for the lamp to remain lit. If either switch is opened, the green lamp will go out. When switches and/or contacts are wired in series, they are said to have an AND relationship. The AND relationship requires that both A *and* B must be closed for the lamp to light. A truth table for this concept is shown in Figure 6–6.

SWITCH	SWITCH	INDICATOR LAMP
A	B	G
OFF	OFF	OFF
OFF	ON	OFF
ON	OFF	OFF
ON	ON	ON

Figure 6–6 Truth Table for Series Devices

8. Devices that perform a *START* function are normally wired in parallel. Figure 6–7 shows two switches (A and B) wired in parallel to control a red indicator lamp. In this configuration, if either switch A *or* B is closed, the red lamp will light.

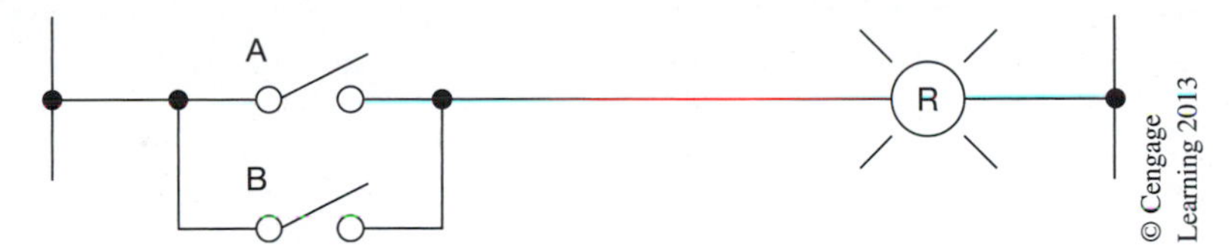

Figure 6–7 Two Switches Wired in Parallel

When switches or contacts are wired in parallel, they are said to have an OR relationship. The OR relationship requires that either A *or* B be closed for the red indicator lamp to light. A truth table for this concept is shown in Figure 6–8.

SWITCH	SWITCH	INDICATOR LAMP
A	B	R
OFF	OFF	OFF
OFF	ON	ON
ON	OFF	ON
ON	ON	ON

Figure 6–8 Truth Table for Parallel Devices

With this understanding of what a ladder diagram is, and the rules that apply to it, a discussion of a basic motor *STOP/START* circuit (shown in Figure 6–2) can begin.

BASIC *STOP/START* CIRCUIT

As stated earlier in this chapter, the wiring diagram in Figure 6–1 is great for showing actual physical locations of the circuit wiring and the components. It does not, however, show the electrical relationships of the devices as simply as the ladder diagram. The wiring diagram is used for original installation and some troubleshooting, whereas the ladder diagram is used to show the electrical relationships of the components, and to speed understanding of how the circuit works.

From viewing the ladder diagram in Figure 6–9, it can be seen that when power is applied to the circuit, the motor starter coil M cannot energize because there is an incomplete path for current flow due to the open *START* button and the N.O. M contacts (2 and 3). The *START* button and the N.O. M contacts are wired in parallel and have an OR relationship. When the *START* button is pushed, a path for current exists from L1 potential through the normally closed *STOP* button, through the now closed *START* button, through the coil of the motor starter (M), and on through the N.C. overload contacts to L2 potential.

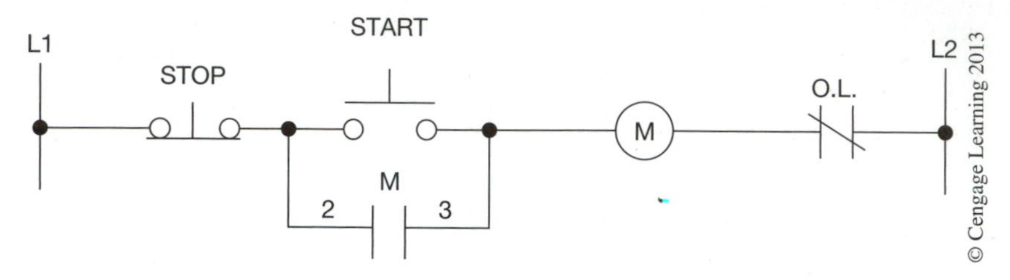

Figure 6–9 Ladder Diagram for Basic *STOP/START* Circuit

When the starter coil M energizes, the M contacts (2 and 3) close, providing an alternate path for current flow. At this point, the *START* button could be released, and the circuit would remain energized, or held in, by the holding contacts (2 and 3) of the motor starter. When contacts from a motor starter or other device are wired in this fashion, they are often referred to as holding, maintaining, or sealing contacts as the circuit is held, maintained, or sealed-in after the *START* button is released.

When the holding contacts (2 and 3) are closed, the main motor contacts of the motor starter are also closed and the motor is started. The operation of the motor is normally taken for granted and is not shown on the ladder diagram. By keeping the ladder diagram as simple and uncluttered as possible, the explanation of the relationships between components and how the control portion of the circuit works is greatly enhanced.

Figure 6–10 again shows the wiring diagram of a motor *STOP/START* circuit. While this diagram looks entirely different from the ladder diagram, both are electrically the same. This comparison shows the electrician or technician why the ladder diagram is preferred.

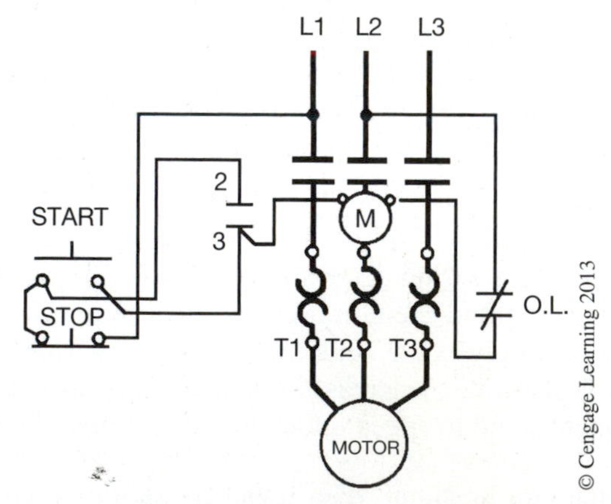

Figure 6–10 Wiring Diagram for Basic *STOP/START* Circuit

The ladder diagram has been the "working language" of electricians and electrical engineers for many years, and helps explain why most programmable controllers are programmed using ladder logic. While this method of programming is welcomed by some, it has frustrated other PLC users who have not been exposed to, or trained in, relay ladder logic.

SEQUENCED MOTOR STARTING

Relay ladder diagrams can become large and complex. It is not the purpose of this text to cover them in great detail, but instead to discuss the basic rules and present some concepts to enhance understanding of circuits that are discussed in later chapters.

Figure 6–11 shows a ladder diagram for a circuit that starts three motors.

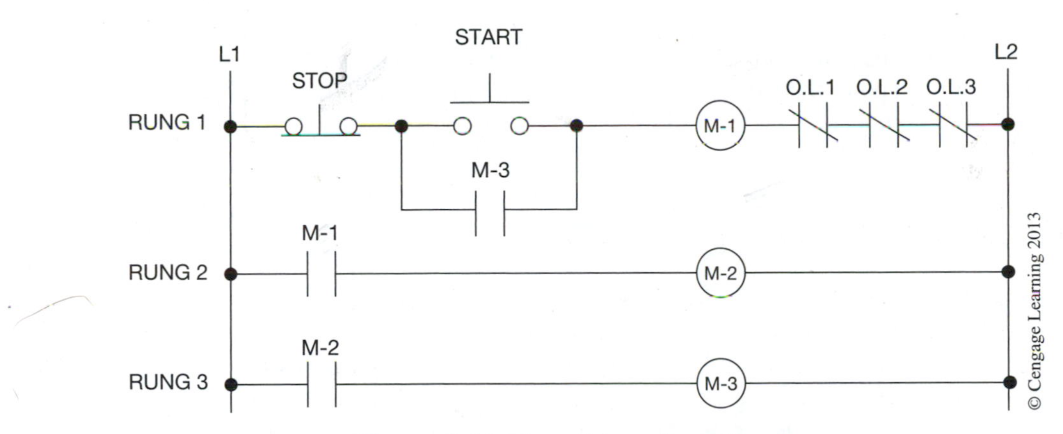

Figure 6–11 Three-Motor Start Circuit

Rung 1 contains the *STOP/START* buttons and the motor starter coil M-1 for motor 1. Notice that the holding contacts wired in parallel with the *START* button are not M-1 contacts, but instead are M-3 contacts. With this arrangement, rung 1 cannot be sealed in, or maintained, unless motor starter 3 energizes and closes its contacts. Additionally, the M-1 contacts in rung 2 must close to energize motor starter 2 (M-2). And M-2 contacts in turn must close in rung 3 to energize motor starter 3 (M-3). When the *START* button of this circuit is pushed, it operates as follows:

1. M-1 energizes, closing the N.O. M-1 contacts in rung 2, and M-2 energizes.
2. M-2 N.O. contacts close in rung 3 and energize M-3.
3. M-3 N.O. contacts in Rung 1 close and act as holding contacts to keep the circuit energized after the *START* button is released.

Note: *This sequence happens almost instantaneously.*

4. Pushing the *STOP* button deenergizes M-1, which deenergizes M-2 in rung 2 when the normally open M-1 contacts go open. M-2's deenergizing opens the M-2 contacts in rung 3 and deenergizes M-3. With M-3 deenergized, the N.O. M-3 contacts in rung 1 open.

5. By wiring all three overload contacts in series with M-1 in rung 1, it is ensured that an overload on any motor would shut down all motors. An open overload contact would have the same effect as pushing the *STOP* button.

It could be said that this circuit consists of basically three elements: inputs, outputs, and logic.

The inputs consist of the *STOP* button, the *START* button, and the overload contacts. The outputs are motor starters M-1, M-2, and M-3. The logic that caused the sequential starting were N.O. contacts M-1, M-2, and M-3.

These three elements—inputs, outputs, and logic—also work well with programmable controllers. The inputs are wired to input modules, the outputs are wired to output modules, and the processor performs the logic functions.

Figure 6–12 shows the wiring diagram for the three-motor circuit just discussed. This diagram further illustrates the point that while wiring diagrams are great for giving the physical locations of components, they do not show the control function of the circuit as clearly as a ladder diagram does.

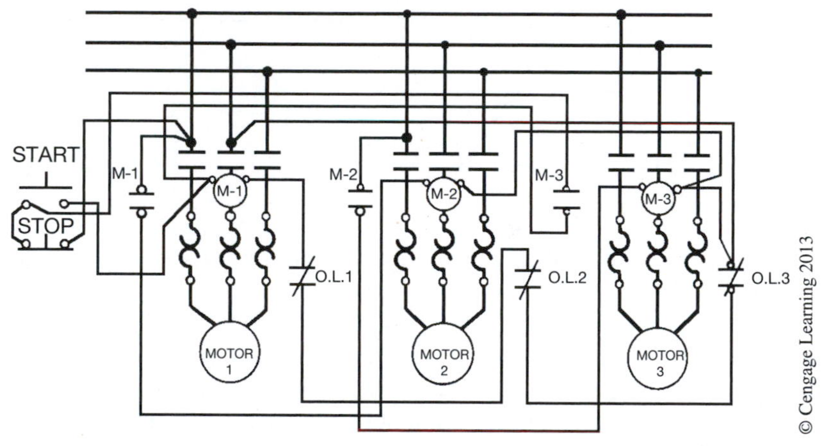

Figure 6–12 Wiring Diagram for Three-Motor Circuit

DIGITAL LOGIC GATES

While the typical PLC is programmed using ladder logic symbols, some older PLCs were programmed using digital logic notations such as AND, OR, NOT, etc. To better understand these digital logic notations and to see how they compare to relay ladder logic, we will cover six basic digital logic gates.

Figure 6–13 shows a two-input AND gate.

From the truth table, we see that both inputs A *and* B must be TRUE, or set to 1, before the output is turned *ON,* or set to 1. The AND gate functions like the two switches that were wired in series to a

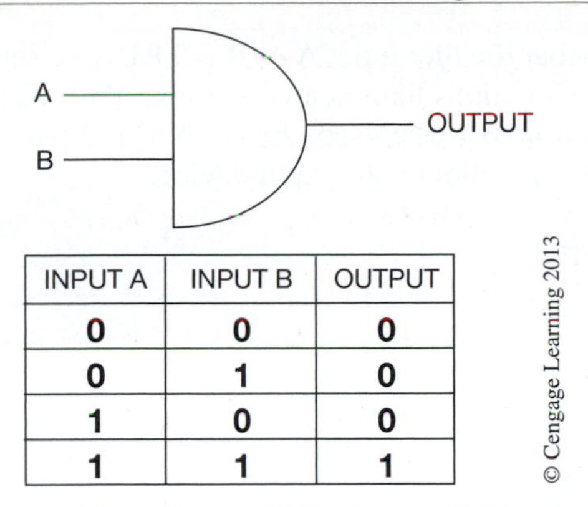

INPUT A	INPUT B	OUTPUT
0	0	0
0	1	0
1	0	0
1	1	1

Figure 6–13 Two-Input AND Gate with Truth Table

lamp in Figure 6–5. Both switch A and switch B had to be closed for the lamp to light. Figure 6–14 shows AND logic for two programmed input devices wired in series to an output device.

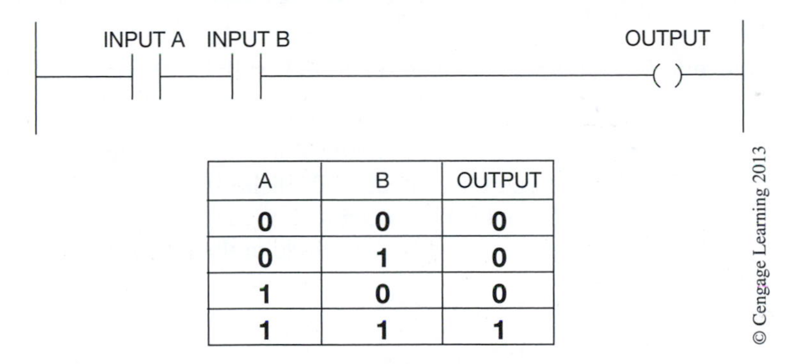

A	B	OUTPUT
0	0	0
0	1	0
1	0	0
1	1	1

Figure 6–14 Two Input Devices Wired in Series with Truth Table

Figure 6–15 shows a two-input OR gate.

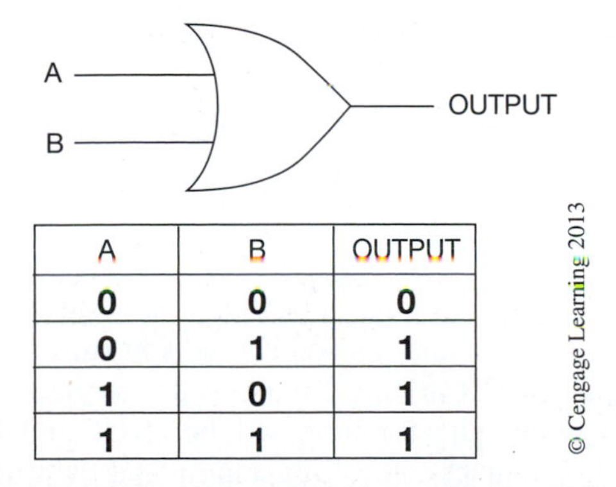

A	B	OUTPUT
0	0	0
0	1	1
1	0	1
1	1	1

Figure 6–15 Two-Input OR Gate with Truth Table

From the truth table, we see that if either input A *or* B is TRUE, or set to 1, the output will be turned *ON,* or set to 1. The OR gate functions like the two switches that were wired in parallel to a lamp in Figure 6–7. If either switch A *or* B was closed, the lamp would light. Figure 6–16 shows OR logic for two input devices wired in parallel to an output device.

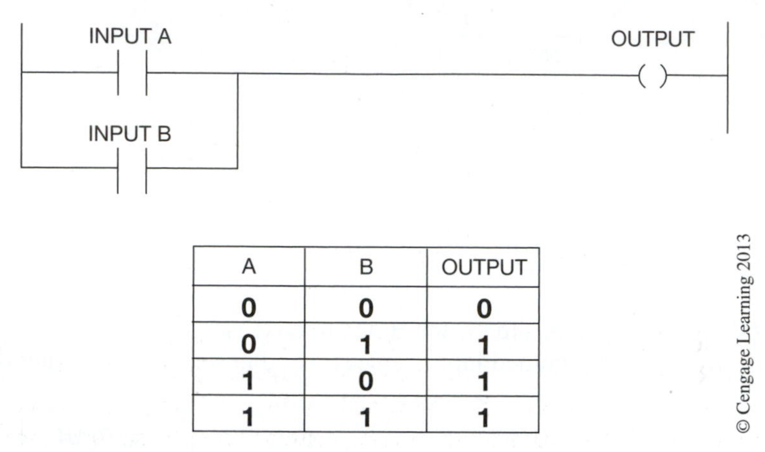

A	B	OUTPUT
0	0	0
0	1	1
1	0	1
1	1	1

Figure 6–16 Two Input Devices Wired in Parallel with Truth Table

The next gate is called a NOT gate and is often referred to as an inverter. The inverter, or NOT gate, will have only one input lead and one output lead. If the input is *OFF,* or set to 0, then the output will be *ON,* or set to 1. If the input is *ON,* or set to 1, then the output will be *OFF,* or set to 0. Figure 6–17 shows a NOT gate with a truth table. The circle in the output line is used to indicate an inverted function.

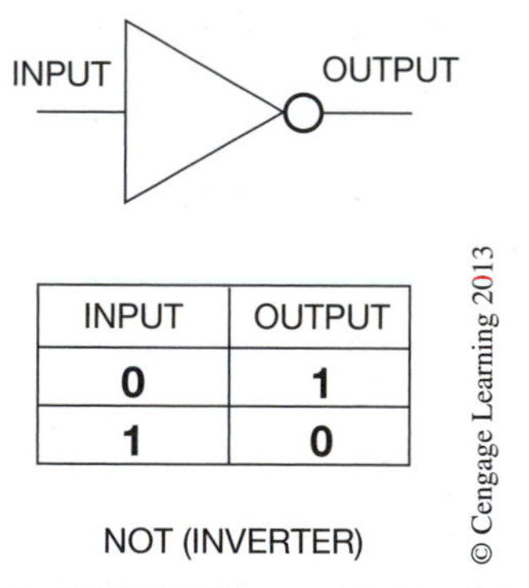

INPUT	OUTPUT
0	1
1	0

NOT (INVERTER)

Figure 6–17 NOT Gate with Truth Table

The NOT gate functions like the normally closed contacts in rung 2 for the three-rung ladder diagram in Figure 6–4. As long as the CR in rung 1 remains deenergized, or *OFF,* the N.C. CR contact in rung 2, which controls the green indicator lamp, will be TRUE and the indicator lamp will be *ON.* Figure 6–18 shows an N.C. CR contact controlling a lamp and the truth table for the circuit.

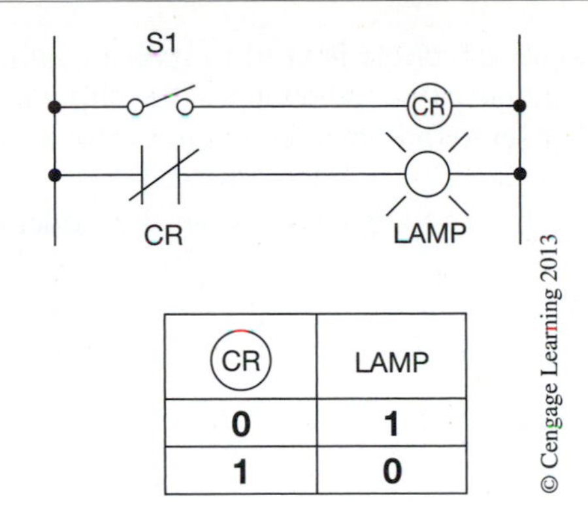

CR	LAMP
0	1
1	0

Figure 6–18 N.C. CR Contact Controlling a Lamp with NOT logic

As long as the single-pole switch (S_1) that is wired in series with the CR coil is open, CR is *OFF,* or set to 0. With CR *OFF,* the logic for the normally closed contacts will be TRUE, and the lamp will be *ON*. When S_1 is closed, CR will energize, turn *ON,* the N.C. CR contacts will open, and the light will be turned *OFF*. The truth table reflects the action of the CR coil that controls the action of the CR contacts. It may help to understand the truth table if we think that the N.C. CR contacts are controlled by the CR coil. If the CR coil is *OFF,* or set to 0, we can think of the CR N.C. contacts also being closed, or set to 1. As the input will be inverted when the CR coil is energized, or set to 1, then the output device controlled by the NC contacts will be FALSE, or set to 0.

The programmable controller will use NOT logic in the same way as described above. If the output device is set to 0, or *OFF,* any N.C. contacts associated with the output device with the same address will also be set to 1. The NOT logic will be used in the next chapter for the Examine Off instruction. When an N.C. contact is addressed with the same address as an output coil, the N.C. contact will be true as long as the output coil is *OFF,* or FALSE.

By combining the NOT gate with the AND gate, we get what is called a NAND gate. Figure 6–19 shows a two-input NAND gate with a truth table.

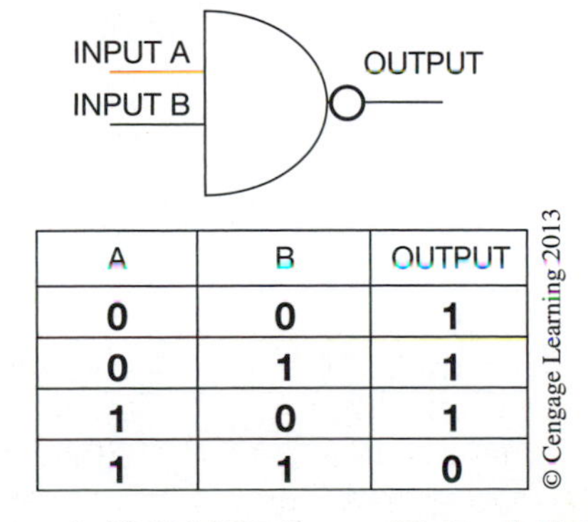

A	B	OUTPUT
0	0	1
0	1	1
1	0	1
1	1	0

Figure 6–19 NAND Gate with Truth Table

As discussed with the NOT logic, the circle is used to indicate an invert function. By placing the invert, or NOT, symbol at the output of the AND gate, the output can only be TRUE when one or both of the inputs are FALSE or set to 0. Figure 6–20 shows the equivalent relay circuit.

To help understand the logic of the NAND gate, consider the ladder diagram in Figure 6–20.

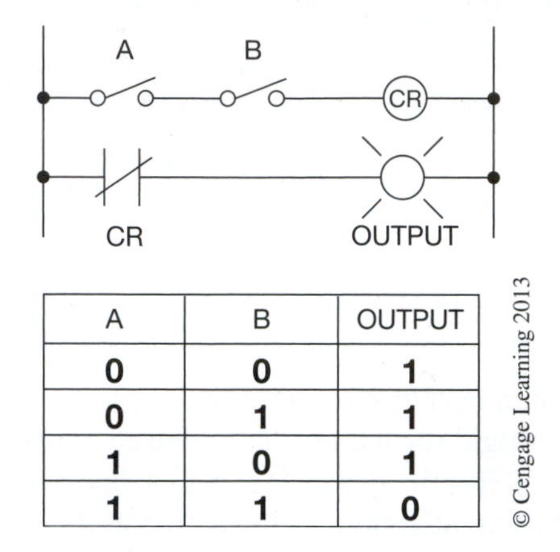

A	B	OUTPUT
0	0	1
0	1	1
1	0	1
1	1	0

Figure 6–20 Ladder Diagram with NAND Logic

With both inputs A and B *OFF,* or set to 0, the inverted output will be set to 1, or be turned *ON*. If only input A is set to 1, the inverted output will remain set to 1 because input B is still open, or set to 0. If input A is opened and input B is closed or is set to 1, the inverted output will again remain set to 1. However, If both A and B are closed, or set to 1, then the inverted output will be set to 0, or *OFF*. From the ladder diagram, we can see that with both input devices A and B open, CR would not be energized and the output would be set to 1, or *ON*. For the output device to go FALSE, or to 0, both input switches A *and* B will be closed, or set to 1, as illustrated in the truth table.

When we combine the NOT gate with an OR gate, we get what is called a NOR gate. Figure 6–21 shows the NOR gate with a truth table.

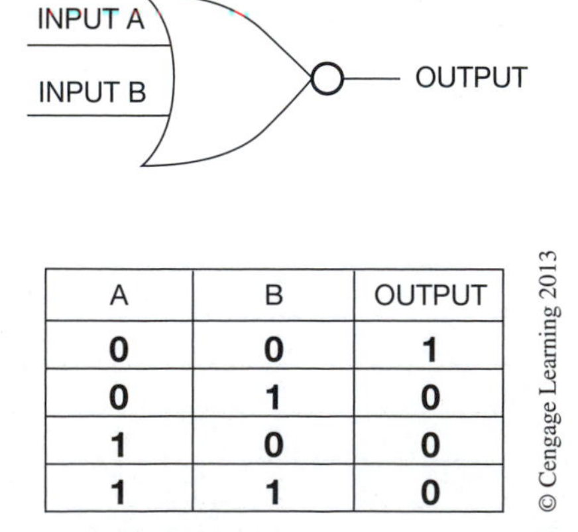

A	B	OUTPUT
0	0	1
0	1	0
1	0	0
1	1	0

Figure 6–21 NOR Gate with Truth Table

To help understand the logic of the NOR gate, consider the ladder diagram in Figure 6–22.

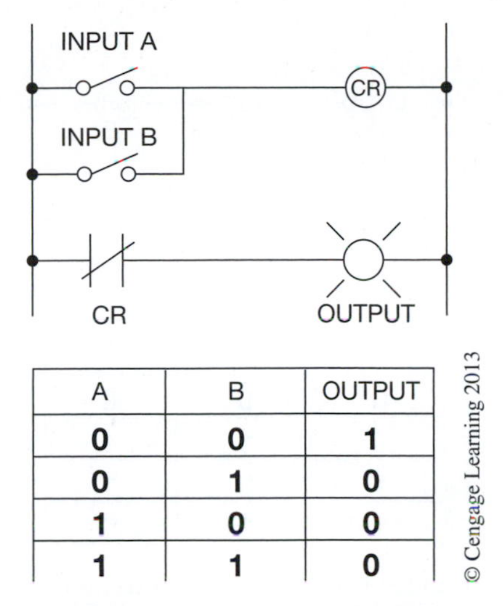

A	B	OUTPUT
0	0	1
0	1	0
1	0	0
1	1	0

Figure 6–22 Ladder Diagram with NOR Logic

From this figure we can see that if either input A *or* B is closed the CR coil will energize and the normally closed contacts in rung 2 will open and turn *OFF* the output. With this configuration, the only time the output lamp will light is if both switches A and B are open, or set to 0.

The final logic gate that will be covered is the exclusive OR gate (XOR). The XOR gate with truth table is shown in Figure 6–23.

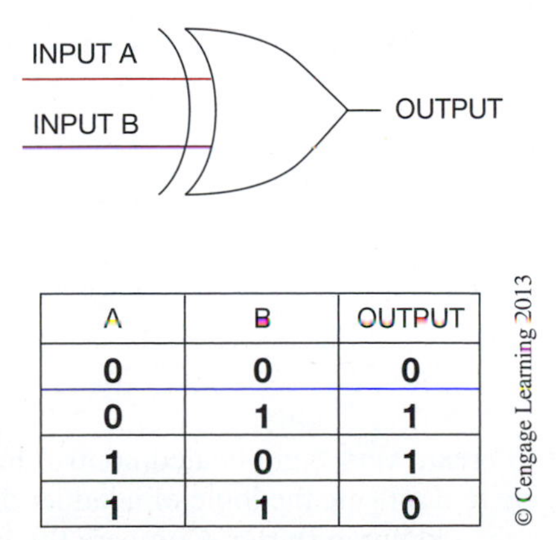

A	B	OUTPUT
0	0	0
0	1	1
1	0	1
1	1	0

Figure 6–23 XOR Logic Gate with Truth Table

The XOR logic gate will only turn the output *ON* when *either* input A *or* B is *ON*, but *not both ON*. This logic gate can be compared to the two double-circuit pushbuttons shown in Figure 6–24a.

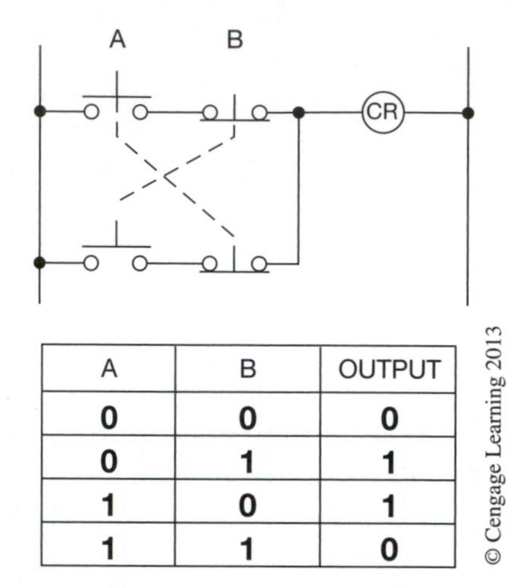

A	B	OUTPUT
0	0	0
0	1	1
1	0	1
1	1	0

Figure 6–24a Ladder Logic Circuit Equivalent to the XOR Gate

This ladder diagram shows us that as long as one pushbutton is pushed, but *not* both, the output device will be turned *ON*. If button A is pushed, a complete path for current now exists, as shown in Figure 6–24b. Similarly, if button B is pushed, a complete path for current now exists, as shown in Figure 6–24c.

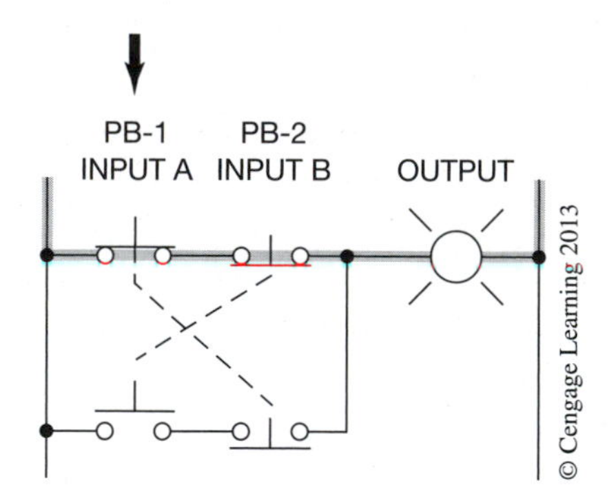

Figure 6–24b Input A Closed and Output is *ON*

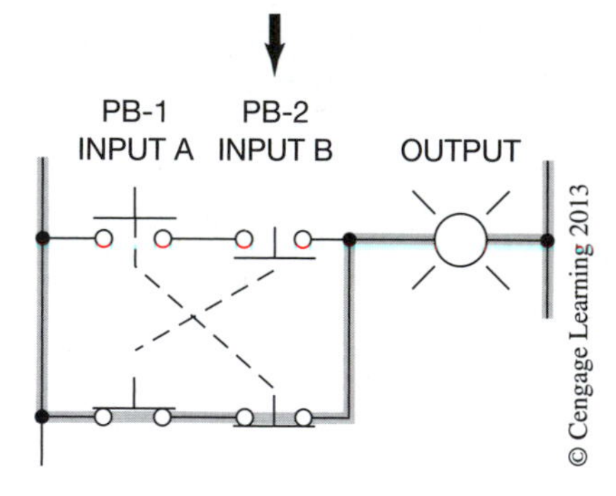

Figure 6–24c Input B Closed and Output is *ON*

Logic gates can be combined to create very complicated control logic. Figure 6–25 shows an OR gate combined with an AND gate to duplicate the logic of a ladder diagram that contains a start button, holding contacts, float switch, and pump starter. Compare the logic of the ladder diagram with the logic gate equivalent circuit and the truth table.

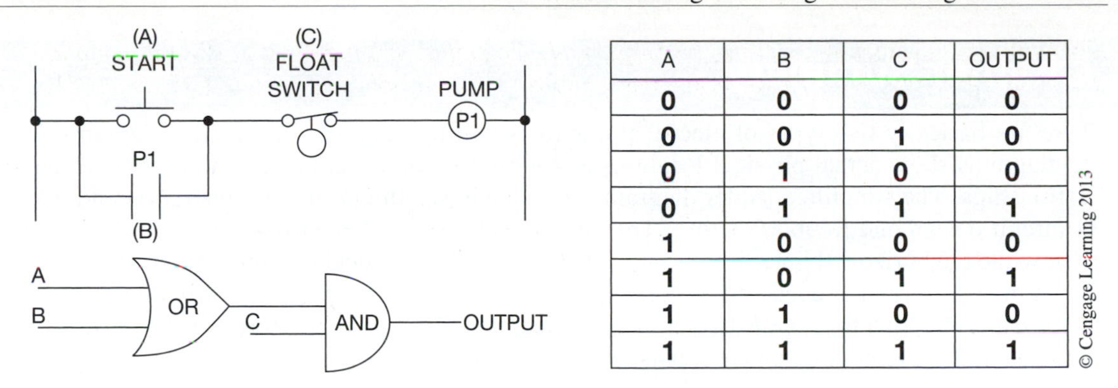

A	B	C	OUTPUT
0	0	0	0
0	0	1	0
0	1	0	0
0	1	1	1
1	0	0	0
1	0	1	1
1	1	0	0
1	1	1	1

Figure 6–25 Combining an OR Gate and an AND Gate

Figure 6–26 shows a ladder diagram with four contacts that control an output device. We can see from the diagram that to turn on the output, any of the following combinations of contacts is required:

A and C
A and D
B and D
B and C

By combining two OR gates and an AND gate, we can duplicate the logic of the four contacts, as shown in Figure 6–26 and verified by the truth table.

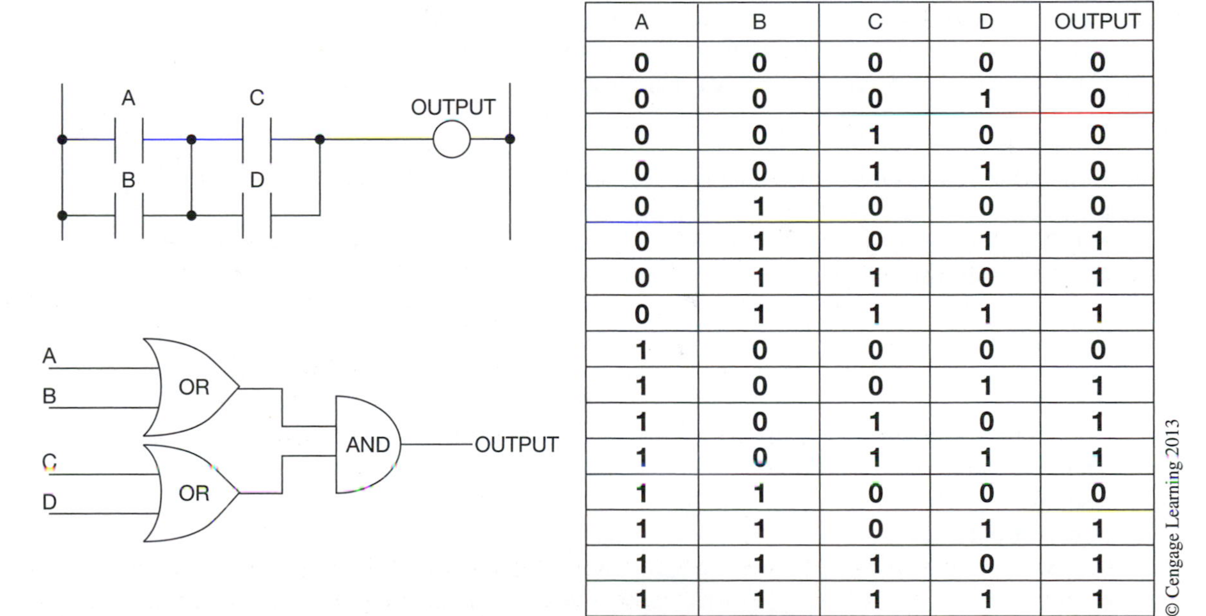

A	B	C	D	OUTPUT
0	0	0	0	0
0	0	0	1	0
0	0	1	0	0
0	0	1	1	0
0	1	0	0	0
0	1	0	1	1
0	1	1	0	1
0	1	1	1	1
1	0	0	0	0
1	0	0	1	1
1	0	1	0	1
1	0	1	1	1
1	1	0	0	0
1	1	0	1	1
1	1	1	0	1
1	1	1	1	1

Figure 6–26 Combining OR and AND Gates

Chapter Summary

There are basically two types of electrical diagrams: wiring diagrams and ladder diagrams. Wiring diagrams show actual physical locations and wiring, whereas ladder diagrams show electrical relationships. The simplified ladder diagram speeds understanding of circuit operation and is used for circuit design and troubleshooting. The vertical sides of the ladder diagram are referred to as *rails*, while the horizontal lines or logic are called *rungs*. On electrical diagrams, devices are always shown in their normal or deenergized condition. When two or more devices are wired in series, they perform an AND function, while two or more devices wired in parallel perform an OR function. The elements of the ladder diagram are inputs, outputs, and logic.

Logic gates can be used that duplicate the logic of the pushbuttons, contacts, and control devices typically used in motor control circuits. Common logic gates are AND, OR, NOT, NOR, NAND, and XOR (exclusive OR). Logic gates can be combined to create complex control circuits. The elements of the ladder diagram are inputs, outputs, and logic. Ladder diagrams are favored over wiring diagrams when a basic understanding of the control circuit is needed. The wiring diagram is used to show actual physical locations and relationships between components, whereas the ladder diagram shows electrical relationships without regard to actual location.

Review Questions

1. Define the terms *normally open* and *normally closed.*
2. Describe the difference between a wiring diagram and a ladder (schematic) diagram.
3. Explain the operation of the circuit in Figure 6–9 if M contacts 2 and 3 do not close.
4. Contacts wired in parallel have what relationship?
   a. AND
   b. OR
5. Contacts wired in series have what relationship?
   a. AND
   b. OR
6. The two main vertical lines of a ladder diagram are often referred to as:
   a. rungs
   b. power ports
   c. rails
   d. tracks
   e. none of the above
7. The horizontal lines of a ladder diagram are referred to as:
   a. rungs
   b. power ports
   c. rails
   d. tracks
   e. none of the above

8. Devices that are intended to perform a *STOP* function are normally wired in ______________ with each other.
9. Devices that are intended to perform a *START* function are normally wired in ______________ with each other.
10. How are contacts that are associated with relays, motor starters, timers, and the like identified?
11. Convert the wiring diagram below into a ladder diagram.

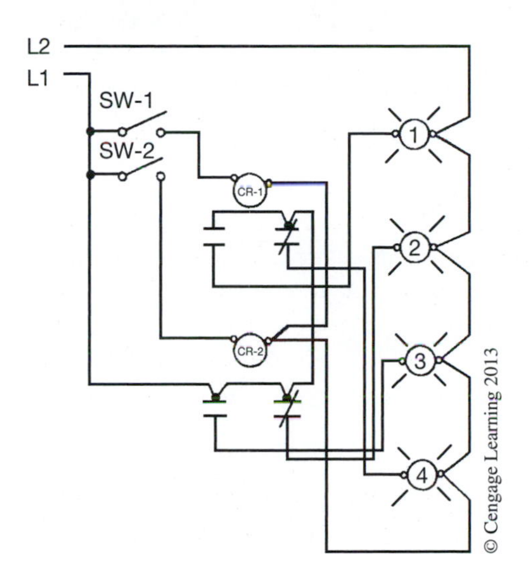

12. Convert the wiring diagram below into a ladder diagram.

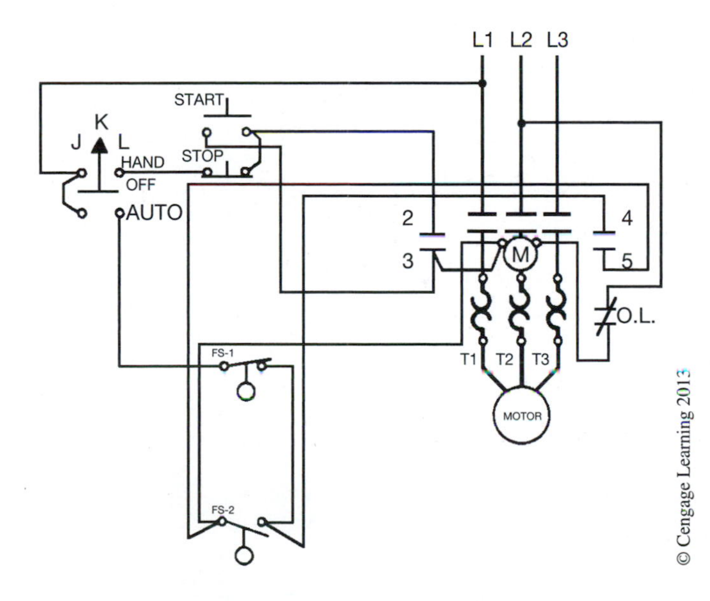

13. Identify the following logic gates.

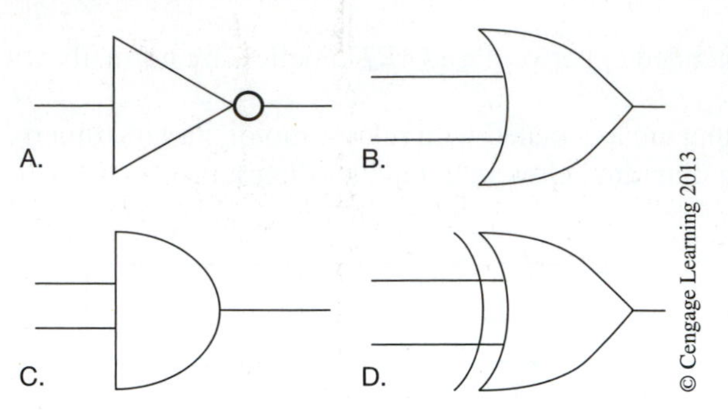

CHAPTER 7 Relay Type Instructions

Objectives

After completing this chapter, you should have the knowledge to:

- Understand the *EXAMINE ON* instruction.
- Understand the *EXAMINE OFF* instruction.
- Write and understand the logic for a standard *STOP/START* motor circuit.

The next step in understanding how the programmable logic controller works is to learn how ladder logic is changed into processor logic. The actual programming is accomplished using either a hand-held programming device or a computer. Figure 7–1a shows the Allen-Bradley MicroLogix PLCs and the Hand-Held Programmer for programming and monitoring. Figure 7–1b shows the keyboard of the hand-held programmer.

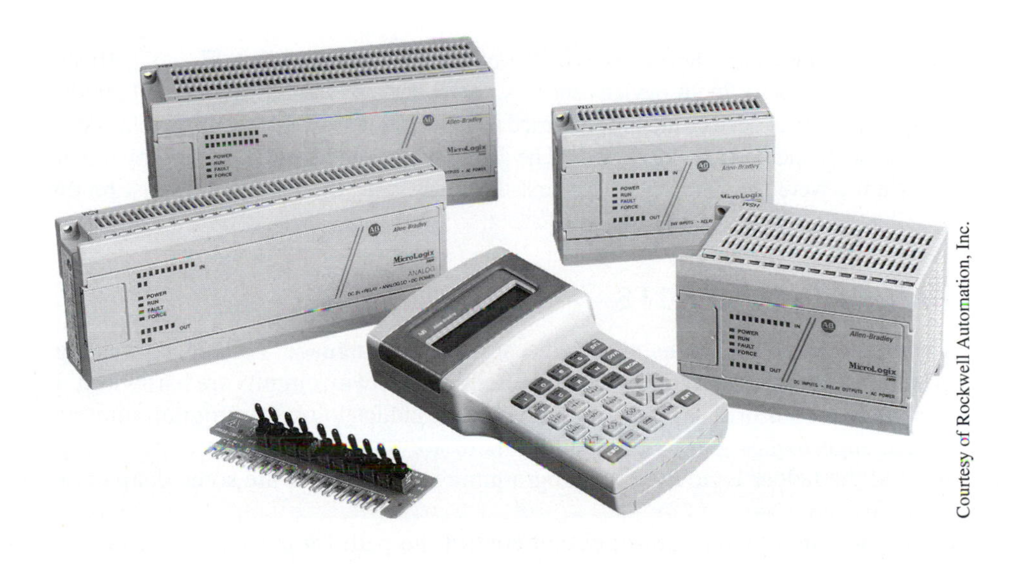

Figure 7–1a Allen-Bradley MicroLogix PLCs and Hand-Held Programmer

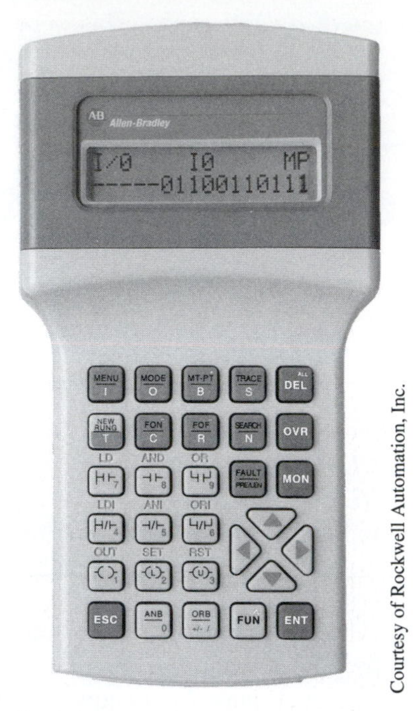

Figure 7–1b Allen-Bradley Hand-Held Programmer

Regardless of the type of programmer used, some common relay symbols are standard. These symbols include normally open contacts, normally closed contacts, and coil or output. Figure 7–1b shows that the keyboard has no symbols for input devices such as *STOP* buttons, limit switches, and pressure switches. The contacts from all input devices are programmed using either the N.O. or the N.C. relay contact symbols (above the numbers 8 and 5 on the keyboard). The actual programming devices used by the different PLC manufacturers are covered in Chapter 8, but first the relay logic used for PLCs must be discussed and *understood.*

PROGRAMMING CONTACTS

A PLC is normally programmed using a ladder logic–type language. Ladder logic is a good choice for a programming language because it closely resembles the way circuits are hardwired. Electricians and technicians feel comfortable with, and understand, ladder logic, so programming with a ladder logic–type language makes good sense. While there are many similarities between standard relay ladder logic and the ladder logic used for programming a PLC, there are some distinct differences.

Hardwired contacts in a motor control circuit control the path for current flow to the output (coil, light, solenoid, etc.). The contact symbols that are used when programming a PLC are actually logic instructions that the processor uses to make decisions.

The PLC N.O. contact symbol is actually an instruction that tells the processor to look for an *ON* condition at the address that corresponds to the symbol. If an *ON* condition exists, the instruction is said to be logically *true,* and logic continuity exists. This is much like saying that if a contact is closed, current can flow.

Note: *As stated earlier in the text, the terms* current flow *and* power flow *are often used by the various PLC manufacturers to indicate that a circuit is complete, or logically true.*

Figure 7–2 shows a simple circuit containing a single-pole switch and a lamp for an output. Figure 7–3 shows the equivalent circuit when programmed with a PLC. Addresses shown are Allen-Bradley PLC-5 format. An "I" preceding a word and bit number indicates an input, whereas an "O" preceding a word and bit address indicates an output.

Figure 7–2 Simple Circuit

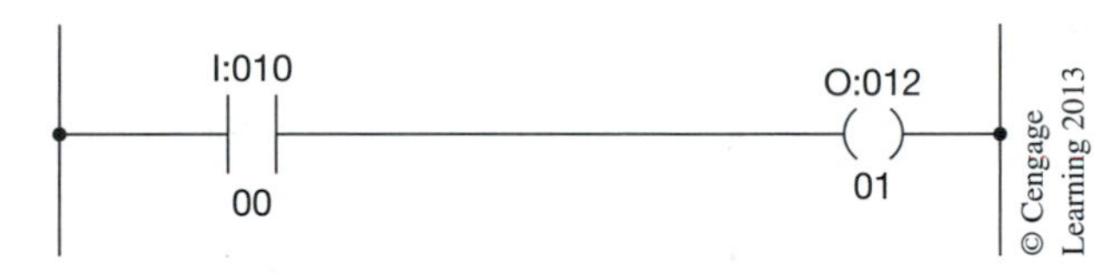

Figure 7–3 Equivalent Circuit Programmed with a PLC

In Figure 7–2, if the switch is closed, current flows through the switch contacts and the lamp lights. In Figure 7–3, the single-pole switch is shown as an N.O. contact symbol with the address I:010/00. The lamp, or output, is shown as a circle, and is given the address O:012/01. The output is actually bit 01 of output image word 012.

Because of the way this program is written, the normally open contact symbol tells the processor to look at address location I:010/00 (input for single-pole switch); if a closed (*ON*) condition is found, then the logic of the circuit is true. When the logic of the circuit is true, the processor is instructed to turn *ON* output O:012/01. There is, in reality, no actual electrical connection between the switch (I:010/00) and the lamp (O:012/01). Instead, it is the processor that turns the lamp *ON* or *OFF* depending on the logic of the program that is written and the status of the input device.

Figure 7–4 shows the switch and lamp as they are wired to their respective I/O modules. It is the N.O. contact symbol in the program that tells the processor to examine the single-pole switch for an *ON* condition. If the switch is open (*OFF*), the program logic is not true and the processor will not turn on the lamp. On the next processor scan, however, if the switch has been closed, it will be *ON,* the logic of the circuit will be true, and the processor will turn *ON* the lamp.

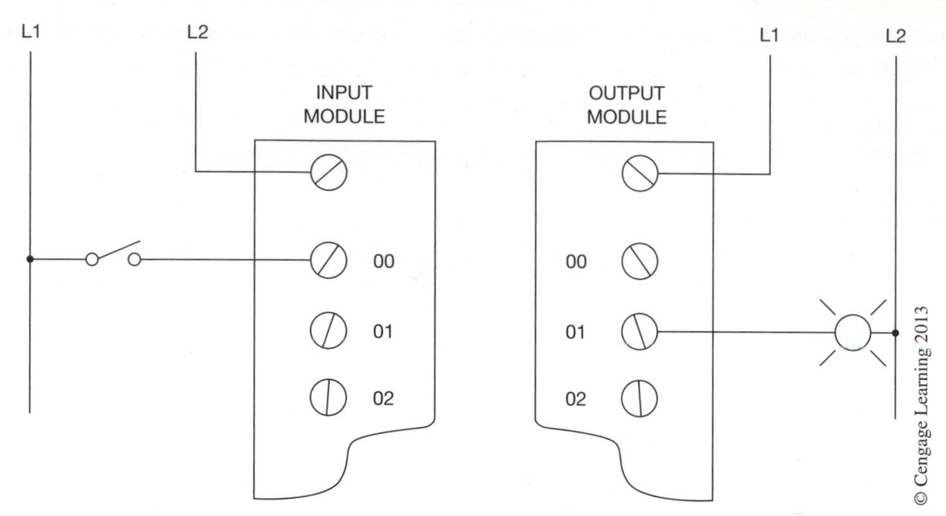

Figure 7–4 Input and Output Devices Wired to I/O Modules

Figure 7–5a shows the bit status of input image word 010 when the switch is open. With the switch open (*OFF*), the logic of the circuit cannot be true so the lamp (output image word 012 bit 01) will not be *ON*. The *OFF* condition of the switch and the lamp is indicated by a 0 in bit location 00 in input word 010 and output image word 012 bit 01. When the switch is closed, the bit that represents the switch (00) will change to a 1. This makes the circuit logically TRUE, the processor will turn the lamp to *ON* (bit 01), and a 1 will be shown in that location (Figure 7–5b).

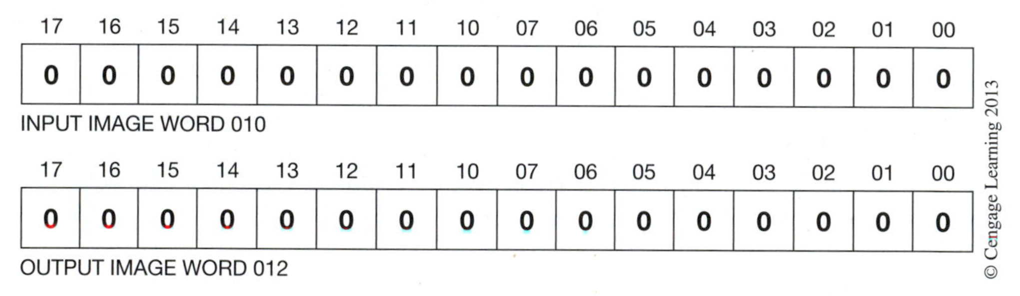

17	16	15	14	13	12	11	10	07	06	05	04	03	02	01	00
0	0	0	0	0	0	0	0	0	0	0	0	0	0	0	0

INPUT IMAGE WORD 010

17	16	15	14	13	12	11	10	07	06	05	04	03	02	01	00
0	0	0	0	0	0	0	0	0	0	0	0	0	0	0	0

OUTPUT IMAGE WORD 012

Figure 7–5a Bit Status for I/O Word 1 with Switch Open

17	16	15	14	13	12	11	10	07	06	05	04	03	02	01	00
0	0	0	0	0	0	0	0	0	0	0	0	0	0	0	1

INPUT IMAGE WORD 010

17	16	15	14	13	12	11	10	07	06	05	04	03	02	01	00
0	0	0	0	0	0	0	0	0	0	0	0	0	0	1	0

OUTPUT IMAGE WORD 012

Figure 7–5b Bit Status for I/O Word 1 after Switch Is Closed

Because the processor views the N.O. contact instruction as a request to examine a given address for an *ON* condition, it is referred to as an **EXAMINE ON** instruction. The opposite instruction, the N.C. contact, is referred to as an **EXAMINE OFF** instruction.

These instructions are also referred to as examine if closed (XIC) and examine if open (XIO), as shown in Table 7–1.

Table 7–1 Examine if Closed (XIC) and Examine if Open (XIO)

Examine On	Examine Off
—] [—	—]/[—
Normally Open	Normally Closed
Examine if Closed	Examine if Open
XIC	XIO

The EXAMINE OFF instruction is only logically true when the device referenced is *OFF,* or open. Figure 7–6 shows the EXAMINE OFF instruction now used for address I:010/00. The processor is asked to examine the address location for an *OFF* (open) condition. If an *OFF* condition is found, then the instruction is logically true and the output O:012/01 would be turned *ON*. If, on the other hand, the switch were found to be *ON* (closed), the logic would be false, and the lamp would not be turned *ON*.

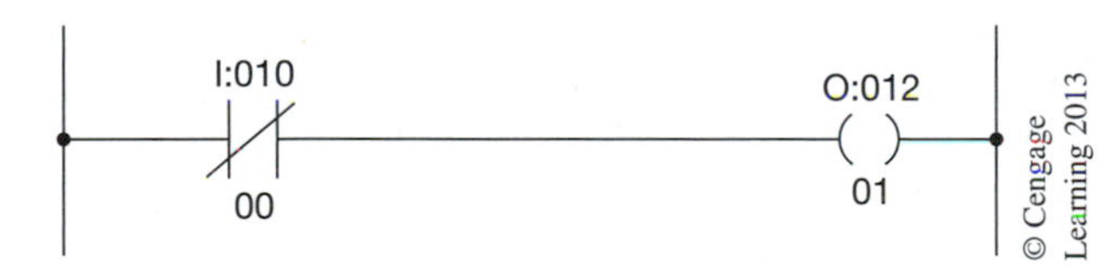

Figure 7–6 EXAMINE OFF Instruction

At first these two instructions may seem to be contrary to the logic of hardwired contacts, so it is important to remember that these are instructions to the processor, and are not hardwired contacts. A review of both instructions seems appropriate.

EXAMINE ON

Whenever the processor sees an N.O. contact in the user program, it views the contact symbol as a request to examine the address of the contact for an *ON* condition. If the N.O. contact has an input address, and if the real-world input device is closed, or *ON,* the processor sets the appropriate bit in the input register to 1, or *ON*. As the EXAMINE ON instruction is looking for an *ON* condition, a bit set to 1, or *ON,* is a true condition, and a logic path exists through the contacts. If the real-world input had been open, or *OFF,* the processor would have cleared the appropriate bit to 0, or *OFF,* and the contact would be false as far as the logic of the ladder diagram was concerned and would not allow a logic path.

EXAMINE OFF

When the N.C. symbol is programmed in a ladder diagram, the processor views it as a request to examine the address of the contact for an *OFF* condition. Any address that is actually *OFF* becomes logically true and power can flow. If an N.C. contact has an input address and the real-world input device is open, or *OFF,* the processor sets the bit to 0, or *OFF*. As the EXAMINE OFF instruction is looking for an *OFF* condition, a bit set to 0, or *OFF,* is a true condition and there would be logic continuity, so power can flow through the contact. If the input device had been closed, or *ON,* the bit would be set to 1, or *ON*. The EXAMINE OFF instruction can only be logically true when an *OFF* condition exists. Any bit set to 1 is viewed as an *ON* condition, which makes an EXAMINE OFF (N.C.) contact false, and no power can flow.

To further reinforce the EXAMINE ON and EXAMINE OFF instruction concepts, a look at a standard *STOP/START* station may be helpful. Figure 7–7a shows a standard *STOP/START* circuit with overload contacts using a standard ladder diagram. Figure 7–7b shows the equivalent circuit program for a PLC.

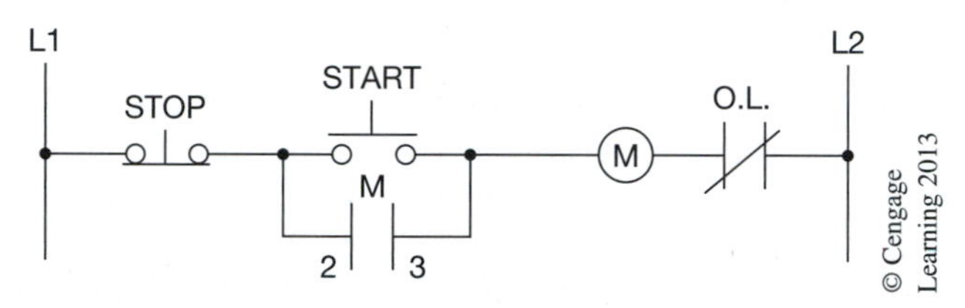

Figure 7–7a Standard *STOP/START* Ladder

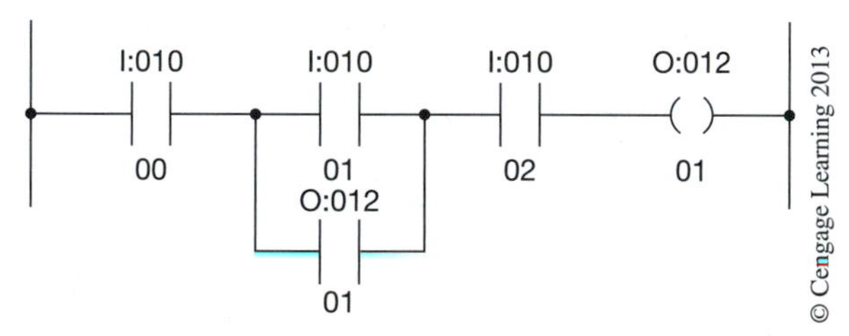

Figure 7–7b Equivalent *STOP/START* Circuit Programmed with a PLC

Note: *Addresses shown are PLC-5 format. An "I" preceding a word and bit number indicates an input, whereas an "O" preceding a word and bit address indicates an output. Since the output must be the last item programmed on a rung, the overload contacts (O.L./I:010/02) are programmed ahead of the motor output.*

Once the input devices are wired to the input module(s), as shown in Figure 7–7c, and the PLC system is "powered up," or turned *ON,* the processor scans the inputs and sets the corresponding bits to 1 or 0 depending on the status of the real-world input devices. If an input is open, the corresponding bit is set to 0, or *OFF,* whereas any bit that represents a closed device will be set to 1, or *ON*.

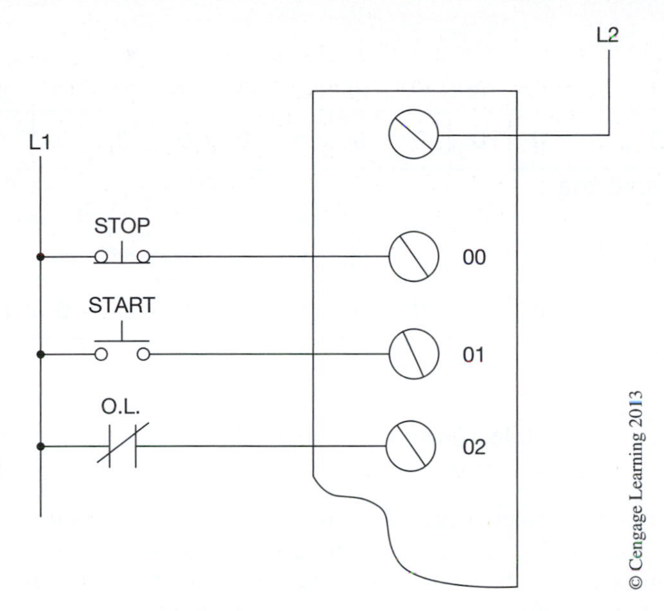

Figure 7–7c Input Devices Connected to an Input Module

In Figure 7–7c, the *STOP* button (I:010/00) and the O.L. contact (I:010/02) are wired in N.C. but programmed as normally open contacts (EXAMINE ON), as shown in Figure 7–7b. Because the *STOP* button and overload contacts are actually closed, bits 00 and 02 of word 010 are set to 1, or *ON*. Figure 7–8a shows the bit status of input image word 010 with the processor in the *RUN* mode.

													O.L.	START PB	STOP PB
17	16	15	14	13	12	11	10	07	06	05	04	03	02	01	00
0	**0**	**0**	**0**	**0**	**0**	**0**	**0**	**0**	**0**	**0**	**0**	**0**	**1**	**0**	**1**

INPUT IMAGE WORD 010

														M	
17	16	15	14	13	12	11	10	07	06	05	04	03	02	01	00
0	**0**	**0**	**0**	**0**	**0**	**0**	**0**	**0**	**0**	**0**	**0**	**0**	**0**	**0**	**0**

OUTPUT IMAGE WORD 012

Figure 7–8a Bit Status for Input Image Word 010 and Output Image Word 012 with the Processor in *RUN* Mode

With the *STOP* button and the O.L. contacts (bits 00 and 02) set to 1, or *ON,* we need only press the *START* button to complete the circuit. When the *START* button (I:010/01) is depressed, bit 01 is set to 1 (*ON*) during the next processor scan (shown in Figure 7–8b), and the circuit logic to the output is complete.

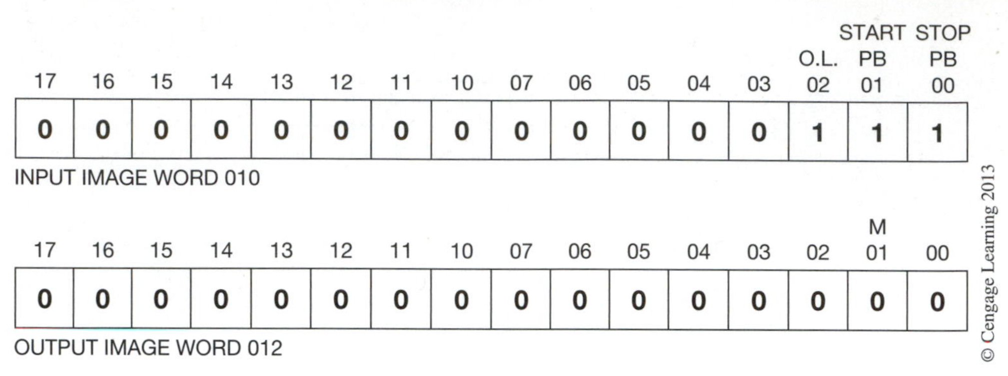

Figure 7–8b Bit Status for Input Image Word 010 and Output Image Word 012 While *START* Button Is Being Depressed

With the circuit logic now complete (true), the processor sets bit 01 of output word 012 to 1, or *ON* (Figure 7–8c), and motor O:012/01 is *energized* and held energized by holding contacts O:012/01. The holding contacts O:012/01 have the same address as the motor (O:012/01) because they both reference the same bit (01) of word 012 in the output register. The holding contacts do not actually exist as hardwired contacts, but are used to maintain the logic path for the circuit. The EXAMINE ON instruction with the address O:012/01 is the equivalent of holding contacts, and when the motor is energized (turned *ON*), bit 01 of word 012 is set to 1, or *ON,* and an alternate logic path is now complete.

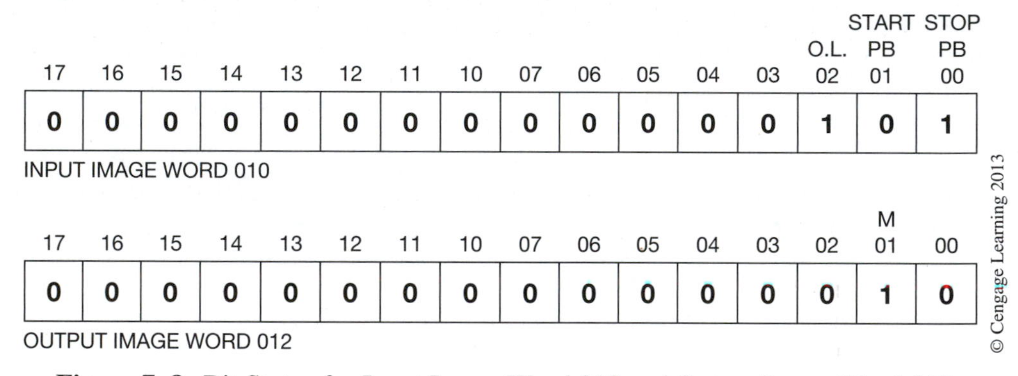

Figure 7–8c Bit Status for Input Image Word 010 and Output Image Word 012 After Output O:010/01 Is Energized and the *START* Button Is Released

When the STOP button (input I:010/00) is depressed, the processor clears bit 00 to 0, and the circuit logic is broken, or goes FALSE. Bit 01 is cleared to 0, and the real-world output device connected to terminal 01 of the output module drops out (turns *OFF*). Words 010 and 012 in the I/O register now appear as shown in Figure 7–8d, with only bit 02, the overload contact, set to 1.

When the STOP button is released or closed again, bit 00 is again set to 1, or *ON* (Figure 7–8e). The output (O:012/01) is not energized, however, as the START button (bit 01) is 0 (*OFF*), and the holding contacts (bit 01 of word 012) are cleared to 0 (*OFF*).

17	16	15	14	13	12	11	10	07	06	05	04	03	O.L. 02	START PB 01	STOP PB 00
0	0	0	0	0	0	0	0	0	0	0	0	0	1	0	0

INPUT IMAGE WORD 010

17	16	15	14	13	12	11	10	07	06	05	04	03	02	M 01	00
0	0	0	0	0	0	0	0	0	0	0	0	0	0	0	0

OUTPUT IMAGE WORD 012

Figure 7–8d Bit Status for Input Image Word 010 and Output Image Word 012 with the *STOP* Button Depressed

17	16	15	14	13	12	11	10	07	06	05	04	03	O.L. 02	START PB 01	STOP PB 00
0	0	0	0	0	0	0	0	0	0	0	0	0	1	0	1

INPUT IMAGE WORD 010

17	16	15	14	13	12	11	10	07	06	05	04	03	02	M 01	00
0	0	0	0	0	0	0	0	0	0	0	0	0	0	0	0

OUTPUT IMAGE WORD 012

Figure 7–8e Bit Status for Input Image Word 010 and Output Image Word 012 with the *STOP* Button Released

Another way to look at the N.O. and N.C. symbols used for programming the PLC is called the relay analogy.

For the sake of discussion, imagine that each input device is connected to an invisible control relay inside the input module, and that each control relay has one normally open and one N.C. contact, as shown in Figure 7–9.

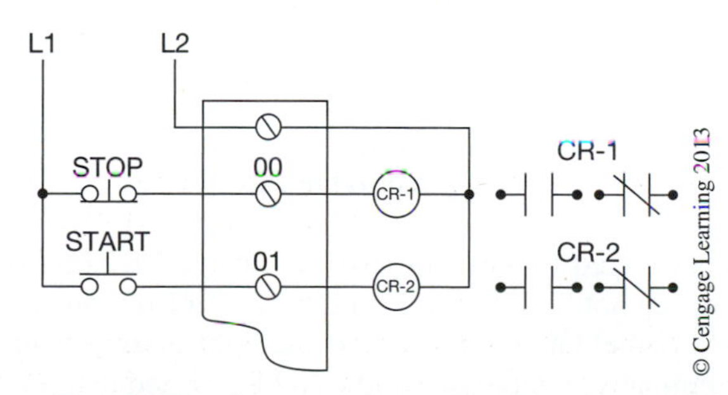

Figure 7–9 Imaginary Control Relays Wired to an Input Module

Connect one lamp to the N.O. contacts, and another lamp to the N.C. contacts of CR-1, as shown in Figure 7–10.

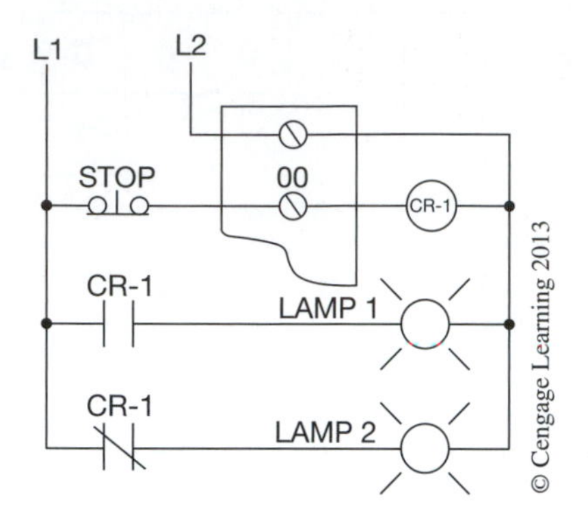

Figure 7–10 Lamps Wired to N.O. and N.C. CR-1 Contacts

When power is applied to L1 and L2, CR-1 energizes through the N.C. contacts of the *STOP* button. With CR-1 energized, the N.O. contacts of CR-1 close, and lamp 1 lights, as indicated in Figure 7–11. The normally closed contacts of CR-1 are now open, so lamp 2 *cannot* light.

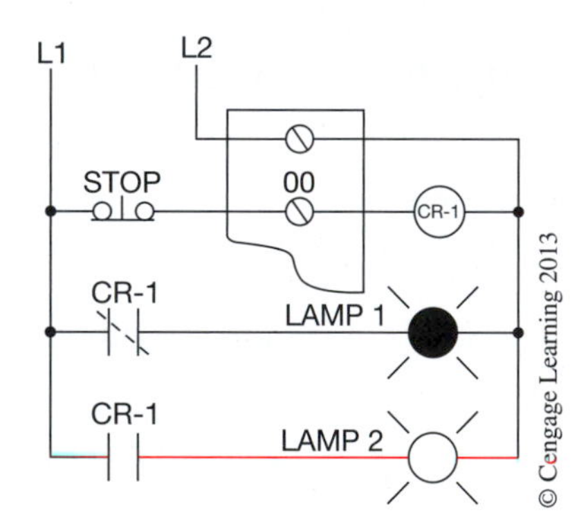

Figure 7–11 Lamp 1 Lights with Power Applied to L1 and L2

Wired in this manner, N.O. CR-1 contacts controlled by a N.C. pushbutton will close or *conduct* when power is applied to the circuit. Likewise, N.O. contacts *programmed* to represent a N.C. pushbutton will conduct when power is applied to the PLC.

An N.O. *START* button connected to an input terminal and an invisible or imaginary control relay (shown in Figure 7–12) would not light Lamp 1 until the *START* button is depressed, and would only stay lit as long as the button is held down. Lamp 2 would light as soon as power is applied, but would go out when the *START* button was depressed and CR-2 energized (Figure 7–13).

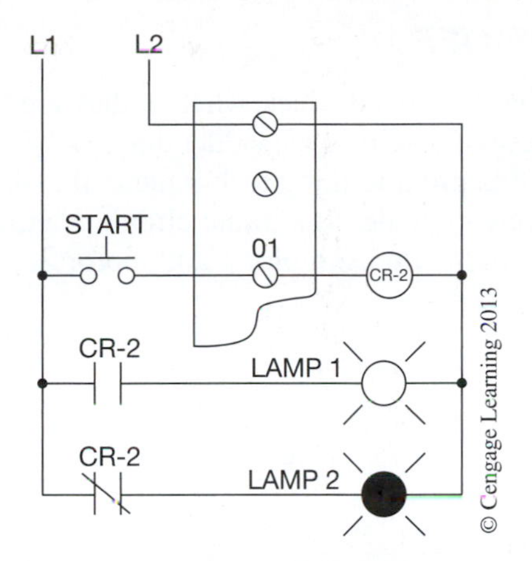

Figure 7–12 Power Applied—*START* Button Not Depressed

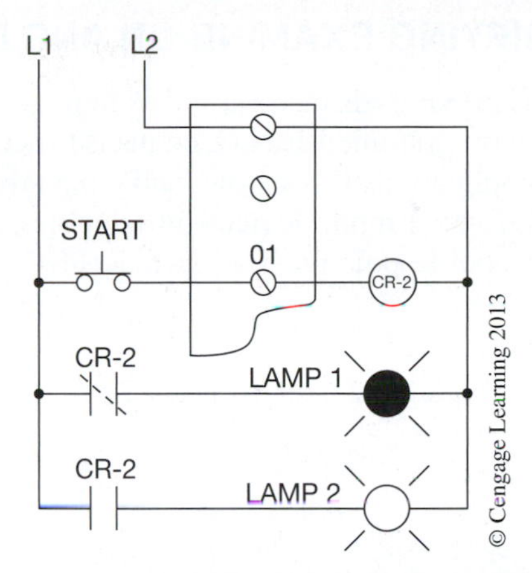

Figure 7–13 Power Applied—*START* Button Depressed

The rules for contacts that represent real-world input devices are shown in Figures 7–14a and 7–14b.

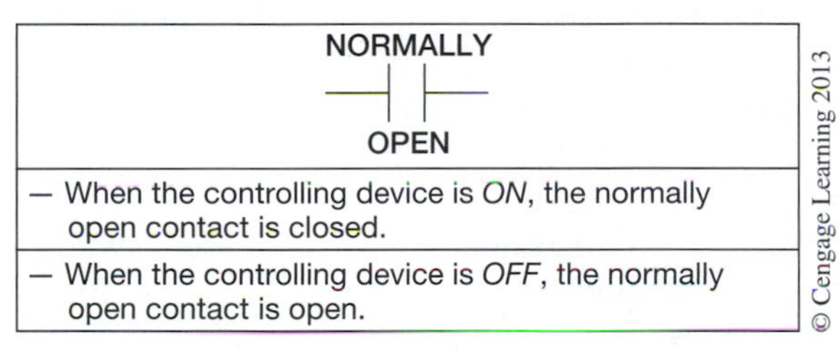

Figure 7–14a Normally Open Contact of an External Input

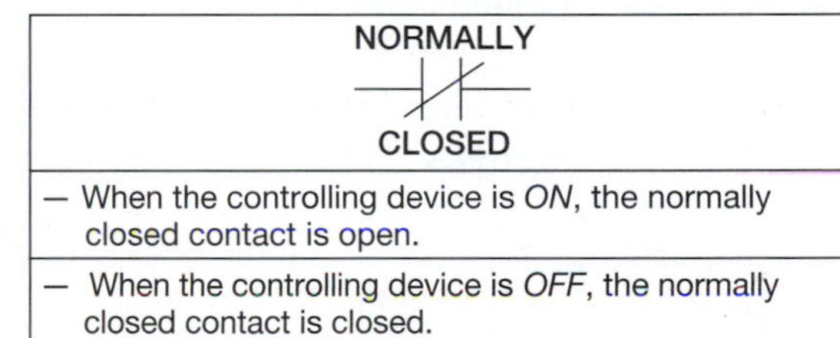

Figure 7–14b Normally Closed Contact of an External Input

There are no invisible control relays in the input modules, and there are also no symbols on the programming device for *STOP* buttons, *START* buttons, limit switches, and the like. As long as relay contact symbols must be used in place of regular input symbols, the relay analogy is an easy way to explain why N.O. contacts are programmed to represent N.C. input devices.

It does not matter which approach you use to understand the logic behind the way that PLCs are programmed, as long as you do understand it. The author believes that the EXAMINE ON and EXAMINE OFF approach is the easiest and clearest way to look at programming. The conversion of ladder diagram to PLC program will be quite simple once you clearly understand the concept of EXAMINE ON and EXAMINE OFF.

CLARIFYING EXAMINE ON AND EXAMINE OFF

From the previous discussion and examples, it appears that all input devices, whether they are N.O. or N.C., are programmed as N.O. contacts to achieve the desired results in the ladder diagram. For many circuit applications this is true, and a big advantage of this programming technique is the ability to turn single-pole input devices into double-, three-, or four-pole devices in the circuit. Figure 7–15 shows a double-pole pressure switch (PS-1) controlling two outputs: motor 1 and motor 2.

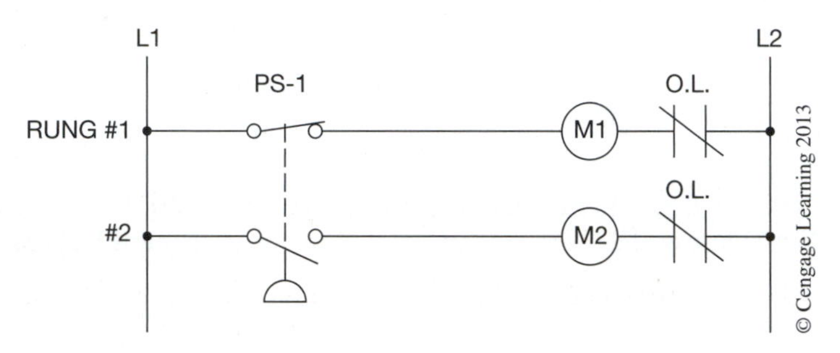

Figure 7–15 Double-Circuit Pressure Switch

When power is applied to the circuit, motor 1 will start through the N.C. contacts of PS-1 in rung 1. Motor 2 cannot start, however, due to the N.O. contacts of PS-1 in rung 2. When PS-1 is actuated, the N.C. contacts in rung 1 will open and motor 1 will go *OFF,* while the N.O. contacts in rung 2 will close, turning motor 2 *ON.*

By programming the same circuit on a PLC, the necessity of buying a double-circuit pressure switch is eliminated. One N.O. and 1 N.C. contact having the same address are used (see Figure 7–16a). The address is actually the address of a discrete N.C. single-circuit pressure switch (illustrated in Figure 7–16b).

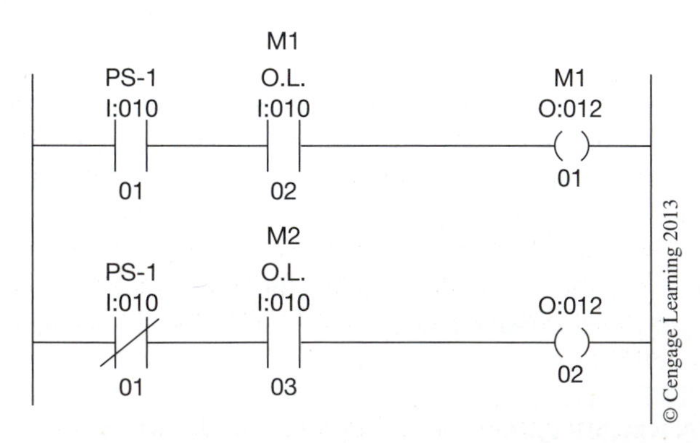

Figure 7–16a Double-Circuit Pressure Switch Circuit Programmed for a Typical PLC

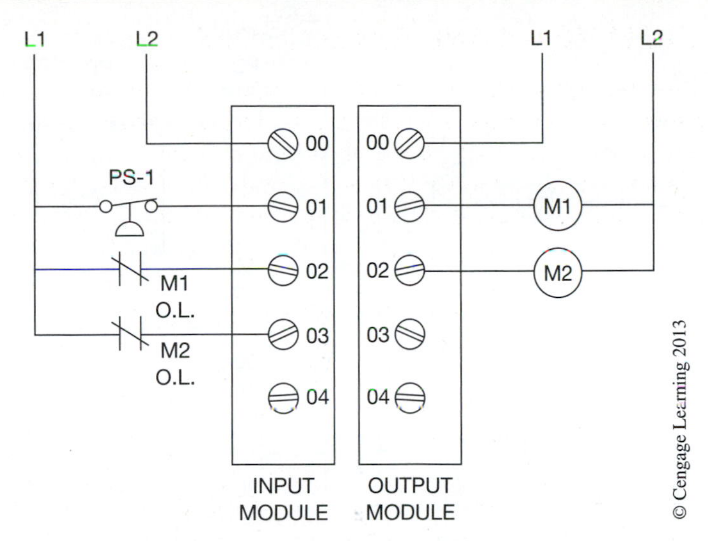

Figure 7–16b Actual Wiring of a Single-Pole Pressure Switch

When the processor is placed in the RUN mode, it examines all N.O. contacts for an *ON* condition, or EXAMINE ON. As PS-1 and both overload contacts are closed (*ON*), bits 01, 02, and 03 of input word 010 are set to 1, making those portions of the ladder diagram true. When the processor EXAMINES OFF the N.C. contact (I:010/01), it sees that bit 01 is set to 1 so that this part of the ladder diagram is false. Motor 1, address O:012/00, is *ON* and motor 2, address O: 012/01, is *OFF*.

When the pressure switch is actuated, the processor continues scanning the user program and examines all N.O. contacts for *ON,* and all programmed N.C. contacts for *OFF*. Since the pressure switch (PS-1) is now actuated, the N.C. contacts are open, and the N.O. contacts (I:010/01) go false, turning *OFF* motor 1 (O:012/00), whereas the N.C. contacts (I:010/01) go true, and turn motor 2 (O:012/01) *ON*.

Even though only double, three, and four poles were mentioned, there is no limit (except user memory size) to the number of times an input device can be addressed and used in a programmed circuit. This programming technique allows for six-pole, seven-pole, eight-pole, and so on, devices to be programmed using only a single-pole discrete device.

There are many more applications and circuits that normally require two- or three-pole devices that now only require single-pole devices when programmed for a PLC. Examples are double-pole limit switches for forward and reversing circuits, and double-pole pressure switches for duplex controllers.

When programming contacts (N.O. or N.C.) are controlled by outputs, the familiar standard relay logic is used. Figure 7–17a shows a standard *STOP/START* station with pilot lights. Lamp 1 (green) indicates power is available, and lamp 2 (red) indicates the circuit is activated. Figure 7–17b shows how the circuit is programmed.

When the processor is placed in the RUN mode, lamp 1 lights due to the N.C. contacts O:012/00. When the *START* button is depressed, output O:012/00 energizes, N.C. contacts O:012/00 open (go false),

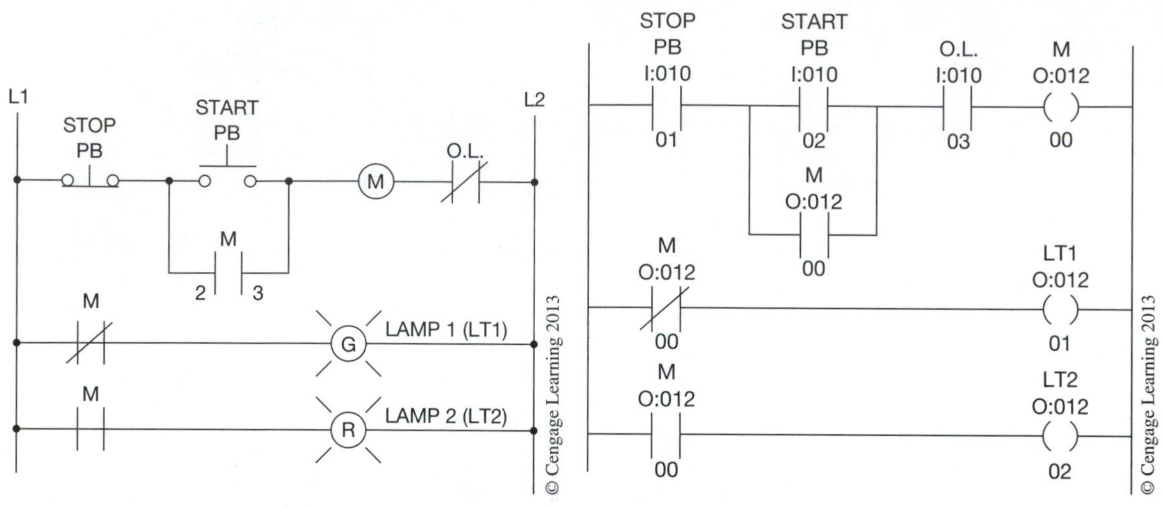

Figure 7–17a Ladder Diagram for *STOP/START* Station with Indicator Lamps

Figure 7–17b Programmed Circuit for *STOP/START* Station with Indicator Lamps

and lamp 1 (O:012/01) goes out. N.O. contacts O:012/00 close (go true), lamp 2 (O:012/02) turns *ON,* and the holding contacts (O:012/00) go true to complete the circuit logic.

Notice that the output, holding contacts, N.C. contact, and N.O. contact for the lamps all have the same address (O:012/00). The address refers to bit 00 of word 012, and this same bit (00) is referenced four times in the ladder diagram. The EXAMINE OFF, or N.C. contact, is logically true when bit 00 is cleared to 0 (*OFF*), whereas the EXAMINE ON, or N.O. contacts, are not logically true until bit 00 is set to 1 (*ON*).

Figure 7–18a shows the bit status for the circuit with power applied; Figure 7–18b, with the *START* button depressed; Figure 7–18c, with the *START* button released; Figure 7–18d, with the *STOP* button depressed; and Figure 7–18e, after the *STOP* button is released.

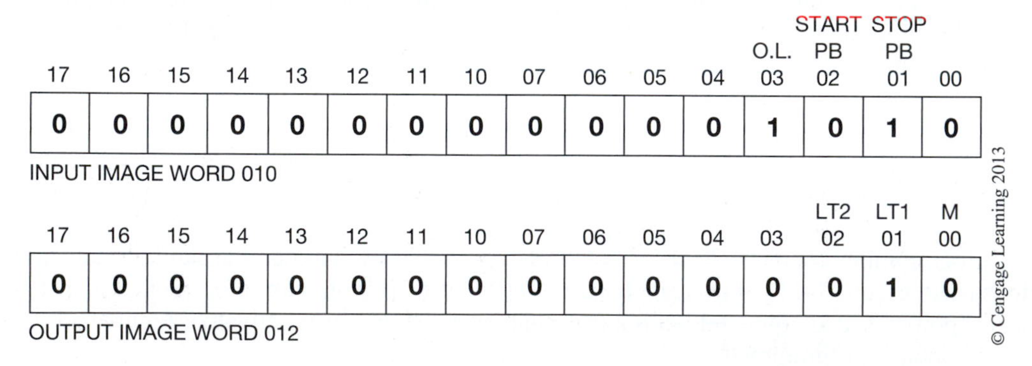

Figure 7–18a Bit Status for Input Image Word 010 and Output Image Word 012 with Power Applied

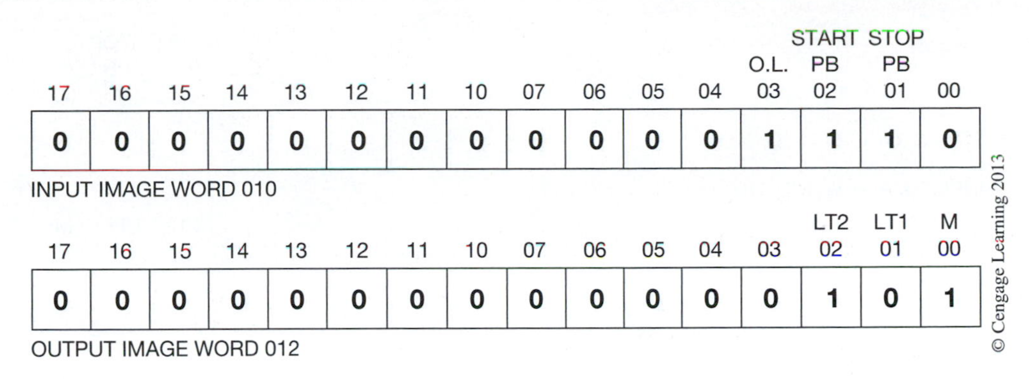

Figure 7–18b Bit Status for Input Image Word 010 and Output Imagc Word 012 with *START* Button Depressed

												O.L.	START PB	STOP PB	
17	16	15	14	13	12	11	10	07	06	05	04	03	02	01	00
0	0	0	0	0	0	0	0	0	0	0	0	1	0	1	0

INPUT IMAGE WORD 010

													LT2	LT1	M
17	16	15	14	13	12	11	10	07	06	05	04	03	02	01	00
0	0	0	0	0	0	0	0	0	0	0	0	0	1	0	1

OUTPUT IMAGE WORD 012

Figure 7–18c Bit Status for Input Image Word 010 and Output Image Word 012 with *START* Button Released

												O.L.	START PB	STOP PB	
17	16	15	14	13	12	11	10	07	06	05	04	03	02	01	00
0	0	0	0	0	0	0	0	0	0	0	0	1	0	0	0

INPUT IMAGE WORD 010

													LT2	LT1	M
17	16	15	14	13	12	11	10	07	06	05	04	03	02	01	00
0	0	0	0	0	0	0	0	0	0	0	0	0	0	1	0

OUTPUT IMAGE WORD 012

Figure 7–18d Bit Status for Input Image Word 010 and Output Image Word 012 with *STOP* Button Depressed

												O.L.	START PB	STOP PB	
17	16	15	14	13	12	11	10	07	06	05	04	03	02	01	00
0	0	0	0	0	0	0	0	0	0	0	0	1	0	1	0

INPUT IMAGE WORD 010

													LT2	LT1	M
17	16	15	14	13	12	11	10	07	06	05	04	03	02	01	00
0	0	0	0	0	0	0	0	0	0	0	0	0	0	1	0

OUTPUT IMAGE WORD 012

Figure 7–18e Bit Status for Input Image Word 010 and Output Image Word 012 with *STOP* Button Released

Chapter Summary

The relay logic used for programming the PLC at first seems to be in conflict with standard ladder logic. But once the concepts of EXAMINE ON and EXAMINE OFF are understood, the logic process is easy to understand. An EXAMINE ON instruction looks for an *ON* condition, and will be logically true when an *ON* condition (a 1) is found. The EXAMINE OFF instruction will be logically true when an *OFF* condition (a 0) is found. Looking at the actual status of the bits within individual I/O words is another way to help understand how the processor logic works. Using the XIC or examine if closed (EXAMINE ON) and the XIO or examine if open (EXAMINE OFF) approach may be helpful. Others find that the relay analogy approach to understanding the N.O. (EXAMINE ON) and N.C. (EXAMINE OFF) symbols used for programming is better. Another approach is to accept the fact that an N.C. *STOP* button, or a similar closed input device, must be programmed using an N.O. contact symbol, and just have the philosophy that logic is relative to application.

Review Questions

1. Briefly describe the action of the EXAMINE ON instruction.
2. When an N.O. limit switch is wired to an input module, and programmed using an N.O. contact symbol (EXAMINE ON), the instruction will be true (check all correct answers):
   a. when power is applied and the key switch is in the RUN position
   b. when the limit switch is closed
   c. as long as the limit switch is open
   d. never

3. If the N.O. limit switch in Question 2 is programmed using an N.C. contact symbol (EXAMINE OFF), the instruction will be true (check all correct answers):
   a. when power is applied and the key switch is in the RUN position
   b. when the limit switch is closed
   c. as long as the limit switch is open
   d. never
4. Briefly describe the action of the EXAMINE OFF instruction.
5. Indicate the logic (T [True] or F [False]) for the following contacts:

CONDITION OF INPUT DEVICE	PROGRAM INSTRUCTION	LOGIC TRUE-FALSE
		T F
		T F
		T F
		T F

6. The XIO instruction is the same as:
   a. EXAMINE ON
   b. EXAMINE OFF
7. The XIC instruction is the same as:
   a. EXAMINE ON
   b. EXAMINE OFF

CHAPTER 8 Programming a PLC

Objectives

After completing this chapter you should have the knowledge to:

- Explain the term *On-Line Programming*.
- Describe basic programming techniques.
- Define the term *mnemonic* and give examples of mnemonic names.
- Describe the *Force On* and *Force Off* features, and the hazards that could be associated with each.

Programmers and programming software can create, modify, monitor, and load programs into user memory, and can also make changes to the program while the processor and driven equipment are running. This feature is often referred to as On-Line Programming. Changing the program while the processor is running must *only* be done by persons with a complete understanding of the circuit operation and the process or driven equipment. To prevent unauthorized On-Line Programming, a key switch is provided, either on the programming device or on the processor, to restrict the access to a monitor-only mode, or to Off-Line and monitor-only mode. When programming using a personal computer and specialized software, a **password** may be used to limit the access to the processor program. Passwords act like a key switch. If the wrong password is entered when requested, the user is denied further access to the program.

Off-Line Programming, which means that the program is being developed off line (without being connected to the process or driven equipment), is the most common method of programming—and, of course, the safest. Since few programs are ever created without mistakes, it is always best to create the program off line. Once the program is complete, it should be tested while still in the Off-Line mode. After testing and verifying the program (circuit) in Off-Line mode, the PLC can be put in the On-Line mode for final verification, testing, and operation.

The function or functions of the processor that can be locked out with the key switch or by using passwords are fairly standard, but vary from PLC to PLC.

Most PLCs are designed so that any input contact (N.O. or N.C.) or output coil can be **Forced On** or **Off.** This feature (and troubleshooting aid) allows the operator to Force On, or make a contact or coil go *ON* regardless of actual status or circuit logic of the input or output device. Similarly, contacts and coils can be Forced Off, or turned *OFF,* regardless of the actual device status or circuit logic.

Caution: The Force On and Force Off feature should *never* be used except by personnel who completely understand the circuit *and* the process machinery or driven equipment. An understanding of the potential effect that forcing a given contact or coil has on machine operation is essential if hazardous and/or destructive operation is to be avoided.

PROGRAMMING WITH A COMPUTER

Many companies have developed software for programming PLCs. For purposes of illustrating how the software works, and the relative ease of programming, the author has selected the RSLogix software created by Rockwell Automation, Inc. for programming the Allen-Bradley family of programmable controllers. This software, in various versions, can be used to program the Logix5000, SLC 500, or the MicroLogix family of processors. The software is Windows® based and very user friendly. For the first programming example, we will use RSLogix 500 to program an SLC 5/03 processor mounted in a 4-slot chassis, as shown in Figure 8–1.

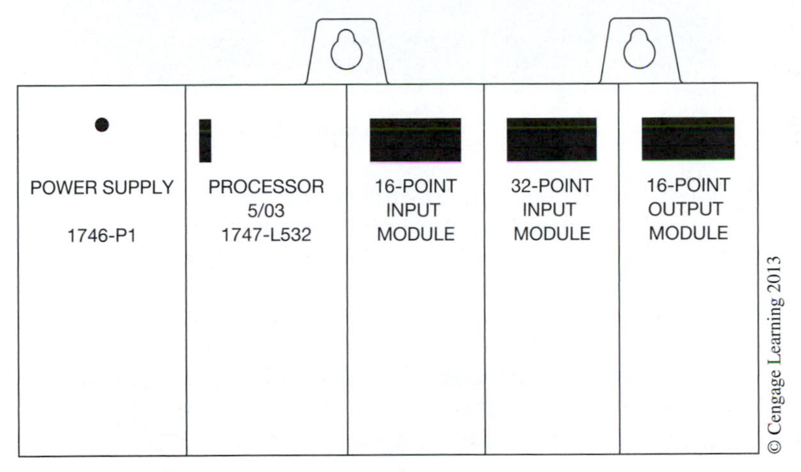

Figure 8–1 SLC 5/03 Processor Mounted in a 4-Slot Chassis

The circuit to be programmed is a simple *STOP/START* circuit that uses a *START* button, a *STOP* button, and holding contacts to control an exhaust fan. The ladder diagram of the circuit is shown in Figure 8–2.

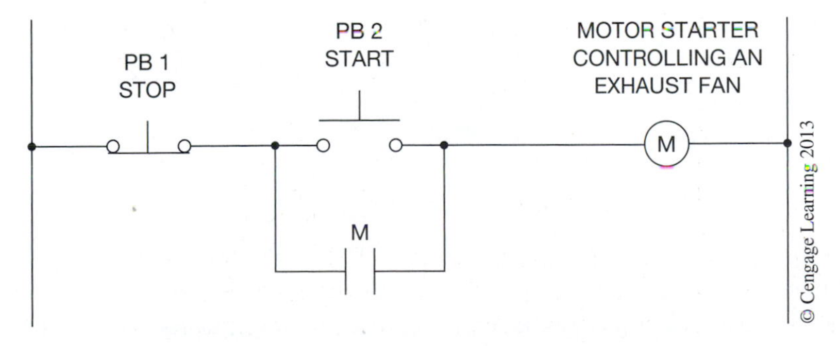

Figure 8–2 Basic *STOP/START* Circuit

To start the RSLogix software, double click on the RSLogix icon for the PLC family you are using. The opening screen for the RSLogix 500 software is shown in Figure 8–3.

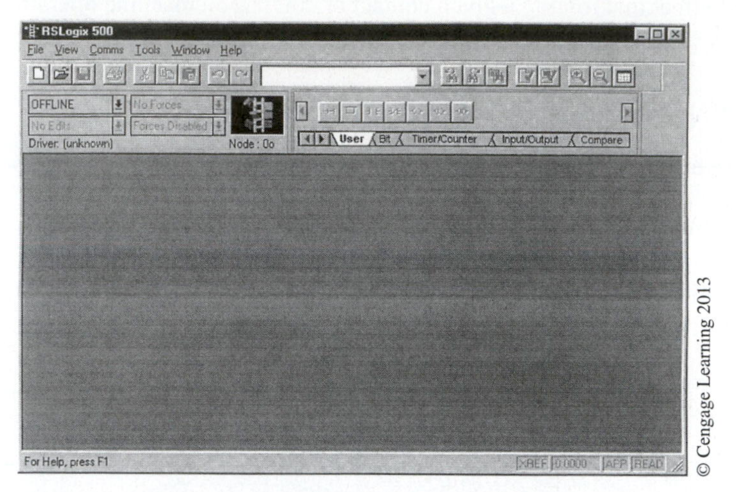

Figure 8–3 RSLogix Opening Screen

Using the computer mouse, left click on *File* and then click on *New*. The window that appears, titled *Select Processor Type,* lists the processors that the RSLogix 500 software can program. Scroll down the list until you find the processor you are using. Figure 8–4 shows the window. Note that an SLC 5/03 processor has been highlighted. Note also that below the 5/03 are listings for MicroLogix 1000 and 1500 processors.

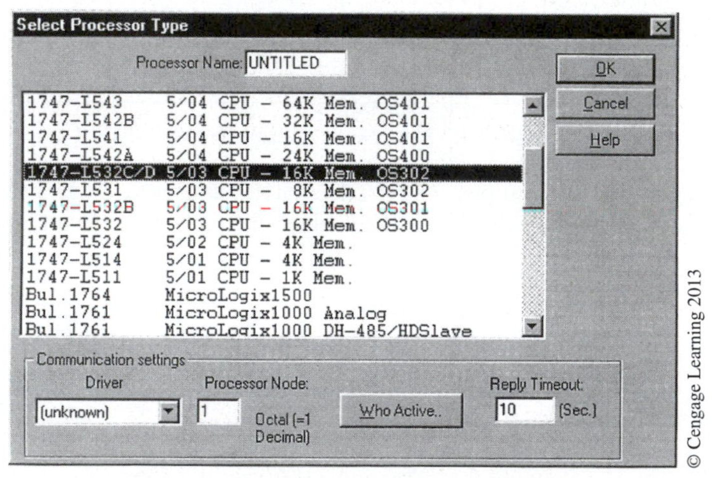

Figure 8–4 *Select Processor Type* Screen

Once the processor has been selected, the processor name needs to be entered. In Figure 8–4, the box where the name of the processor is entered, *Processor Name:,* shows the name as UNTITLED.

The name given the processor is typically the name of the process that the processor controls. Examples are Lumber Stacker 1, Log Sorter, Palletizing Unit, etc. For this example, the processor name will be TEST. Figure 8–5 shows the window after the processor name has been entered.

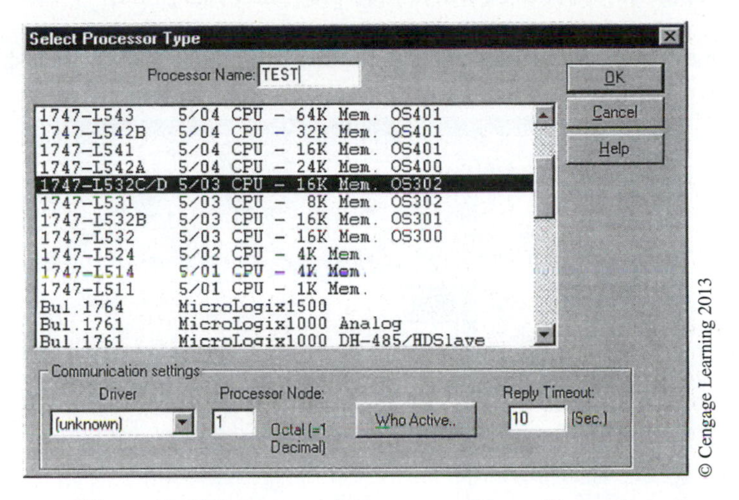

Figure 8–5 Screen with Processor Name Entered

The next step is to single click on the *OK* button that causes the software to enter the information for the processor that has been selected (SLC 5/03). Figure 8–6 shows the new screen.

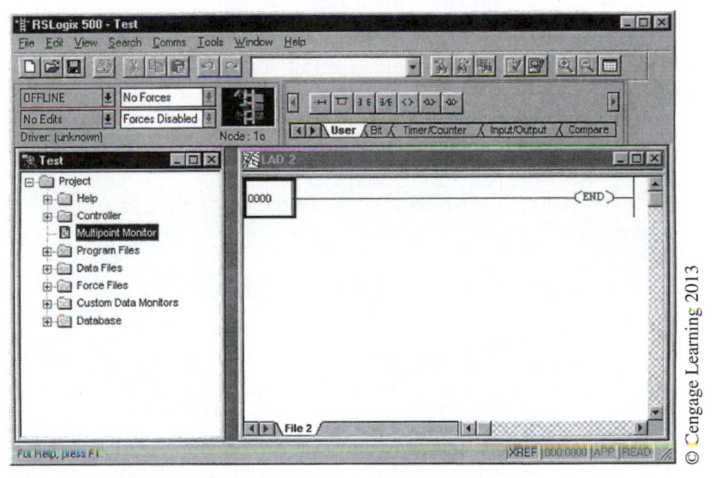

Figure 8–6 Screen After Processor Information Has Been Stored

The left side of the screen shows the *Project Tree* while the right side of the screen is the *Programming Area*. Either area can be increased in size, minimized, or closed by left clicking the mouse on the appropriate symbol.

Under the project tree, the main folders are *Help, Controller, Multipoint Monitor, Program Files, Data Files, Force Files, Custom Data Monitors,* and *Data Base.* Not all of these folders will be discussed, since the intent of this section is merely to illustrate the relative ease of programming using the RSLogix software, not to be a definitive programming guide.

Double clicking on the *Controller* folder opens the folder and produces the screen shown in Figure 8–7.

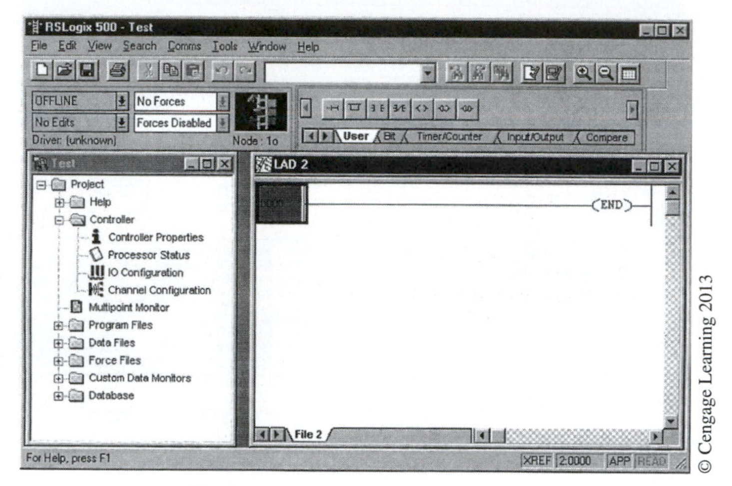

Figure 8–7 Controller Screen

Double clicking on *Controller Properties* produces the screen shown in Figure 8–8. From this screen, we can see the number of program files and data files, and also determine the memory used and the memory left. In the screen shown there is an asterisk (*) in both the Memory Used and Memory Left areas because no program has been developed yet and no memory has been used.

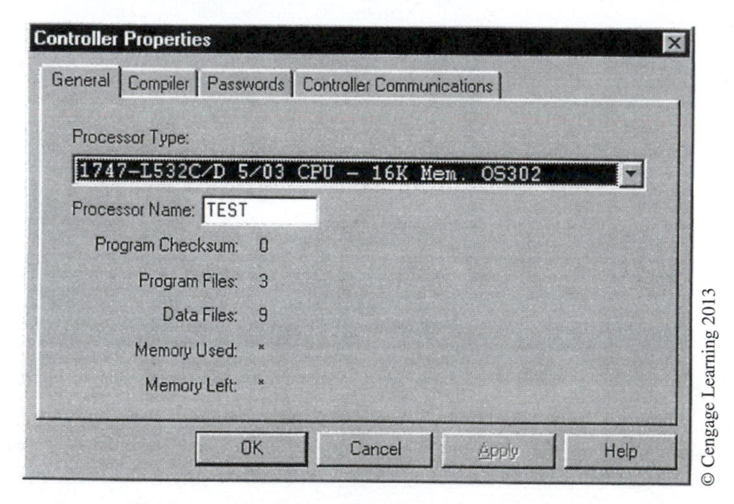

Figure 8–8 *Controller Properties* Screen

Closing *Controller Properties* and double clicking on the *IO Configuration* symbol produces the screen shown in Figure 8–9. From this screen, we can configure the I/O. As stated earlier, the SLC 5/03 processor is installed in a 4-slot chassis. From this window select either a 4-slot, 7-slot, 10-slot, or a 13-slot chassis from the pull-down list for Rack 1.

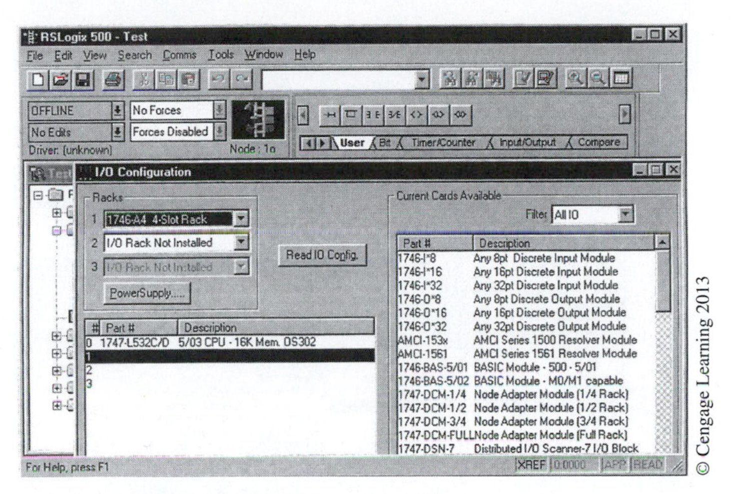

Figure 8–9 *IO Configuration* Screen

Notice that in Figure 8–9 the 5/03 processor is shown installed in slot 0. The processor will always be installed in slot 0 when using modular I/O. To enter the correct I/O module that is installed in slot 1 of our 4-slot chassis, find the module on the list to the right and double click. Figure 8–10 shows the window after double clicking on 1746-I*16, which is a 16-point discrete input module installed in slot 1 of the 4-slot chassis (Figure 8–1).

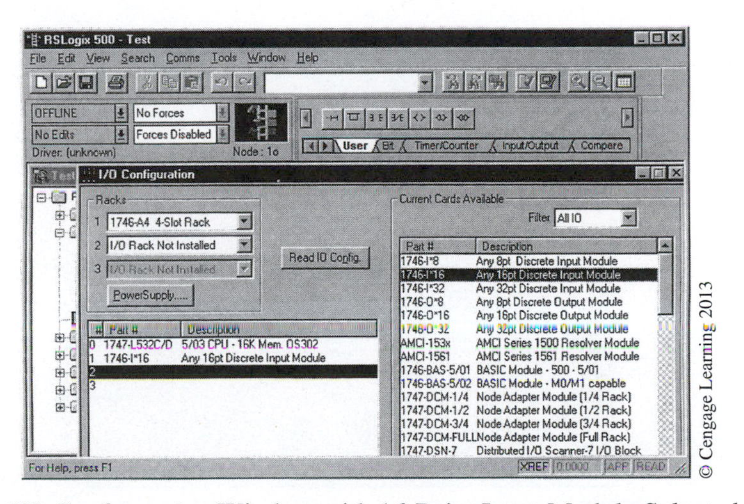

Figure 8–10 *I/O Configuration* Window with 16-Point Input Module Selected for Slot 1

To complete the I/O Configuration, select a 32-point input module (1746-I*32) for slot 2 and a 16-point output module (1746-O*16) for slot 3. The completed I/O Configuration is shown in Figure 8–11. If the processor was in the On-Line mode, the software could look at the I/O modules installed in the chassis and automatically enter the appropriate I/O module type and number into this window by pressing the "Read IO Config" button.

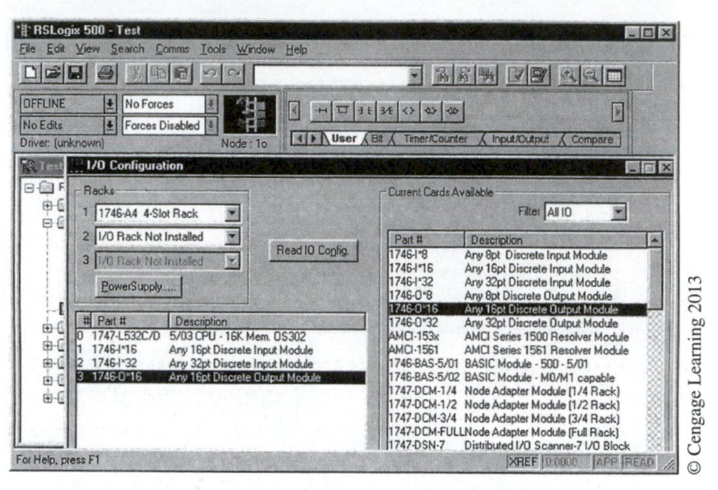

Figure 8–11 Completed *IO Configuration* Window

The software has the ability to indicate the minimum size of power supply that should be used based on the number and type of I/O modules that have been selected. To determine the correct power supply to use, single click on the *Power Supply* button. The *Power Supply Loading* window is shown in Figure 8–12. This window shows that our SLC 5/03 system consists of 1 Rack, and the minimum power supply recommended is a 1746-P1.

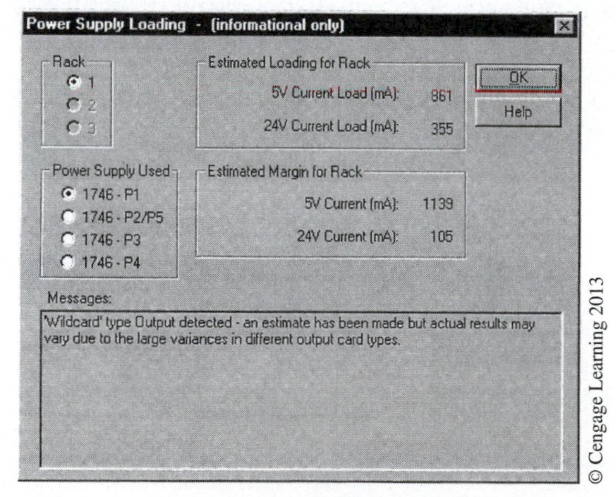

Figure 8–12 *Power Supply Loading* Window

This window estimates the loading of the rack. In this illustration, the estimated 5-volt current load in mA (milliamps) is 861 and the 24-volt current load is estimated at 355 mA. This window also estimates the margin of mA load that is provided by a 1746-P1 power supply beyond the mA load for the mix of I/O currently installed. The estimate for the 5-volt current is 1139 mA. We could add an additional 1139 mA of load before the P1 power supply would be at full capacity. The 24-volt current load margin is given as 105 mA. Note the message at the bottom of the window that says:

> *"Wildcard" type output detected—an estimate has been made but actual results may vary due to the large variances in different output card types.*

This message informs the programmer that the current values given are only estimates and the type of discrete output cards used will determine the actual mA load. Common sense dictates that if the estimated mA margin for the rack is low, the next size power supply should be used.

The RSLogix software gives an **OVERLOAD—Use a Larger Power Supply** warning when the mix of I/O requires a larger power supply than has been selected. To determine the correct power supply, once an overload warning has been displayed, click on the next larger power supply and see if the warning disappears. If not, click on the next larger power supply listed. Continue to click on the next larger power supply until the OVERLOAD display disappears.

After closing the *Controller Properties* screen, open the *Data Files* folder by double clicking on the folder. The contents of this folder are shown in Figure 8–13.

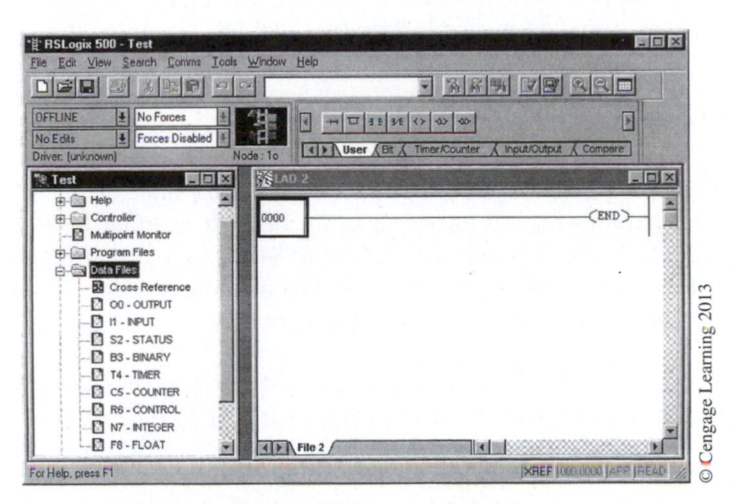

Figure 8–13 *Data File* Folder Contents

The Data File folder allows the user to determine the status of I/O files, as well as the status file (S2), binary file (B3), timer file (T4), counter file (C5), control file (R6), integer file (N7), and floating point file (F8).

When the processor is in the On-Line mode, the actual status of the input and output devices, 1 or 0, is determined by looking at the bit status in either the *Input* or the *Output* file. Double clicking on *I1—Input* produces the *Data File I1 (bin)—Input* window shown in Figure 8–14.

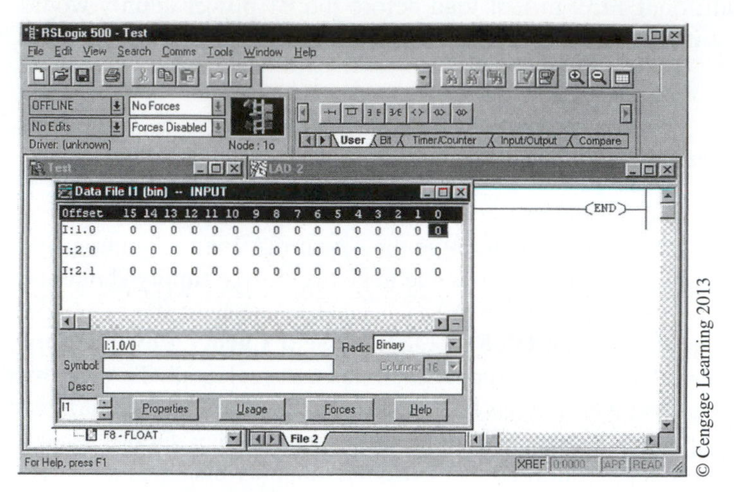

Figure 8–14 *Data File* Window

This window shows the status of the first three words of the input file. The first address I:1.0 indicates that this is an input module installed in slot 1. If we refer back to Figure 8–1, the 16-point input module was indeed installed in slot 1. Also, in Figure 8–11, an input module 1746-I*16 was shown installed in slot 1 of the 4-slot chassis. The individual input devices are represented by bits 0 through 15. Again note that the bit addresses are in decimal format, not in octal, like the Allen-Bradley PLC-5 family.

The next two addresses are I:2.0 and I:2.1. These addresses indicate that an input module is installed in slot 2 but requires two 16-bit words of memory. If we refer back to Figure 8–1 again, a 32-point input module was installed in slot 2. Figure 8–11 showed that input module 1746-I*32 was installed in slot 2 of the 4-slot chassis. Because this input device has 32 points, or terminals, it requires two 16-bit words of memory, as shown in Figure 8–14.

From this window, each bit (input device) can be assigned a *Symbol* and a *Description*. The symbol tells what the device is and the description explains what it does. The symbol and the description are entered by typing the information into the appropriate boxes. For the example in Figure 8–15, for address I:1.0/0, the symbol was identified as *PB 1* (pushbutton 1), and the description as *Stop*. This is the address that will be used for the *STOP* button in the *STOP/START* circuit to be programmed. The information for each connected input device can be entered in this same manner. Information also can be assigned to input and output devices as the program is being written. If this is the case, the symbol type and description will be shown automatically when a *Data File* window is opened.

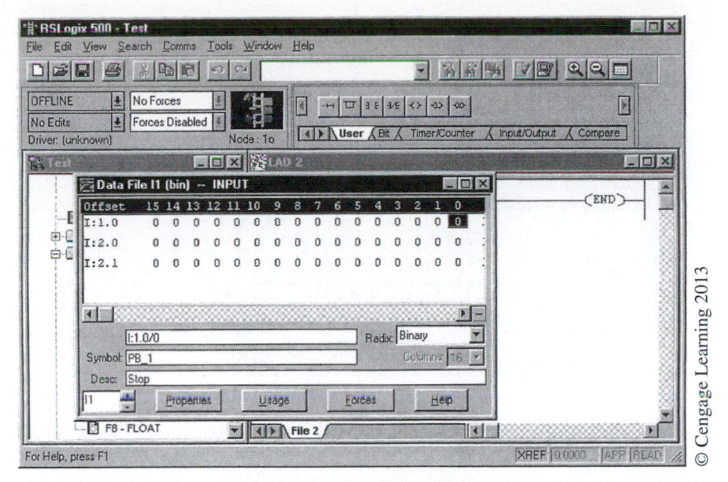

Figure 8–15 Input Data File with Symbol and Description of Address I:1.0/0 Entered

If we close the *Data File Input* screen and double click on *O0—Output,* the *Output Data File* will open, as shown in Figure 8–16.

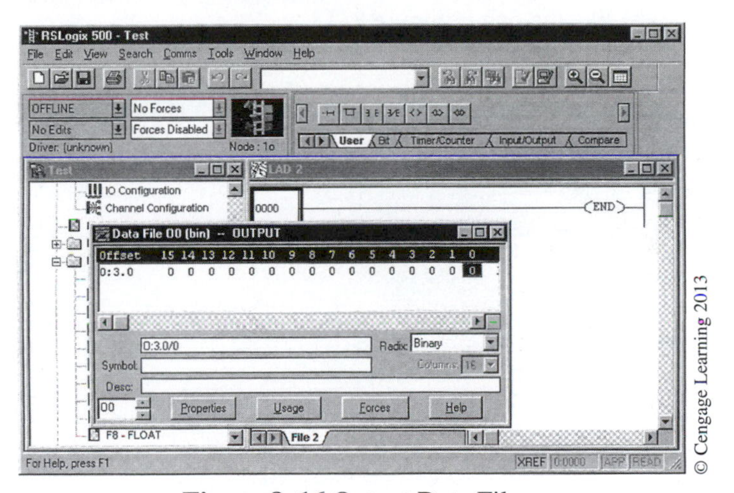

Figure 8–16 Output Data File

From this screen, we can add the symbol and description for each of the 16 outputs represented by bit 0 through 15. Figure 8–17 shows the screen after the symbol and description for output address O:3.0/0 have been typed in. This is the address that will be used for the motor starter that controls the exhaust fan in our *STOP/START* circuit. As with the *Input Data File*, each output device can be given a symbol and a description from this screen.

Again, this brief discussion of the program tree and the information that can be obtained is intended to show the power of the RSLogix software and is not a complete guide to the software.

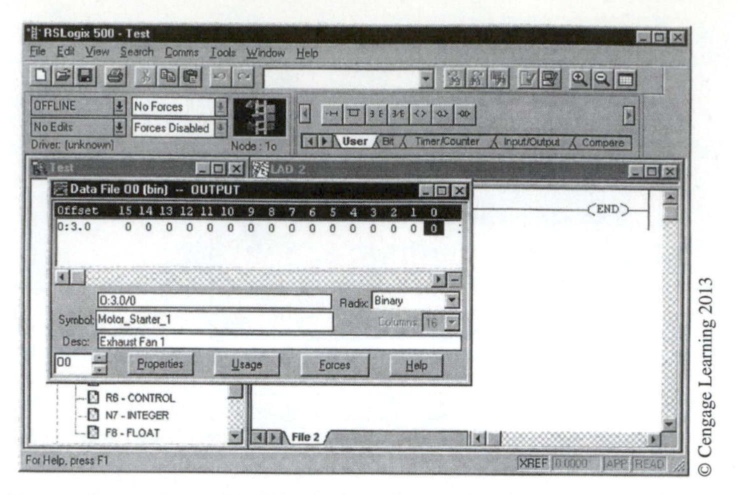

Figure 8–17 Output Data File with Symbol and Description of Address O:3.0/0 Entered

Closing the *Output Data Table* screen and maximizing the programming side of the display, we see that the screen now looks like the one in Figure 8–18.

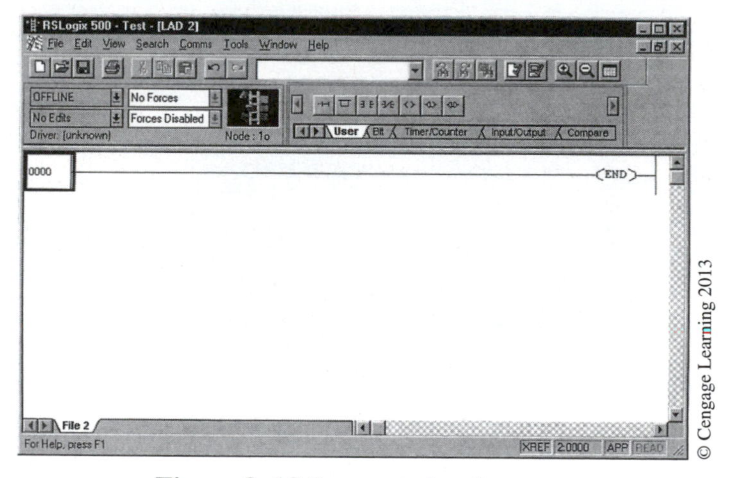

Figure 8–18 Programming Screen

The box in the upper left-hand corner of the programming screen has four zeros (0000) inside the box. The four zeros indicate Rung 0000. To start to develop a program, single click on the *New Rung* button above the program area. The *New Rung* button is the first button and uses this symbol: --|—|. Figure 8–19 shows the screen with the new rung added.

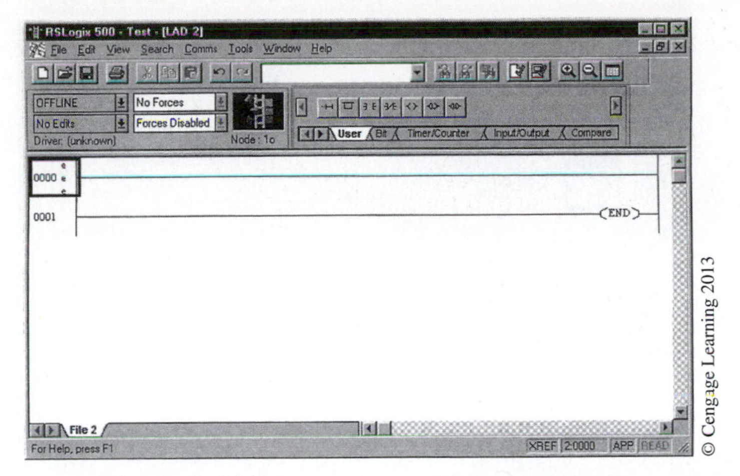

Figure 8–19 Program Area with Rung Added

As shown in Figure 8–19, two rungs now show on the screen. The first rung is 0000; the second rung is 0001. Note that there are three small letter *e*'s at Rung 0000. The *e*'s indicate that this rung is being edited.

Figure 8–20 shows the simple *STOP/START* circuit without overloads that will be programmed.

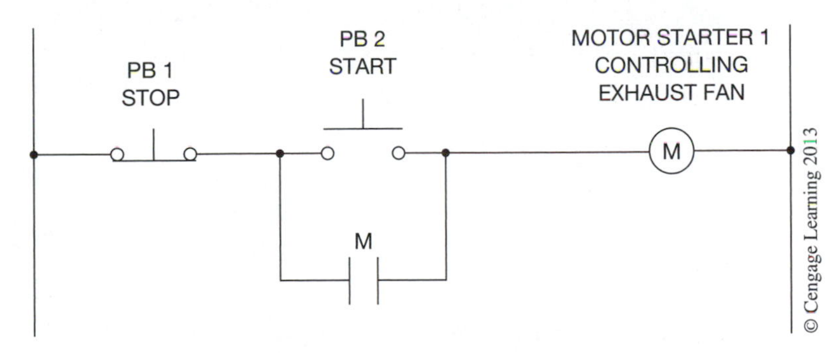

Figure 8–20 Simple *STOP/START* Circuit without Overloads

The first step is to program the *STOP* button. This is accomplished by single clicking on the *Normally Open* contact symbol (—] [—). This is the third symbol in the group of seven symbols. This N.O. contact symbol is also referred to as XIC, or Examine if Closed. Figure 8–21 shows the screen with the N.O. contact.

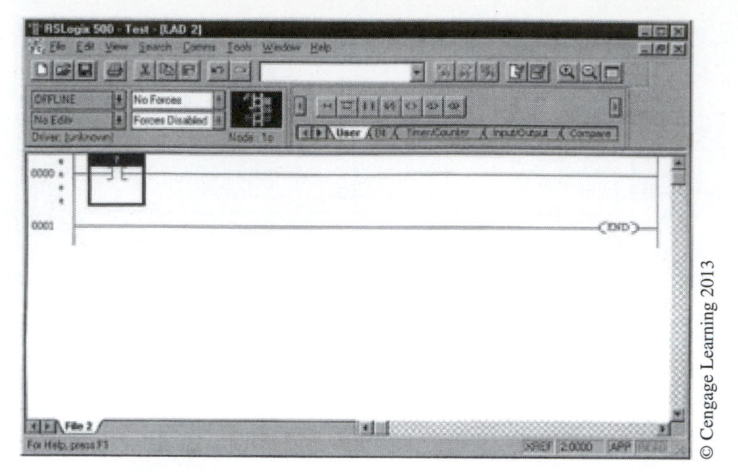

Figure 8–21 Screen with N.O. Contact

Double click on the question mark (?) above the contact and enter the address of the *STOP* button. We are using address I:1.0/0. Figure 8–22 shows the screen with the address I:1.0/0 entered above the contact.

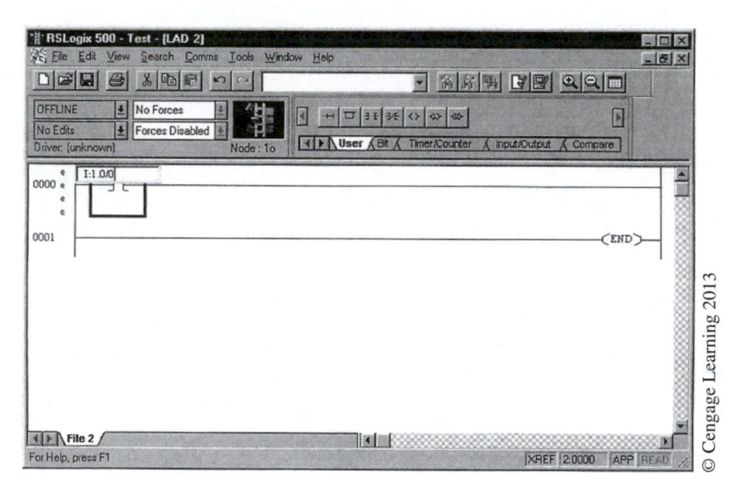

Figure 8–22 Address Entered above the N.O. Contact

After typing in the address, hit the *Enter* key. Figure 8–23 shows how the screen looks now.

Remember that earlier, the address I:1.0/0 was assigned the symbol *PB 1* and the description of *STOP*. When the address was entered, the information entered for address I:1.0/0 was automatically added to the programmed N.O. contacts. Notice also that the address is I:1/0. The program eliminates any unnecessary zeros. So address I:1.0/0 becomes I:1/0 (input device in slot 1 connected to terminal 0).

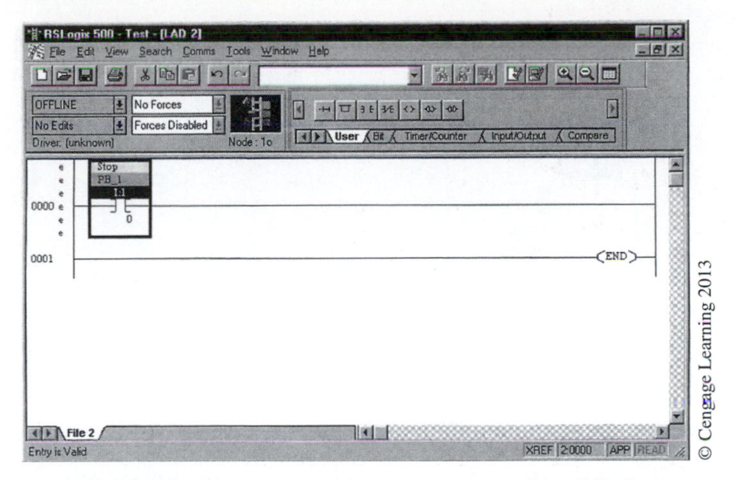

Figure 8–23 Screen with *STOP* Button Identified

Note: *When entering the addresses, the semicolon (;) can be used instead of the colon (:). The software will automatically convert the ";" to a ":" in the address.*

The next step in the programming process is to add the *START* button. Since the *START* button is in parallel with the holding contacts, single click on the *BRANCH START* symbol button. The button symbol is ⊓ .

After the *BRANCH START* symbol is clicked, the screen now appears as shown in Figure 8–24.

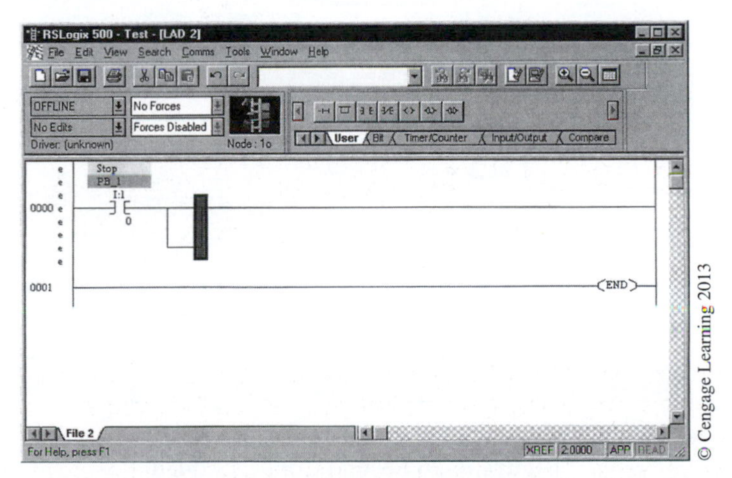

Figure 8–24 Screen with *BRANCH START* Symbol

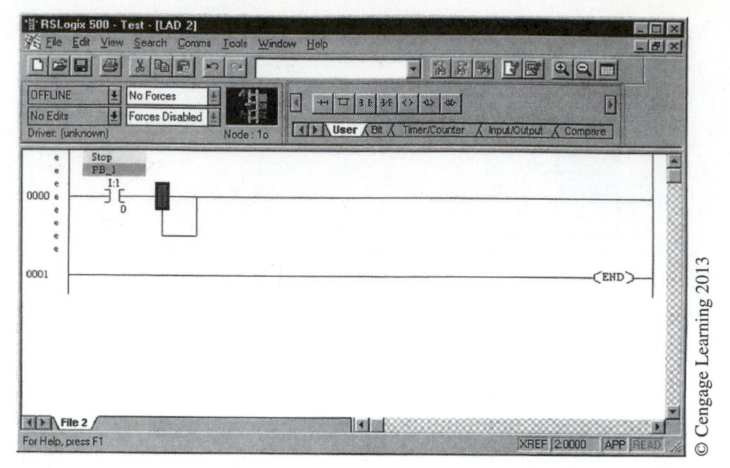

Figure 8–25 Heavy Line Moved to Upper Left-Hand Side of Symbol

Notice that a dark vertical line appears on the right side of the *BRANCH START* symbol. It is necessary to move this line to the upper left-hand corner of the symbol before contacts can be added. To move the line, point the arrow of the mouse at the upper left-hand corner of the *BRANCH START* symbol and single click. The heavy line now moves to the upper left portion of the symbol, as shown in Figure 8–25.

Next, single click on the N.O. symbol to add the next contact. Figure 8–26 shows the screen after the next contact has been added.

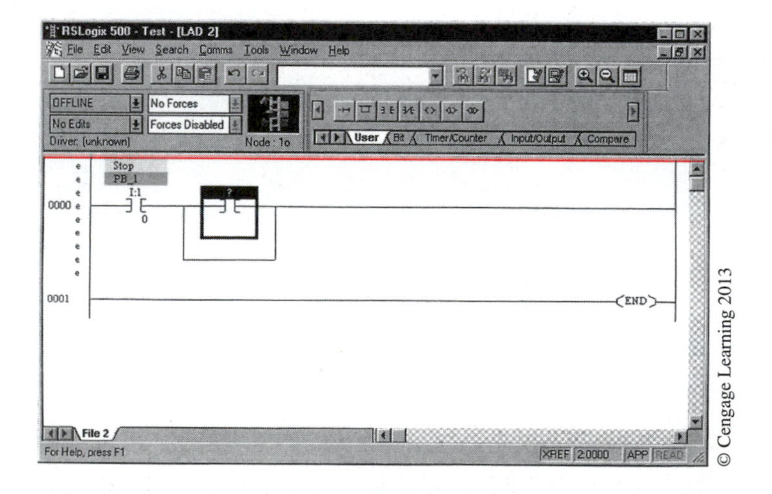

Figure 8–26 Second Contact Added

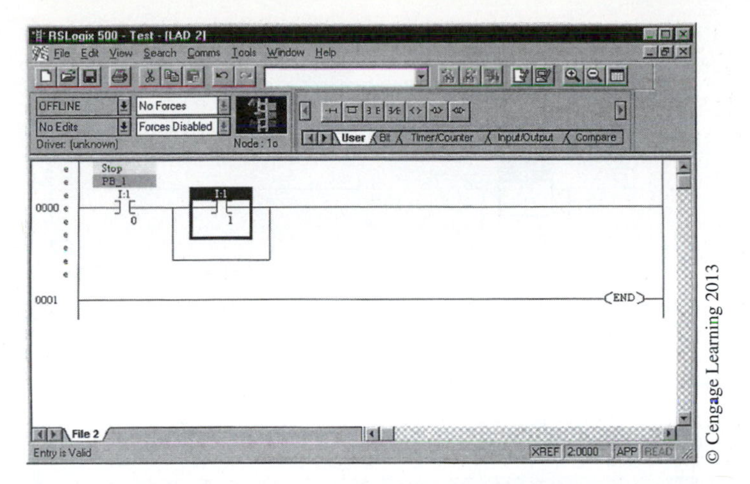

Figure 8–27 *START* Contacts Addressed

Because all unnecessary zeros will not be displayed, double click on the (?) and enter the address I:1/1 for the *START* button. Figure 8–27 shows the *START* button contacts with the address I:1/1.

Notice that there is no symbol or description for the address I:1/1 as was the case for address I:1/0. The symbol and description can be entered by opening the Input Data File as was done earlier. However, another way to enter the information is to place the cursor on the N.O. contact symbol (*START* button) and single click the right mouse button. From the menu that appears, select *Edit Symbol.* The screen is shown in Figure 8–28.

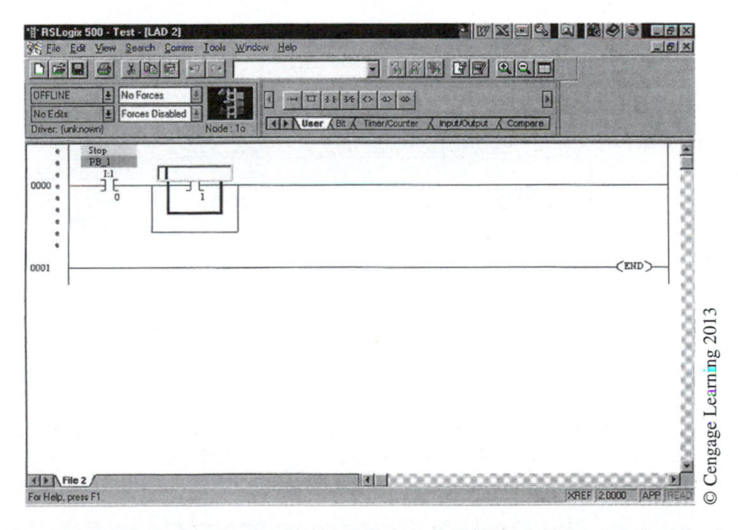

Figure 8–28 Box above *START* Button for Entering the Symbol

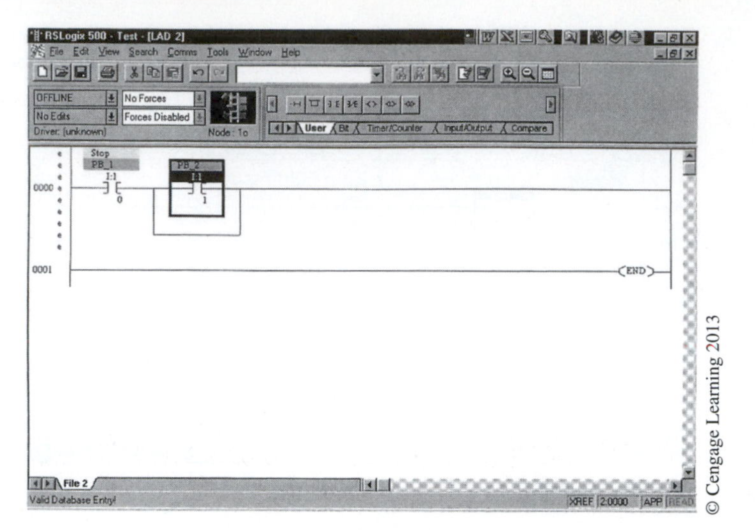

Figure 8–29 PB 2 Symbol Added to *START* Button

In the box above the *START* button, type in the assigned symbol *PB 2* and press the Enter key. Figure 8–29 shows the screen with the symbol PB 2 added above the *START* button.

The description of the input device (*START* button) can be added by placing the cursor on the *START* button again, single clicking the right mouse button, and then selecting *Edit Description* from the menu that appears. Type *Start* in the box for the description and then click on *OK* to enter the information. Figure 8–30 shows the completed *START* button.

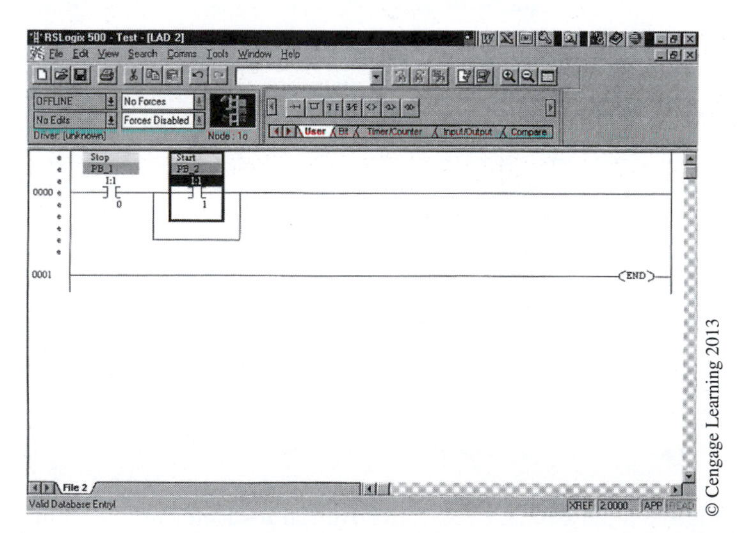

Figure 8–30 *START* Button with Symbol and Description Notations

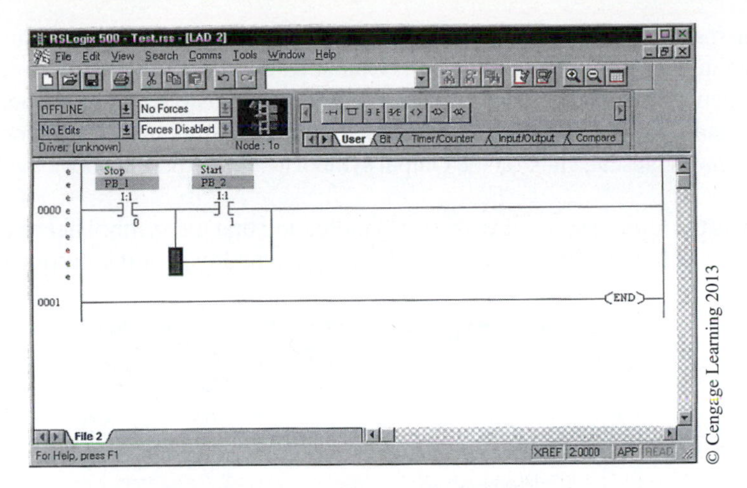

Figure 8–31 Heavy Line in Lower Left Corner

To add the holding contact from the Motor Starter, point the arrow at the lower left-hand corner of the *BRANCH START* instruction and click. This causes the heavy line to appear as shown in Figure 8–31.

Select and left click on the N.O. symbol above the program area to add the holding contacts that are in parallel with the *START* Button. As the holding contacts are controlled by the output, which in this case is a motor starter connected to output terminal 0 of the Output module in slot 3, the holding contacts will have the same address as the Motor Starter, O:3.0/0. As before, the program ignores all unnecessary zeros, so the address can be entered as O:3/0. Figure 8–32 shows the screen after the holding contacts have been added and addressed. Notice also that the symbol notation and the description notation added earlier for address O:3.0/0 are automatically added to the holding contacts. The symbol tells what it is, in this case contacts from Motor Starter 1, that controls Exhaust Fan 1.

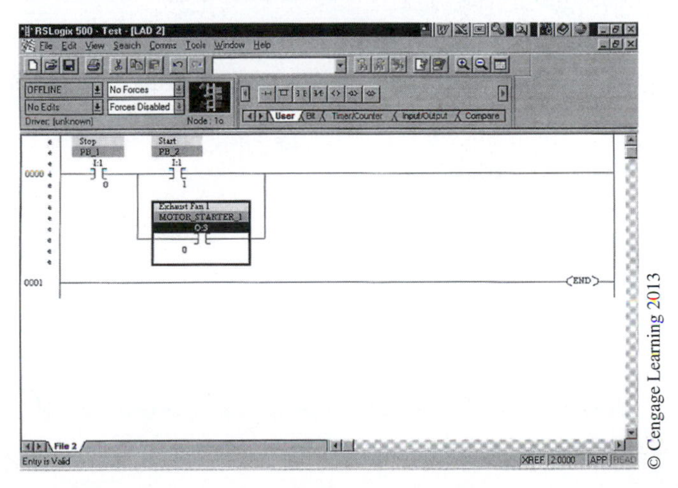

Figure 8–32 Holding Contacts with Address, Symbol, and Description Notation

The last procedure necessary to complete our simple *STOP/START* circuit is to program the output device. The output device in this case is Motor Starter 1. Output address O:3.0/0 was given the symbol notation Motor Starter 1 when the Output Data File was opened earlier. To finish our program, point the mouse arrow at the upper right-hand corner of the parallel contacts and click the left button. This produces a heavy vertical line. Once the line is present, click on the Output symbol (—( )—) button above the programming area.

Figure 8–33 shows the Output symbol with the (?) indicating that the symbol needs an address. Double click the left mouse button on the (?) and enter the assigned address for the output O:3/0. Remember,

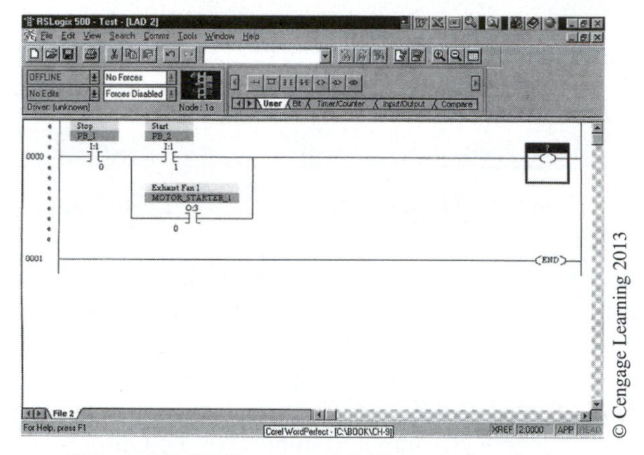

Figure 8–33 Screen with Output Symbol

there is no need to add any unnecessary zeros because the program will not display them. Figure 8–34 shows the completed *STOP/START* circuit with all addresses, symbols, and description notations added.

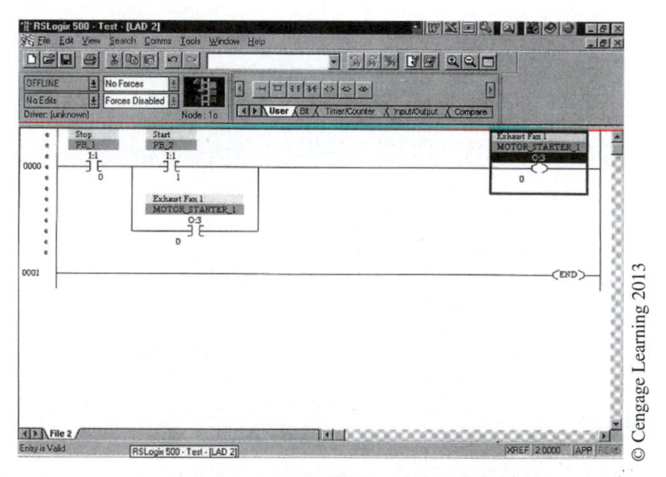

Figure 8–34 Completed *STOP/START* Circuit

The RSLogix software has been designed to check the finished program to make sure that all addresses and instructions are correct. To verify the correctness of the project (*STOP/START* Circuit), click on the *Verify Project* button at the top of the program area. The icon for the Verify Project button is a computer with a check mark. Single click on the icon and the verification process begins. If the project has no errors or faults, the program will place a message across the lower left corner of the screen. The message will read "Verify has completed, no errors found." Had there been errors in the program, the errors would be identified by program rung number. Figure 8–35 shows the screen with the "verify completed" message.

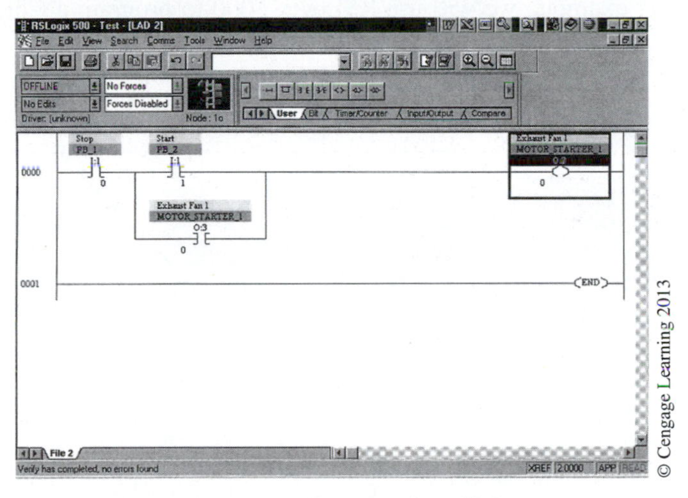

Figure 8–35 "Verify Completed" Screen

Remember that earlier we opened a Controller Properties screen (Figure 8–8) and one of the things noted was that no Memory Used or Memory Left values were given. That was because we had not developed a program yet. Now that a program has been developed, we can open that screen again and see the number of memory words that were used. Figure 8–36 shows the Controller Properties screen and indicates that our program, or project "TEST," used seven Instruction Words and 100 Data Table words. The screen shows there are 12,281 instruction words left.

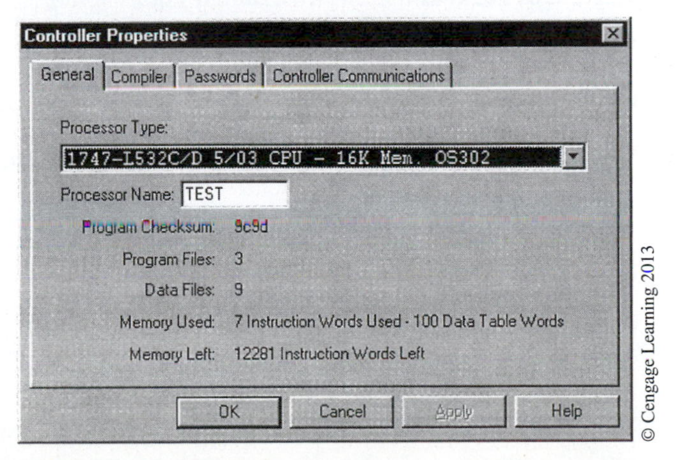

Figure 8–36 Controller Properties Screen Showing Memory Words Used

The seven (7) program words used were for the first contact (*START*), then a *BRANCH START* instruction, another contact (*STOP*), another *BRANCH START,* another contact (holding contacts), a *BRANCH CLOSE* instruction, and finally an *OUTPUT* instruction.

PROGRAMMING WITH THE LOGIX5000 SOFTWARE

For this programming example, we will use RSLogix 5000 to program a ControlLogix 1756-L62 processor mounted in a 4-slot chassis, as shown in Figure 8–37.

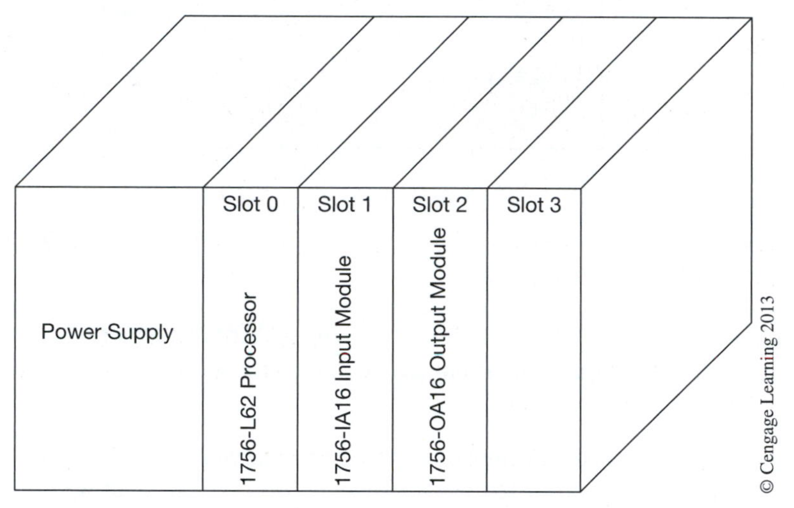

Figure 8–37 ControlLogic Processor Mounted in a 4-Slot Chassis

The circuit to be programmed will be the same simple *STOP/START* circuit that was used in the first example. The ladder diagram of the circuit is shown again in Figure 8–38.

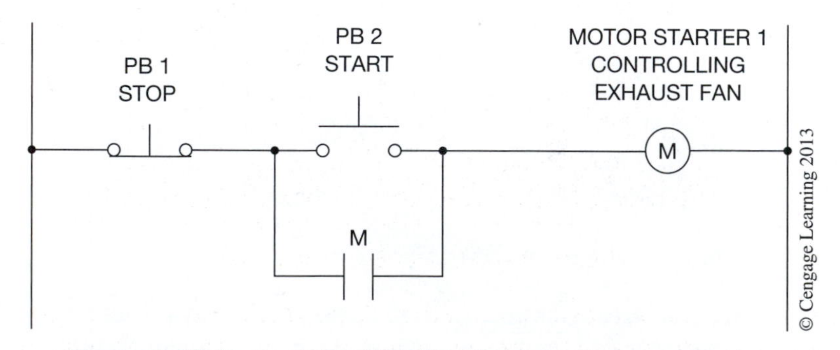

Figure 8–38 Basic *STOP/START* Circuit without Overloads

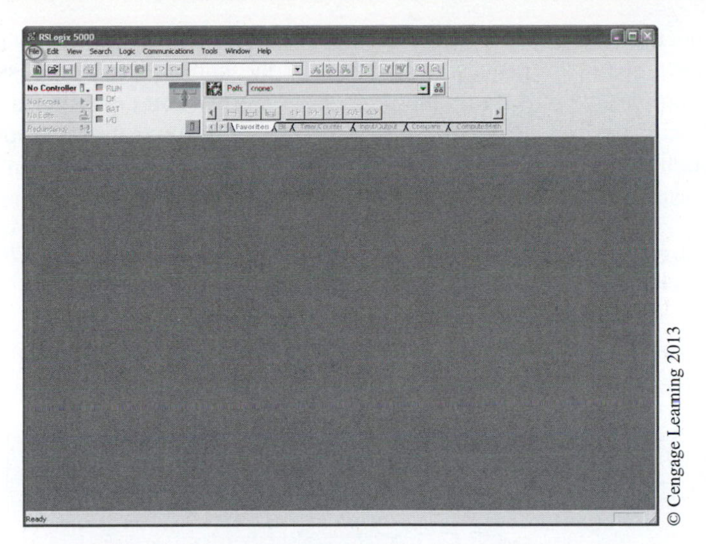

Figure 8–39 RSLogix 5000 Opening Screen

Start the RSLogix software by double clicking the RSLogix 5000 icon. The opening screen for the RSLogix 5000 software is shown in Figure 8–39.

Using the computer mouse, left click on *File* and then click on *New*. The window that appears, titled *New Controller*, lists the processor type, revision, name, description, chassis type, slot, and create in. Figure 8–40 shows the window.

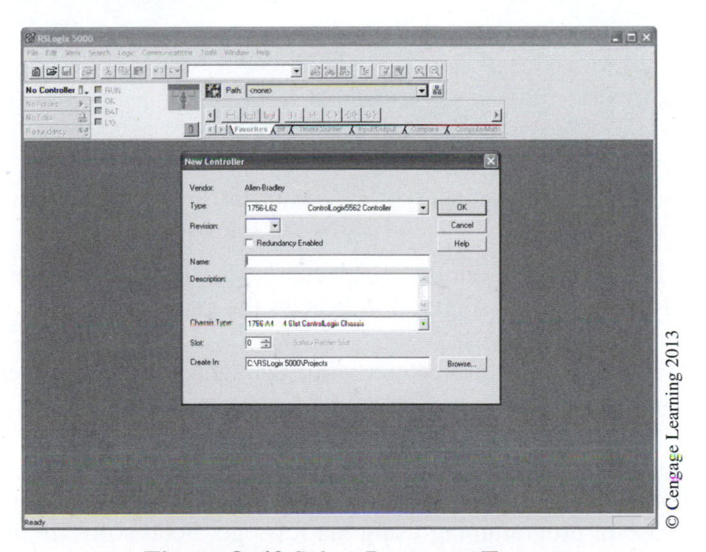

Figure 8–40 Select Processor Type

The first step is to select the processor type, left click the black arrow to the right of the box, and use the pull-down menu to choose the processor you are using. In our example we are using the 1756-L62 processor. After selecting the processor type, select the major revision of the firmware for which this project is configured. Next, enter the processor name in the *Name* box. As mentioned previously, the name given the processor is typically the name of the process that the processor controls. In our example, we will name the processor "Test". You can enter a short description in the *Description* box if desired. From the *Chassis Type* pull-down menu, choose the appropriate chassis type in which the controller will reside. In our example, we have chosen the 1756-A4, 4-slot ControlLogix Chassis. In the *Slot* field, enter the slot number for the controller.

Note: *In ControlLogix, controllers can be placed in any slot. It is also possible to place multiple controllers in the same chassis.*

After entering the processor slot number, enter a directory in which you want to store the project file in the box *Create In*. Figure 8–41 shows our completed example.

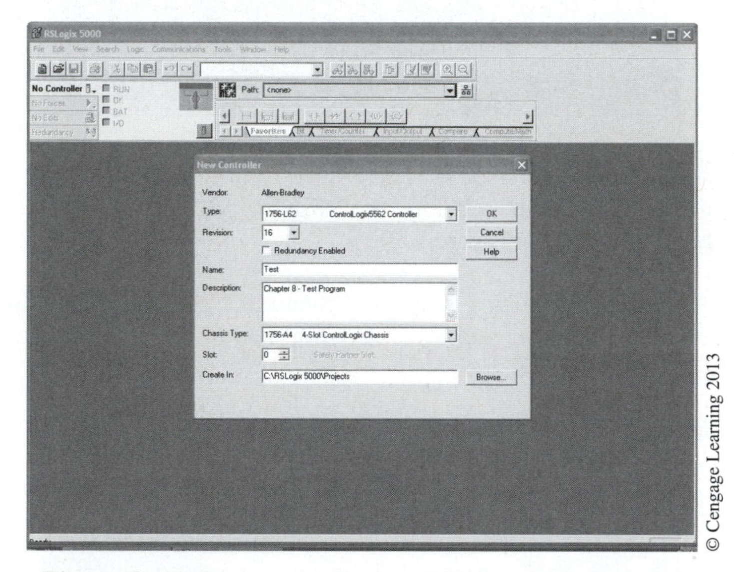

Figure 8–41 Screen with Processor Information Entered

The next step is to click on the *OK* button, which causes the software to enter the information for the processor that has been selected. Figure 8–42 shows the new screen.

The left side of the screen shows the project tree while the right side of the screen is the display area, which displays various screens depending on your selection in the project tree.

Under the project tree, the main folders are *Controller Test, Task, Motion Groups, Data Types, Trends,* and *I/O Configuration*. Not all of these folders will be discussed, since the intent of this section is merely to illustrate ease of programming using the RSLogix 5000 software, not to be a definitive programming guide.

In studying Figure 8–42, you will notice that under the *I/O Configuration* folder there is a picture of a chassis titled "1756 Backplane, 1756-A4". Under the chassis is a picture of a processor that

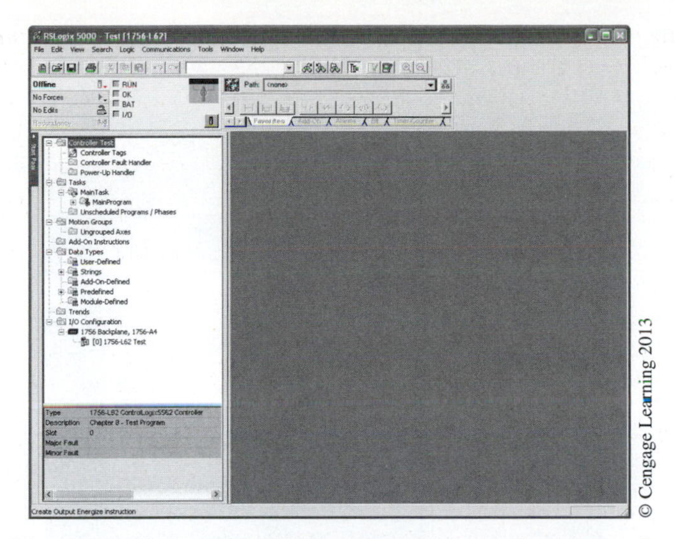

Figure 8–42 Screen after Processor Information Entered

is titled "[0] 1756-L62 Test". As you recall, we configured the processor to reside in slot [0] of the 4-slot chassis and gave it the name "Test".

The next step in the configuration process is to configure the I/O modules that are installed in the chassis with the processor. Reviewing Figure 8–37, you will notice that we showed a 1756-IA16 Input module located in slot 1 and a 1756-OA16 Output module located in slot 2. To enter the correct I/O modules that are installed in slots 1 and 2 of our 4-slot chassis, right click on the picture of a chassis titled "1756 Backplane, 1756-A4" and left click on "*New Module*".Figure 8–43 shows the window after clicking on "*New Module*".

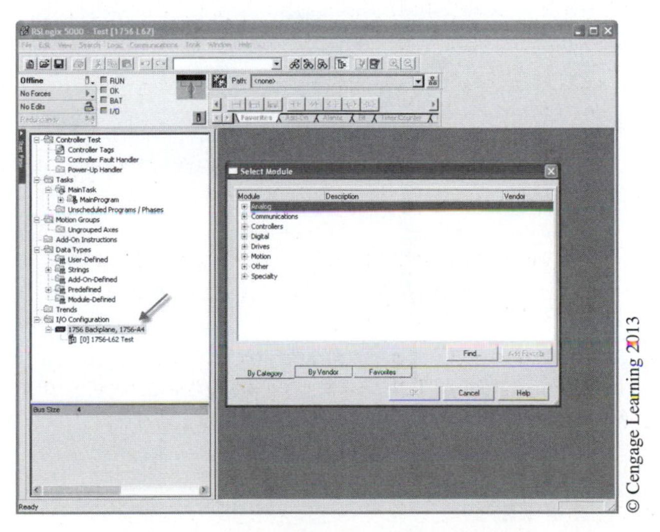

Figure 8–43 I/O Configuration Screen

Now left click on the + sign to the left of where it says "*Digital*" and you should see the screen shown in Figure 8–44.

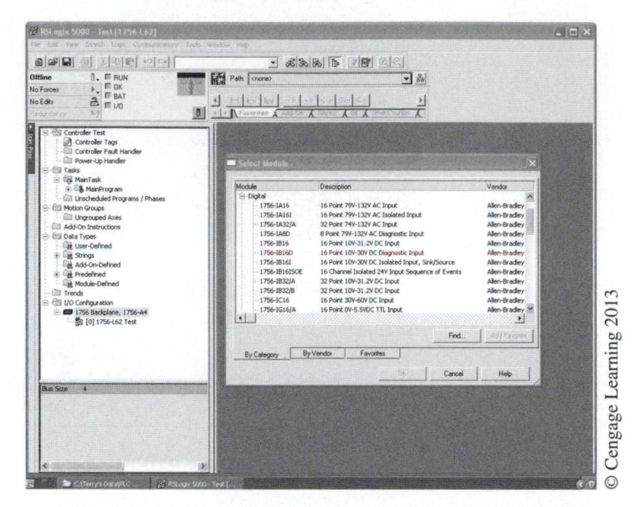

Figure 8–44 I/O Configuration Screen Showing I/O Modules

Using the scroll button on the right side of the window, scroll down until you find the digital input module in slot 1 of our chassis (1756-IA16) and then highlight the module by left clicking on it. Once it is highlighted, left click the *OK* button at the bottom of the screen. If you are creating a module for which more than one major revision is allowed, you will be prompted to choose which revision you would like to create. Simply choose the appropriate revision from the pull-down menu and then click *OK*. If you are not sure which revision to use, then select the highest-numbered revision. After clicking *OK*, you should see the screen shown in Figure 8–45.

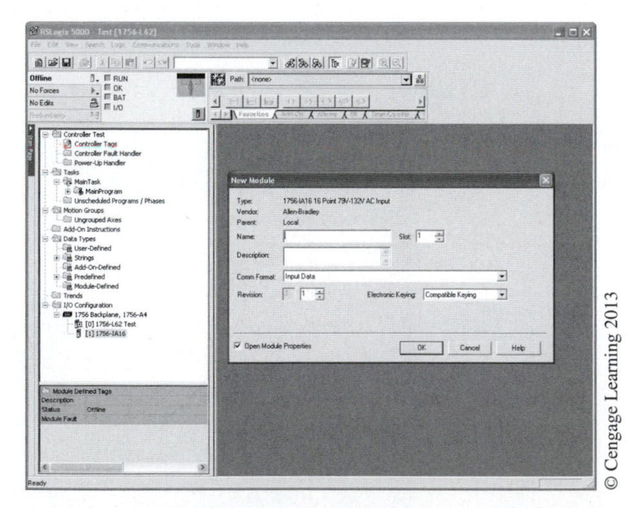

Figure 8–45 I/O Module Configuration Screen

To complete the configuration of the new module, enter the slot number of the module in the box titled *Slot*. The *Name* and *Description* boxes are optional text boxes that you can use to add additional information about the new module. The other boxes can be left in their default settings. Once you are finished, click *OK* and you should see the popup window shown in Figure 8–46. Click *OK* to leave the module properties at their default settings.

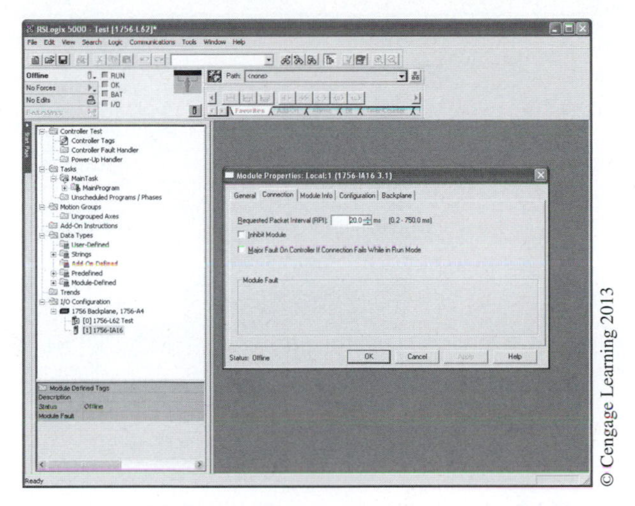

Figure 8–46 I/O Module Properties Dialog Window

Note: *The ControlLogix family of PLCs are very advanced PLCs and have many features that others do not. There are configuration options available so the user can customize the operation of the processor for a given control or process application. This book is not intended to show all of the features or configuration options. One should consult the many programming and reference manuals available from the manufacturer to better understand all of the features.*

After selecting *OK* you should see the 1756-IA16 module located under the picture of the chassis in the left window pane seen in Figure 8–47.

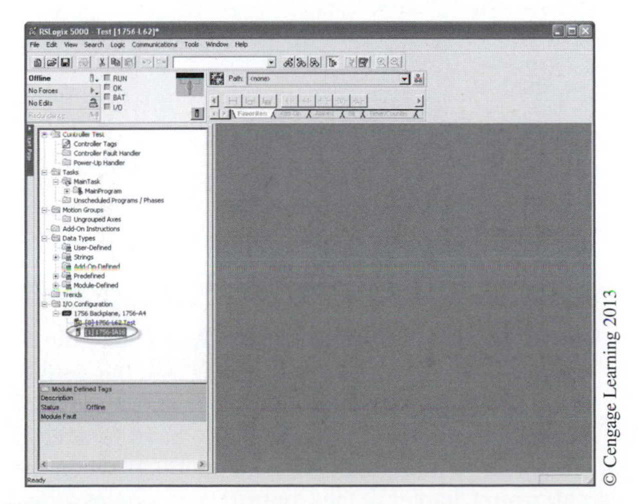

Figure 8–47 Project Tree Showing 1756-IA16 Module

Now, repeat the same process for the 1756-OA16 Output module located in slot 2. When you are finished you should see both I/O modules listed under the 4-slot chassis in the left window pane of the project tree seen in Figure 8–48.

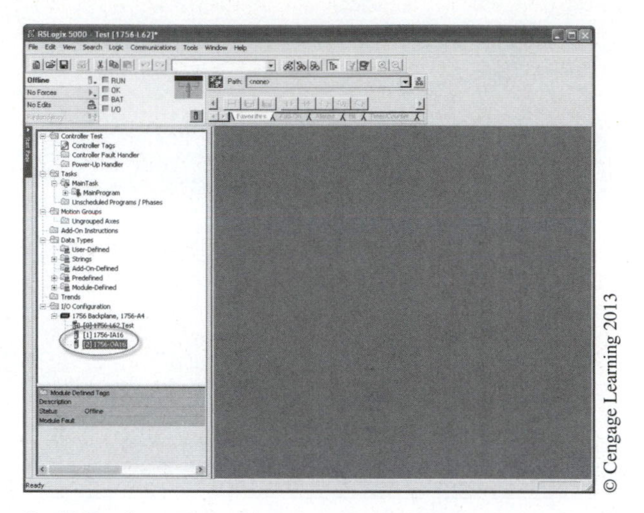

Figure 8–48 Project Tree Showing Both I/O Modules Configured

Open the controller scoped tags window by double clicking on the *Controller Tags* folder in the left window pane. After the controller scoped tags window opens, left click on the + sign to the left of where it says "Local:1I". The "Local" stands for local chassis, the "1" stands for slot 1, and the "I" stands for input. If you recall, we configured slot 1 for the digital input module 1756-IA16, which has 16 digital inputs. Now click on the + sign to the left of "Local:1:I.Data" and you should see the screen in Figure 8–49.

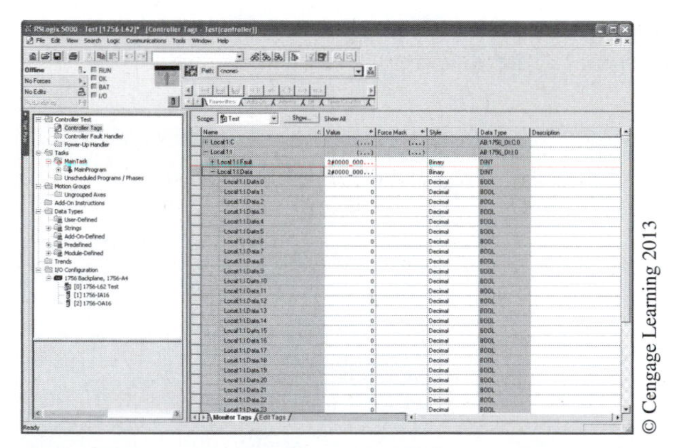

Figure 8–49 Input Module 1756-IA16 Data Points

From this screen you can see all of the input data points (tags) for the 1756-IA16 Input module located in slot 1. After studying Figure 8–49 you probably have noticed that there are more than 16 input data tags. How can there be more than 16 input data tags when the input module only has 16 input points? If you recall from Chapter 4, the ControlLogix processor is a 32-bit word processor and allocates memory in full word increments. Since the input module only has 16 inputs, the first 16 input data tags are used (Data.0…15) and the remainder are unused data tags.

When the processor is in the *On-Line* mode, the actual status of the input devices, 1 or 0, is determined by looking at the bit status in the column titled "*Value*". From this window, each input data tag point can be assigned a description in the "*Description*" column to the right of the input data tag name.

In our example, we will assume that the *STOP* pushbutton has been wired to digital input point 0 (Local:1:I.Data.0) and the *START* pushbutton to input point 1 (Local:1:I.Data.1). In Figure 8–50, we have added the descriptions for each button in the description field for each input tag.

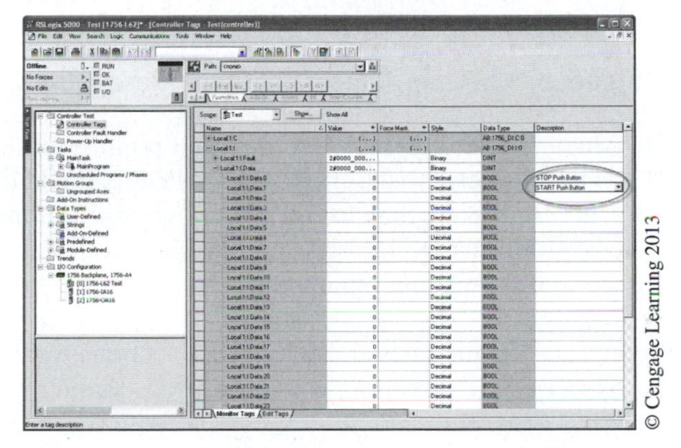

Figure 8–50 Digital Input Tags with Descriptions

Now minimize the input tags for slot 1 by clicking on the (–) sign. The screen should look like Figure 8–51.

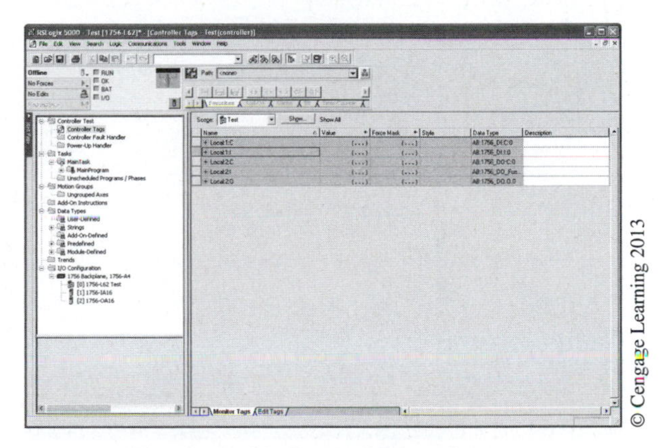

Figure 8–51 Control Tags Screen

In Figure 8–52, we have expanded the output data tags associated with the 1756-OA16 Output module located in slot 2 of the chassis. Since this is an output module, we have expanded "Local:2:O" to reveal the output data tags for each output point on the module.

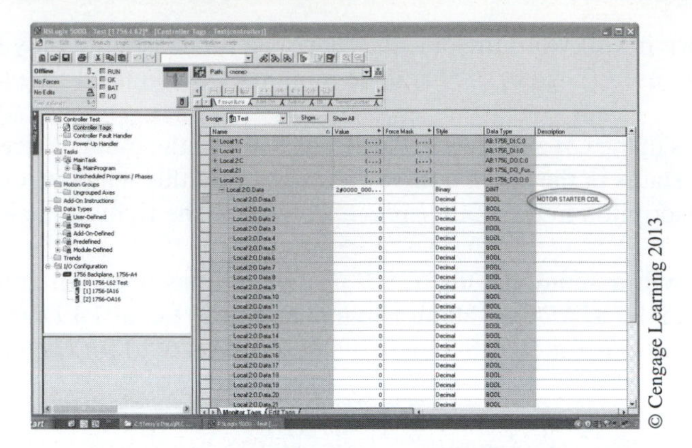

Figure 8–52 Digital Output Tags with Descriptions

We will assume that our Motor Starter coil is wired to output point 0 on the 1756-OA16 Module (Local:2:O.Data.0) and has been given the description "MOTOR STARTER COIL" as shown in Figure 8–52. It is good practice to take the time to add all of your descriptions for the I/O points that are used before starting to program the ladder logic. In this way, your description will already be there when you enter the tag address while programming.

Now that we have configured our I/O modules and entered descriptions for the I/O points to be used, it is time to program our *START/STOP* circuit. If you recall from Chapter 4, the ControlLogix processor divides logic into *Tasks, Programs*, and *Routines*. By default, when you create a new project, a main task, main program, and main routine are automatically created. The *Routine* is where the ladder logic is entered. When you double click on the "*MainRoutine*" icon in the project tree, the screen should look like Figure 8–53.

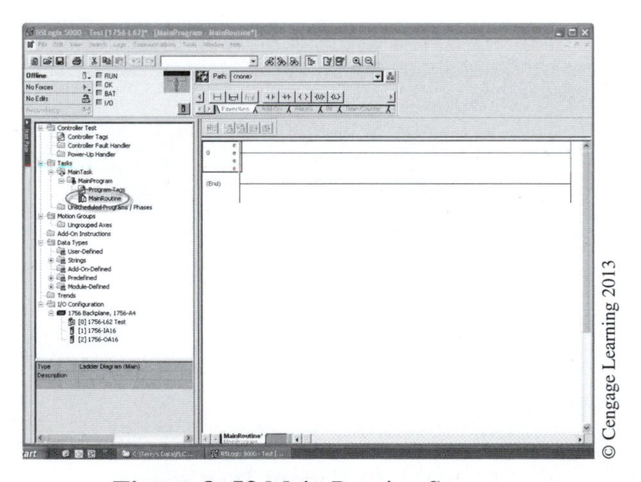

Figure 8–53 Main Routine Screen

Note: *Refer to Chapter 4 and the manufacturer's literature for a detailed description of creating and using Tasks, Programs, and Routines.*

The "*MainRoutine*" ladder logic programming window (Figure 8–53) is where the ladder logic program will be entered for our *STOP/START* circuit. The box in the upper left-hand corner of the programming screen has a zero (0) inside the box indicating that *Rung* 0 is selected. Note that there are four small e's at Rung 0. The e's indicate that this rung is being edited.

Just like in the first example, the first step will be to program the *STOP* button. This is accomplished by single clicking on the *Normally Open* contract symbol (—] [—) in the instruction toolbar located above the programming window. Ladder programming elements can also be dragged from the instruction toolbar to a valid placement location (green target circles). The green target circles indicate where a ladder logic element will be inserted when the mouse button is released. Figure 8–54 shows the screen with the N.O. (—] [—) contact inserted.

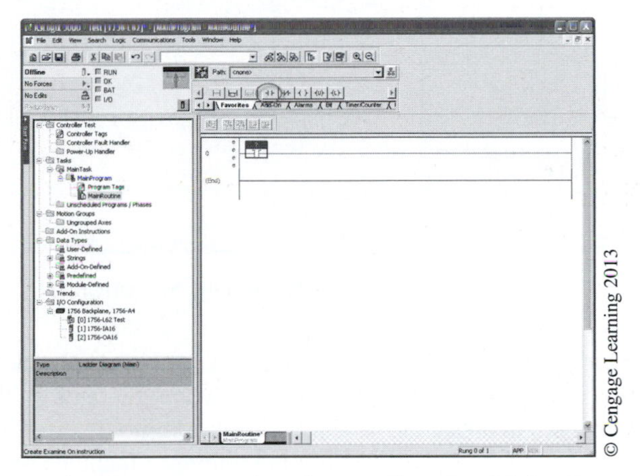

Figure 8–54 Programming Screen Showing N.O. Contact

Double click on the question mark (?) above the contact, enter the tag for the *STOP* button, and then press the *Enter* key. If you recall, the tag for the *STOP* button was Local:1:I.Data.0. Figure 8–55 shows the tag entered above the N.O. contact. Notice that your description "STOP Pushbutton" also appeared above the tag. This is a good indicator that you have entered the tag correctly.

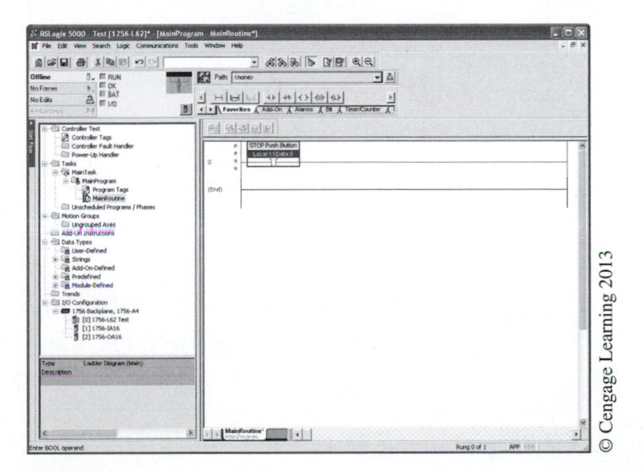

Figure 8–55 Programming Screen with STOP Button Identified

Already created tags can also be selected from a drop-down list in the text box by clicking on the black arrow to the right of the text box. You must then expand the tag structure as necessary to find the tag you wish to select. Figure 8–56 shows the tag structure of the Input module and a bit grid that you can use to select the appropriate Input bit.

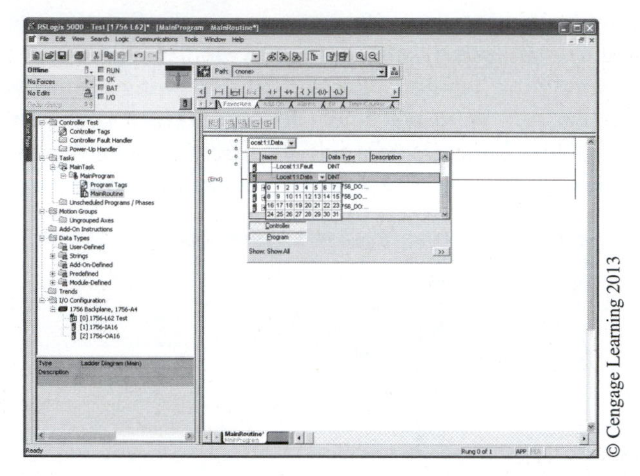

Figure 8–56 Drop-Down Tag Selection

If a tag is not already created, it can be created from the tag text box by simply entering in a name for the tag and pressing the *Enter key*. The tag will be "*Undefined*" until you right click on the tag box, at which time a menu will appear allowing you to create the new tag. This is commonly done with internal memory tags rather than I/O tags. See the manufacturer's programming reference manuals for details on creating tags.

The next step is to add the *START* button. Since the *START* button is in parallel with the holding contacts, single click on the *BRANCH START* symbol button on the toolbar. After the *BRANCH START* symbol is clicked, the screen now appears as shown in Figure 8–57.

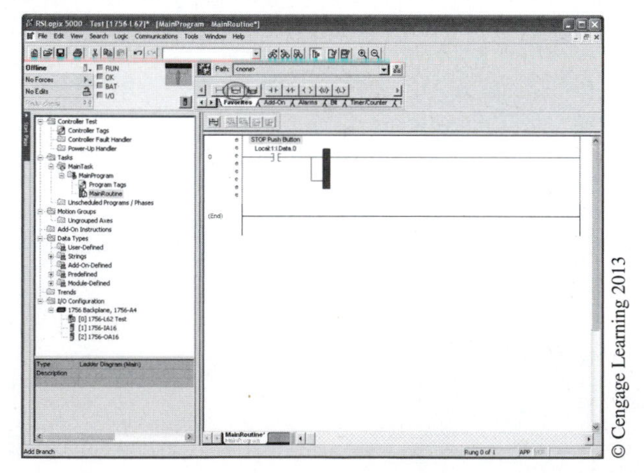

Figure 8–57 Programming Screen with *BRANCH START* Symbol

Notice that a dark vertical line appears on the right side of the *BRANCH START* symbol. It is necessary to move this line to the upper left-hand corner of the symbol before contacts can be added. To move the line, point the arrow of the mouse at the upper left-hand corner of the *BRANCH START* symbol and click. The heavy line now moves to the upper left portion of the symbol, as shown in Figure 8–58.

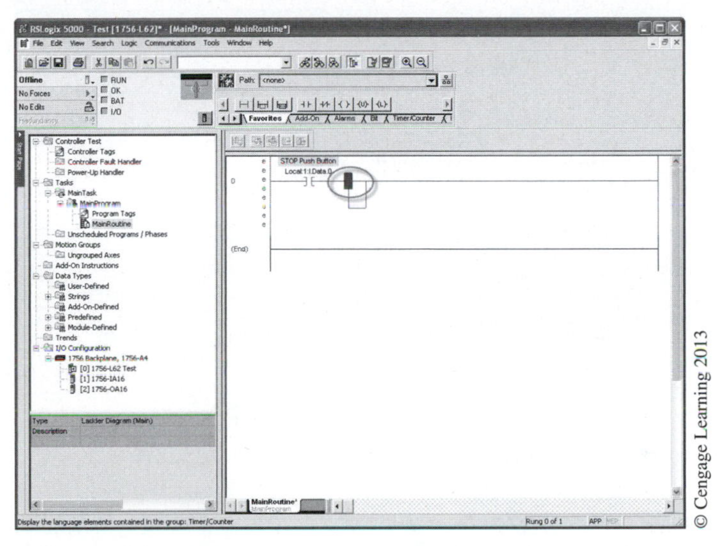

Figure 8–58 Heavy Line Moved to Upper Left-Hand Corner

Next, single click on the N.O. symbol (—] [—) to add the next contact. Figure 8–59 shows the screen after the N.O. contact has been added.

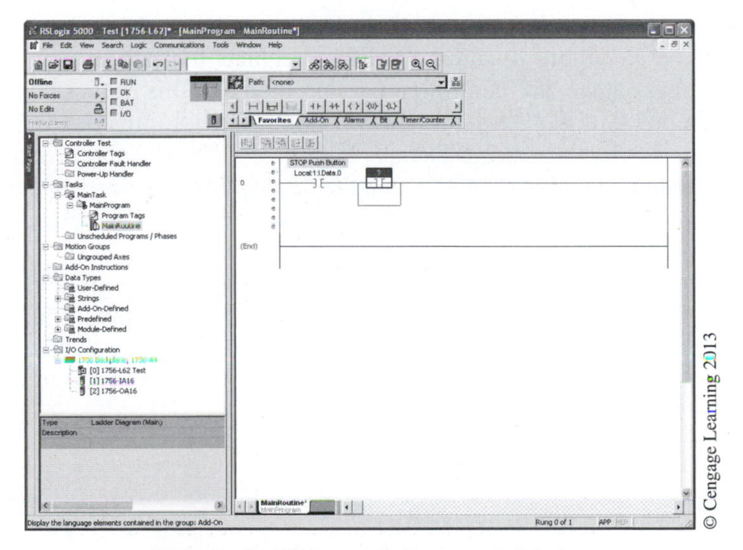

Figure 8–59 Second Contact Added

Enter the tag for the *START* button by double clicking on the question mark (?) above the contact and entering the tag. When finished, press the *Enter* key. If you recall, the tag for the START button was Local:1:I.Data.1. You can also use the drop-down list in the text box. Figure 8–60 shows the screen with the tag entered above the N.O. contact. Notice that your description "START Pushbutton" also appeared above the tag.

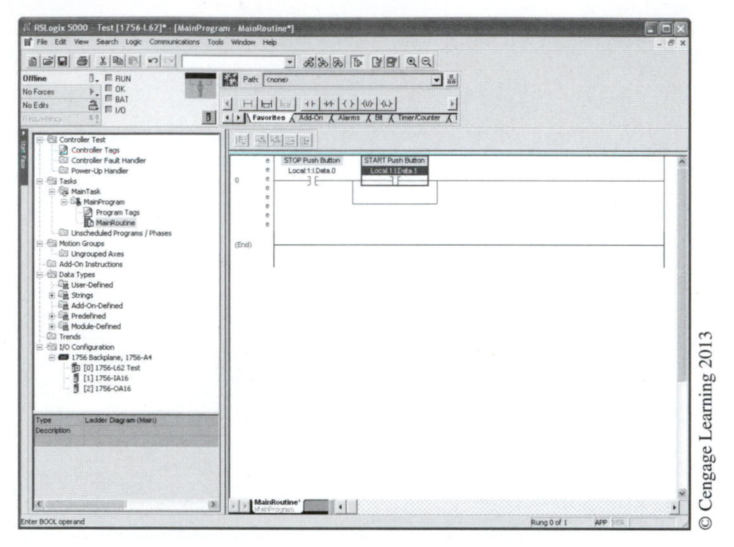

Figure 8–60 Tag Added to Start Button

To add the holding contact from the Motor Starter, point the arrow at the lower left-hand corner of the *BRANCH START* symbol and click. This causes the heavy line to appear in the lower left corner of the symbol.

Select and click the N.O. symbol to add the holding contacts that are in parallel with the *START* button. As the holding contacts are controlled by the output, in this case a motor starter coil connected to output terminal 0 of the Output module in slot 2, the holding contacts will have the same address as the Motor Starter, Local:2:O.Data.0. Figure 8–61 shows the screen after the holding contacts have been added and tagged. Notice that the description you entered earlier appears above the tag for the Motor Starter.

The last procedure necessary to complete our simple *STOP/START* circuit is to program the output device. The output device in this case is the Motor Starter coil. To finish our program, click the upper right-hand corner of the *BRANCH START* symbol. This produces a heavy vertical line. Once the line is present, click on the Output Coil symbol (—( )—) button in the toolbar above the programming window.

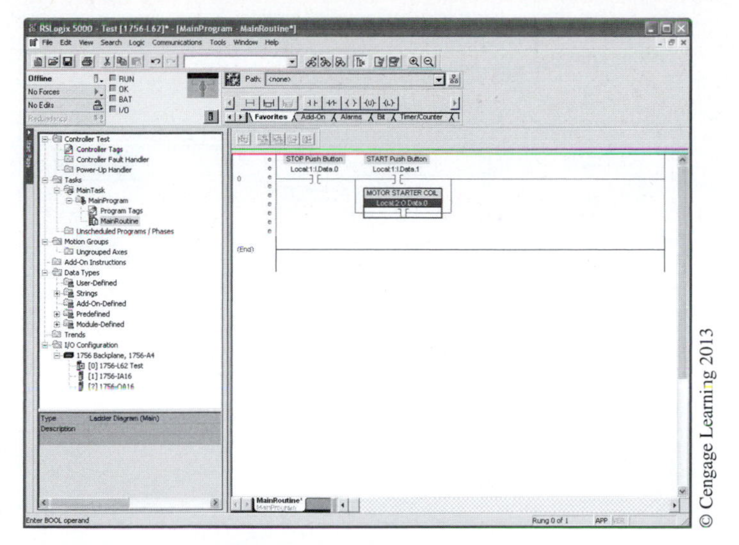

Figure 8–61 Holding Contacts with Tag and Description

Figure 8–62 shows the Output Coil symbol with the (?) indicating that the symbol needs a tag. Double click on the (?) and enter the tag for the Motor Starter coil, Local:2:O.Data.0. Figure 8–63 shows the completed *STOP/START* circuit with all tags and descriptions added.

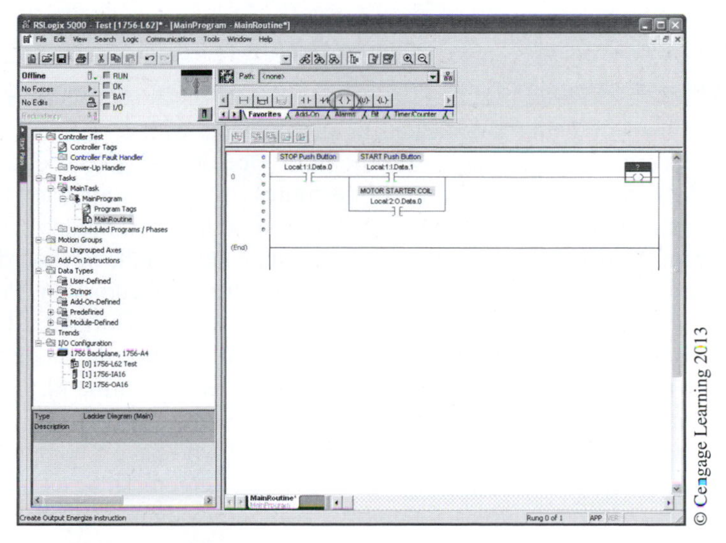

Figure 8–62 Programming Screen with Output Coil

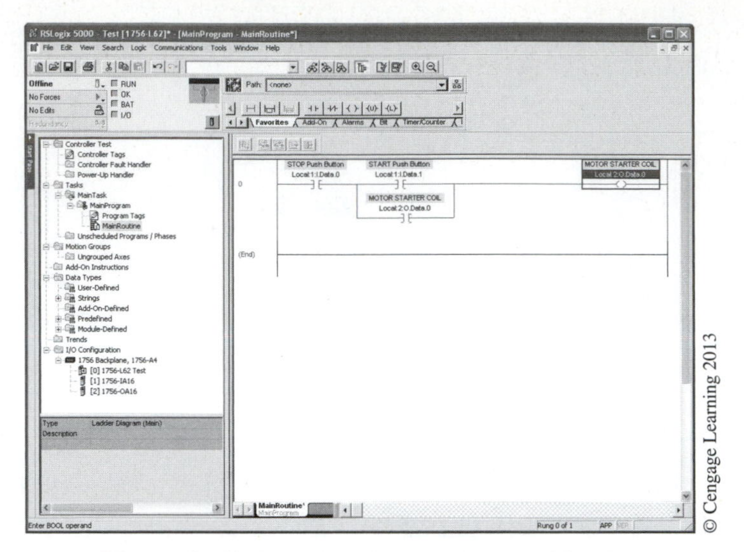

Figure 8–63 Completed *STOP/START* Circuit

If there are no errors, the four small *e*'s to the left of the rung will disappear. The ControlLogix software can check for errors and provide a description by right clicking on the rung number to the left of the finished rung and selecting "*Verify Rung*". If there are any errors, an error window will appear below the programming window describing the errors.

Chapter Summary

The PLC is programmed by using a dedicated programmer or with a personal computer that uses software that has been created to program a specific PLC. Operational (OP) codes are used during the programming to tell the processor what to do, while addresses and tags are used to tell the processor where to do it. The programming device (programmer) is used to enter, modify, and monitor the user program. The program (ladder diagram) is entered by pushing keys on the keyboard in a prescribed sequence so that the results can be displayed on the screen of a personal computer. Programs are usually developed while the PLC is in the Off-Line mode. Once the program is complete, it can be tested in the Off-Line mode to ensure correct operation before it is placed in the On-Line mode for final testing and verification. Keys or passwords are used to prevent unauthorized use of the PLC. From the programming device, contacts and coils can be forced *ON* or *OFF* while the circuit is operational. The FORCE ON, FORCE OFF capability should be restricted to personnel who have a complete understanding of the circuit and the driven equipment. The programming device screen can be used as a troubleshooting aid to test the circuit prior to entry into user memory, or after the circuit is entered into memory and is operational. Contacts and coils are either intensified or displayed in reverse video to indicate true logic or power flow.

Programming a PLC is not difficult, but time must be spent to become familiar with the programming device, the software, and the programming techniques used by the various PLC manufacturers.

Review Questions

1. What does the term *On-Line Programming* mean?
2. What is the function of the cursor?
3. What is the FORCE feature used for?
4. Timers and counters use words of memory, but contacts, coils, and *BRANCH START* instructions do not.
   T F
5. When is *Off-Line Programming* normally used?
6. What are the mnemonic names of the following instructions?
   EXAMINE ON ____________
   NEXT BRANCH ____________
   TIMER ON-DELAY ____________
   BRANCH END ____________
7. The RSLogix software can tell you if the right power supply has been used, based on the number and mix of I/O modules.
   T F
8. Describe briefly the shorthand method of programming using the RSLogix software.
9. ControlLogix processors use what in place of addresses as found on RSLogix processors?
10. In a ControlLogix processor, ladder logic is entered in:
    Tasks, Routines, or Programs

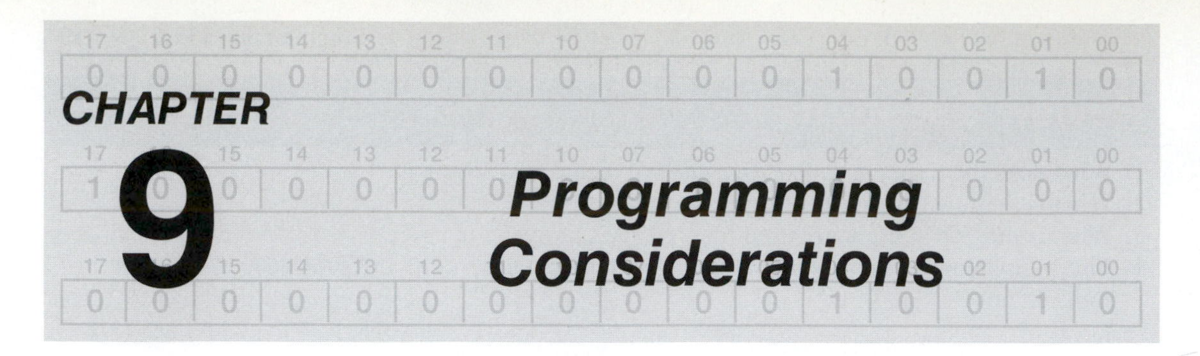

CHAPTER 9 Programming Considerations

Objectives

After completing this chapter, you should have the knowledge to:

- Define a *network.*
- Describe the term *dummy relay*.
- Understand the horizontal and vertical contact limits.
- Define the term *nesting*.
- Correctly wire and program *STOP* buttons.
- Describe the difference between logical and discrete holding contacts.

NETWORK LIMITATIONS

A **network** is defined as a group of connected logic elements used to perform a specific function. Figure 9–1 shows a typical network consisting of seven series contacts and three parallel branches. A network also constitutes one rung of a ladder diagram that starts at the left rail and ends at the right rail.

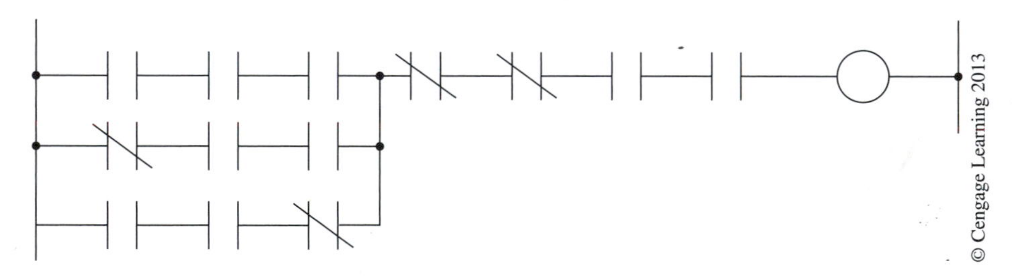

Figure 9–1 Network (Rung)

Some PLC manufacturers have virtually no network limitations, whereas other PLC systems are limited by the number of contacts or other logic symbols that can be included on the horizontal line of a network, and the parallel branches (lines) that make up one network. A typical PLC network limitation of ten series contacts per line and seven parallel lines, or branches, is shown in Figure 9–2. Additionally, some PLCs are further limited because they only allot one output per rung or network, and the output must be on the first line.

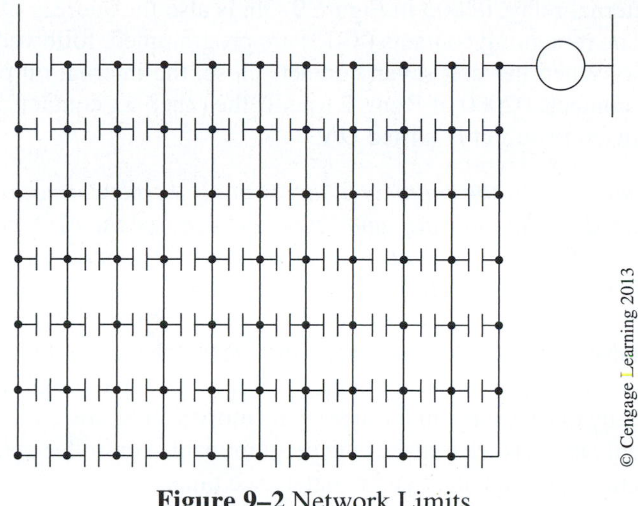

Figure 9–2 Network Limits

Note: *Special functions and other logic symbols alter the network limitations and requirements; check the operation or program manual of the specific PLC for additional information.*

While the number of elements and lines within a network is limited, only the size of user memory limits the number of networks or rungs.

When a circuit requires more series contacts than the network allows (Figure 9–3a), the contacts are split into two rungs (Figure 9–3b). The first rung contains part of the required contacts and is programmed to an internal, or "dummy," relay. Internal relays are actually a bit and word location in storage memory.

EXCEEDS HORIZONTAL LIMIT
DISCRETE OUTPUT
1 2 3 4 5 6 7 8 9 10 11 12 13

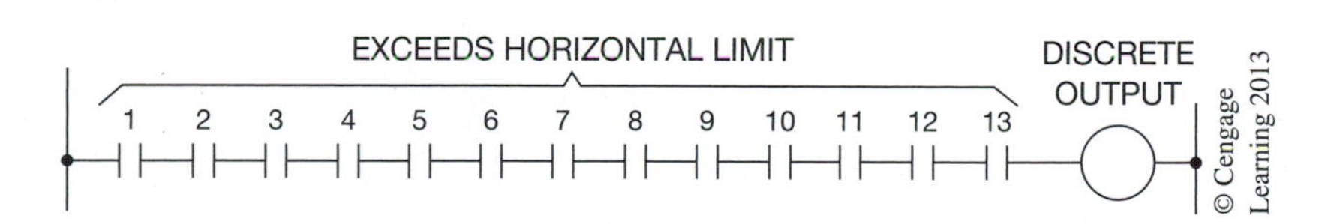

Figure 9–3a Contacts Exceed Horizontal Limit

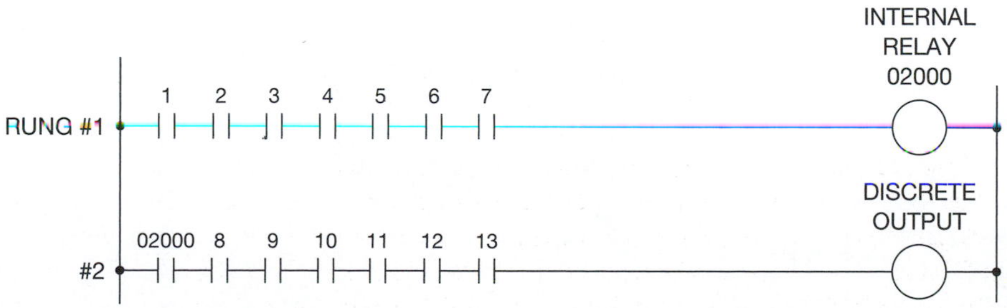

INTERNAL RELAY 02000
RUNG #1
1 2 3 4 5 6 7
DISCRETE OUTPUT
#2
02000 8 9 10 11 12 13

Figure 9–3b Contacts Split into Two Rungs

The address of the internal relay, 02000 in Figure 9–3b, is also the address of the first N.O. contact on the second rung. The remaining contacts (8–13) are programmed, followed by the address of the discrete output device. When the first seven contacts close, the internal output, 02000, is set to 1. This makes the N.O. contacts 02000 of Rung 2 true. If the other six contacts (8–13) are closed, the rung is true, and the discrete output is turned *ON*.

Note: *It is not necessary to split the contacts in any ratio. If the network allows 10 horizontal contacts, 10 could be placed on the top rung, and three could follow the N.O. contacts of the internal relay (02000) on Rung 2. This technique applies not only to N.O. contacts, but to N.C. (or combinations of N.O. and N.C.) contacts as well.*

The internal relay just used does not exist as a real-world device that has to be hardwired, but is merely a bit in the storage memory that performs the logic of a relay. In actual programming, internal control relays that do not actually exist, except in the storage memory as bits, are extensively used. The use of these internal relay equivalents is what makes the programmable controller unique, eliminating hours of hardwiring and shortening installation and maintenance time.

When a program requires more parallel branches than the network allows, the circuit can be split into two networks, or rungs. The first six parallel contacts are programmed to an internal or dummy relay, as shown in Figure 9–4. A contact with the same address as the internal, or dummy, relay is then programmed in parallel with the remaining contacts to control the output.

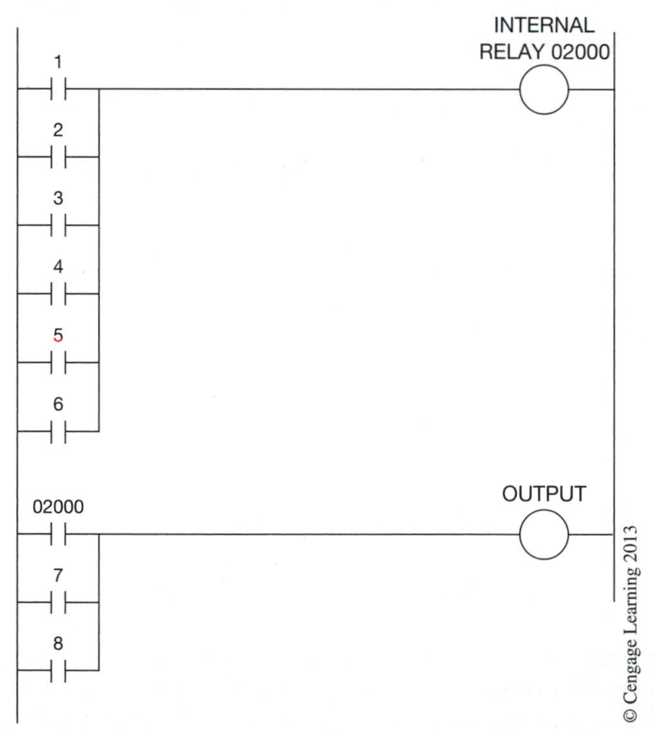

Figure 9–4 Parallel Contacts Split into Two Rungs (Networks)

Sometimes it is not the PLC that limits the number of horizontal or parallel contacts, but the display capability of the monitor. The screen may only be able to display 10 horizontal contacts, even though the PLC can be programmed with an unlimited number of horizontal contacts. While the processor may allow more contacts to be programmed than the monitor can display, it is not a good idea to do so. If the contacts cannot be seen on the screen, it is difficult to troubleshoot the circuit later. The ability to view each contact on the monitor and to know the status (*ON* or *OFF*) of each contact, as well as the status of the output devices, is what makes the PLC such a powerful tool.

Note: *When using a computer as a programming device, most software packages allow the electrician or technician to "shift" the screen to monitor all the instructions, even when the normal screen is filled with instructions.*

Many PLCs have networks that allow for more than one output (parallel outputs) (Figure 9–5). With this parallel output configuration, all of the outputs are *ON* or *OFF* at the same time, based on the network logic.

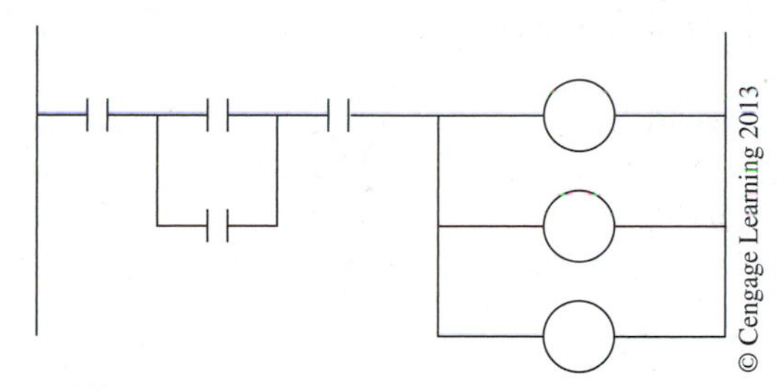

Figure 9–5 Parallel Output Format

Other PLCs allow multiple outputs that can be *ON* or *OFF* at different times, depending on the network logic (illustrated in Figure 9–6).

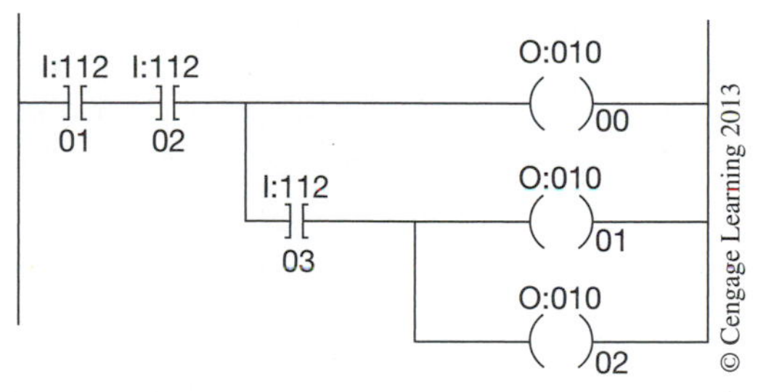

Figure 9–6 Multiple Outputs

PROGRAMMING RESTRICTIONS

In addition to the number of horizontal contacts on one line, and the number of lines in a network or rung, the PLC does not allow for programming vertical contacts (Figure 9–7). In the real world, one could wire the circuit as shown in the figure, but programming restrictions would not allow the PLC to be programmed in this manner.

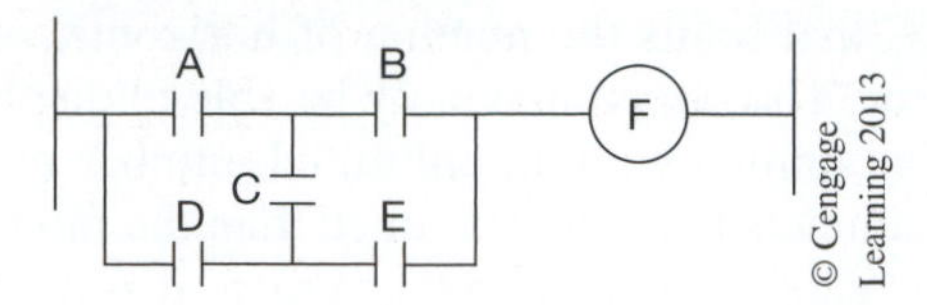

Figure 9–7 Vertical Contacts

If one analyzes the logic of the circuit in Figure 9–7, the circuit logic shows that output F can be energized by any of the following contact combinations: A, B (Figure 9–8a); A, C, E (Figure 9–8b); D, C, B (Figure 9–8c); and D, E (Figure 9–8d).

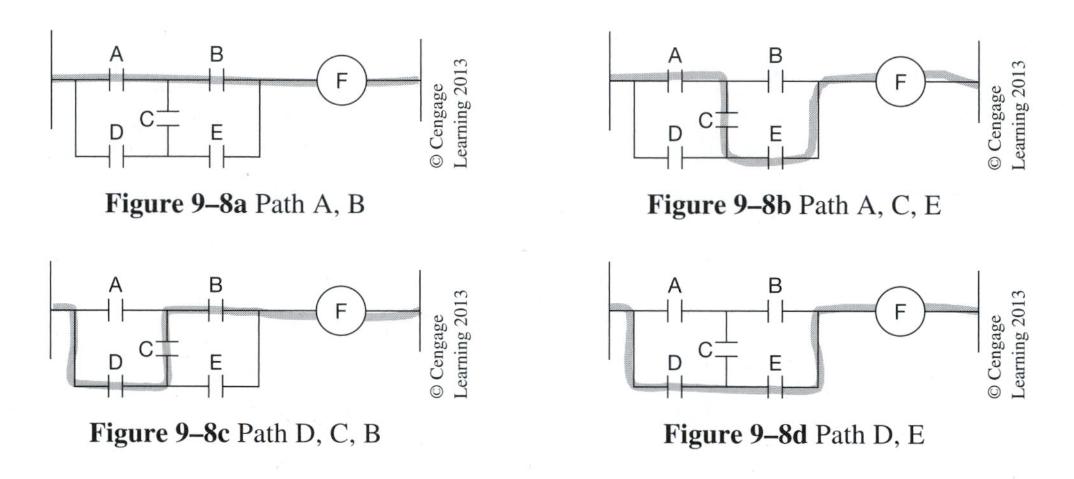

Figure 9–8a Path A, B

Figure 9–8b Path A, C, E

Figure 9–8c Path D, C, B

Figure 9–8d Path D, E

To duplicate the logic, the circuit could be programmed as shown in Figure 9–9.

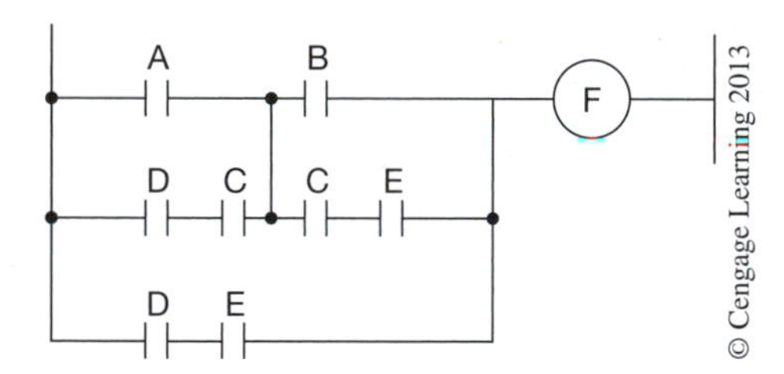

Figure 9–9 Equivalent Circuit without Vertical Contacts

This circuit maintains the circuit logic. Contact combinations A, B; A, C, E; D, C, B; and D, E all energize output F.

Another limitation to circuit programming is the way in which the processor considers power flow, or logic continuity, when it scans a rung of logic. Flow is from left to right *only,* and vertically *up* or *down*. The processor *never* allows logic continuity (power flow) from right to left.

Normally, relay logic for the circuit shown in Figure 9–10 would indicate the following possible contact combinations to energize output G: A, B, C; A, D, E; F, E; and F, D, B, C.

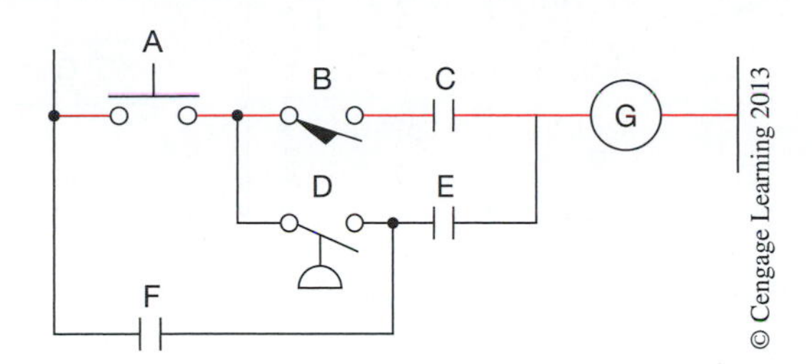

Figure 9–10 Hardwired Circuit

If the circuit shown in Figure 9–10 was programmed into user memory as shown in Figure 9–11a, the processor would ignore contact combination F, D, B, C because it would require power flow (logic continuity) from right to left. If combination F, D, B, C was required, the circuit would be reprogrammed as shown in Figure 9–11b.

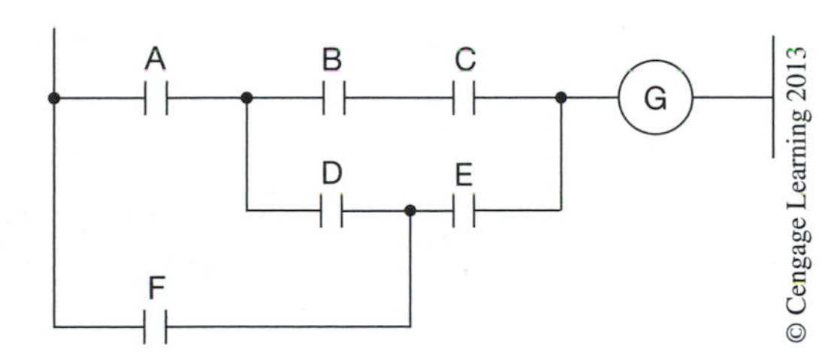

Figure 9–11a Circuit Improperly Programmed

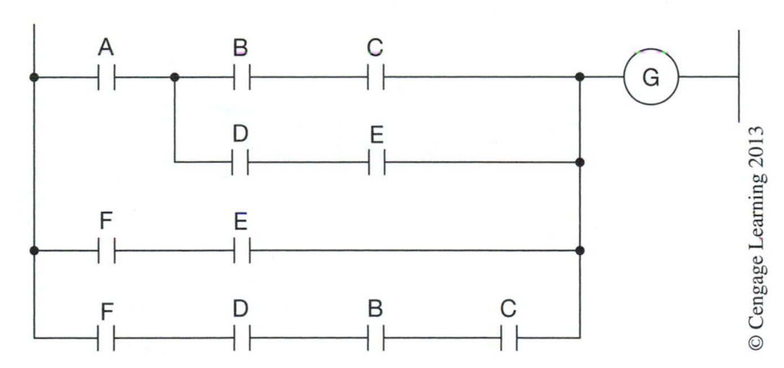

Figure 9–11b Circuit Properly Programmed

The last restriction placed on the programming of circuits into user memory by some—*but not all*—PLCs is the use of "a branch circuit within a branch circuit," or the **nesting** of contacts. Figure 9–12a is an example of a circuit that has nested contacts (L and G) or "a branch within a branch." To obtain the required logic, the circuit is programmed as shown in Figure 9–12b. The duplication of contacts J and K eliminates the nested contacts L and G.

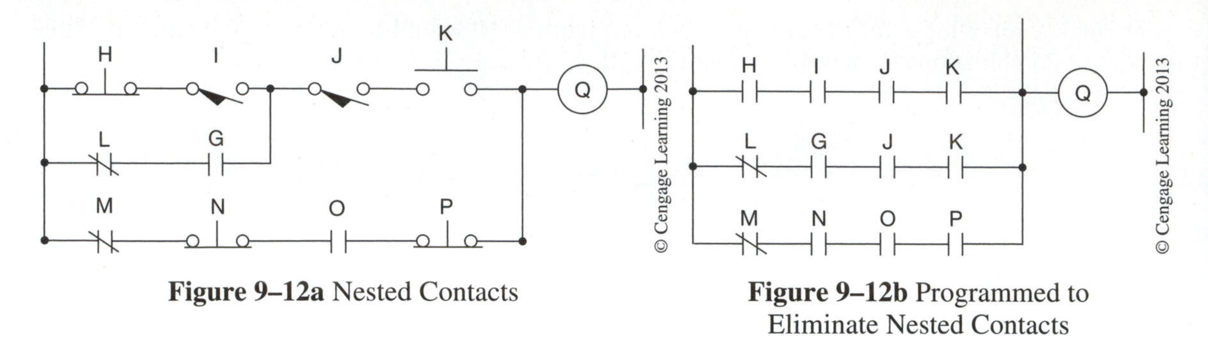

Figure 9–12a Nested Contacts

Figure 9–12b Programmed to Eliminate Nested Contacts

Figures 9–13a and 9–13b are other examples of circuits with "a branch within a branch" (in this case "branches within a branch"), and how the circuits are programmed to maintain circuit logic.

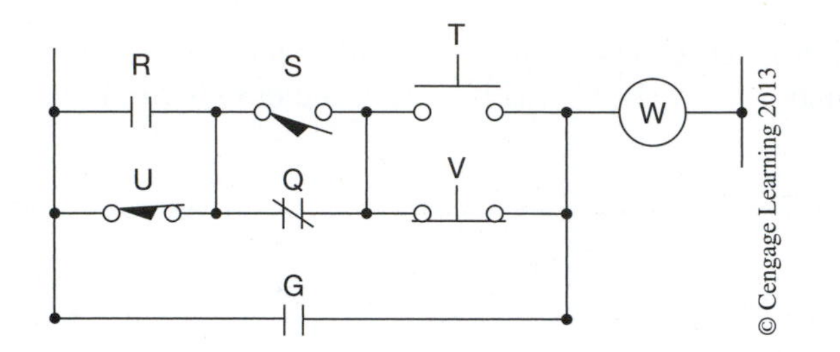

Figure 9–13a Branches within a Branch

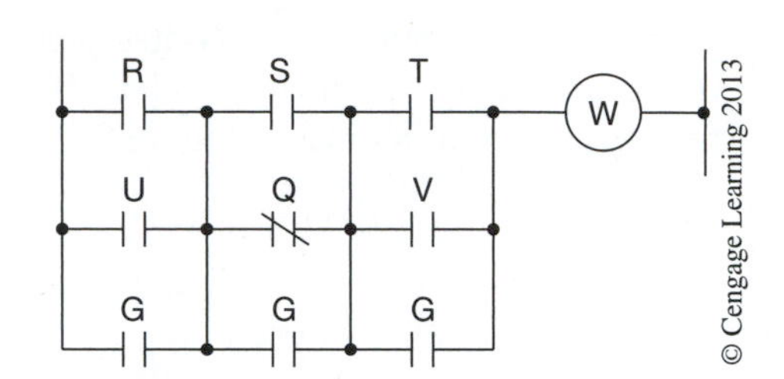

Figure 9–13b Programmed to Eliminate Branches within a Branch

The easiest way to avoid nesting is to remember that all branches must start at a common point, and all of the branches must end at the same location. Figure 9–13b is actually three parallel branches in series as illustrated in Figure 9–14.

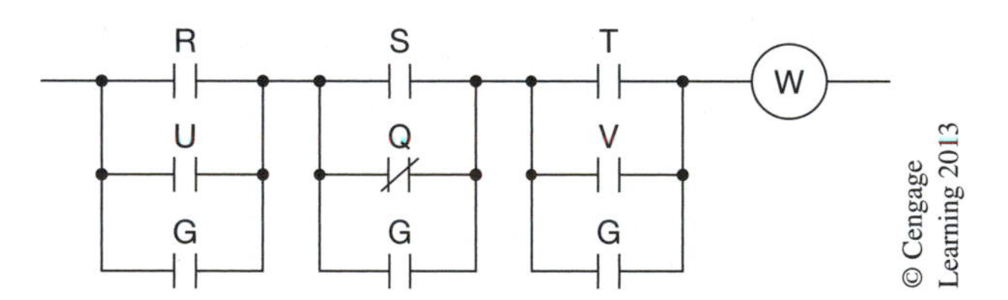

Figure 9–14 Parallel Series Combination

PROGRAM SCANNING

As discussed in Chapter 3, the processor first determines the status of the input devices, then it scans the user program, and then updates (turns *ON* or *OFF*) the outputs. The way the processor scans the program varies from PLC to PLC. One common method is to scan the program from left to right and top to bottom, similar to the way in which a book is read. In this method, the processor scans the first rung of the program from left to right, then the second rung from left to right, and continues in this fashion until all the rungs have been scanned. In the next scan, the processor returns to the first rung and starts all over again, scanning each rung in order from top to bottom. Figure 9–15 illustrates the order of scanning for a processor that scans from left to right and top to bottom. This is the scanning method used by Allen-Bradley for their family of PLCs.

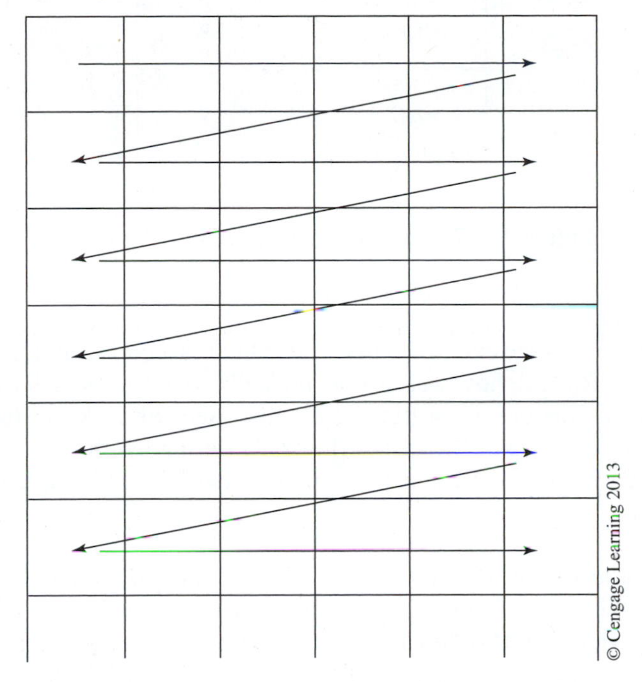

Figure 9–15 Processor Scan Left to Right, Top to Bottom

Look at the circuit shown in Figure 9–16. If S1 is closed, or true, the logic of Rung 1 is true, making the logic of Rung 2 true, which in turn makes the logic of Rung 3 true. Lamps 1, 2, and 3 are turned *ON* at the end of the first scan.

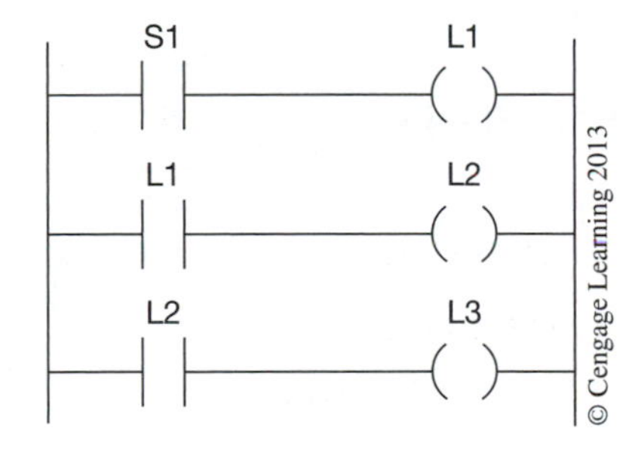

Figure 9–16 One Scan Turns On L3

If, however, the circuit was programmed as shown in Figure 9–17, L3 would *not* turn *ON* until the third scan was completed.

In the first scan, Rung 1 is not yet logically true because the processor does not know the status of L2. Rung 2 would also not have logic continuity because the processor does not yet know the status of L1.

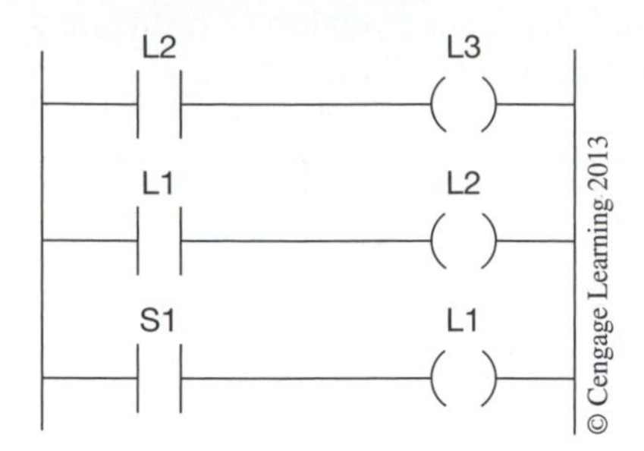

Figure 9–17 Three Scans Required to Turn On L3

When the processor scans Rung 3, however, it now is logically true because S1 is closed. The processor, therefore, turns *ON* L1 at the end of the first scan. On the second scan, L2 is still false, so Rung 1 has no logic continuity. Rung 2, however, is now logically true because L1 was turned *ON* after the first scan, and L1 contacts are closed, or true. S1 is still closed, which keeps the third rung true, so at the end of the second scan, L1 and L2 are *ON*. It is only during the third scan that Rung 1 becomes true. With L2 now *ON,* the logic of Rung 1 is complete and L3 will be turned *ON* at the end of the third scan.

Each scan took only a matter of milliseconds (msecs) to complete, so the delay in turning on L3 is not perceptible to the naked eye. However, in certain high-speed processes, the time lost due to poor programming may be significant and must be considered.

Another example of how the processor scans is illustrated by duplicating output addresses. If the same output address is inadvertently used twice in one program, the last rung in which the address is used will indicate the status of the output. For example, if the address is used in Rung 5 and the logic of Rung 5 is true (which would tell the processor to turn *ON* the output), and the address is used again in Rung 13 and the logic of Rung 13 is false, the output will not be turned *ON* because the processor scanned Rung 13 last, and the last state of the output (true or false) will be based on the last rung scanned.

PROGRAMMING *STOP* BUTTONS

During the early days of programmable controllers, it was common for salespeople to use a demonstrator model that had the *STOP* buttons wired N.O. This technique was used so the salespeople did not have to explain why a *STOP* button was shown as N.O. in a PLC program.

Figure 9–18a shows a standard *STOP/START* station ladder diagram, and Figure 9–18b shows the equivalent diagram used by some PLC salespeople.

From an understanding of how a *STOP/START* circuit works, and an understanding of the EXAMINE ON and EXAMINE OFF instructions, it is easy to see that the only way the circuit could be logically true would be for the *STOP* button to be wired open. By using an N.O. *STOP* button, the EXAMINE OFF instruction is true and the circuit energizes when the *START* button is pressed. The problem is that once the circuit is energized, the only way it can be stopped, or turned *OFF,* is if the *STOP* button is pushed and the contacts close. While this circuit will work, there is a built-in *danger* that must be considered.

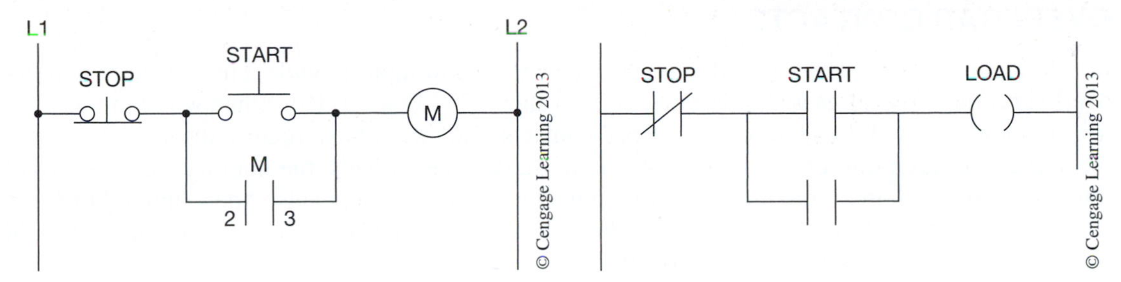

Figure 9–18a Standard *STOP/START* Ladder Diagram

Figure 9–18b PLC Programmed *STOP/START* Circuit

If a *STOP* button is wired in an N.O. position, the switch is impossible to close if it becomes jammed, and it would be impossible to deenergize the circuit. Similarly, if a wire breaks on the *STOP* button, it is possible to complete the logic of the circuit and energize the equipment, but *impossible* to deenergize the equipment. With the wire broken, changes in the status of the *STOP* button cannot be conveyed to the processor, and the circuit and/or equipment cannot be deenergized.

Safety Note: *All* STOP *buttons must be wired so that a failure of the switch or a broken wire will automatically break logic continuity and turn the circuit* OFF. *A good programmer will always wire the devices and program the circuit so that if the real-world device fails, it creates a safe condition, not a safety hazard.*

This practice of wiring *STOP* buttons in the N.O. position was common during the 1980s. If an electrician or technician finds equipment wired in this fashion, he or she should change the *STOP* buttons to N.C. and the PLC program from EXAMINE OFF to EXAMINE ON.

The *START* button should be wired N.O. and programmed with an EXAMINE ON (XIC) instruction. As a general rule, all input devices, except *STOP* buttons, are wired N.O. and given EXAMINE ON instructions in the program.

LOGICAL HOLDING INSTRUCTIONS

In previous programming examples we have used the output address to address the holding contacts. This method of providing holding logic works well in many applications and eliminates the need to actually wire the holding contacts on the motor starter. When the output address is also used for the holding contacts (logic), the circuit is maintained logically because the output point has been turned *ON*. This is no guarantee, however, that the actual motor starter connected to the output module has been energized.

DISCRETE HOLDING CONTACTS

The only real way to know that the starter has energized is to actually wire the holding contacts of the motor starter to an input module point, and use that address when programming the holding contacts or when programming motor fault logic. This method has many advantages, and, short of installing a motor sensor, is the best way to verify that the motor starter has been energized.

OVERLOAD CONTACTS

It is common practice *not* to wire the overload contacts to an input module, but instead to wire the overload contacts in series with the starter coil (as shown in Figure 9–19). When wired in this manner, a motor overload that opens the overload contacts also opens the circuit to the starter coil, and the starter will drop out, or deenergize. When the starter deenergizes, the holding contacts (which must be wired to an input module and programmed in the PLC circuit) also open, and the PLC circuit loses logic, which, in turn, turns *OFF* the point on the output module that is connected to the starter coil. This arrangement only works if the holding contacts are wired to an input module and programmed into the PLC program.

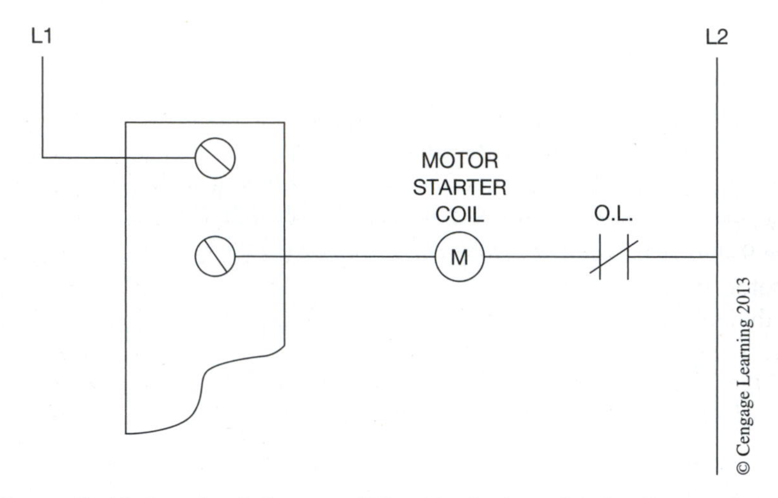

Figure 9–19 Overload Contacts Wired in Series with the Starter Coil

If the holding contacts are not wired to an input module, but instead the output address is used for holding logic, the overloads trip and interrupt the circuit to the starter coil and the starter deenergizes, but the PLC logic is *not* broken. Without the holding contact to open the logic in the PLC program, the output module point would remain *ON,* even though the starter coil circuit is open. This wiring scheme can cause a safety hazard. With the logic remaining true, the motor restarts automatically when the overload is reset.

Some applications require the addition of a hardwired backup control circuit for PLC-controlled motors. These hardwired backup control circuits provide a means to control a motor or other output independent of the PLC, if for any reason the PLC is unable to control its outputs. Many of these hardwired backup control systems can be found in critical control systems such as water and wastewater systems, environmental systems, emergency power, cooling systems, etc. When a hardwired backup control circuit is used in conjunction with a PLC to control a motor, then the overload contact on the motor starter must be wired in series with the motor coil to provide protection for the motor under both means of control, PLC *and* hardwired. Figure 9–20 shows a motor controlled from a PLC *and* hardwired control circuit.

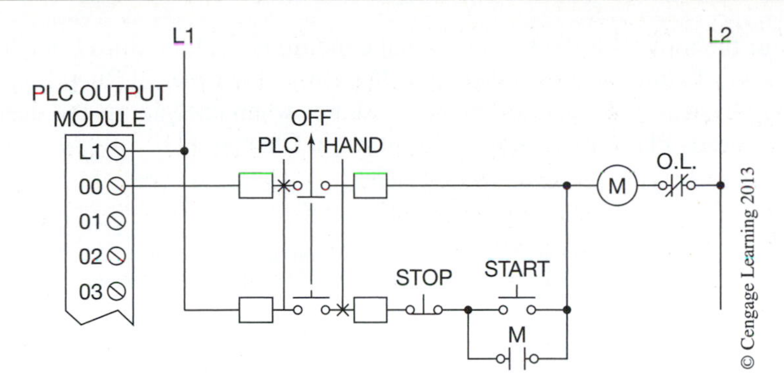

Figure 9–20 PLC and Hardwired Controlled Motor

By applying common sense and good PLC programming practices, many of the previously mentioned problems concerning overload contacts can be solved. One method that the author and many control engineers use to solve these problems and provide additional benefits is as follows:

Step 1. Hardwired the overload contacts in series with the motor starter coil, as was shown in Figure 9–19. This will protect the motor regardless of the method being used to control the motor, PLC *or* hardwired, and will not depend upon software for motor protection.

Step 2. Wire the holding contacts to an input module (as shown in Figure 9–21). The holding contact input can be used to monitor the motor starter for two abnormal conditions: failure of the motor starter to pull in when the output to the starter coil is *ON,* and failure of the motor starter to drop out when the output to the starter coil is *OFF or* deenergized.

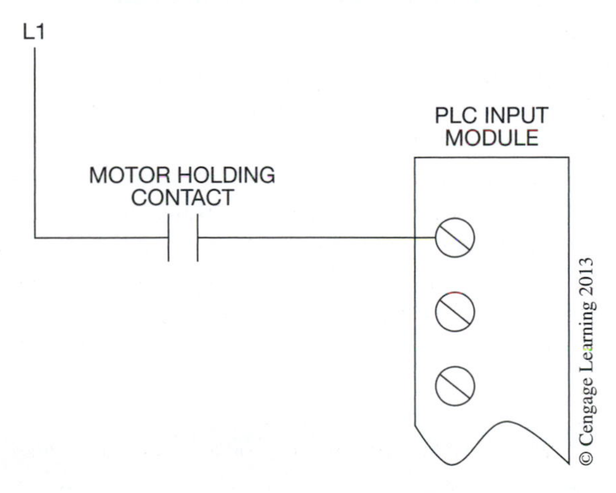

Figure 9–21 Motor Holding Contacts Wired to Input Module

Step 3. Write a rung of PLC logic that will monitor the motor starter for either of the two abnormal conditions described above, failure to pull in *or* drop out (see Figure 9–22, Rung 2). If an abnormal condition is detected, then Rung 2 will turn *ON* an internal memory bit labeled "Motor Fault" that will seal in and remain *ON* until an operator or maintenance person reattempts to start the motor. The first two branches in Rung 2 monitor the motor for the two abnormal conditions, and the third branch acts as a holding circuit if a motor fault is detected. The On-Delay timer in Rung 2 is used to allow for the physical movement of the motor starter when energizing and deenergizing the motor starter (PLC timers will be discussed in Chapter 11).

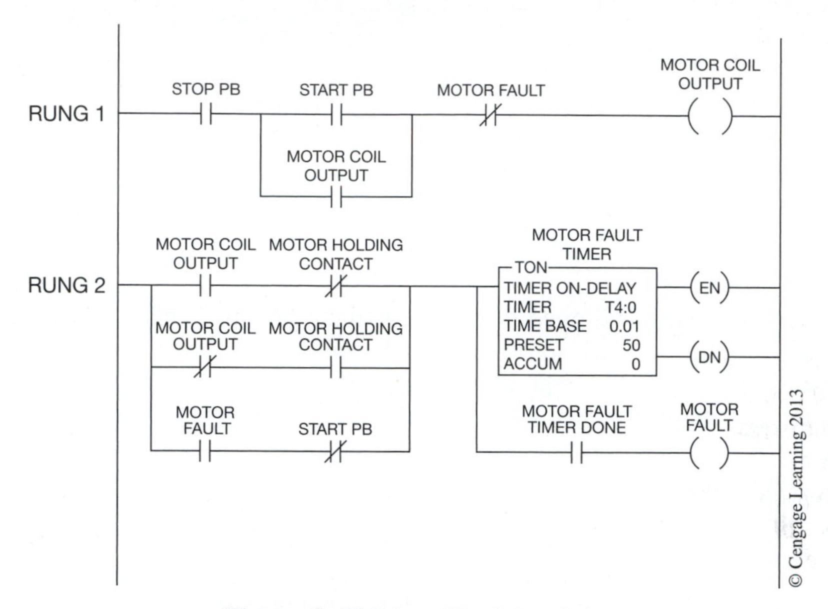

Figure 9–22 Motor Fault Logic

Step 4. When one is writing the logic to control the motor coil, an N.C. contact should be inserted in series with the motor coil output. This N.C. contact should have the same PLC address as the "Motor Fault" memory bit described above. Figure 9–22, Rung 1 shows an example of this logic. Now, whenever the motor starter fails to pull in or drop out for any reason, the condition is detected and the motor coil output is turned *OFF* or deenergized until the motor is again restarted.

Note: *The "Motor Fault" memory bit can also be used to flash a light, such as the motor's run light, to alert the operator and maintenance personnel of the problem. See Chapter 22, "Programming Examples," for examples of this logic.*

Chapter Summary

Each PLC has a maximum network size, or matrix, that limits the number of horizontal and vertical contacts for any one network, or rung. The only limitation to the number of rungs (networks) is memory size. Since the processor reads power flow (logic) from left to right *only,* and vertically either *up* or *down,* the logic of a relay circuit must be examined carefully to ensure that the logic is maintained when the circuit is programmed into the user memory. The program device does not allow contact to be programmed vertically, but the logic of a ladder diagram with vertical contacts may be duplicated by adding additional contacts. Depending on the PLC, contacts may or may not be programmed as a "branch-within-a-branch," or nested.

Consideration must be given to the way that the processor scans the program to eliminate unnecessary scans before a line or rung of logic goes true. *STOP* buttons must always be wired as normally closed and use an EXAMINE ON (XIC) instruction to work correctly and safely. Holding contacts can be either logical or discrete (real world). The discrete method has the added advantage of verifying that the motor starter has indeed energized. Holding contacts wired to a PLC input provide the added advantage of monitoring the motor starter for abnormal conditions.

Review Questions

1. Define the term *network.*
2. Draw and label a diagram to show how a network that requires 14 series contacts to control a discrete output can be programmed on a PLC that limits series logic elements to 10 per line.
3. Draw a circuit with nested contacts.
4. Draw a circuit that retains the logic of the accompanying figures, which has a vertical contact (D), so the circuit could be programmed into a PLC.

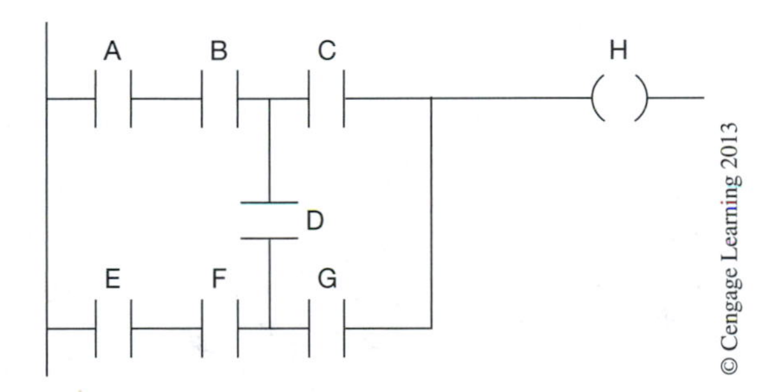

5. Power flow, or logic, in a PLC is considered to be (check all correct answers):
   a. up to down only
   b. up or down only
   c. left to right
   d. right to left
   e. up or down and from left to right
   f. up, down, and from right to left
   g. up to down only and left to right
   h. up to down only and right to left
   i. up or down and left to right or right to left

6. Write a program for a PLC that does *not* allow nested contacts, for the Hand-Off-Auto circuit shown in the accompanying figure.

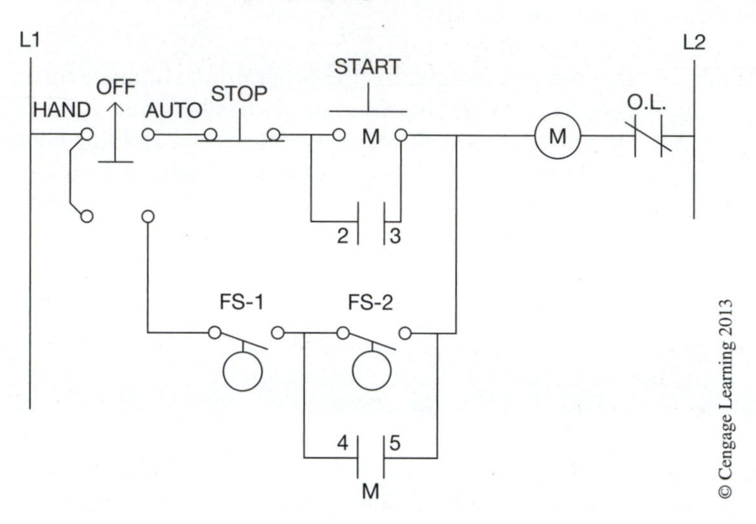

7. Explain how a *STOP* button must always be wired, and why.
8. List two ways that holding contacts can be programmed.
   a. ____
   b. ____
9. Explain two advantages of wiring the holding contacts to an input module and then programming the holding contact address into the PLC program.

CHAPTER 10 Program Control Instructions

Objectives

After completing this chapter, you should have the knowledge to:

- Write a program using a *latching relay*.
- Understand the term *retentive*.
- Write a program using a *master control relay*.
- Understand the importance of a *safety circuit*.
- Describe how an *immediate input instruction* could be used.
- Describe how an *immediate output instruction* could be used.
- Write a program using the *jump and label instructions*.
- Give a reason for using the *temporary end instruction*.

MASTER CONTROL RELAY INSTRUCTIONS

In standard relay control systems, a master control is often used to control power to the entire circuit or just to selected rungs. This allows selected rungs, or the whole circuit, to be deenergized by turning off the master control relay (MCR). Figure 10–1 shows a typical hardwired master control relay that controls power for the whole circuit.

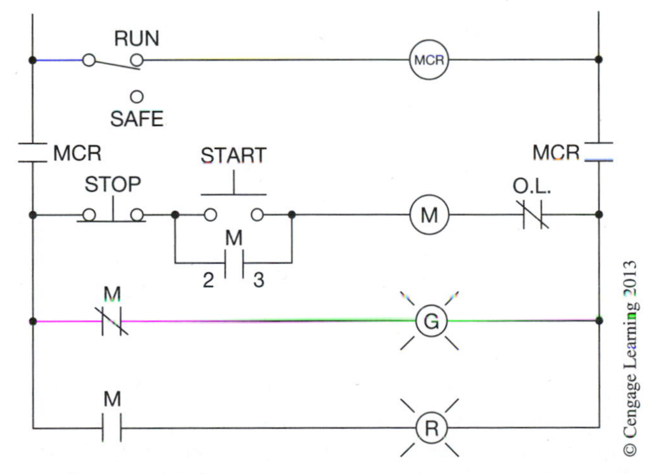

Figure 10–1 Hardwired Master Control Relay

MCRs are often used with circuits that have off-delay timers so the circuit can be shut down completely *without* waiting for the timers to time out.

With PLCs, an MCR function can be programmed to control an entire circuit or just selected rungs of a circuit. When the MCR instruction is programmed as shown in Figure 10–2, any rungs that follow can only become energized if the MCR instruction is energized, or set to a True condition.

Note: *MCRs are sometimes called Master Control Resets because they reset the output to zero or* OFF.

Allen-Bradley Logix5000, SLC 500, and MicroLogix MCR Instruction

As shown in Figure 10–2, when the MCR instruction is *true,* the outputs in the rungs that follow are controlled in a normal fashion by the logic programmed for each rung. If the MCR instruction is *false,* the rungs below the MCR are also *false* and cannot energize even if the programmed logic for each rung is *true*. The exception is retentive outputs such as latching relay instructions. Latching relay instructions will remain *ON,* or *true,* even when the MCR instruction is *false*. Latching relay instructions are covered later in this chapter.

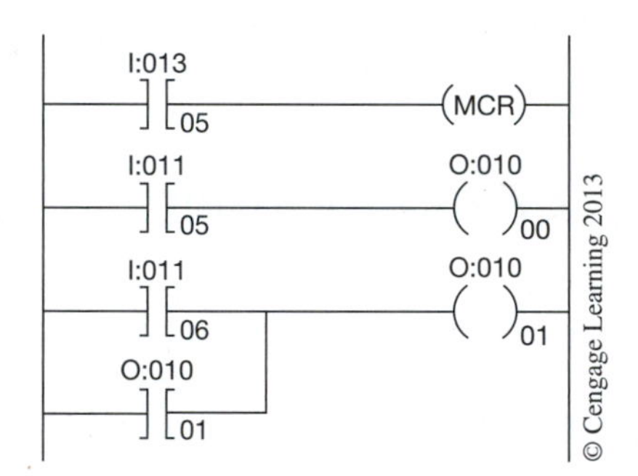

Figure 10–2 Allen-Bradley MCR Instruction

To create an area, or zone, within a program that will turn *OFF* all nonretentive outputs within the area, or zone, two MCR instructions are used. Figure 10–3 shows how the two instructions are programmed. Rung 1 has the first MCR instruction that is controlled by input device I:012/00. Note that the second MCR instruction has been programmed in Rung 5 with no logic element preceding the instruction. When an instruction is programmed in this manner, it is said to be programmed ***unconditionally.*** This second MCR instruction is used to end the zone controlled by the MCR instruction in Rung 1.

When input device I:012/00 goes *true,* the MCR instruction goes *true,* and the rungs between the two MCRs—Rungs 2, 3, and 4—are controlled by the first MCR. As long as the MCR in Rung 1 remains *true,* the normal logic of the rungs below the instruction will control output devices O:010/00, 01, and 02. When the MCR instruction in Rung 1 goes *false,* all output devices between the MCR in Rung 1 and the MCR in Rung 5 will be turned *OFF,* regardless of the logic of the individual rungs

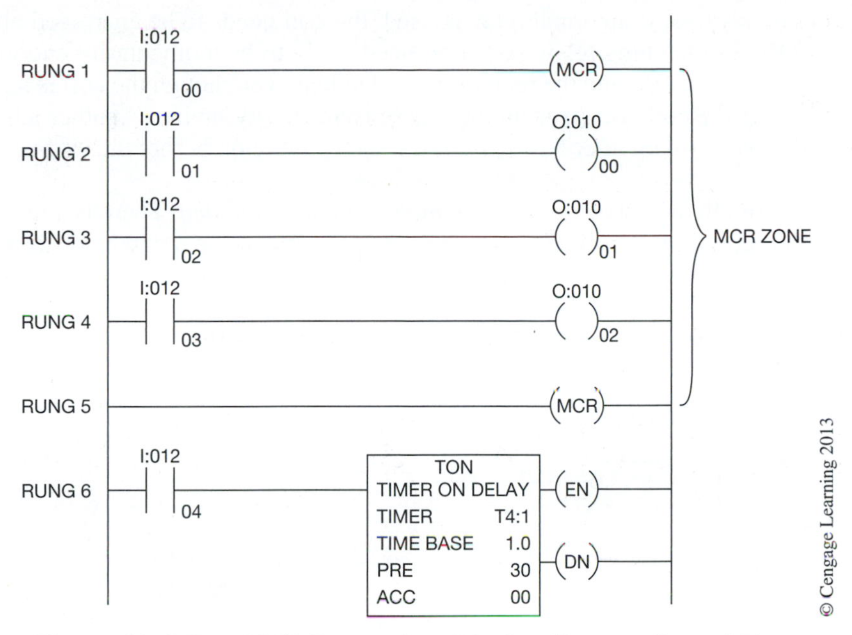

Figure 10–3 Two MCR Instructions Used to Create a Control Zone

(Rungs 2, 3, and 4). Rung 6 is outside the area, or zone, controlled by the MCR instructions and works independently of the MCRs.

Additional MCR zones can be created within the program using pairs of MCR instructions. An MCR zone cannot be nested within another MCR zone.

Safety Note: *A programmed MCR must never be used to replace a hardwired emergency stop or master control relay that provides emergency shutdown. You should still install a hardwired master control relay to provide output power shutdown.*

LATCHING RELAY INSTRUCTIONS

Before discussing how a latching relay function is programmed with a PLC, it may be helpful to review traditional hardwired latching relays.

Latching relays are used when it is necessary for contacts to stay open and/or closed even though the coil is only energized for a short time (40 milliseconds) by a momentary signal. Latching relays are often used for lighting applications where multiple circuits are needed for the lights in a large room or auditorium. Instead of having a switch for each circuit, a multiple contact latching relay is used. Each lighting circuit is wired to a set of contacts on the latching relay. One switch now controls the latching relay, which in turn controls several circuits of the lighting load. The latch and unlatch feature of the relay only requires three wires, so wiring is greatly reduced, and controlling the latching relay from multiple locations is quite simple. Another advantage of using the latching relay occurs when lighting is normally turned *ON* at the start of the work day, and left *ON* until quitting time.

Under these circumstances, if a normal relay is used, the coil needs to be energized all day long to keep the lights *ON*. The latching relay, however, needs only to be momentarily energized to latch, or turn the connected load *ON*, and the relay remains latched even though the coil is no longer energized. By not having the coil energized all day, there is an energy saving. Another advantage is the fact that the lights will turn on after a power outage if they were on before the outage.

Latching relays normally use two coils: one to *latch* and one to *unlatch*. There is a mechanical linkage that holds the relay in the latched, or closed, position. When the unlatch coil is energized, the coil action disengages the mechanical latch and allows the relay to open.

Figure 10–4 shows the wiring diagram for a mechanical latching relay.

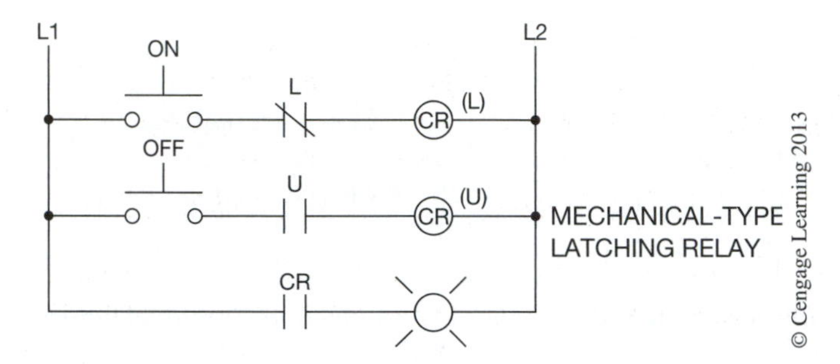

Figure 10–4 Mechanical Latching Relay

When the *ON* button is pushed, the latch coil energizes and opens the N.C. latch (L) contacts and closes the unlatched (U) contacts. Opening the N.C. L contacts deenergizes the L coil. The length of time it took to push the *ON* button and to energize the latch coil (which opened the N.C. L contacts and deenergized the L coil) was only a fraction of a second. During the short time the latch coil energized, it closed the N.O. CR contacts, completing the circuit to the lamp. The CR contacts remain closed even though the latch coil deenergized because of the mechanical latch mechanism. To open the mechanically latched contacts to turn the light *OFF* requires the *OFF* button be pushed. The U contacts in the unlatch coil circuit are now closed. Pushing the *OFF* button energizes the unlatch coil, which in turn closes the N.C. L contacts, and opens the U contacts, which deenergizes the unlatch coil. For the brief instant that the unlatch coil is energized, it releases the mechanically latched CR contacts so they can open and turn the light *OFF*.

Mechanical latching relays can be replaced by programming internal latching instructions. Like the dummy relays discussed earlier, the programmed internal latching relays do not exist as real-world devices but can perform all the logic of an actual latching relay.

Programmed latch and unlatch instructions, like their physical real-world counterparts, are **retentive** during a power failure. When the processor loses power, is switched to either the TEST or the PROGRAM modes, or detects a major fault, discrete outputs are turned *OFF;* the state of the latch instruction is retained in memory, however, and when power is restored or the processor is switched back to the RUN mode, the outputs that were *ON* previously return to their *ON* state.

Figure 10–5 shows latch and unlatch rungs as they would be programmed using the Allen-Bradley Output Latch (OTL) and Output Unlatch (OTU) instructions for their Logix5000, SLC 500, and MicroLogix PLCs.

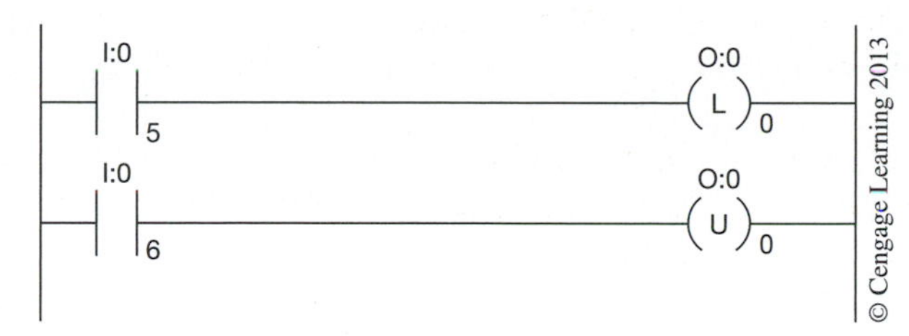

Figure 10–5 Programmed Latch and Unlatch Rungs

The OTL instruction is a retentive output instruction that can be programmed to turn *ON* an output device. This instruction cannot turn an output device *OFF*. Once an output device has been turned *ON* by an OTL instruction, an OTU (output unlatch) instruction must be used to turn the device *OFF*. As these two instructions must be used in pairs, it follows that they will use the same address. Note in Figure 10–5 that both the latch (L) and unlatch (U) coils have the same address (O:0/0).

The address, which is bit 0 of output image table word 0, will be set to 1, or *ON,* when the latch rung is true (input I:0/5 closed), and will be cleared to 0 or turned *OFF* when the unlatch rung is true (input I:0/6 closed). Like normal latching relays, only a momentary closure of input device I:0/5 latches output coil (L) O:0/0, and the output remains latched, or *ON,* until the unlatch coil (U) rung is true by closing input I:0/6.

Normally, an internal storage bit (dummy relay) is used for the latch and unlatch address, rather than an actual discrete output address. If a discrete output address is used, the output, once latched, remains *ON,* even if programmed after an open MCR rung. When an internal storage bit is used for the latch and unlatch address, the bit is still retentive, but turns *OFF* if programmed *after* an MCR rung that is open.

Although all PLCs are designed and manufactured to the highest standards and quality, a latching or MCR instruction should not be depended on for machine safety. A hardwired safety circuit should always be added. A safety circuit is recommended by most PLC manufacturers to ensure maximum safety rather than depending on a programmed MCR or latching relay alone.

SAFETY CIRCUIT

The concept of safety circuits has been discussed earlier in the text, and it is an important enough subject to be covered again. The National Electrical Manufacturing Association (NEMA) standards for programmable controllers recommend that consideration be given to the use of emergency stop functions that are independent of the programmable controller. The standard reads in part:

> When the operator is exposed to the machinery, such as loading or unloading a machine tool, or where the machine cycles automatically, consideration should be given to the use of an electromechanical override or other redundant means, independent of the controller, for starting or interrupting the cycle.

Figure 10–6 shows how a control relay (CR) and *safe-run switch* is added to interrupt L1 and L2 to the discrete output devices of an automatic machine or process.

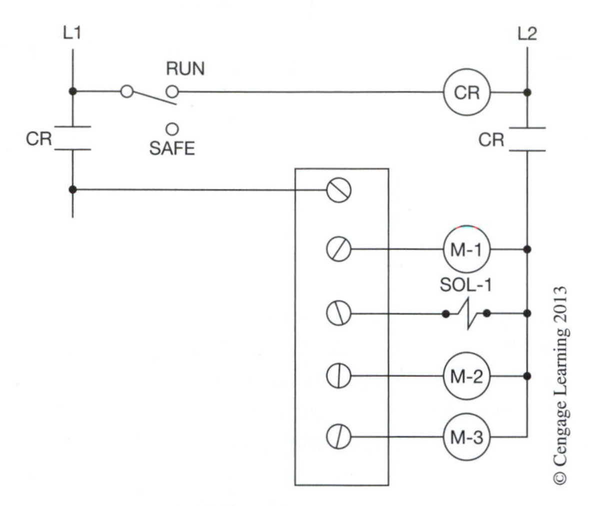

Figure 10–6 Safety Circuit

JUMP AND LABEL INSTRUCTIONS

Used in combination, these two instructions allow for skipping over portions of the program to save program scan time. If there is a portion of the program that is not operational during certain portions of the process, the portion that is not used and/or needed can be jumped over or bypassed until it is needed again. By jumping over parts of the program, one decreases the scan time and more scans can be completed in a given period of time, which, in turn, means more frequent updating of information in the program. The jump instruction (JMP) tells the processor to jump over a portion of the program. Where to jump to is controlled by the label instruction (LBL). Figure 10–7 shows a jump and label instruction in Allen-Bradley PLC-5 format. When the JMP instruction is true, the processor will jump over Rungs 3 and 4 and go directly to Rung 5, as shown.

The JMP instruction is assigned a three-digit number from 000 to 255, and the rung that the processor is to jump to is given an LBL of the same number. In Figure 10–7 the JMP and LBL instruction is also enabled, and the processor is instructed to jump all successive rungs until it reaches the rung that contains the label instruction with the number 20. In this illustration, only two rungs are jumped. In actual practice, any number of rungs can be jumped.

The jump and label instruction can be used to jump forward or backward in the program, depending on need. Jumping backward adds to the total scan time.

Note: *Jumping backward an excessive number of times could increase the scan time to a point where the watchdog timer will time out (the processor has a watchdog timer that is reset on each*

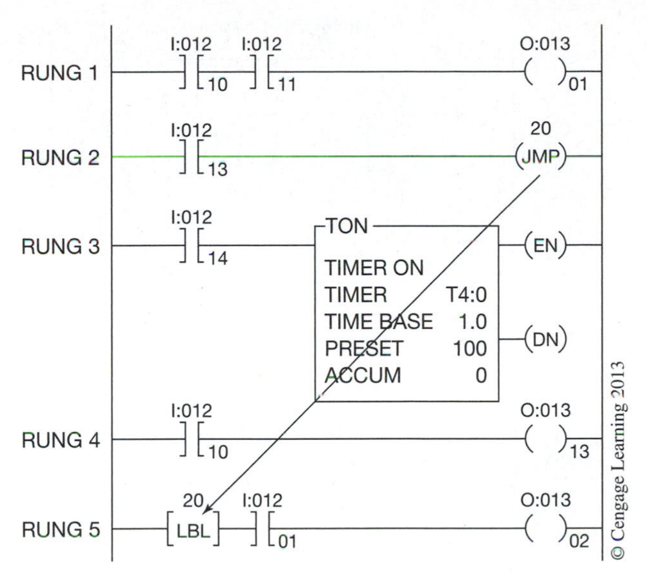

Figure 10–7 Programming the Jump (JMP) and Label (LBL) Instructions

scan). If the scan time exceeds the watchdog timer's preset time, the processor goes into a fault *condition.*

Safety Note: *When a portion of the program is jumped over, the outputs located within that portion will remain in their last state until scanned again by the processor.*

JUMP TO SUBROUTINE, SUBROUTINE, AND RETURN INSTRUCTIONS

The jump to subroutine, subroutine, and return instructions are used to direct the processor to go to a different routine (subroutine), scan it, return to the routine, and continue to scan. The formats used by the various PLC manufacturers to program these instructions vary widely, and for this reason, only instruction blocks are shown in Figure 10–8. The blocks are identified using the Allen-Bradley mnemonics, or labels: jump to subroutine (JSR), subroutine (SBR), and return (RET). Subroutines are very valuable for program organization, and for using blocks of programming logic over and over by simply changing the variables used in the subroutine. To use this group of instructions, consult the programming guide for the PLC system that is being used.

TEMPORARY END INSTRUCTION

The temporary end instruction (TND) is used to place a temporary end to the routine. When the instruction is inserted into the routine, the processor stops scanning the routine and moves to the end of the current routine. This instruction is often used when a new program is being debugged for the

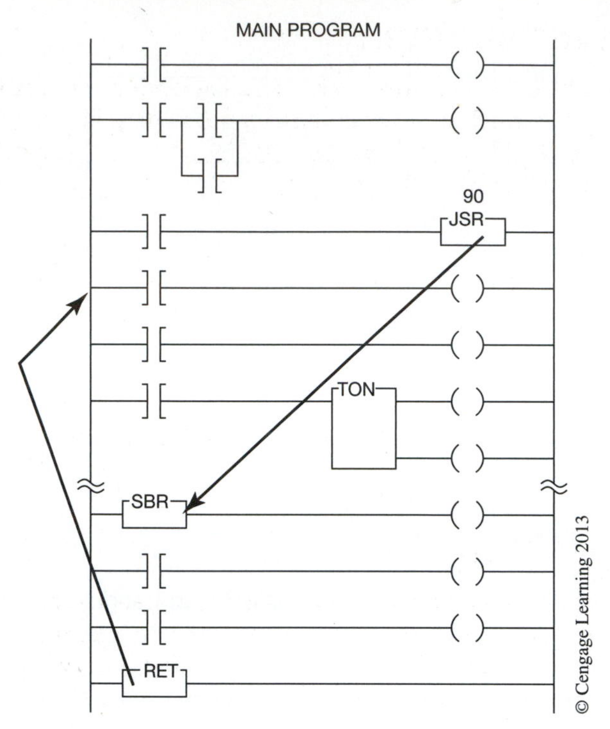

Figure 10–8 Jump to Subroutine, Subroutine, and Return Instructions

first time, because it allows for portions of the routine to be checked out without running the entire routine. Figure 10–9 shows a TND using the Allen-Bradley format. When the TND instruction is *true,* the processor stops scanning at Rung 3 and does not scan Rungs 4 and 5.

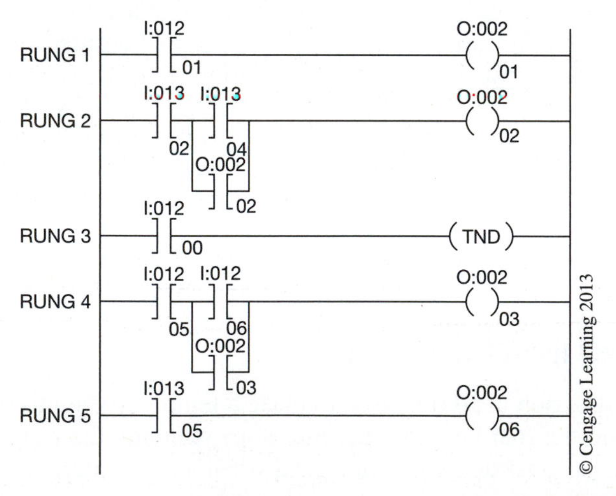

Figure 10–9 Allen-Bradley PLC—Temporary End Instruction (TND)

ALWAYS FALSE INSTRUCTION

The always false instruction (AFI) is also used when debugging a new or modified program. By inserting the always false instruction in a rung, it is ensured that the rung will always be false, regardless of the status of other instructions in the rung. Figure 10–10 shows a rung of logic with the AFI instruction programmed at the start of the rung.

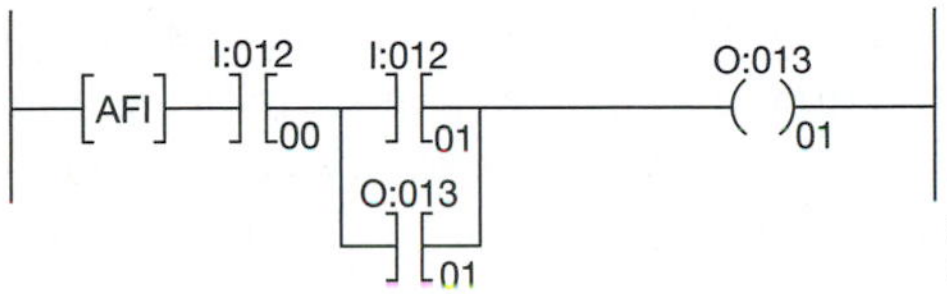

Figure 10–10 Always False Instruction

ONE-SHOT INSTRUCTION

The one-shot instruction (ONS) is an input instruction that makes the rung true for just one program scan, based on a false-to-true transition of the instruction that precedes the one-shot instruction. Figure 10–11 shows a rung of logic with the one-shot instruction programmed after input device I:011/04, which is controlling output O:010/12.

Figure 10–11 One-Shot Instruction

When the rung is programmed as shown, the output is turned *ON* for one scan, and one scan *only,* when input I:011/04 is true (makes a false-to-true transition). The output cannot be turned *ON* again until the input device is first opened, then closed again, making a false-to-true transition. With the next false-to-true transition, the output device is again only turned on for one scan. This is a beneficial instruction when an output signal or operation is wanted for only one scan.

Math operations, data or word moves, and the like, are completed only once if a one-shot instruction is put in series or "anded" with the instruction. The one-shot instruction is often used with timers and counters for changing preset and accumulated values. This technique is discussed further in Chapter 13.

Note: *This chapter has covered some of the basic instructions that are available for programming with a PLC. As the instruction sets vary with each manufacturer, it is necessary that the programming manuals be consulted to determine what instructions are available, what their mnemonics or designations are, and how to properly use them in a program.*

Chapter Summary

Latching and master control instructions can be programmed to serve the same control functions as their real-world counterparts. Where personnel safety is a factor, a hardwired safety circuit should be added, instead of depending on latching or MCR instructions alone. Through the use of various PLC instructions, the programmer can cause the processor to jump between specific rungs of logic, jump to subroutines then back to the original rung by using special instruction blocks, place a temporary end statement in the program to limit the amount of program that the processor will scan, use an always false instruction to keep a rung of logic from going true, and program one-shot instructions that limit activity to only that one program scan. Although the mnemonics used by the various manufacturers will differ for each of their instruction sets or blocks, the main purpose of each instruction is the same. Once the electrician or technician has mastered programming one type of PLC, the transition to other types becomes easier.

Review Questions

1. Will both programmed and real-world latching relays, if latched, remain latched if power is lost and then restored?
2. Latching relays are normally used when it is necessary for:
   a. contacts to open and/or close only while the coil is energized.
   b. contacts to open and/or close every 30 seconds.
   c. contacts to stay open and/or closed even though the coil is only energized a short time.
   d. none of the above.
3. Explain why an MCR is often used with off-delay timers.
4. An MCR can be used to control (check all correct answers):
   a. selected circuit rungs (networks).
   b. entire circuits.
   c. individual contacts within a rung (network).
   d. all of the above.
5. Define the term *unconditional.*
6. When using PLCs, NEMA recommends that consideration be given to stop functions independent of the PLC. Explain briefly why this recommendation is made.
7. Define the term *retentive.*
8. What does the *jump and label instruction* do?
9. Give one reason why you might use a *temporary end instruction.*
10. What is the function of the *always false instruction?*
11. What is the function of a *one-shot instruction?*

CHAPTER 11 Programming Timers

Objectives

After completing this chapter, you should have the knowledge to:

- Describe how *pneumatic time delay relays* work.
- Write a program using *ON delay* and *OFF delay* timers.
- Describe the difference between an *ON delay timer* and a *retentive timer.*
- Explain how to extend the time range of timers by *cascading*.

PNEUMATIC TIMERS (GENERAL)

To fully understand how a PLC can be programmed to replace pneumatic time-delay relays, both the basic pneumatic time-delay relay and the standard symbols used must be understood.

Figure 11–1 shows a complete Allen-Bradley pneumatic timing relay, and Figure 11–2 shows a cutaway view of the contact and timing mechanism.

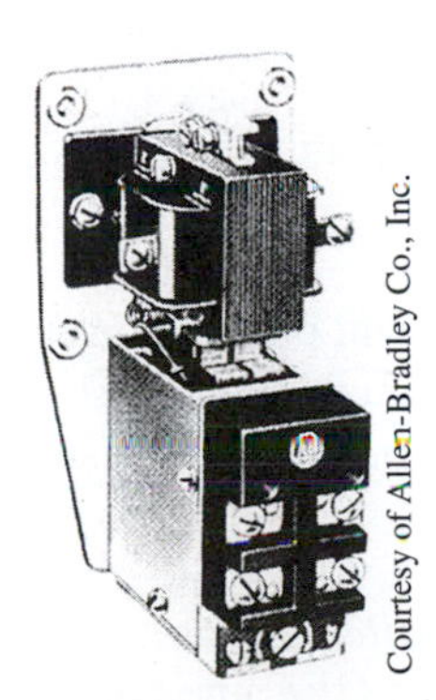

Figure 11–1 Pneumatic Timing Relay

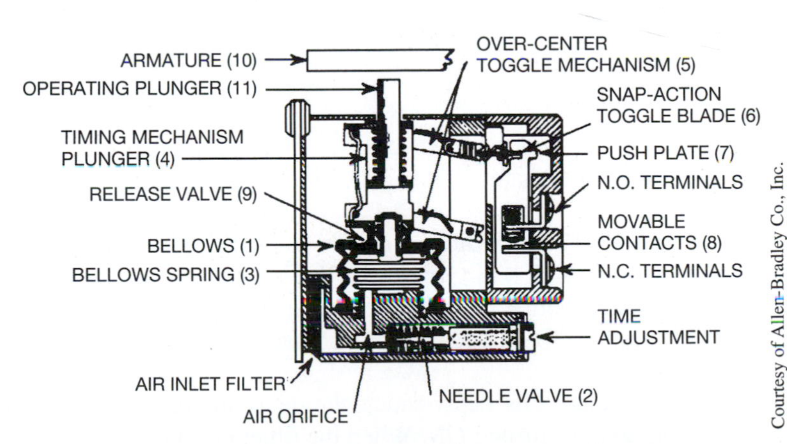

Figure 11–2 Cutaway View of Contact Unit and Timing Mechanism

For the timer to time when power is applied (coil energized), the solenoid unit—coil, core piece, and armature—is mounted so that the natural weight of the armature (10) pushes down on the operating plunger (11). This causes the bellows (1) and bellows spring (3) to collapse the bellows, and dispel the air out through the release valve (9). When the coil is energized, the armature is attracted magnetically to the pole pieces, and lifts up and off the bellows assembly. Air now comes in through the air inlet filter, past the needle valve (2), and fills the bellows with air. The incoming air expands the bellows upward, pushing on the timing mechanism plunger (4). As the plunger rises, it causes the over-center toggle mechanism (5) to move the snap-action toggle blade (6) upward. This picks up the push plate (7) that carries the movable contacts (8) to open the N.C. contact and close the N.O. contact. The time it takes for the bellows to fill with air and activate the contact mechanism is controlled by adjusting the needle valve in the air orifice. The valve is adjusted with a screwdriver as shown in Figure 11–3. A counterclockwise rotation moves the needle valve further into the air orifice, restricting airflow into the bellows, slowing the airflow, and increasing the time it takes for the bellows to expand and operate the contact mechanism. Conversely, clockwise adjustment of the needle valve decreases the time it takes the bellows to fill with air and activate the contacts after the armature has been lifted off the bellows mechanism.

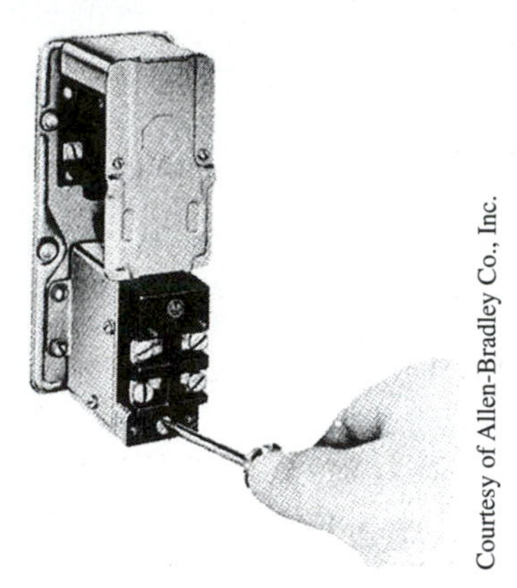

Courtesy of Allen-Bradley Co., Inc.

Figure 11–3 Pneumatic Timer Adjustment

When the contact action is delayed after the coil has been energized and the armature is lifted up and off the bellows mechanism, it is called ON delay.

When the coil of an **ON delay timer** is de-energized, the armature drops down, pushing on the operating plunger, which in turn pushes down on the bellows expelling air through the release valve. The downward motion of the bellows causes the snap-action toggle blade to instantaneously snap the N.C. contact closed and the N.O. contact open.

To summarize the ON delay timer, the delay in contact operation begins *after* the timer coil has been energized, or turned *ON*. When the timer coil is de-energized, or turned *OFF,* the contacts go back to their normal condition instantly. Figure 11–4 shows a pneumatic timer with the solenoid unit mounted for ON delay.

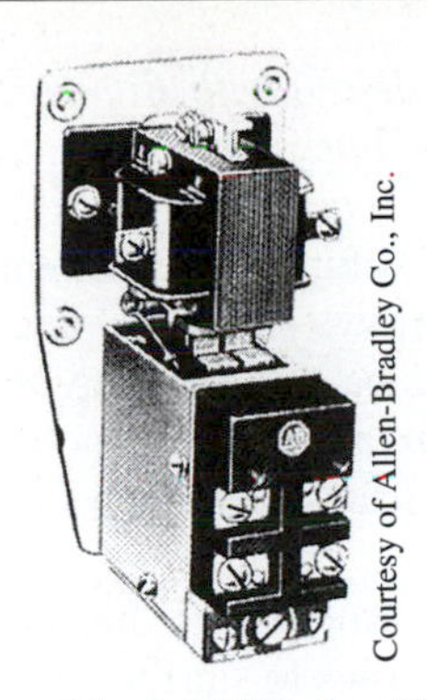

Figure 11–4 ON Delay Timer

Figure 11–5 illustrates the electrical symbols used to indicate ON delay contacts.

The arrowhead indicates that movement is up. Since ON delay contacts can only time *after* the armature has lifted up off the bellows, this method of identifying timed contacts is easy to remember. Another common method of identifying timed contacts is shown in Figure 11–6.

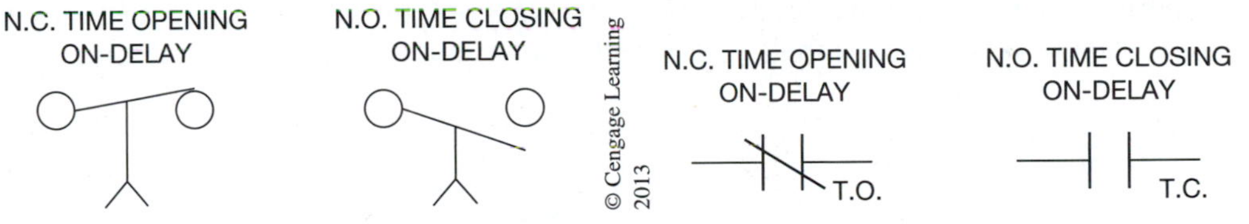

Figure 11–5 ON Delay Symbols

Figure 11–6 ON Delay Symbols

Note: *Remember that "normal" for contacts is how they are open or closed, with the coil of the relay de-energized and time expired.*

For a pneumatic timer to time when power is removed from the relay coil **(OFF delay),** the solenoid unit is mounted as shown in Figure 11–7. With a spring holding the armature up, no weight is applied to the bellows assembly, and the bellows are filled with air in a fully extended position.

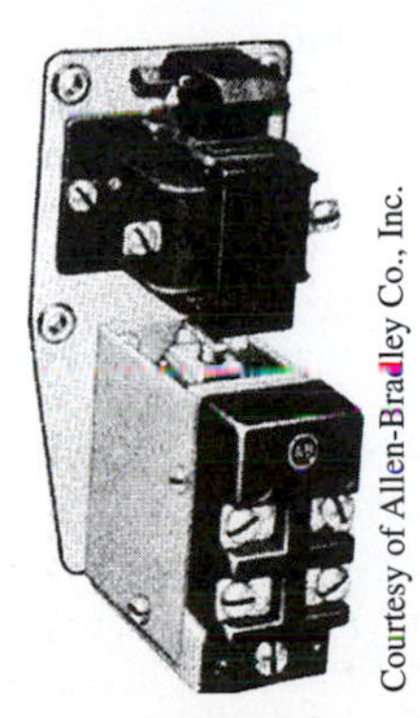

Figure 11–7 OFF Delay Timer

Note: *Compare Figure 11–4 (the ON delay) with Figure 11–7 (the OFF delay) to clearly see the difference in the mounting of the solenoid assemblies.*

When the coil is energized and the armature moves down, the armature pushes on the operating plunger. The plunger pushes on the bellows assembly, and all air is immediately forced out of the bellows through the release valve. This causes the snap-action contact assembly to instantly open the N.C. contact and close the N.O. contact. The contacts stay in this configuration as long as the coil is energized and the armature is holding the bellows mechanism down (compressed).

When the relay coil is de-energized, or turned *OFF,* the spring on the armature lifts it up and off the operating plunger, which allows the bellows to start to fill with air. The N.C. contact remains open, and the N.O. contact remains closed until the bellows are filled with enough air to activate the snap-action contact mechanism. When the contact mechanism has been activated, the N.C. contacts go closed and the N.O. contacts go open.

Figure 11–8 shows the electrical symbols for OFF delay contacts.

To avoid confusion when reading electrical drawings with OFF delay contacts, it must be remembered that normal refers to the coil *after* it has been de-energized (turned *OFF*), and the time set for the timer has elapsed. The other symbols used for OFF delay contacts are shown in Figure 11–9.

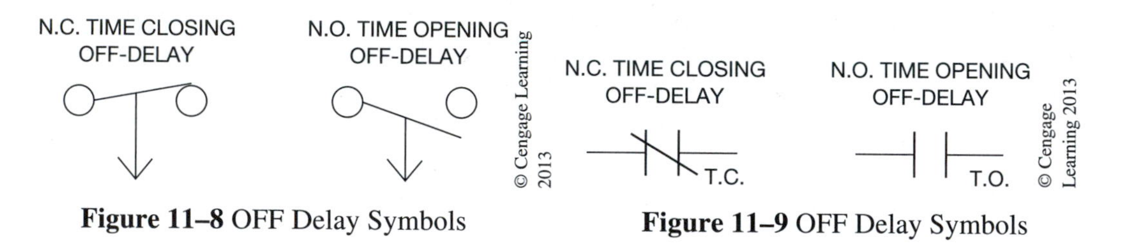

Figure 11–8 OFF Delay Symbols

Figure 11–9 OFF Delay Symbols

Figure 11–10 compares both types of symbols used for ON delay and OFF delay timer relays.

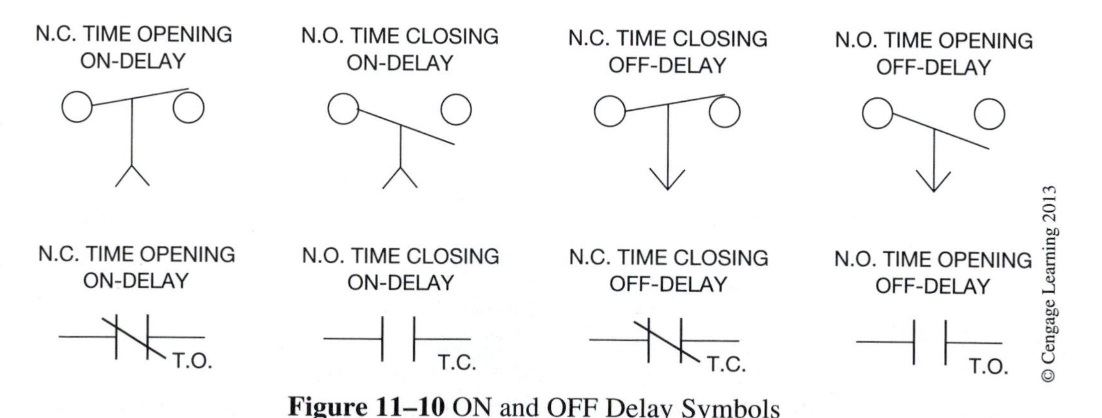

Figure 11–10 ON and OFF Delay Symbols

Reviewing the two types of symbols commonly used in motor control diagrams, an electrician or technician should have no trouble determining the type of timing relay (ON delay or OFF delay) used, or what is normal (open or closed) for the timed contacts.

The basic pneumatic timing relay is designed so that additional instantaneous contacts may be added, as shown in Figure 11–11. The instantaneous contacts operate when the coil is energized or de-energized independent of the timing mechanism. Figure 11–12 shows the electrical symbol for contacts with an asterisk (*), which is sometimes used to indicate instantaneous contacts of a timing relay.

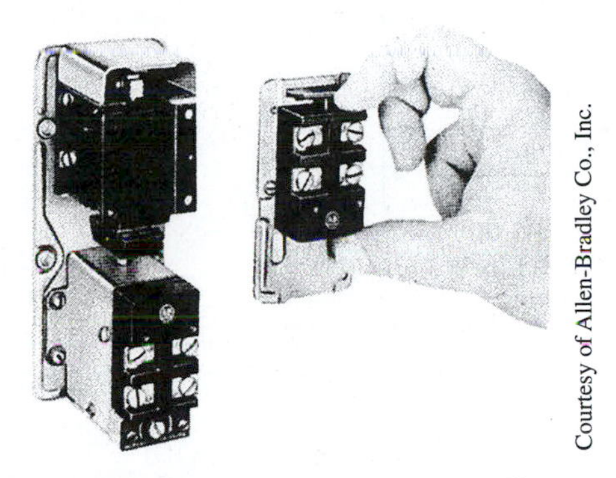

Courtesy of Allen-Bradley Co., Inc.

Figure 11–11 Adding Instantaneous Contacts

Figure 11–12 Instantaneous Contact Symbols

Figure 11–13a shows a simple light circuit controlled by an ON delay timer set for five seconds. The amount of delay is written near the timer coil on the diagram for understanding and for troubleshooting. Figure 11–13b shows that when S^1 is closed, the coil of the pneumatic timer energizes, lifts the armature up and off the bellows, and the timing starts. Figure 11–13c shows the circuit after three seconds have elapsed (not enough time for the timer to time out) with the lamp circuit still open. After five seconds have elapsed (Figure 11–13d), the N.O. time closing contacts close, and the lamp lights. As long as S^1 remains closed, the timer coil is energized, and the timed contacts stay closed. When S^1 is opened (Figure 11–13e), the coil circuit is broken, and the coil deenergizes. This causes the timed contacts to open, thereby turning *OFF* the lamp. The timed contacts will open the instant the coil deenergizes because they are timed only when power is applied to the coil.

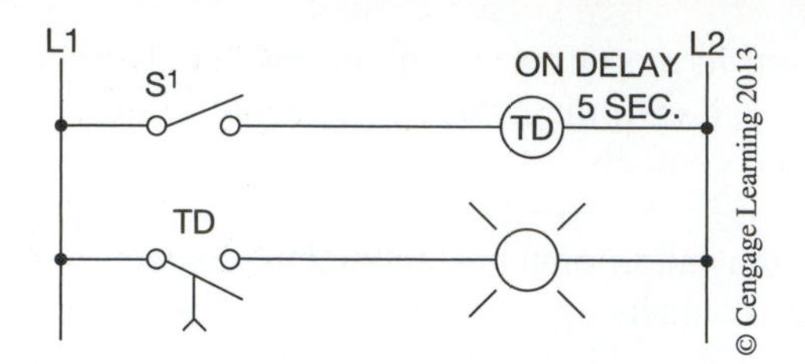

Figure 11–13a ON Delay Timer Circuit

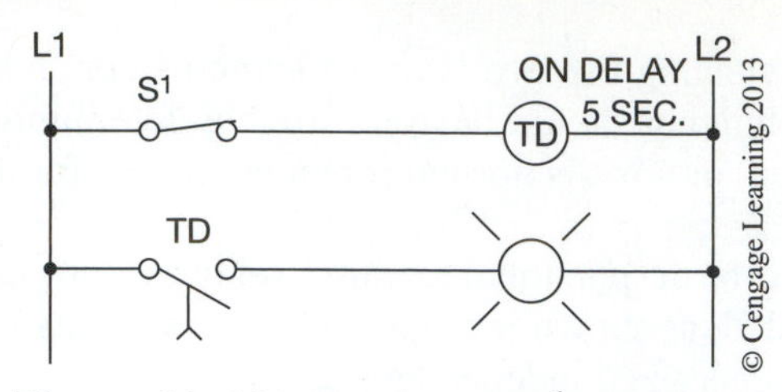

Figure 11–13b The Instant S^1 is Closed

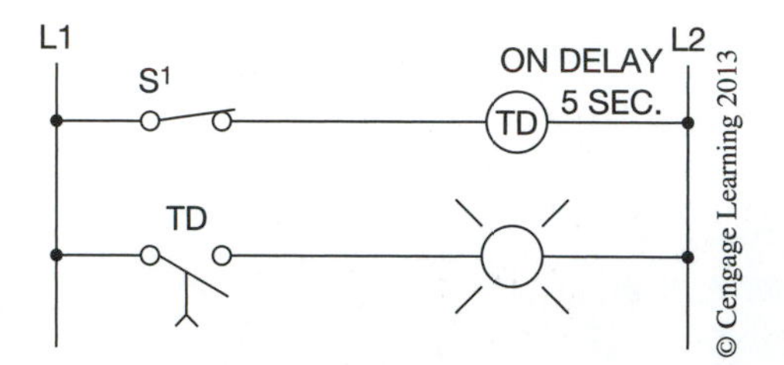

Figure 11–13c Three Seconds After S^1 is Closed

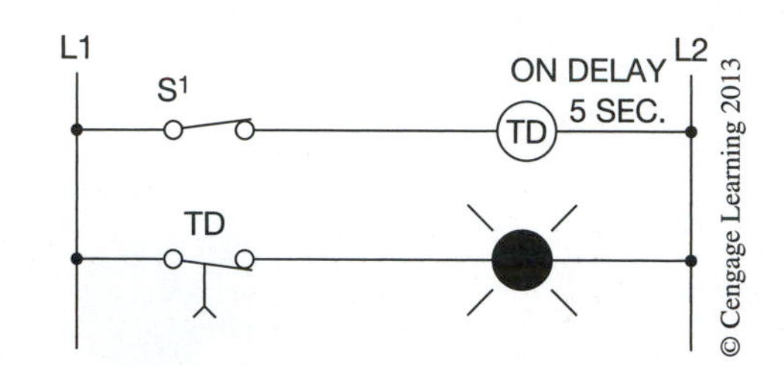

Figure 11–13d Five Seconds After S^1 is Closed

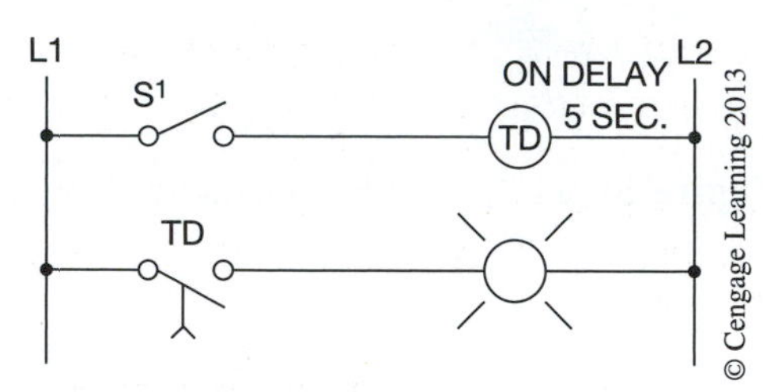

Figure 11–13e The Instant S^1 is Opened

Figure 11–14a shows the same circuit but with an OFF delay timer. When S^1 is closed (Figure 11–14b), the TD coil energizes, drawing the armature down and compressing the bellows. This causes the N.O. OFF delay contacts to go closed instantly, and the lamp lights. When S^1 is opened (Figure 11–14c), the TD coil is de-energized, the spring-loaded armature is lifted up and off the bellows, and the five-second timing begins. Figure 11–14d shows the circuit after three seconds have elapsed. The lamp remains energized until the full five seconds have elapsed, and the N.O. contacts time out and open (Figure 11–14e).

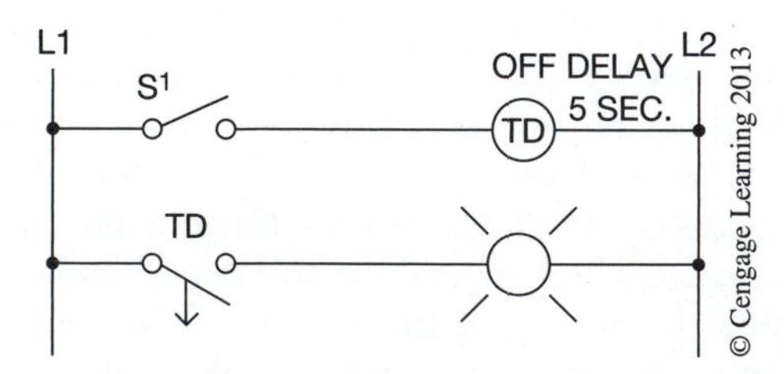

Figure 11–14a OFF Delay Timer Circuit

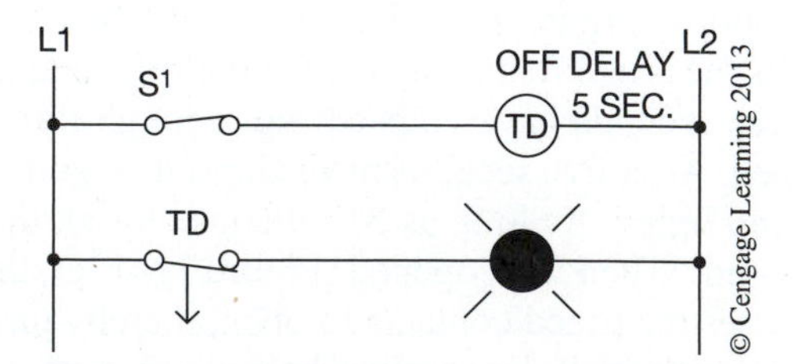

Figure 11–14b The Instant S^1 is Closed

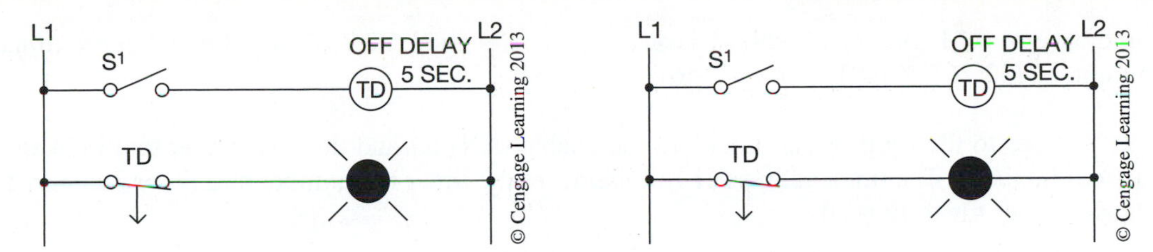

Figure 11–14c The Instant S^1 is Opened

Figure 11–14d Three Seconds After S^1 is Opened

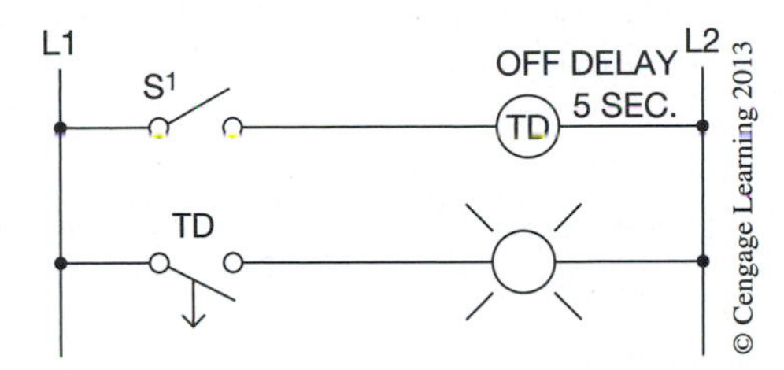

Figure 11–14e Five Seconds After S^1 is Opened

Instead of a bellows assembly, like the pneumatic time delay relay, PLC timers use internal solid state circuitry (clocks) for timing intervals or time base. The various PLC manufacturers use varying approaches for the actual programming of timers. Several methods that are typical for most PLCs will be discussed. Because it is an easy transition from pneumatic timer concepts to programming concepts, the Allen-Bradley approach to programming timers is discussed first.

ALLEN-BRADLEY PLC-5, SLC 500, AND MICROLOGIX TIMERS

Figure 11–15 shows the timer format used by Allen-Bradley. The timer consists of a timing block containing the timer number (address), time base (1 second or 0.01 seconds), and the preset and accumulated times. The preset time can be programmed with any value from 0 to 32,767. If a time base of one second was assigned, 32,767 would equal 9.1 hours (32,767 ÷ [60 × 60] = 9.1); if

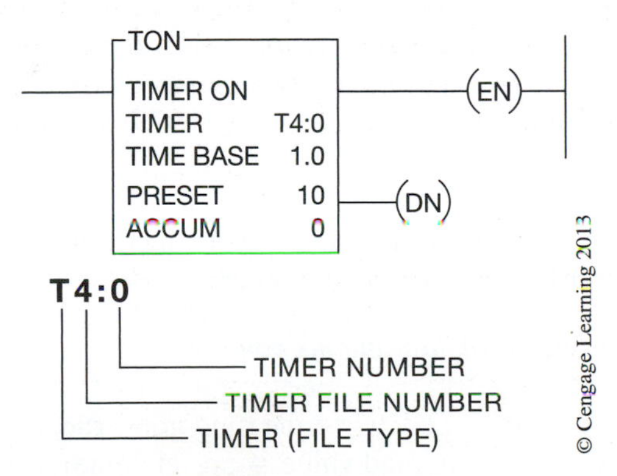

Figure 11–15 Allen-Bradley Timer Format

a time base of 0.01 (one hundredth of a second) was assigned, 32,767 would equal approximately 5.5 minutes ($32,767 \times 0.01 \div 60 = 5.46$).

The two lines to the right of the block are the enable (EN) bit and done (DN) bit that indicate the status of the timer. The timer address (T4:0) identifies the timer file number and timer number. T4:0 indicates timer file 4, timer 0.

File 4 is the default file from the data table for timers using the PLC-5, SLC 500, and the MicroLogix family. The PLC-5 can be programmed to use files 3-999 for additional timer files. The SLC 500 can be programmed to use files 9-255 for additional timer files. The MicroLogix 1000 is limited to one timer file which is the default file, file 4.

By only having one timer file, the MicroLogix 1000 is limited to 40 timers (timer 0 through 39) whereas the SLC 500 can use timers 0 through 255. The PLC-5 can have timer numbers from 0 through 999.

The EN bit is set to 1 (or is true) whenever there is a logic path to the timer block. The DN bit is set to 1 (or is true) when the accumulated value equals the preset value, and the timer has timed out. Figure 11–16 shows how the information for a timer is stored. Three words of memory are used for each timer programmed. The first word of memory uses the first 8 bits for internal use and uses bit 13 for the DN bit, bit 14 for the timer timing bit (TT), and bit 15 for the EN bit. The next two words store the preset and accumulated values of the timer.

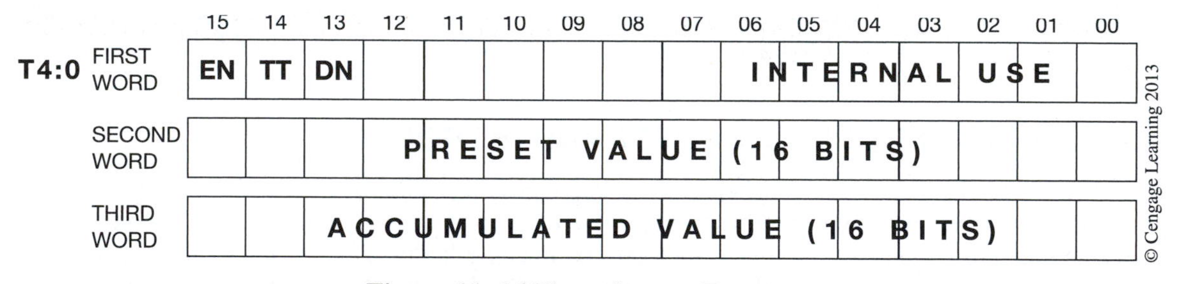

Figure 11–16 Timer Storage Format

Note: *There is no need to remember or memorize the timer bit numbers because the programming software accepts the mnemonics DN, TT, and EN. When addressing timer contacts, enter the timer number first, followed by the timer bit. For example, T4:1/TT or T4:1.TT addresses the TT bit of timer 1, file 4.*

The timer enable bit, bit 15, is set to 1, or turned *ON,* when the rung goes true, and remains set until the rung goes false or a reset instruction resets the timer.

Note: *The EN bit can be used as an instantaneous contact.*

The TT bit, bit 14, is set to 1, or turned *ON,* when the rung goes true, and remains *ON* until the rung goes false; the DN bit is set to 1 (accumulated value = preset value); timing is completed; or a reset instruction resets the timer.

Note: *The TT bit can be used to control a timer timing light that is only* ON *when the timer is actually timing. Figure 11–17a shows how the TT bit is used to control an indicator light, and Figure 11–17b shows the equivalent circuit using a pneumatic timer.*

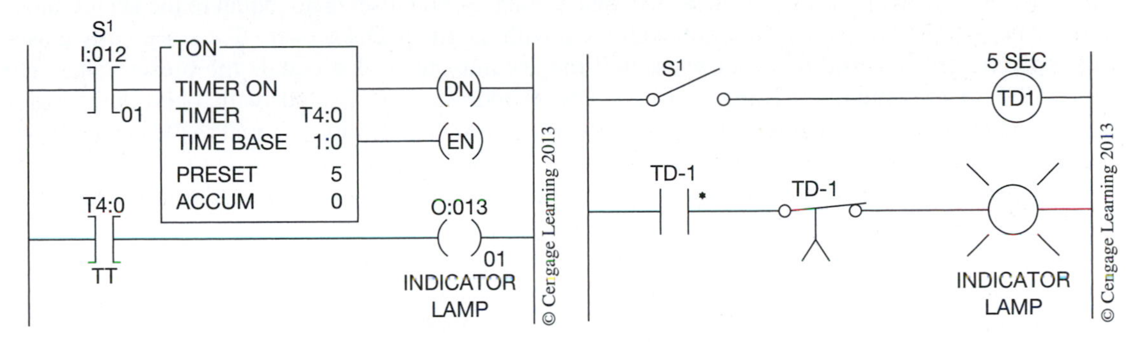

Figure 11–17a TT Bit Used to Control an Indicator Lamp

Figure 11–17b Pneumatic Timer Circuit Used to Control an Indicator Lamp

The DN bit, bit 13, is set to 1 when the accumulated value is equal to the preset value. The DN bit remains set to 1, or *ON,* until the rung goes false or a reset instruction resets the timer.

Note: *The DN bit can be used to control an output, or for other logic within a program.*

Figure 11–18 shows an ON delay timer and how it is programmed to control outputs O:013/01, O:013/02, O:013/03, and O:013/04.

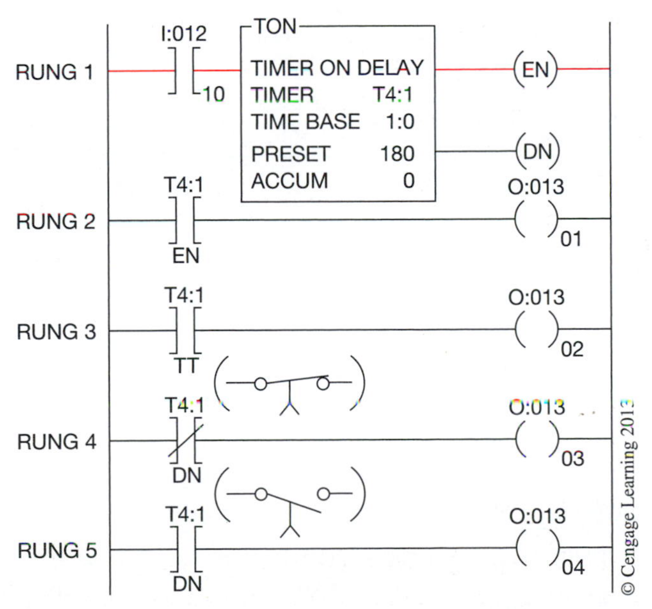

Figure 11–18 Programmed TON Timer

When bit I:012/10 (input device) is true, or set to 1, the timer rung is true, and the processor starts timer T4:0 timing and sets the EN and TT bits to 1. This turns *ON* outputs O:013/01 and O:013/02 in Rungs 2 and 3. The accumulated value increases in one-second intervals. The output in Rung 4, controlled by an EXAMINE OFF instruction, is true as long as the preset is not equal to the accumulated value. The EXAMINE OFF instruction addressed with the timer DN bit acts like a normally closed time-opening contact, and does not open until the accumulated value equals the preset value. The EXAMINE ON instruction in Rung 5 with the DN bit address acts like a normally open time-closing timer contact, and does not close (or go true) until the accumulated value is equal to the preset. When the accumulated time does equal the preset time, the DN bit is set to 1, and output O:013/04 is turned *ON* and output O:013/03 is turned *OFF*. Once the timer instruction has completed timing, the TT bit is reset to 0 and the output (O:013/02) of Rung 3 is turned *OFF*.

Like a pneumatic ON delay timer, when power is removed, the timer is reset to 0. The PLC-5 timer instruction is reset when the input device (I:012/10) is opened.

Figure 11–19 shows a typical timing chart. Notice that when the Rung condition is true (*ON*), the timer will time, but if the Rung goes false (*OFF*), the timer resets to 0, as illustrated, during the first two minutes of the timing diagram.

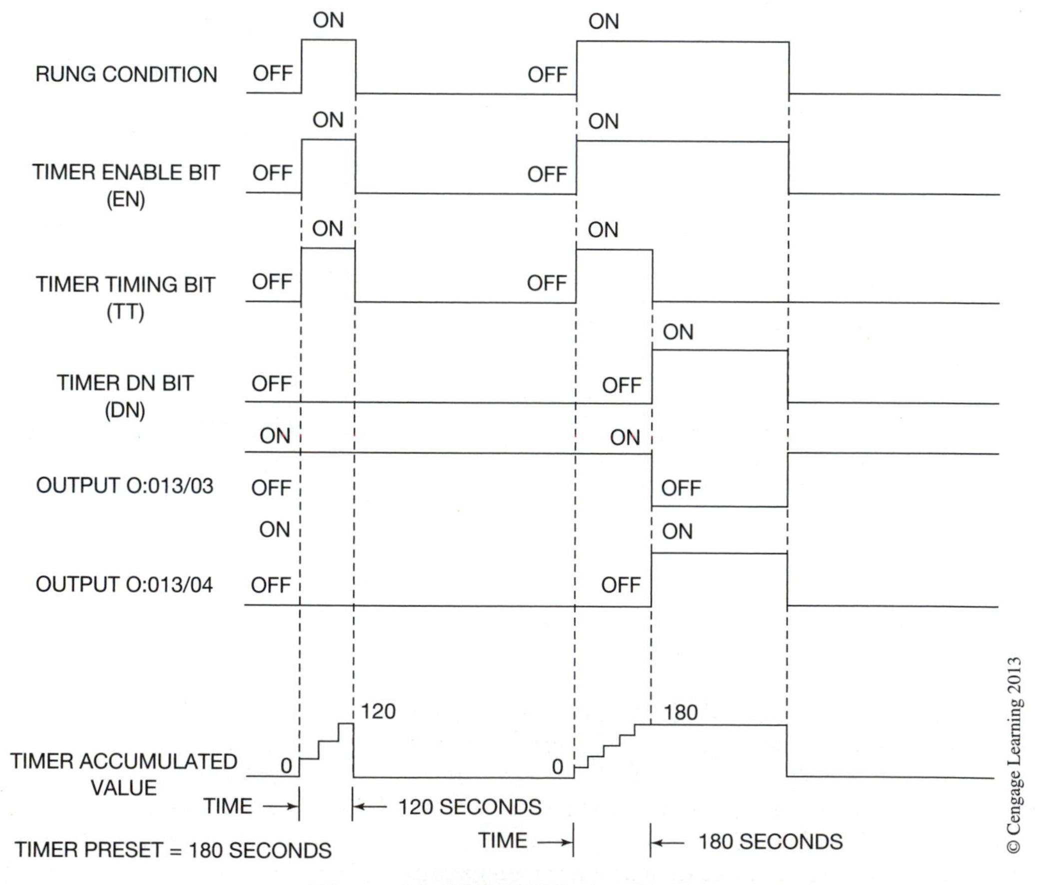

Figure 11–19 TON Timing Chart

Figure 11–20 shows how an OFF delay timer (TOF) is programmed.

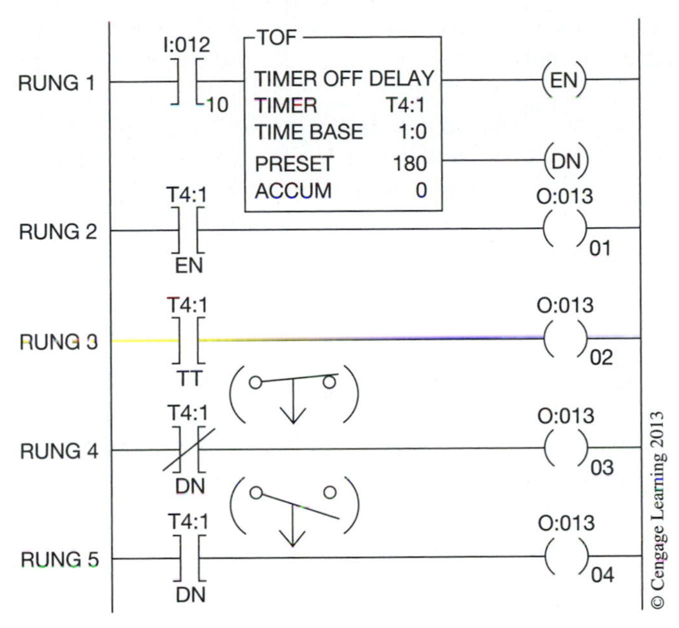

Figure 11–20 Programming an OFF Delay Timer (TOF)

In an OFF delay timer, when bit I:012/10 is set to 1, the DN and EN bits are also set to 1. The DN bit acts like the OFF delay contacts of a pneumatic timer, and the EXAMINE OFF (N.C.) instruction in Rung 4 goes false, or open, while the EXAMINE ON instruction (N.O.) in Rung 5 goes true. When input device I:012/10 is reset, or set to 0, Rung 1 goes false, and the timer starts to accumulate time in one-second intervals as long as the rung remains false. When the accumulated value equals the preset value (180) the timer stops. T4:1.TT was set to 1 while the timer was timing and output O:013/02 in Rung 3 was *ON*. When the accumulated value equaled the preset value and the timer stopped timing, the TT bit was reset to 0 and output O:013/02 was turned *OFF*. When the TT bit is reset to 0, the DN bit (bit 13) is also set to 0, and output O:013/03 in Rung 4 is turned *ON* and output O:013/04 in Rung 5 is turned *OFF*. The TOF instruction is reset by each open-to-closed transition of input device I:012/10. Figure 11–21 shows a typical timing chart for an OFF delay timer.

During the first timing cycle, the timer was only *OFF* for 120 seconds. That was not long enough for the timer to time out, so the outputs controlled by the DN bit did not change. During the second timing cycle, the timer was allowed to time out and the outputs changed states.

Most PLCs also offer a timer that replaces the standard motor-driven timer. A typical motor-driven timer consists of shaft mounted cam(s) that are driven by a synchronous motor. Rotating cam(s) activate (open or close) limit or micro switches. Once power is applied, the motor turns the shaft and cam(s). The positioning of the lobes of the cam(s) and the gear reduction of the motor determine the time it takes for the motor to turn the cam far enough to activate the switches. If power is removed from the motor, the shaft stops. When power is reapplied, the motor continues turning the shaft until the switches are activated. When the timing of a device is not reset due to a loss of power, the timing is said to be retentive.

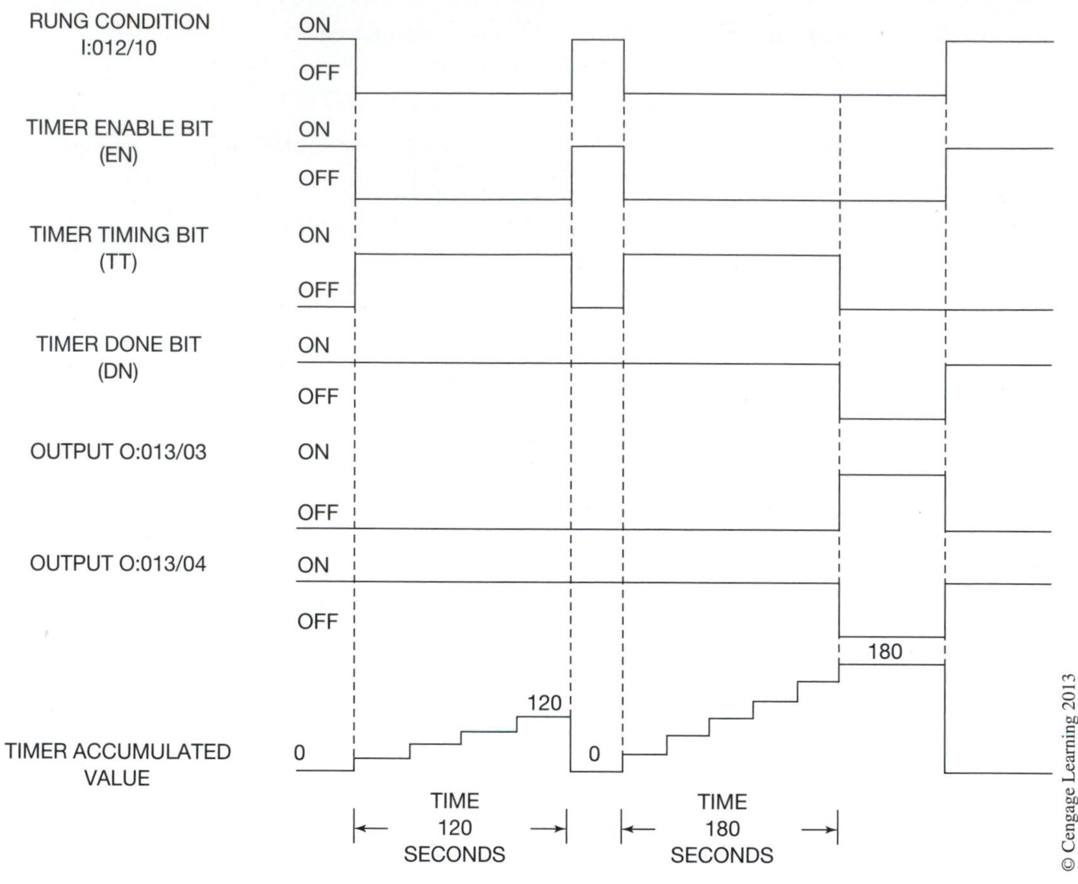

Figure 11–21 Timing Chart for a TOF Timer

Retentive timers (RTO) can be programmed to replace motor-driven timers. The retentive timer lets the timer START and STOP without resetting the accumulated value to 0. The bits associated with the timer EN, TT, and DN function the same as with the TON instruction. The RTO instruction begins timing when the rung goes true. As long as the rung remains true, the timer continues to time until the accumulated value reaches the preset value. If the timer rung goes false, the timer holds the accumulated time, rather than resetting the accumulated value to 0. When the timing rung goes true again, the count picks up from where it was, and continues to accumulate time. Once the accumulated time is equal to the preset time, the processor will set bit 13 (the DN bit) to 1. The DN bit remains *ON* (or set to 1) as long as the accumulated value is equal to or greater than the preset value. Because the retentive timer does not reset to 000 when the timer is de-energized, a reset rung (RES) instruction must be added. The reset instruction must be given the same address as the retentive timer it is intended to reset.

A common problem in programs that have retentive timers is that the timer is not accumulating time, even though the timer rung is true. More often than not, the problem is a reset instruction that is true, which prevents the timer from timing. Figure 11–22 shows how the RTO timer is programmed, including the reset rung (Rung 3), and shows a typical retentive timer timing chart.

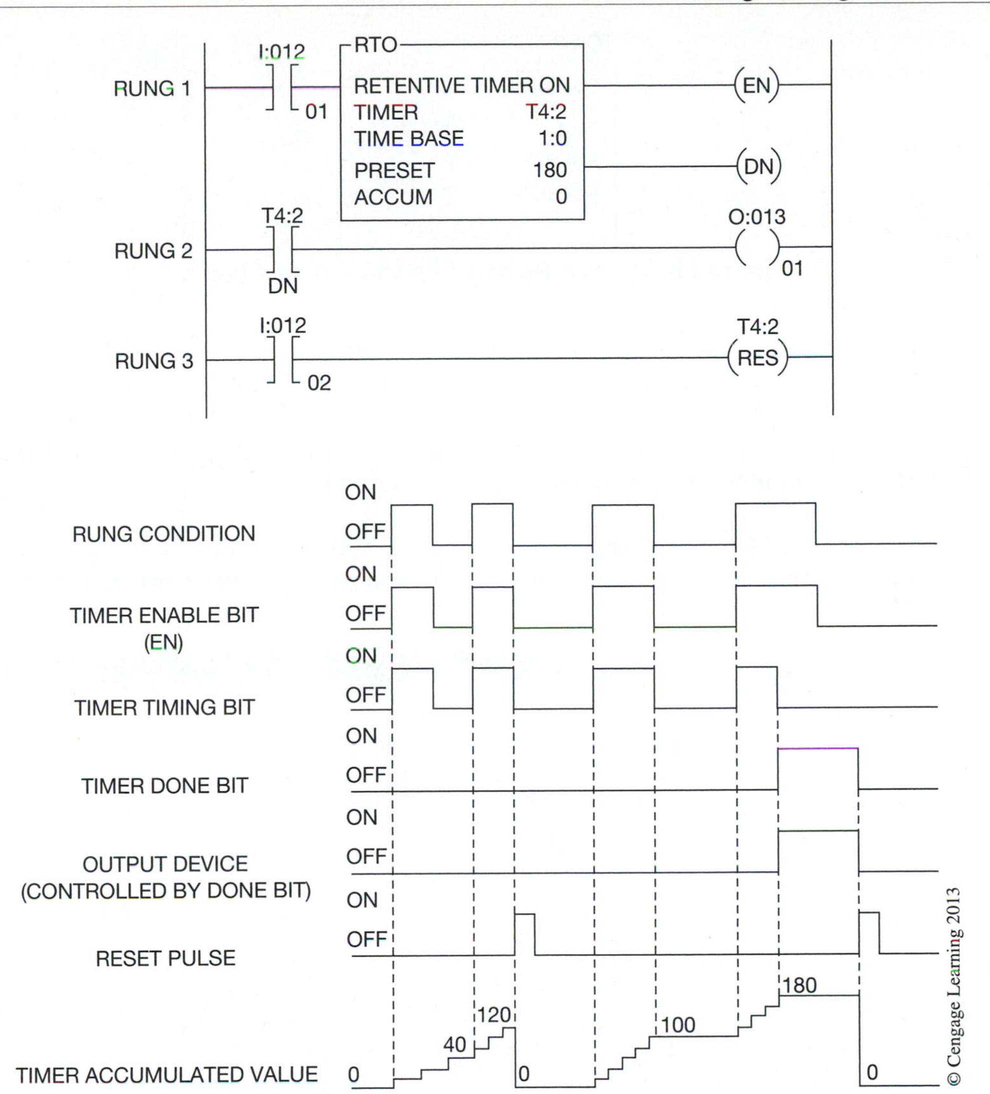

Figure 11–22 Programmed Retentive Timer (RTO) and Timing Chart

ALLEN-BRADLEY LOGIX5000 TIMERS

Figure 11–23 shows the timer format used by Allen-Bradley's Logix5000 controllers. The timer consists of a timing block containing the timer tag, the preset, and accumulated times. The time base is always 1 millisecond for Logix5000 timers. For example, for a 1-second timer, you would enter 1000 for the preset value. The preset time can be programmed with any value from 0 to 2,147,483,647 (DINT).

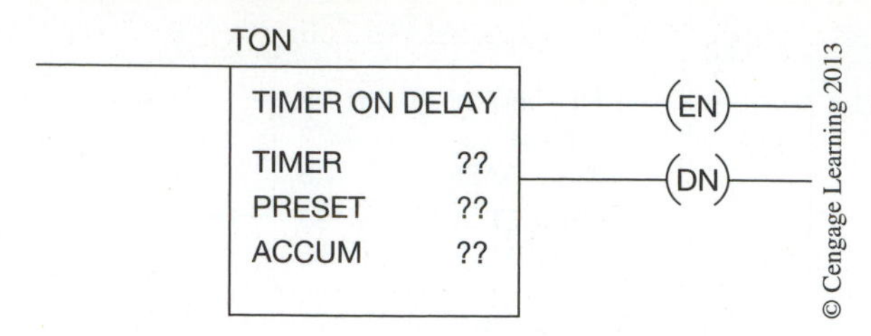

Figure 11–23 Allen-Bradley Logix5000 Timer Format

The same EN, DN, and TT bits are available on the Logix5000 timers as were available on the other Allen-Bradley timers discussed previously. The operation of the TON, TOF, and RTO timers are the same in both types of processors, and the timing charts in Figures 11–19, 11–21, and 11–22 can be applied to both. The only difference between the timers is that the Logix5000 timers have a fixed time base of 1 msecs and use a tag structure to address the timer.

Figure 11–24 shows a TON timer programmed in a Logix5000 controller. Note that the *Timer* parameter contains the tag (address) for the timer. In this example tag name "Flasher" was assigned to

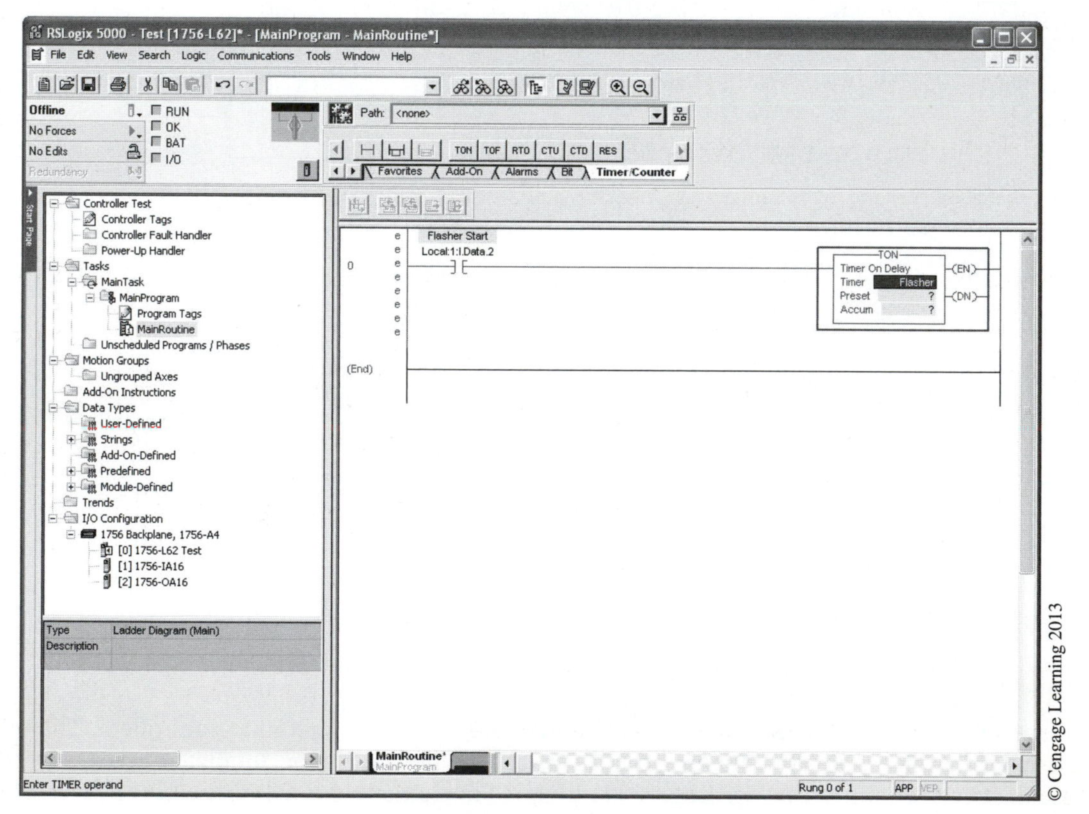

Figure 11–24 Logix5000 TON Instruction

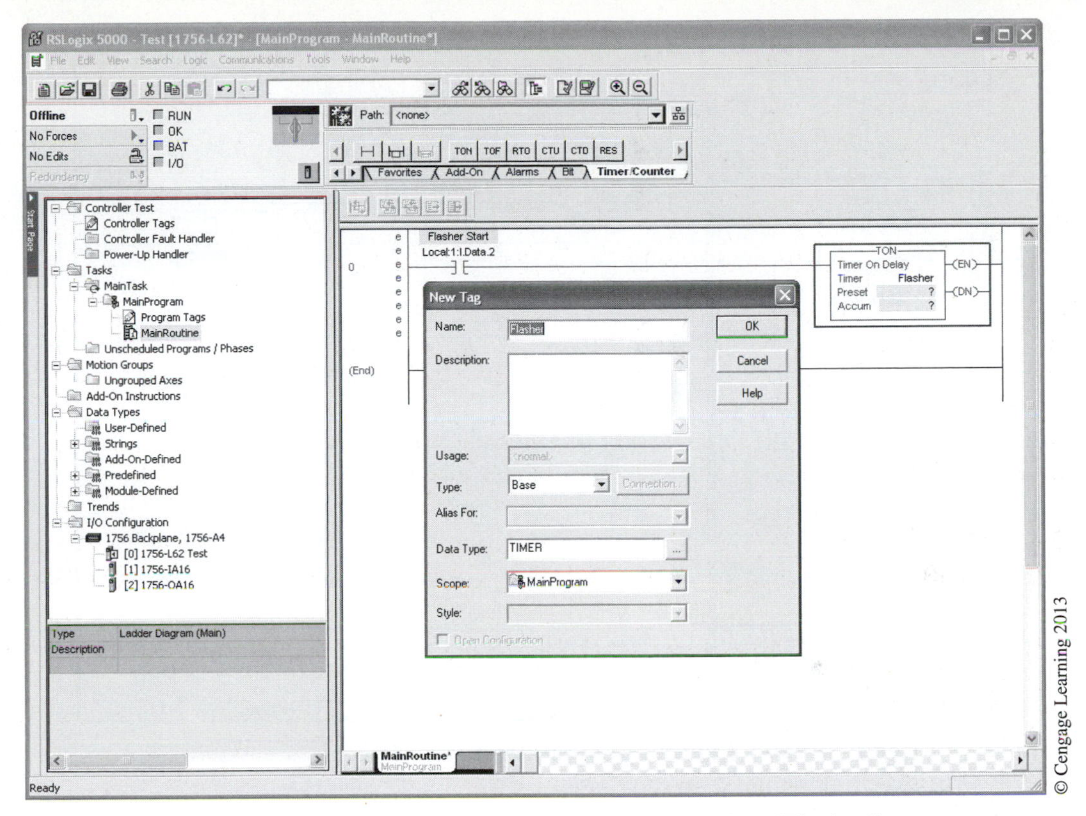

Figure 11–25 New Tag Dialog Box for Timer Tag "Flasher"

the timer. Figure 11–25 shows the new tag window. By default, when creating new tags they will be configured as program-scoped tags.

After the tag has been created, the preset and accumulated values can be entered. The finished TON timer is shown in Figure 11–26.

The Logix5000 controllers have three additional timer instructions—TONR, TOFR and RTOR—that have a built-in reset function. These timers are only available with Function Block programming, which will be covered in Chapter 18.

CASCADING TIMERS

When circuit requirements demand more time than is available from a single timer, two or more timers can be programmed together, as shown in Figure 11–27. Programming two or more timers together to extend the timing range is called **cascading.**

In this circuit, the first timer is controlled by input device I:012/01. When the device is true, the timer starts to time. When the accumulated time is equal to the preset time, the timer done bit is set

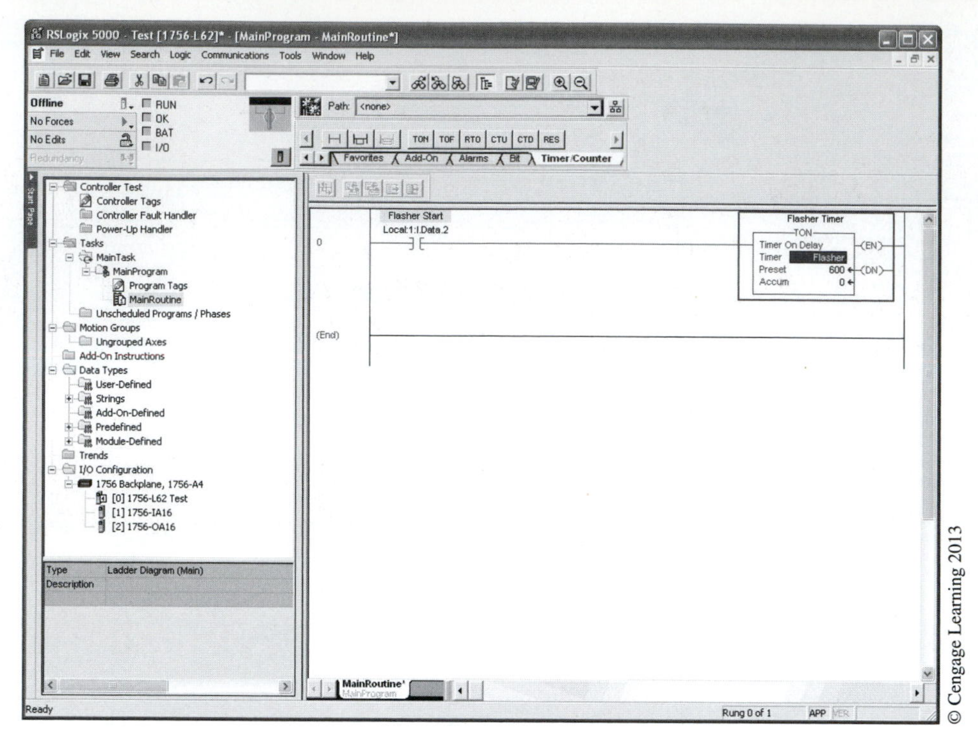

Figure 11–26 Finished TON Timer

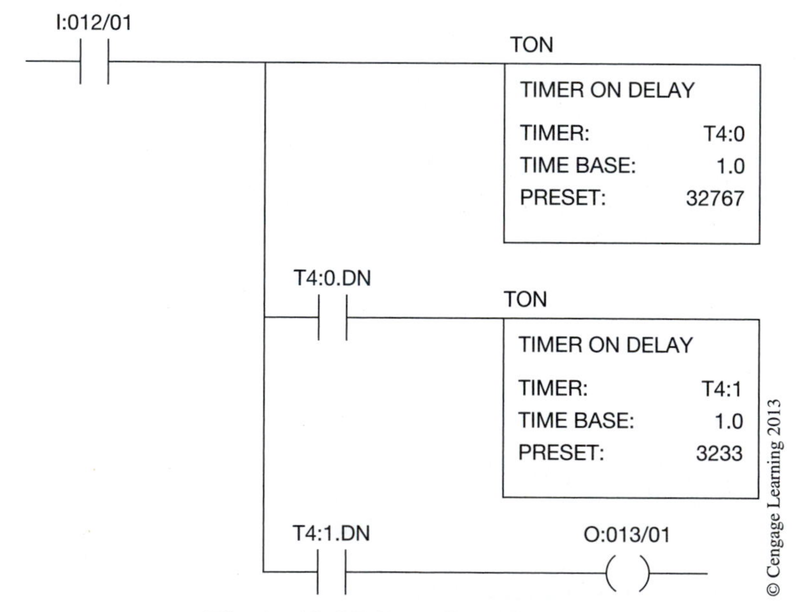

Figure 11–27 Cascading Timers

to 1, or *ON*. When the timer done bit (T4:0.DN) is set to 1, the second timer is enabled and starts to time. When the second timer has timed out, output O:013/01 is turned *ON*. The total time to turn *ON* output O:013/01 after input I:012/01 was true is 36,000 seconds (32,767+3,233), or 600 minutes.

Chapter Summary

Although the format is different for different PLCs, the basic principles are the same. Preset and accumulated times are stored and compared on each processor scan. When the accumulated value equals the preset value, discrete output devices or internal outputs can be turned *ON* or *OFF*. Timers can be programmed for *ON delay* or *OFF delay,* or as *retentive* timers. The only limit to the number of timed and instantaneous contacts that can be programmed is memory size. Programmed timers offer a wider range of time settings and greater accuracy than is possible with hardwired pneumatic timers.

Review Questions

1. Match the standard time delay symbols.

a.

b.

c.

d.

1. N.O.T.O.
2. N.O.T.C.
3. N.C.T.C.
4. N.C.T.O.

2. The amount of time for which a timer is programmed is called the:
   a. preset
   b. set point
   c. desired time (DT)
   d. all of the above
3. As scan time increases, so does the accuracy of any programmed timers.
   T F
4. When the timing of a device is not reset due to a loss of power, the timer is said to be:
   a. holding
   b. secured
   c. retentive
   d. continuous
5. When more time is needed than can be programmed with one timer, two or more timers can be programmed together. This programming technique is called:
   a. stacking
   b. cascading
   c. doubling
   d. synchronizing

6. When programming timers with Allen-Bradley format, which bit will act as an instantaneous contact?
   a. DN
   b. TT
   c. EN
   d. IN
7. When the accumulated time is equal to the preset time, which bit in the Allen-Bradley PLC-5 family will be true?
   a. DN
   b. TT
   c. EN
   d. IN
8. When programming a Logix5000, what preset value would be entered to create a 23-minute timer?

CHAPTER 12 Programming Counters

Objectives

After completing this chapter, you should have the knowledge to:

- Write a program using up and down counters.
- Define the terms *increment* and *decrement*.

Programmed counters serve the same function as the mechanical counters used in the past. Programmed counters can count up, count down, or be combined to count up and down. Counters are similar to timers, except they do not operate on an internal clock but instead are dependent on external or program sources for counting.

ALLEN-BRADLEY PLC-5, SLC 500, AND MICROLOGIX COUNTERS

Allen-Bradley offers two types of counters: up counters (CTU) and down counters (CTD). Both counters are retentive until reset by a reset instruction. Figure 12–1 shows a typical Allen-Bradley up counter.

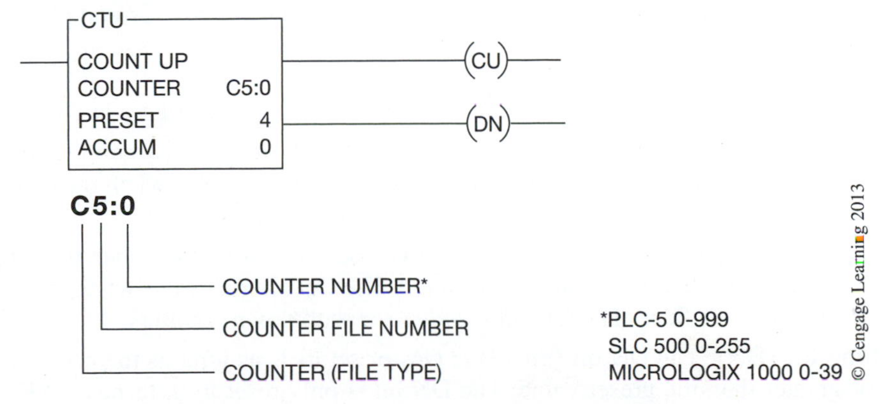

Figure 12–1 Allen-Bradley PLC-5 Counter Format

The Allen-Bradley up counter format is similar to the timer format. The up counter consists of a counter block that contains the up counter address, the preset value, and the accumulated count value, which can be any number from 0 to +32,767. The counter address consists of C for counter, the file number (5 [the default number]), a colon (:), and the counter number. File 5 is the default file from the data table for counters using the PLC-5, SLC 500, and the MicroLogix 1000. The PLC-5 can be programmed to use files 3-999 for additional counter files. The SLC 500 can be programmed to use files 9-255 for additional counter files. The MicroLogix is limited to one counter file, which is the default file, file 5.

By only having one counter file the MicroLogix 1000 is limited to 32 counters (counters 0 through 31), whereas the SLC 500 can use 0 through 255 counters per counter file. The PLC-5 can have counter numbers from 0 through 999. Each counter requires three words of memory, as shown in Figure 12–2.

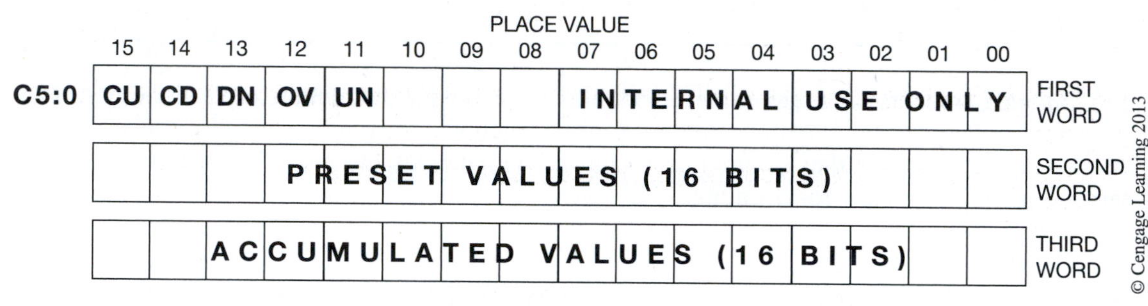

Figure 12–2 Storage Format for Counters

The first word stores the status bits of the counter (bits 11 through 15). The second word holds the preset values, or count, and can range from –32,768 to +32,767. The positive numbers are stored in 16-bit binary, while the negative numbers are stored in the 2s complement. The third word stores the accumulated count, and can be any number from –32,768 to +32,767. (2s complement is covered in Chapter 5.)

The status bits for up and down counters that are stored in the first word are as follows:

Count Up Enable Bit (CU) The CU bit (bit 15) is set to 1, or is true, when the rung is true, and remains true as long as the up counter is enabled. The CU bit goes false when the counter is reset or the counter rung goes false. The CU bit is *only used* with up counters.

Count Down Enable Bit (CD) The CD bit (bit 14) is set to 1, or is true, when the rung is true, and remains true as long as the down counter is enabled. The CD bit goes false when the counter is reset or the counter rung goes false. The CD bit is *only used* with down counters.

Count Done Bit (DN) The DN bit (bit 13) is *ON,* or set to 1, as long as the accumulated value is equal to or greater than the preset value. The DN bit is only reset to 0, turned *OFF,* when the accumulated count is less than the preset value.

Count Up Overflow Bit (OV) The OV bit (bit 12) is set by the processor to 1, or *ON,* when the accumulated count exceeds the *upper limit* of (+)32,767. When this limit is reached, the count wraps around to (–)32,767, and the up counter increments from there.

Count Down Underflow Bit (UN) The UN bit (bit 11) is set by the processor to 1, or *ON,* when the accumulated count exceeds the *lower limit* of (–)32,768. It wraps around to (+)32,767, and the CTD instruction counts down from there.

Figure 12–3 shows a CTU counter and how it is programmed to control outputs O:013/01 and O:013/02. Rung 5 is the reset rung that resets the counter's accumulated value to 0000. An output instruction is used to reset the counter. The reset (RES) command must have the same address as the counter to enable it to be reset.

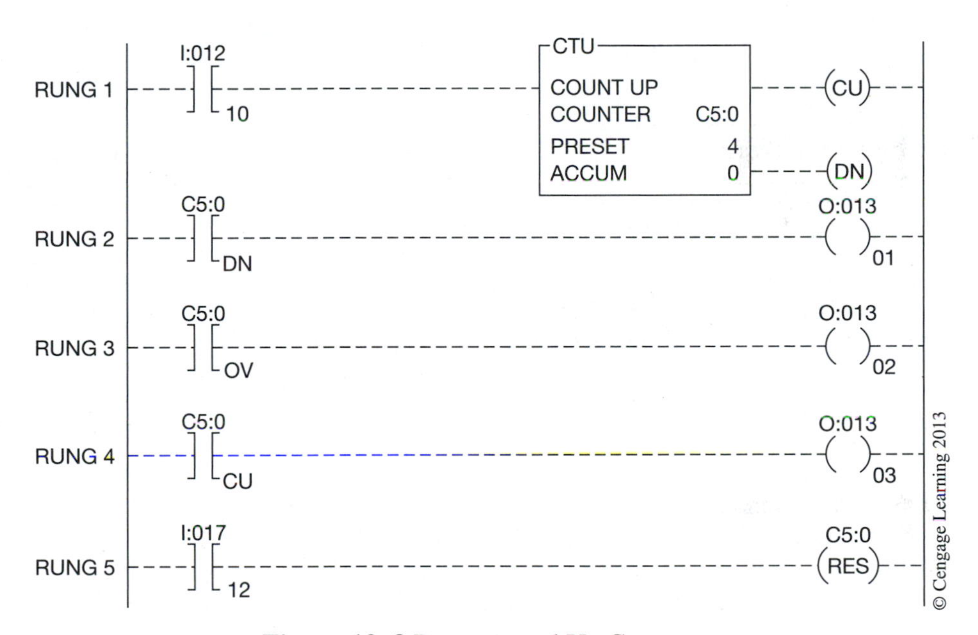

Figure 12–3 Programmed Up Counter

Each time input device I:012/10 in Rung 1 makes a transition from false to true, the counter increments, or counts up by 1. When the accumulated value (count) is equal to or greater than the preset count, the done (DN) bit is set to 1 by the processor, and Rung 2 becomes true, turning *ON* output O:013/01. Rung 3 is not true unless the count exceeds the counter's upper limit of (+)32,767. If the count exceeds the limit, output O:013/02 comes *ON* and remains *ON* until the counter is reset by closing and then opening input device I:017/12 in Rung 5. Bit 15, the count up bit (CU), can be programmed and used to indicate that the counter is enabled and that Rung 1 is true. The CU bit in Rung 4 is set to 1 by the processor any time that input device I:012/10 is true, thereby enabling the counter. Bit 15 is reset to 0 when Rung 1 goes false, or the timer is reset.

Figure 12–4 shows a typical counting chart for a CTU timer.

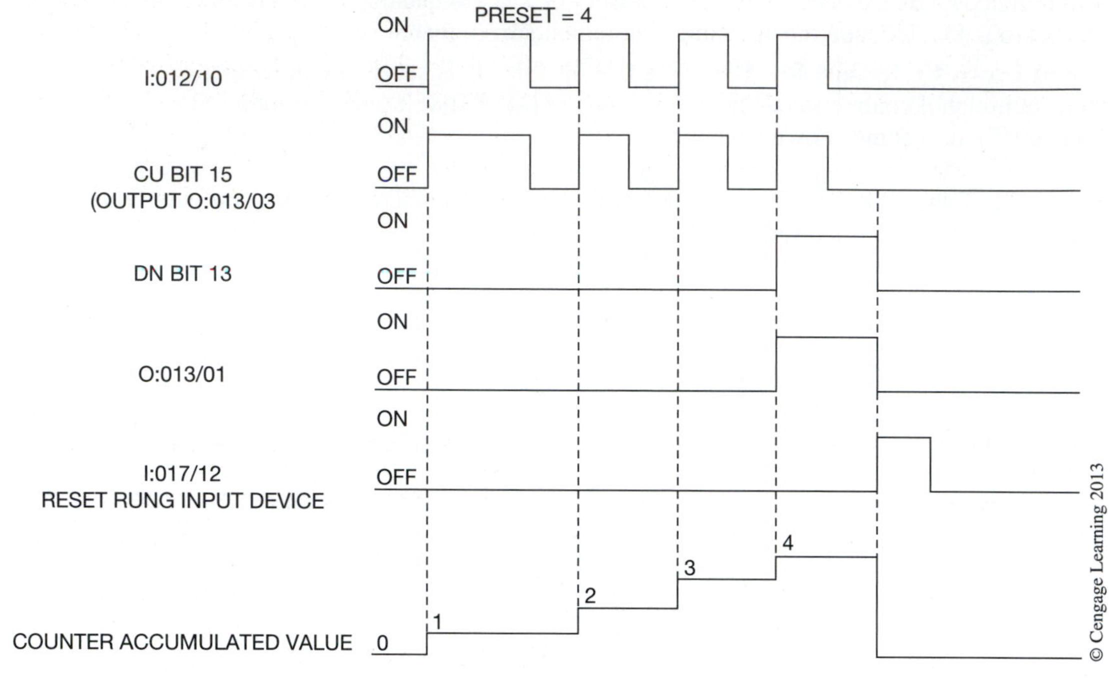

Figure 12–4 CTU Counting Chart

Figure 12–5 shows how a PLC-5 down counter (CTD) is programmed.

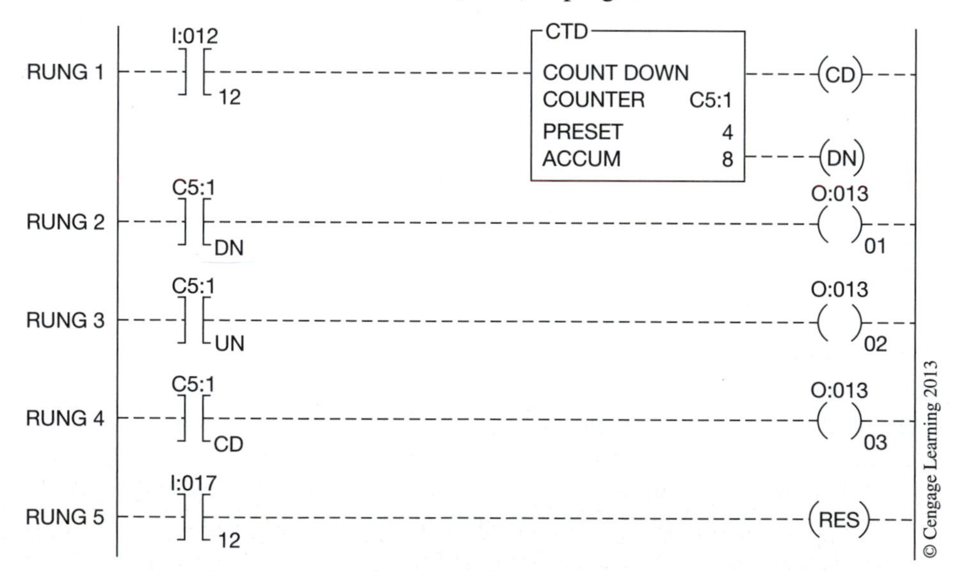

Figure 12–5 Programmed CTD Counter

The CTD counter counts down each time input device I:012/12 in the counter rung (Rung 1) goes from false to true. As long as the accumulated count is equal to or greater than the preset count, the output device (O:013/01) in Rung 2 remains *ON*. When the accumulated count falls below the preset count of 4, output O:013/01 is set to 0, or *OFF*. Rung 3 contains the underflow bit, which is opposite the overflow bit used with the CTU counter, and is only set to 1 when the count goes below (–)32,768. Rung 4 contains the CD bit and is *ON,* or true, any time the counter is enabled. The CD bit mirrors the status of input device I:012/12. Rung 5 is the reset rung and uses input device I:017/12 for resetting the counter. Figure 12–6 shows the counting chart for a count down timer (CTD).

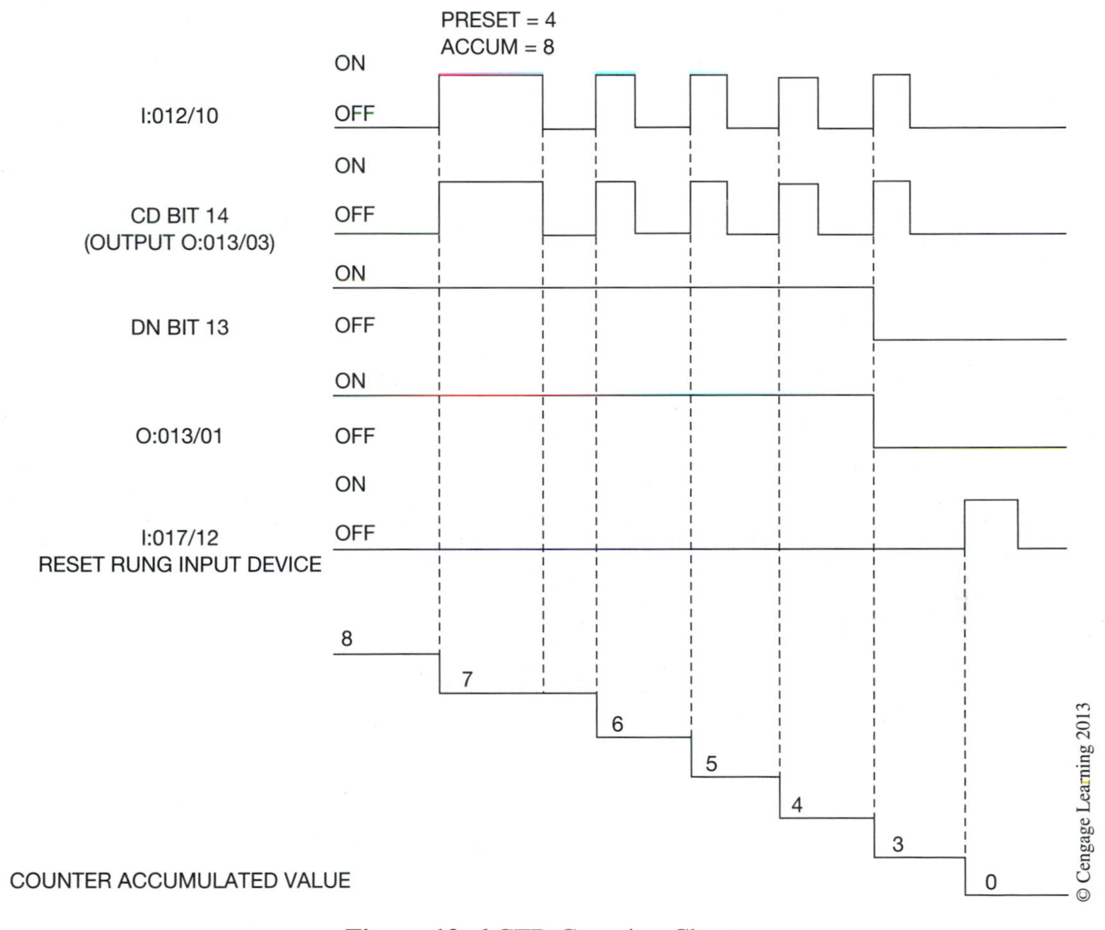

Figure 12–6 CTD Counting Chart

Many times a CTD instruction is combined with a CTU instruction as shown in Figure 12–7a. Figure 12–7b is an example counting chart for the CTD and CTU combination. When combining up and down counters, the same counter file and counter number are used for both counters as well as for the reset instruction in Rung 6.

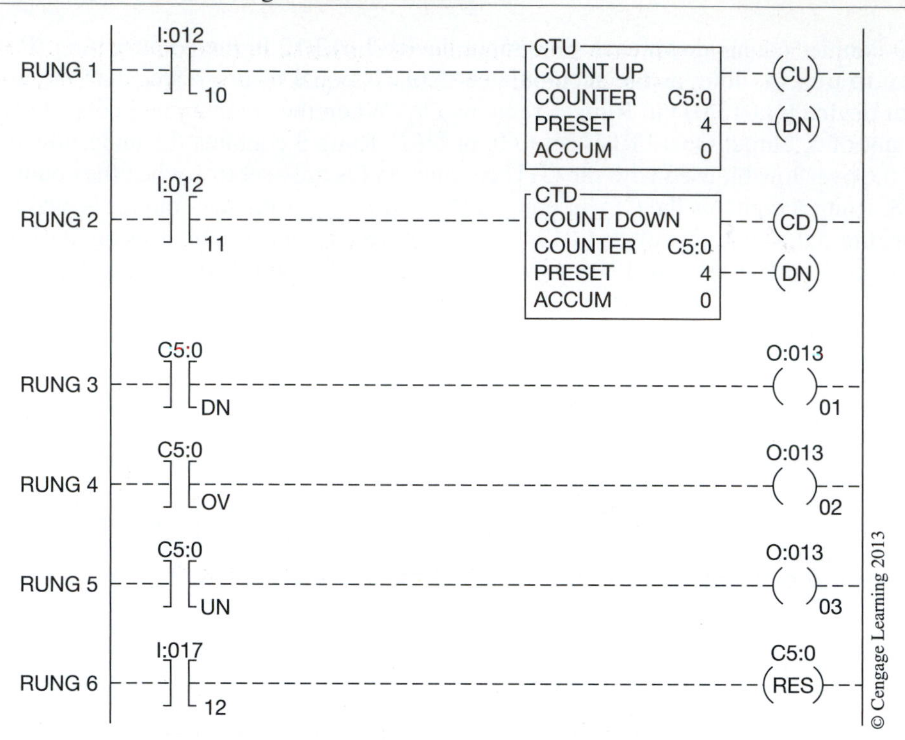

Figure 12–7a Combining CTU and CTD Instructions

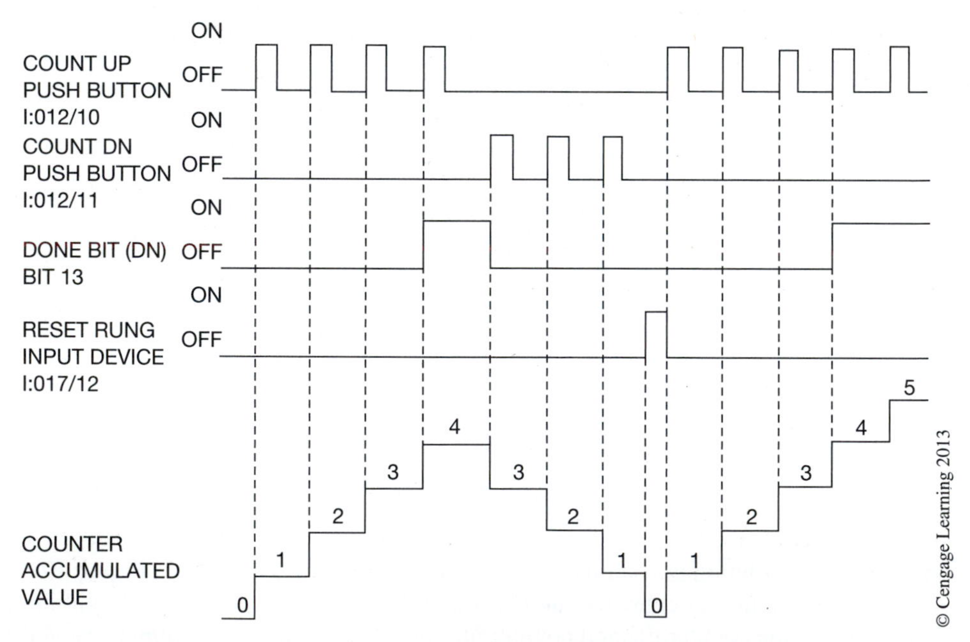

Figure 12–7b Combined CTU and CTD Counting Chart

Up and down counters can be programmed together as shown in Figures 12–8a and 12–8b to count products as they enter a conveyor line (count up) and as they leave the line (count down).

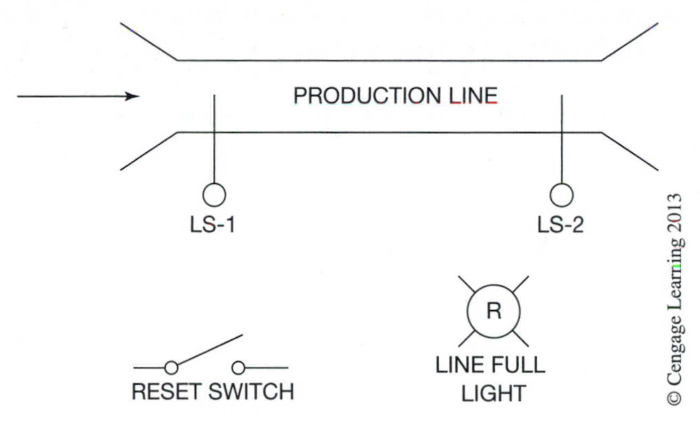

Figure 12-8a Applying Up and Down Counters

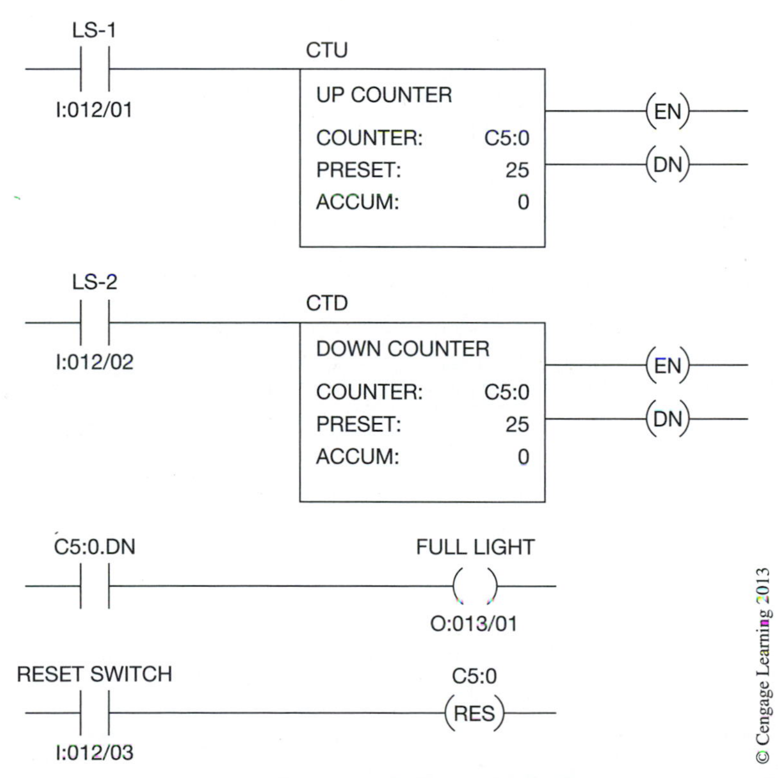

Figure 12-8b Up and Down Counter Logic

As a product enters the conveyor line, input I:012/01 (LS-1) is activated (false-to-true), and the actual count in C5:0 increments from 0 to 1. The next product increments C5:0 to 2, and so on. If eight products entered the line before any were removed, the accumulated count in C5:0 would be 8. The first product to leave the line and activate I:012/02 (LS-2) decrements the actual count in C5:0 (from 8 to 7). The next product that left the line and activated I:012/02 (false-to-true) would again decrease the accumulated count in C5:0 from 7 to 6.

The indicator lamp, output O:013/01 (red), indicates the condition of the production line. When the line is full, the up counter accumulated value is equal to or greater than the preset (25), and the red lamp (output O:013/01) is ON.

Input device I:012/03 is a RESET switch. When the switch is closed, the accumulated value in C5:0 is reset to 0. Up and down counters are retentive and retain their values during power failures.

COMBINING TIMERS AND COUNTERS

Timers can be combined with counters when it is necessary to extend the time of the timer beyond its normal limits. An example of combining a timer with a counter is shown in Figure 12–9.

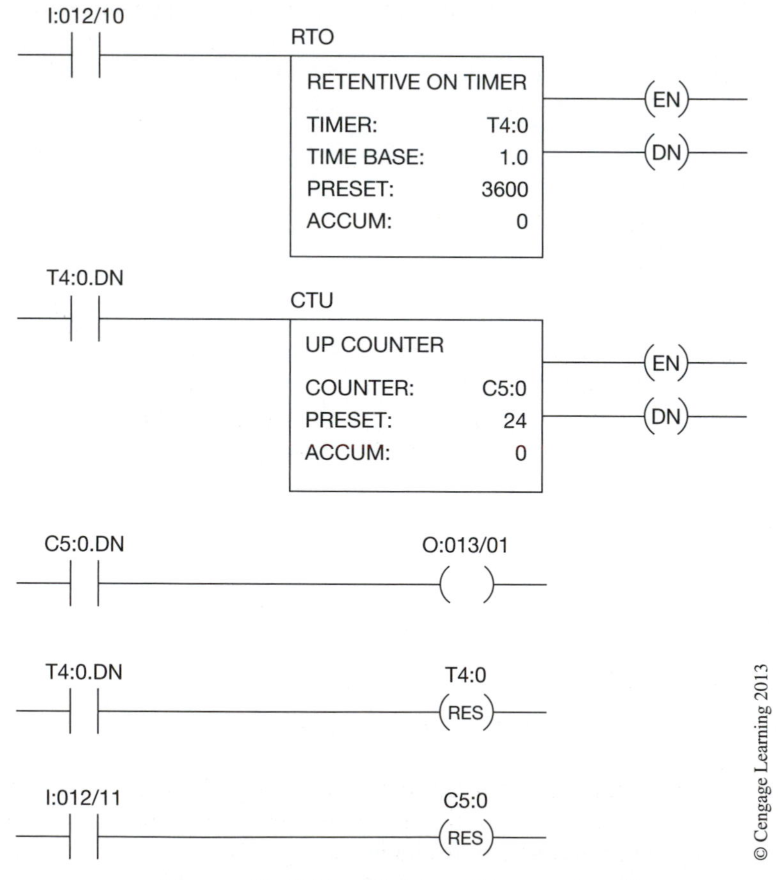

Figure 12–9 Combining a Timer with a Counter

The timer (T4:0) has a time base of 1.0 seconds and a preset value of 3600. The 3600-second preset value is equal to 1 hour. When input device I:012/10 is closed, the timer starts to time in 1-second increments. When the accumulated time is equal to the preset value, the DN bit is set to 1 and the CTU counter counts, or increments, by one. The DN bit also resets the timer and the timer starts to accumulate time again. When the accumulated time on the timer has reached 3600 seconds, the timer increments counter C5:0 again and resets itself. The counter continues to count each time the DN bit makes a false-to-true transition (every 3600 seconds, or one hour) until the accumulated count equals the preset value of 24. When the counter has counted to 24 (24 hours), the counter DN bit is set to 1 and output O:013/01 is turned *ON*. Input I:012/11 is used to reset the counter.

Note: *Remember that the length of the program affects scan time, which in turn affects timer accuracy and total time. The actual time it takes for the counter to count to 24 may be 24 hours plus or minus a few minutes.*

ALLEN-BRADLEY LOGIX5000 COUNTERS

Figure 12–10 shows an up counter format used by Allen-Bradley's Logix5000 controllers. The counter instruction, CTU or CTD, contains the counter tag, and the preset and accumulated values. The counter instruction preset value can be set for any value between 0 and 2,147,483,647 (DINT). If the accumulated value should exceed the upper limit, the overflow bit (OV) will be set and the counter then rolls over to –2,147,483,648 and begins counting up again.

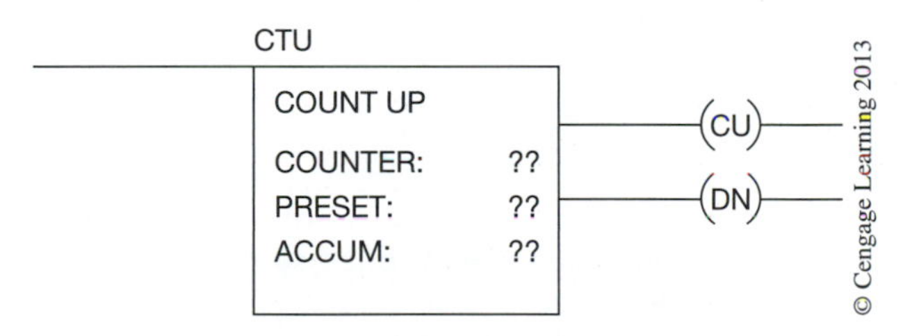

Figure 12–10 Allen-Bradley Logix5000 UP Counter Format

The same CU, CD, DN, OV, and UN bits are available on the Logix5000 counters as were available on the other Allen-Bradley counters discussed previously. The operation of the CTU and CTD counters are the same in both types of processors and the counting charts in Figures 12–4 and 12–6 can be applied to both. The only difference between the counters is the Logix5000 counters use a tag structure to address the counter and can have preset or accumulated values up to ±2,147,483,647 (DINT).

Figure 12–11 shows a CTU counter being programmed in a Logix5000 controller. Note that the *Counter* parameter contains the tag (address) for the counter. In this example tag name "Counter1" was assigned to the counter. Figure 12–12 shows the new tag window. By default, when creating new tags they will be configured as program-scoped tags.

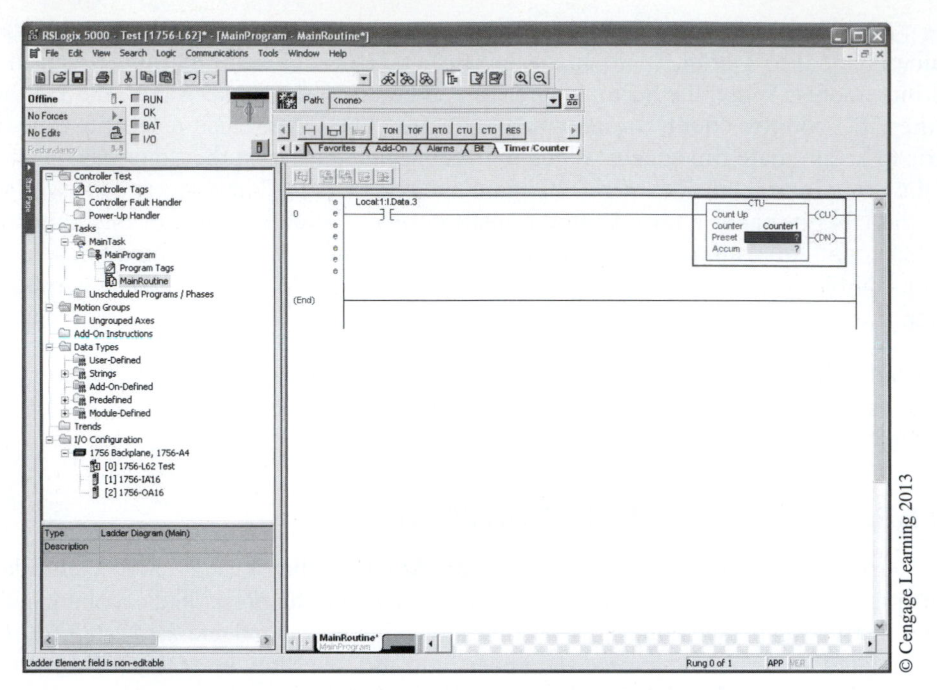

Figure 12–11 Logix5000 CTU Instruction

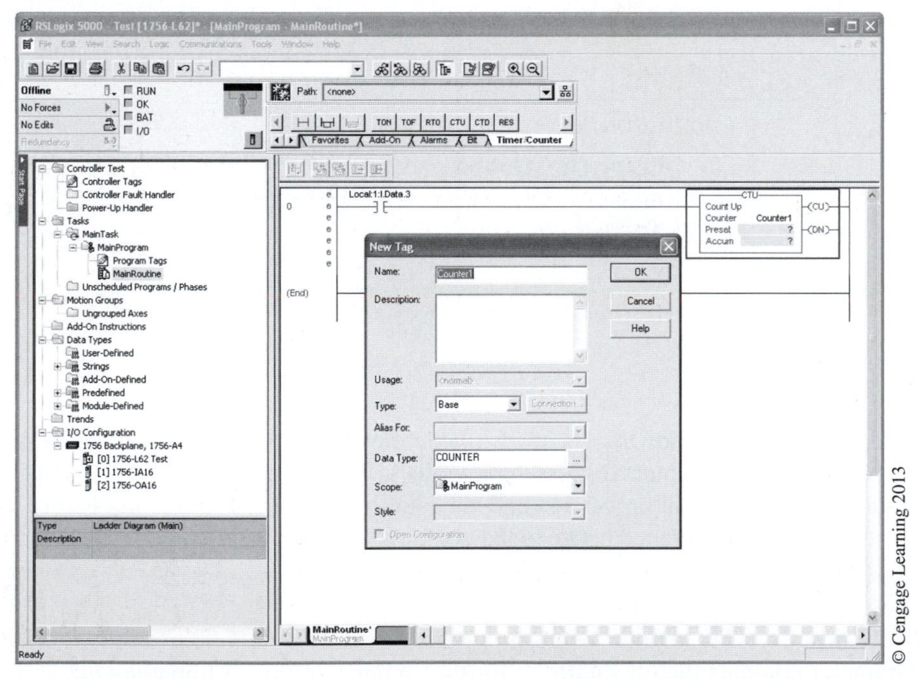

Figure 12–12 New Tag Dialog Box for Counter Tag "Counter1"

After the tag has been created, the preset and accumulated values can be entered. The finished CTU counter is shown in Figure 12–13.

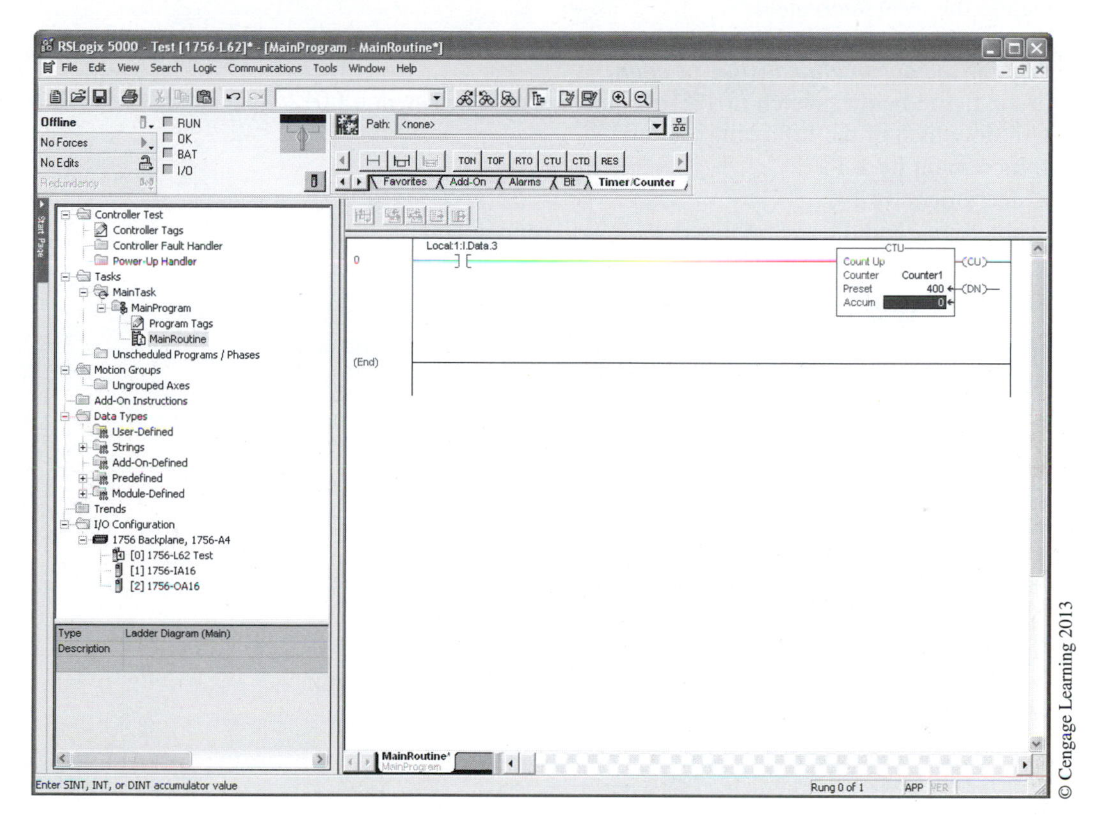

Figure 12–13 Finished CTU Counter

The Logix5000 controllers have one additional counter instruction, CTUD, that is both an up and down counter in one instruction with a built-in reset function. This counter is only available with Function Block programming, which will be covered in Chapter 18.

Chapter Summary

Programmed counters give added flexibility and control to electrical process equipment and/or driven machinery. Similar to timers, counters store values in binary format for the preset and accumulated counts. The processor compares the preset and accumulated values on each scan of the rung, and updates the counter's status bits as appropriate.

Review Questions

1. Define the term *increment.*
2. Define the term *decrement.*
3. What is the *preset value* or *count?*
4. What is the *accumulated value* or *count?*
5. In the figure below, switch I:012/10 is now open. When switch I:012/10 is closed, counter C5:0 will do which of the following?
   a. increment by 1
   b. decrement by 1
   c. not count, and the accumulated value will remain at 0

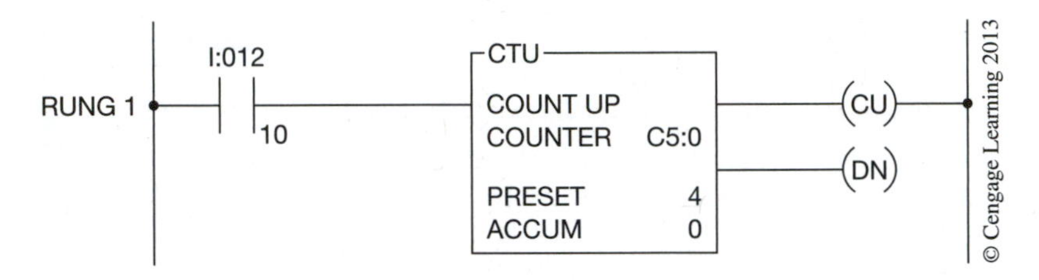

6. Output O:013/01, shown in Rung 2 of the figure below, is true:
   a. only when the count is equal to the preset value
   b. when the count is equal to or greater than the preset value
   c. when the count is less than the preset value
   d. when the accumulated value reaches +32,767 and overflows
   e. when the count goes to 011

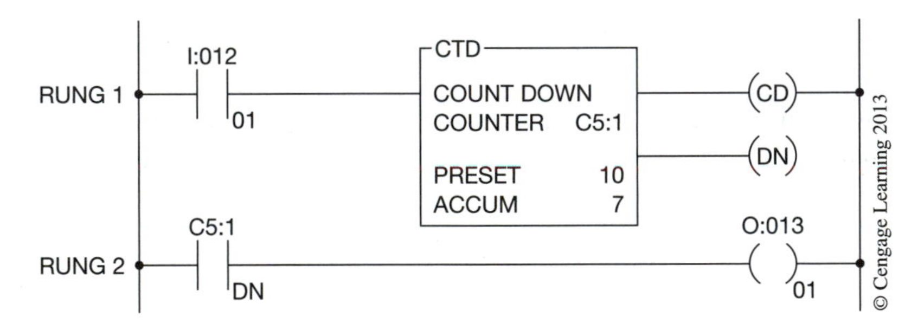

7. When the accumulated count is equal to or greater than the preset count, which bit in the Allen-Bradley SLC 500 family will be true?
   a. CU
   b. CD
   c. OV
   d. DN
8. When is the OV status bit set on an Allen-Bradley SLC 500 up counter?

9. Up and down counters can be programmed together to count up and down.
   T F
10. The reset rung shown in the figure below resets counter C5:0:
    a. automatically when the count reaches 010
    b. automatically when the count reaches 011
    c. only when the count reaches 32,767
    d. only when switch I:017/12 is closed
    e. only when switch I:017/12 is closed and then opened

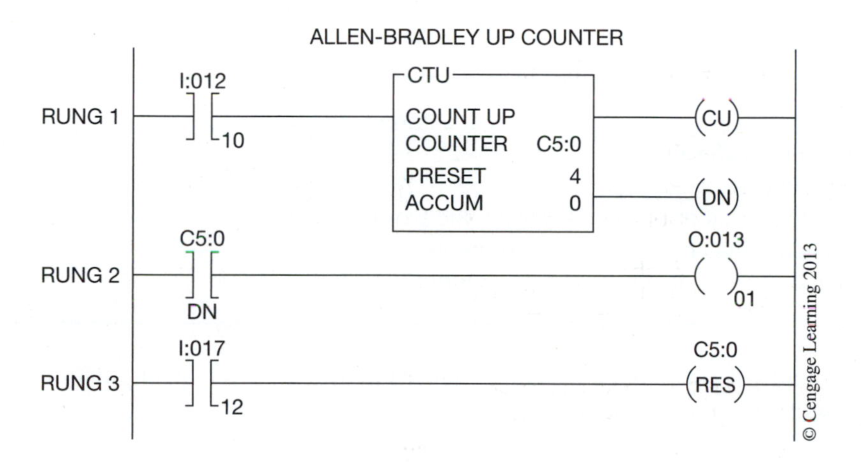

11. Define the term *overflow*.
12. Define the term *underflow*.
13. When an up counter accumulated value equals the preset value, the counter will:
    a. reset itself
    b. stop counting
    c. continue to count
    d. continue to count but go into an overflow condition as soon as the accumulated value exceeds the preset value

CHAPTER

13 Data Manipulation

Objectives

After completing this chapter, you should have the knowledge to:

- Explain what data transfer is.
- Define the term *writing over*.
- Write a rung of logic that transfers data from one word to another.
- Identify the standard data compare instructions.
- Write logic that compares data to control an output.

Most PLCs now have the ability to manipulate data that is stored in memory. Data manipulation can be placed in two broad categories: data transfer and data compare.

DATA TRANSFER

Data transfer consists of moving or transferring numeric information stored in one memory word location to another word in a different location. Words in the user memory portion of the processor may be referred to as data table words, holding registers, internal memory tags, and/or storage register words, depending on the PLC.

Figures 13–1a and 13–1b illustrate the concept of moving numerical data from one word location to another word location. Figure 13–1a shows that numeric (binary) data is stored in word 0 of file N7, and that no information is currently stored in word 1 of file N7.

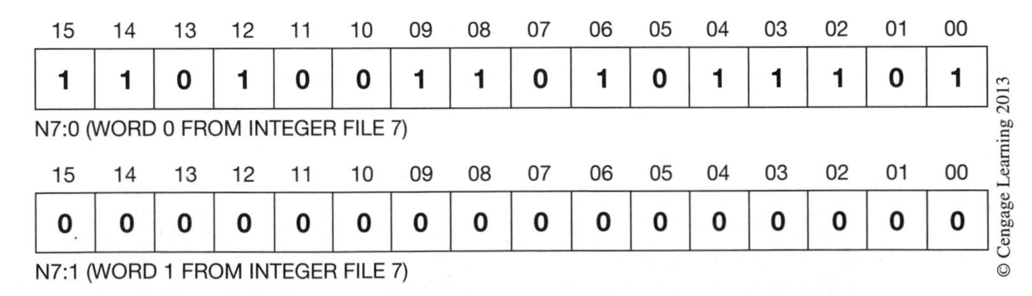

Figure 13–1a Numeric Data Stored in Words 0 and 1 of File N7

15	14	13	12	11	10	09	08	07	06	05	04	03	02	01	00
1	1	0	1	0	0	1	1	0	1	0	1	1	1	0	1

N7:0 (WORD 0 FROM INTEGER FILE 7)

15	14	13	12	11	10	09	08	07	06	05	04	03	02	01	00
1	1	0	1	0	0	1	1	0	1	0	1	1	1	0	1

N7:1 (WORD 1 FROM INTEGER FILE 7)

Figure 13–1b Data Transferred from Word 0 of File N7 into Word 1 of File N7

After the data transfer (Figure 13–1b), word 1 of file N7 now holds the exact or duplicate information that is in word 0 of file N7. If word 1 had information already stored, rather than all 0s, the information would have been replaced. When new data replaces existing data in a word after a transfer, it is referred to as **writing over** the existing data.

ALLEN-BRADLEY PLC-5, SLC 500, AND MICROLOGIX DATA TRANSFER INSTRUCTIONS

The PLC-5, SLC 500, and MicroLogix use a move (MOV) instruction for moving data from one word to another. MOV is an output instruction that copies a value from one word (source address) to another word (destination address). When the rung that holds the MOV instruction is true, the instruction moves data from the source address into the destination address on each processor scan. Figure 13–2 shows the MOV format.

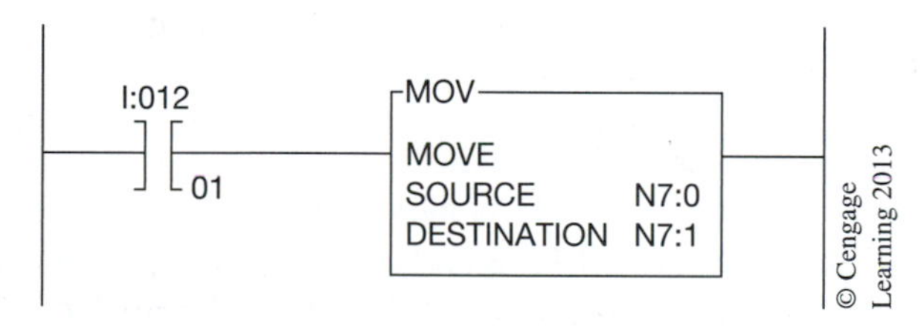

Figure 13–2 PLC-5 MOV Format

When input device I:012/01 is closed and the rung is true, the MOV instruction reads the data from the source address and copies it to the destination address. In this case, the source is integer file N, file number 7, and Word 0. The destination is integer file N, file 7, Word 1. As long as the rung stays true, the values found in Word 0 of N7 are transferred (copied) into Word 1 of N7 on each program scan. The integer file (as discussed in Chapter 4) is used for storing whole numbers.

To illustrate how an MOV instruction could be used, consider a machine that produces two types of products. Product A requires a time delay of 10 seconds during the processing and Product B requires a 20-second delay. The ON delay (TON) timer in the process program would be programmed as shown in Figure 13–3.

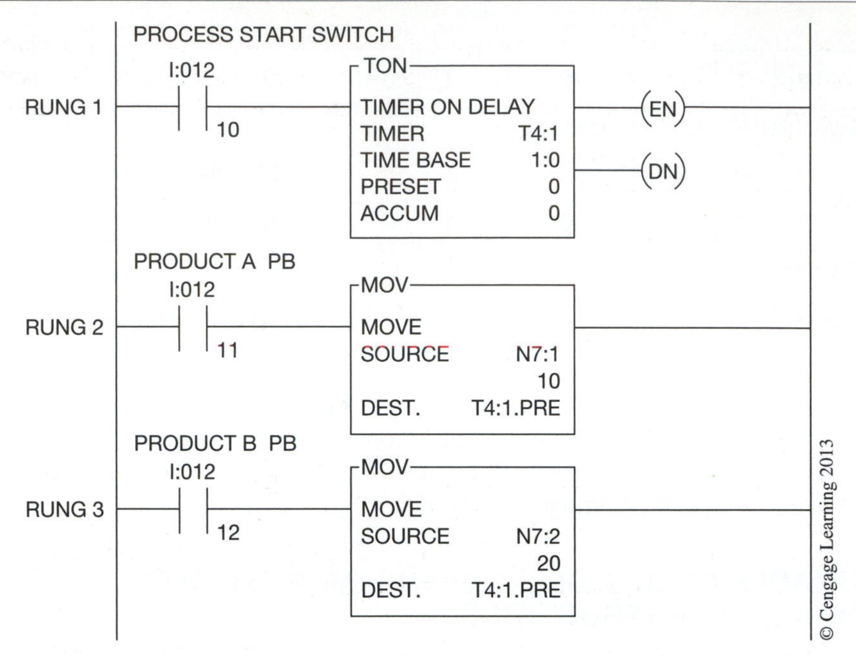

Figure 13–3 MOV Instruction Used to Change Timer Preset Values

The ON delay timer T4:1 is programmed with no preset value as shown in Rung 1. Rung 2 has input device I:012/11 controlling an MOV instruction programmed to move the numeric value found in Word 1 of integer file N7 (Source) into the destination word, which is shown as T4:1.PRE. The destination word is the word that holds the preset value for timer T4:1. Rung 3 is programmed so that when the rung is true, the MOV instruction moves the numeric value in Word 2 of file N7 into the word that holds the preset value for timer T4:1.

When Product A is being processed, input device I:012/10 and I:012/11 are activated, and the numeric value of 10 from Word 1 of file N7 is moved into the word that holds the preset value of ON delay timer T4:1. This gives timer T4:1 a preset value of 10 seconds. As long as input device I:012/11 remains true, the preset value of T4:1 is 10. When Product B is to be run, input device I:012/11 is opened and input device I:012/12 in Rung 3 is closed. With Rung 3 now true, the value found in Word 2 of file N7 (20) is moved into the word that holds the preset value for ON delay timer T4:1. The value 20, from the Source (N7:1), is moved into the word that holds the preset time for timer T4:1 and overwrites the previous information, which was a preset of 10.

In the Allen-Bradley MOV instruction, constants such as 250, 400, 5.5, etc., can be entered into the source location of the MOV instruction rather than a memory address containing the value to be moved. Only use constants in the source location if the source value is to remain unchanged. In fact, entering constant values into the source of the MOV instruction insures that only that value gets moved into the destination address, and is called *hard coding* the source value.

MOV instructions can be used to change preset values of timer preset or accumulated values of counters, as well as for transferring data between any two words to meet program requirements. An example of how an MOV instruction can be used to change preset values of a counter is a program that counts boards in a saw mill. When the mill is producing 2 × 4s, it wants 400 in a stack. However, when the mill is producing 2 × 6s, only 250 boards are needed for a full stack. Figures 13–4a and 13–4b show how to change the preset value of an up counter for each different lumber size by using pushbuttons.

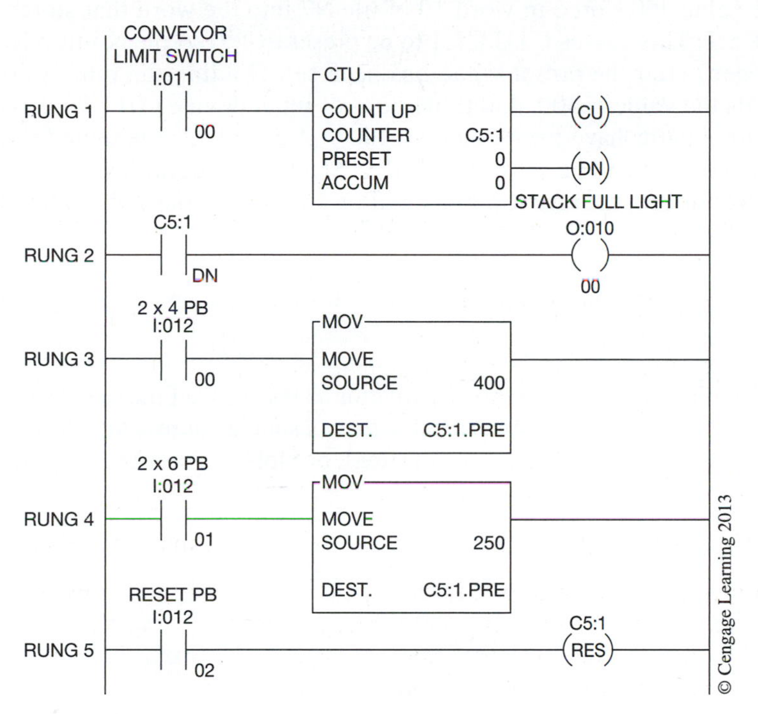

Figure 13–4a Changing Preset Values with an MOV Instruction

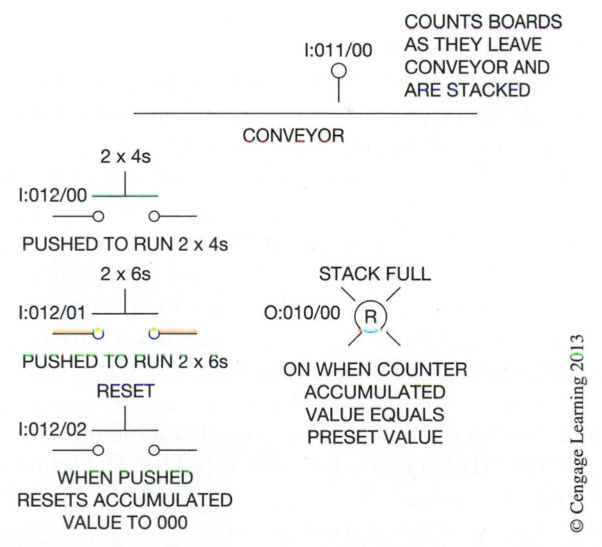

Figure 13–4b Pushbuttons Used to Change Preset Values

Up counter C5:1 is initially programmed with no preset (0). The preset is determined by whichever pushbutton is depressed. If 2 × 4s are to be counted, pushbutton I:012/00 is pushed, enabling the MOV instruction in Rung 3. When this rung is enabled, or true, it tells the processor to move the value 400 stored in word 10 of file N7 into the word that stores the preset value for up counter C5:1. This causes CTU C5:1 to be preset to 400. A pushbutton is used so the rung will go false (open) after the preset value has been set. Holding the button down and keeping Rung 3 true holds the value at 400, and transitions of input device I:011/00 do not increment the counter. After 400 boards have been counted (PR = AC), bit 13 (the done bit) of the first word of C5:1 in the up counter will be set to 1, and the "Stack Full" light, O:010/00, comes *ON*. After the stack has been moved, counter reset pushbutton I:012/02 is pushed to clear the accumulated value back to 0.

To change the preset value of up counter C5:1 from 400 to 250, the pushbutton for 2 × 6s (I:012/01) is depressed.

Another Allen-Bradley data manipulation instruction is the masked move (MVM) instruction. The MVM is an output instruction that copies a value from a source address to a destination address, but in addition allows portions of the data to be **masked,** or blocked from being copied. The format for the MVM instruction is shown in Figure 13–5.

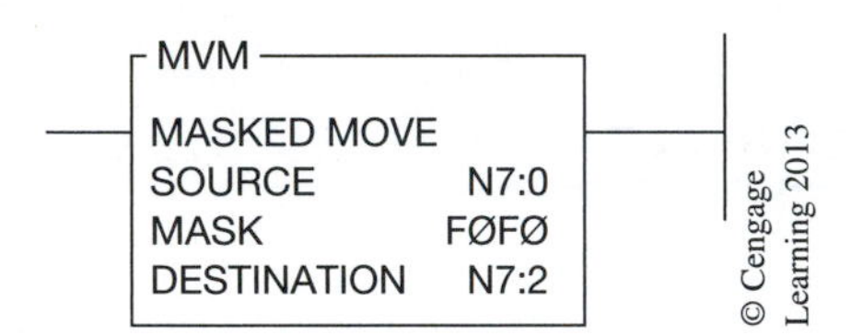

Figure 13–5 Masked Move (MVM) Instruction Format

To program an MVM instruction, a source address and a destination address are required, just as in the MOV instruction. The additional requirement for the MVM instruction is the mask data. For each bit of the destination word that is to be masked, or not copied, a 0 is used. If, on the other hand, it is desired that the data from the source word be written into specific bits of the destination word, a 1 is placed in that bit location. Figure 13–6 clarifies the operation of the MVM instruction.

For each location in the destination word that you want to be overwritten by the data from the source word, a 1 is used. In Figure 13–6, only bits 4 through 7, and bits 12 through 15, are set to 1 in the mask, so only these bits of the destination word will have data transferred in from the source word. Those bits in the destination word that had 0s in the mask (bits 0 through 3 and bits 8 through 11) remain unchanged when the MVM instruction is true.

The bit status for the mask is entered by addressing a word and file that has the desired bit order that is wanted; for example, B100:0 (Binary file, file 100, Word 0). The value can also be entered into the instruction using the hexadecimal format. The mask bit pattern shown in Figure 13–6 is F0F0 in hexadecimal.

Figure 13–6 Mask Bits Used to Block Transfer of Data from Source Address into Destination Address

DATA COMPARE

Data compare opens a new realm of programming possibilities and demonstrates why PLCs are rapidly replacing most, if not all, hardwired control systems.

Data compare instructions, as the name implies, compare the data stored in two or more words and make decisions based on the program instructions. Numeric values in two words of memory can be compared for *less than* (<), *equal to* (=), *greater than* (>), *less than or equal to* (≤), *greater than or equal to* (≥), and *not equal to* (≠) conditions, depending on the PLC.

Data compare concepts were previously used when timers and counters were discussed. The ON delay timer turns *ON* an output when the accumulated value equals the preset value (AC = PR). What happens is that the accumulated numeric data in one memory word is compared to the preset value in another word on each scan of the processor, and when the accumulated value equals the preset value (AC = PR), the output is turned *ON*. Additional programming instructions can compare memory words and turn *ON* outputs when the values are less than (<), equal to (=), greater than (>), and so on.

ALLEN-BRADLEY PLC-5, SLC 500, AND MICROLOGIX DATA COMPARE INSTRUCTIONS

The Allen-Bradley family of programmable controllers has a set of data compare instructions that include *equal* (EQU), *greater than or equal* (GEQ), *greater than* (GRT), *less than or equal* (LEQ), *less than* (LES), and *not equal* (NEQ). Figure 13–7 shows how the EQU instruction is programmed.

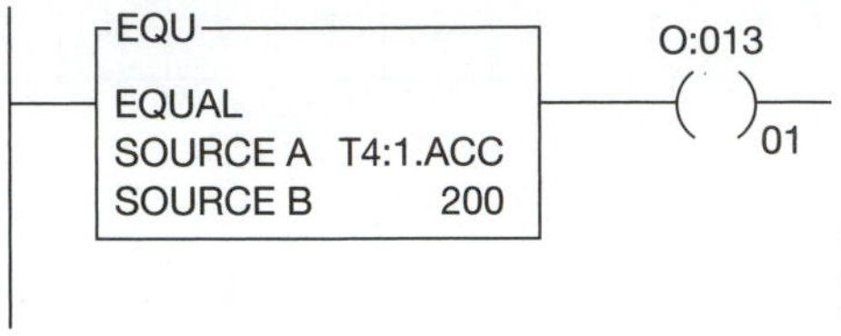

EQU
EQUAL
SOURCE A T4:1.ACC
SOURCE B 200
O:013
01

Figure 13–7 Allen-Bradley PLC-5 Equal To Instruction

The EQU instruction is true, and turns *ON* output O:013/01, when the value in Source A is equal to the value in Source B. Source A and B can be either numeric values or addresses that contain values. The value in Source A is the value of address T4:1.ACC (timer file 4, timer 1, accumulated value), whereas the value in Source B is the numeric value 200. To use either the accumulated or preset values of timers and counters, a period is entered after the timer number, followed by ACC or PRE.

Figure 13–8 illustrates how the GEQ instruction operates.

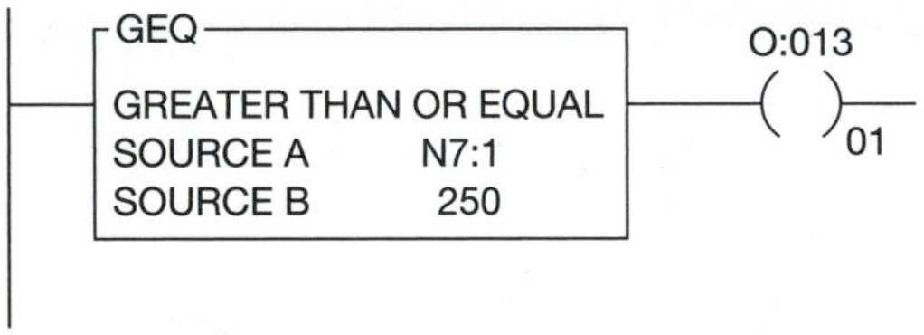

GEQ
GREATER THAN OR EQUAL
SOURCE A N7:1
SOURCE B 250
O:013
01

Figure 13–8 Allen-Bradley PLC-5 Greater Than or Equal To Instruction

This instruction becomes true and turns output O:013/01 *ON* when the value in Source A is greater than or equal to the value in Source B. Again, the value that is in Source A or B can be numeric values or addresses that contain values. In this illustration, the value in Source A is the value stored in integer File 7, Word 1. The value in Source B is the numeric value of 250.

The GRT (greater than) instruction is programmed as shown in Figure 13–9. This instruction is true when the value in Source A is greater than the value in Source B.

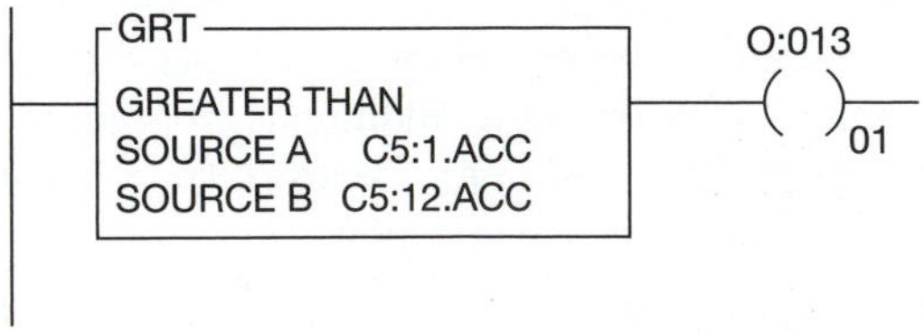

GRT
GREATER THAN
SOURCE A C5:1.ACC
SOURCE B C5:12.ACC
O:013
01

Figure 13–9 Allen-Bradley PLC-5 Greater Than Instruction

The instruction is true as long as the value in Source A is greater than the value in Source B. In Figure 13–9, output O:013/01 is turned *ON* whenever the accumulated value of counter C5:1 is greater than the accumulated value in counter C5:12. As in the earlier example, the preset and accumulated values of timers and counters can be referenced by typing a period followed by either ACC or PRE after the timer or counter address. In Figure 13–9, Source A is the accumulated value of Counter 1, in counter file 5 (C5:1.ACC). Source B is the ACC value of timer 12, in counter file 5 (C5:12.ACC).

The less than or equal instruction (LEQ) is programmed as shown in Figure 13–10. This instruction is true whenever the value in Source A is less than or equal to the value stored in Source B.

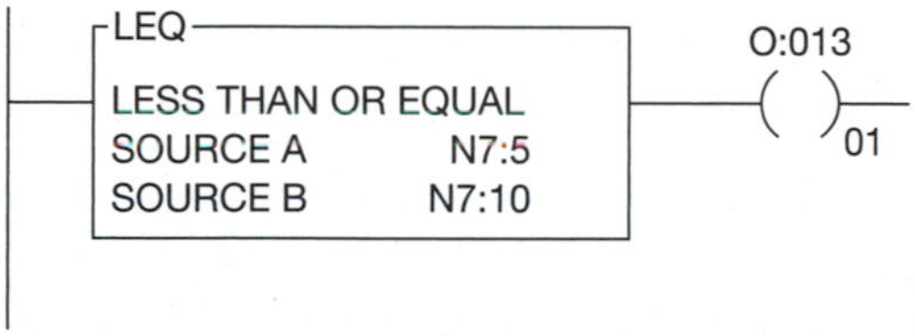

LEQ
LESS THAN OR EQUAL
SOURCE A N7:5
SOURCE B N7:10
O:013
01

Figure 13–10 Allen-Bradley PLC-5 Less Than or Equal To Instruction

The LEQ instruction is true as long as the value in N7:5 is less than or equal to the value in N7:10. When the value in Source A is less than or equal to the value in Source B, output O:013/01 is turned *ON* by the processor.

The LES (less than) instruction is logically true when the value in Source A is less than the value in Source B. Figure 13–11 shows an LES instruction.

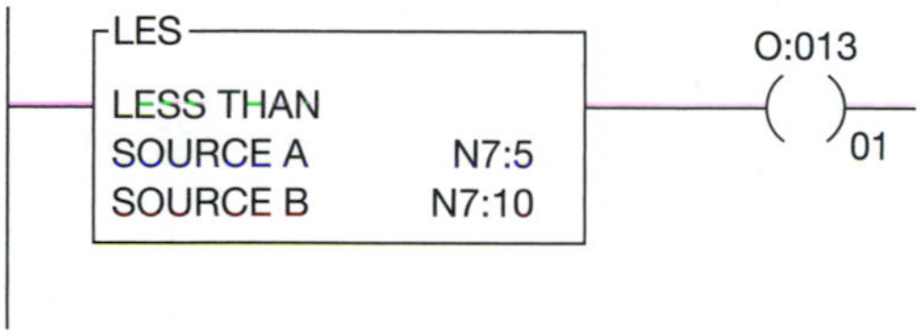

LES
LESS THAN
SOURCE A N7:5
SOURCE B N7:10
O:013
01

Figure 13–11 Allen-Bradley PLC-5 Less Than Instruction

In Figure 13–11, output O:013/01 is *ON* whenever the value in N7:5 (Source A) is less than the value in N7:10 (Source B). When the value in Source A is equal to or larger than the value in Source B, the instruction is not logically true, and output O:013/01 is set to 0, or *OFF*.

The not equal to instruction (NEQ) is programmed as shown in Figure 13–12. This instruction is true whenever the value in Source A is not equal to the value stored in Source B.

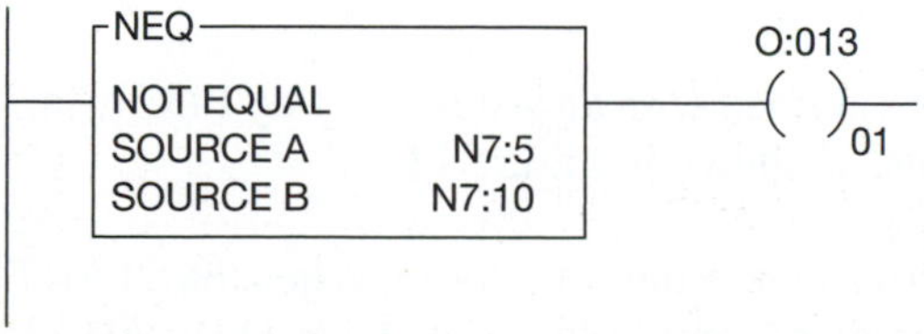

NEQ
NOT EQUAL
SOURCE A N7:5
SOURCE B N7:10
O:013
01

Figure 13–12 Allen-Bradley PLC-5 Not Equal To Instruction

The NEQ instruction will be logically true any time the value in N7:5 (Source A) is not equal to the value in N7:10 (Source B). As long as the values are not equal, output O:013/01 will be turned *ON*. This instruction is logically false only when the value in Source A is equal to the value in Source B.

To graphically demonstrate how data compare instructions can be used, consider the hardwired circuit in Figure 13–13. This circuit uses three pneumatic time delay relays to start up a 4-motor conveyor system in inverse order (4–3–2–1).

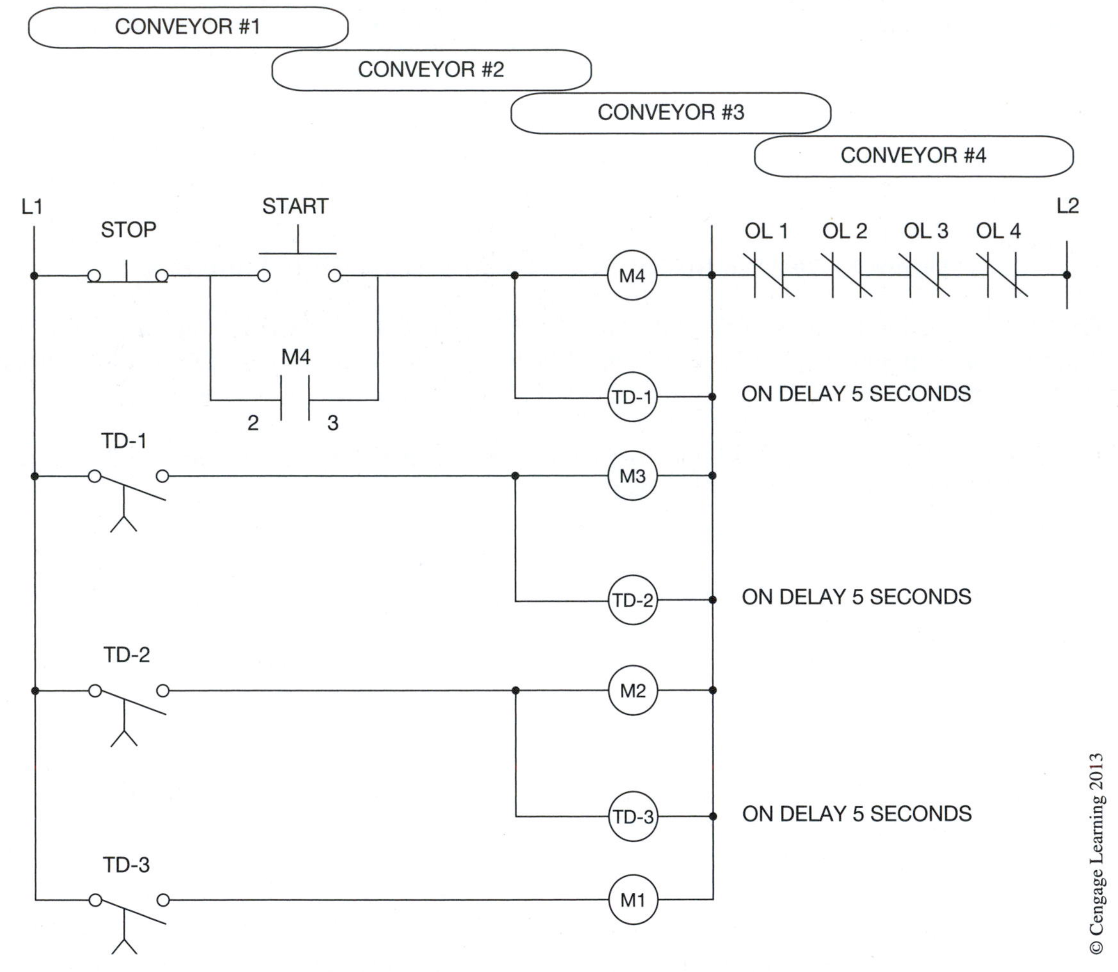

Figure 13–13 Hardwired Conveyor System

The same circuit can be programmed with an Allen-Bradley PLC using only one internal timer and two data compare statements, as shown in Figure 13–14.

Assume that STOP button I:012/00 in Rung 1 is closed. When the START button (I:012/05) is pushed, Output M-4 (O:010/04) energizes, and holding contacts O:010/04 close and hold the circuit in.

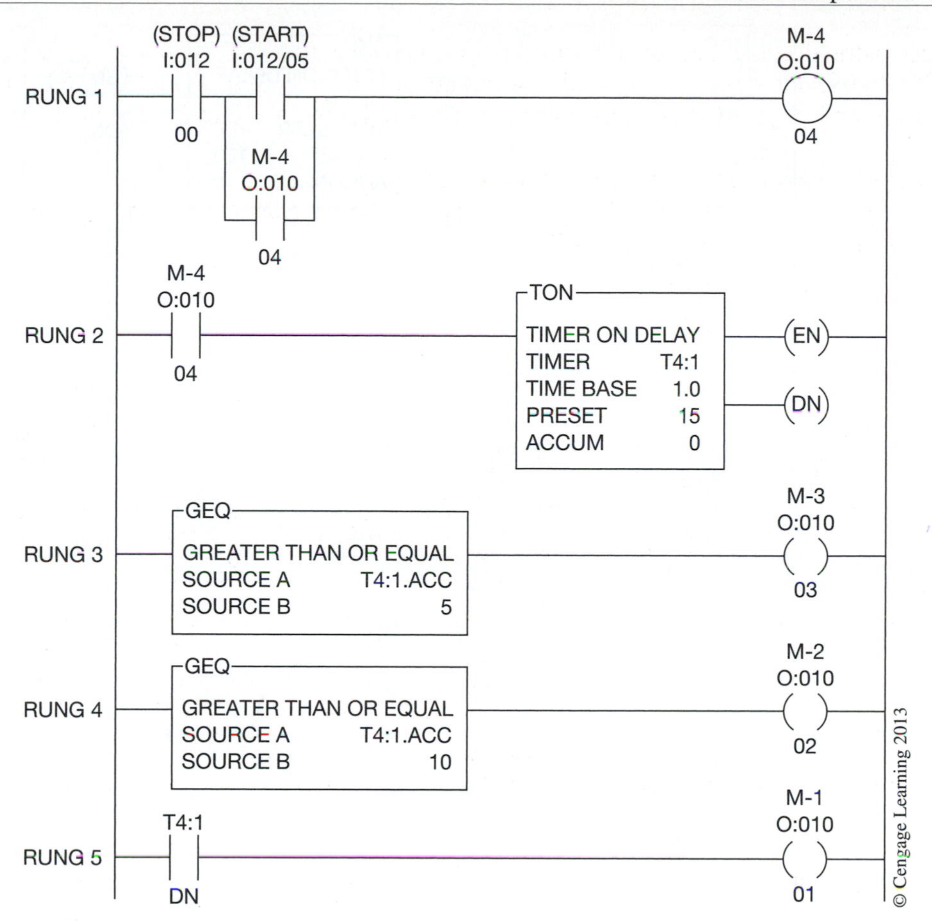

Figure 13–14 Allen-Bradley Data Compare Format

M-4 contacts O:010/04 also close in Rung 2 and enable the timer. The timer has been preset to 15 seconds (1.0 second time base). The accumulated time is stored in timer file T4:1 (remember that timers use three words of memory with the third word holding the accumulated value of the timer).

The GEQ instruction in Rung 3 that controls output O:010/03 (Motor 3) is logically true when the accumulated value of timer T4:1 (Source A) is equal to or greater than the constant in Source B (5 seconds). When the accumulated value reaches 5, output O:010/03 (M-3) is turned *ON*. Similarly, when the accumulated value reaches 10, the logic of Rung 4 will be true and output O:010/02 will be turned *ON*. When the accumulated value of the timer reaches 15 and is equal to the preset value of 15, the done bit (DN), bit 13, will be set to 1 and the last motor, Motor 1, will be turned *ON*.

To further illustrate how the data compare instructions work, consider the program in Figure 13–15 and the time chart for data comparisons in Figure 13–16.

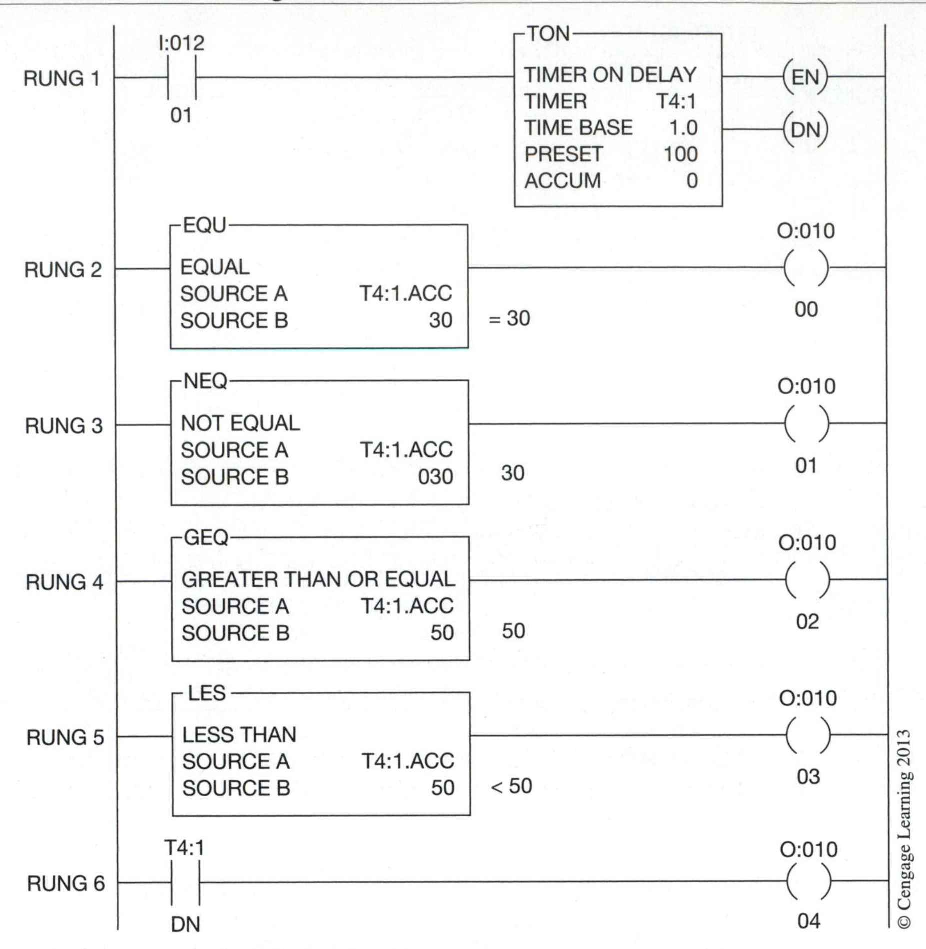

Figure 13–15 Data Compare Instructions

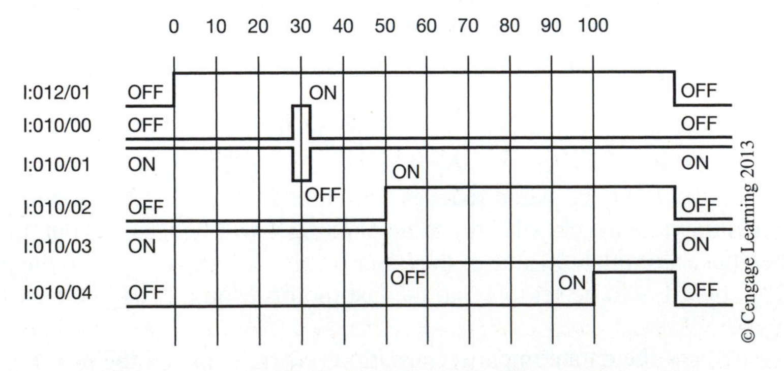

Figure 13–16 Time Chart for Data Comparisons

When power is applied, but before I:012/01 is closed to activate the timer, outputs O:010/01 (Rung 3) and O:010/03 (Rung 5) are energized. The NEQ instruction preceding output O:010/01 is true if the value in Source A is not equal (≠) to 30. With I:012/01 open, timer T4:1.ACC value is 00 and is *not* equal to 30. Output O:010/03 (Rung 5) is energized because the LES instruction is true whenever the accumulated value of T4:1 is less than (<) 50.

When input I:012/01 closes, the timer is enabled and starts to time. At time 30, the EQU instruction preceding O:010/00 goes true because the accumulated value in T4:1 is equal to (=) 30, and output O:010/00 in Rung 2 is energized. This is only true when the accumulated value of timer T4:1 is 30. When it advances to 31, the EQU instruction goes false and O:010/00 goes *OFF*. Output O:010/01, which was *ON* because of the not equal to (≠) instruction, now goes *OFF* for one second because the NEQ instruction was false when the accumulated value of T4:1 was equal to (=) 30.

When the accumulated time reaches 50, output O:010/02 (Rung 4) turns *ON* because the GEQ instruction preceding it goes true when T4:1.ACC is equal to or greater than (≥) 50. The rung is true when the accumulated value in T4:1 is equal to (=) 50 and remains true as long as the accumulated value is 50 or greater. Output O:010/02 remains *ON* until the timer is turned *OFF* and the accumulated value is reset to 00.

Output O:010/03, which was *ON,* now goes *OFF* when the accumulated value of T4:1 reaches 50 because the LES instruction that precedes it is only true when the value of Source A is less than (<) 50.

Output O:010/04, the timer done bit (DN), comes *ON* at 100 when the accumulated value equals the preset value. The time chart in Figure 13–16 illustrates the *ON* and *OFF* states of the outputs in relation to time and the data compare instructions.

The Allen-Bradley PLC-5 has an instruction that is a combination of the previous instructions, it is the **compare instruction,** or **CMP.** The CMP instruction is an input instruction that compares values from addresses or files. Figure 13–17 shows a CMP instruction and the compare expression is designated as T4:0.ACC = N7:2.

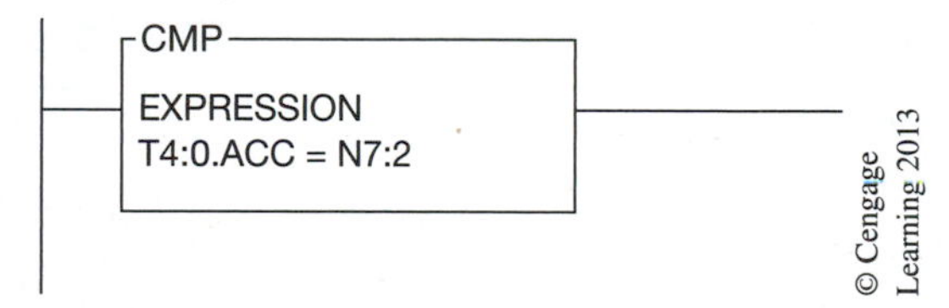

Figure 13–17 Allen-Bradley PLC-5 Compare Instruction

The CMP instruction in this case is true only when the accumulated value of T4:0 is equal to the value found in integer file 7, word 2.

The table in Figure 13–18 shows the different operators (symbols) that the CMP instruction uses. Because standard computer keyboards do not have keys for not equal, less than or equal to, or greater than or equal to, the CMP instruction uses variations, as shown in Figure 13–19.

OPERATOR	DESCRIPTION	EXAMPLE
=	EQUAL TO	TRUE IF A = B
< >	NOT EQUAL TO	TRUE IF A < > B
<	LESS THAN	TRUE IF A < B
< =	LESS THAN OR EQUAL TO	TRUE IF A < = B
>	GREATER THAN	TRUE IF A > B
> =	GREATER THAN OR EQUAL TO	TRUE IF A > = B

Figure 13–18 Available Operators (Symbols) for CMP Instruction

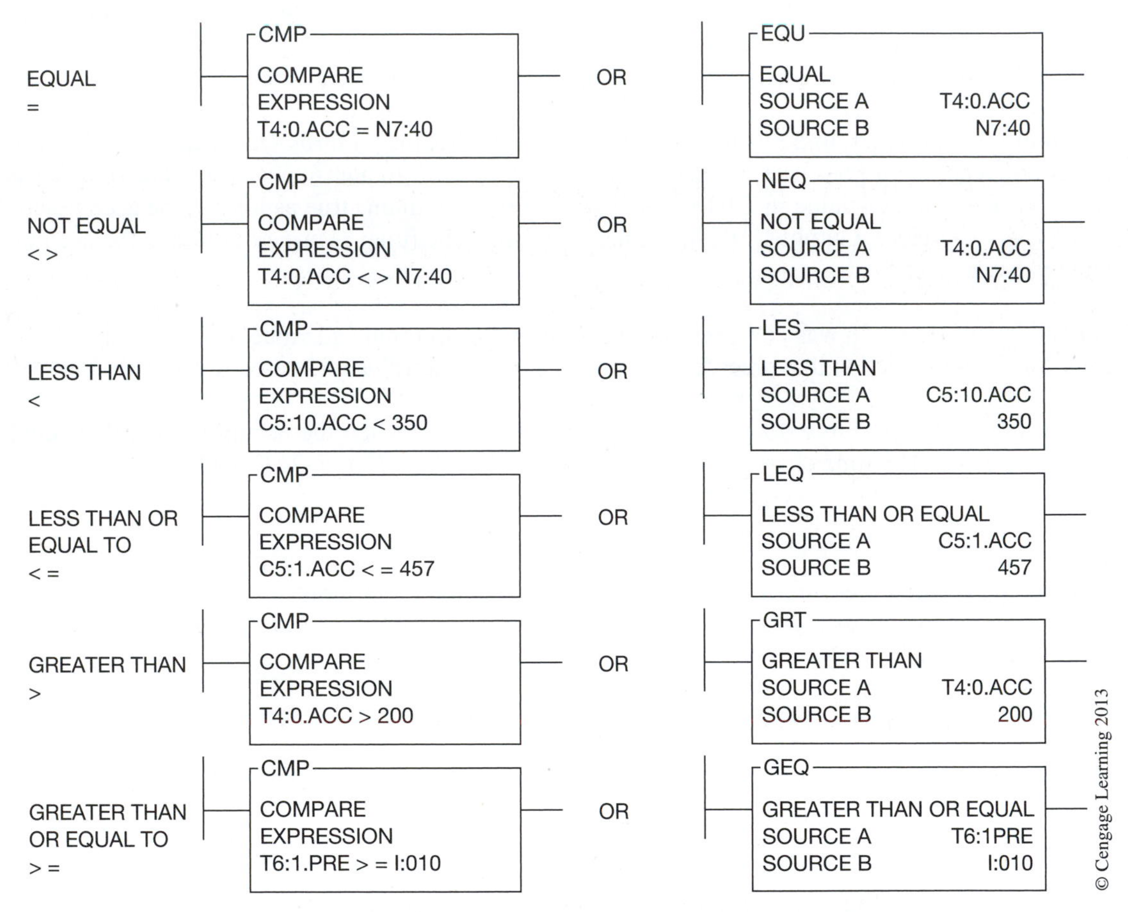

Figure 13–19 Comparing the PLC-5 CMP Instruction to Data Compare Instructions

Note: *While the CMP instruction duplicates the other data compare instructions, the execution time for the CMP instruction is longer than the execution time for equivalent comparison instructions (for example, GRT, LEQ, etc.). A CMP instruction also uses more words per instruction than the equivalent comparison instructions. The advantage, however, is that the CMP instruction* *can also perform math operations as part of the compare and also multiple compare operations. Figure 13–20 shows an example of a CMP instruction with embedded math operation.*

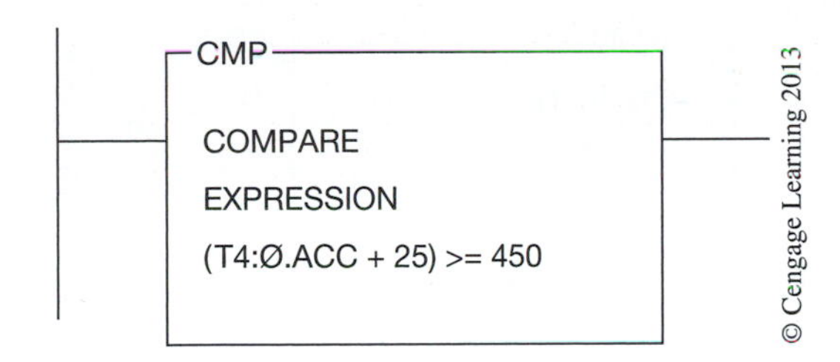

Figure 13–20 CMP with Embedded Math Operation

The LIM, or limit test instruction, is an input instruction used by Allen-Bradley to test for values inside or outside a specific range. The instruction is false until it detects that the test value is within certain limits. It then goes true. When the instruction detects that the test value has again gone outside the prescribed limits, the instruction goes false. This instruction is perfect for monitoring analog signals and making program decisions based on the analog value(s). Figure 13–21 shows the format for an LIM instruction.

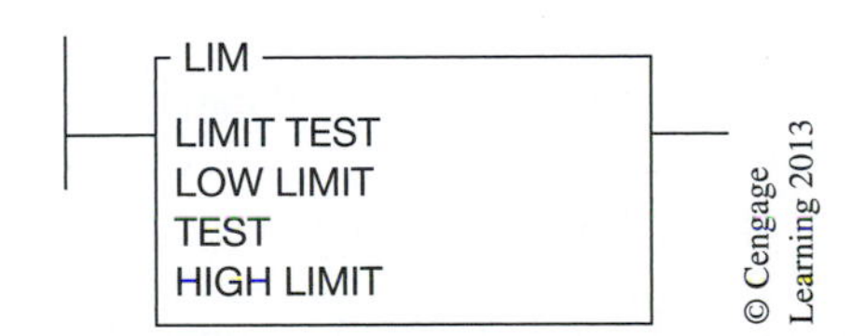

Figure 13–21 PLC-5 Limit Test Instruction Format

To program an LIM instruction, the following information must be provided:

Low Limit

Low limit is a constant, or an address that determines the lower limit of the test range. The value in the low limit can be either integer (whole number) or floating point (number[s] and a decimal).

Test Value

The test value is the address that contains the value examined to determine if it is inside or outside the specified range.

High Limit

High limit is a constant (numeric value) or an address that determines the upper limit of the test range. The value in the high limit can be either integer or floating point value.

An example of how an LIM instruction is used is shown in Figure 13–22. In the example, the LIM instruction is used to turn *ON* an indicator lamp whenever the accumulated value of T4:1 is between the values stored in N7:10 and N7:20.

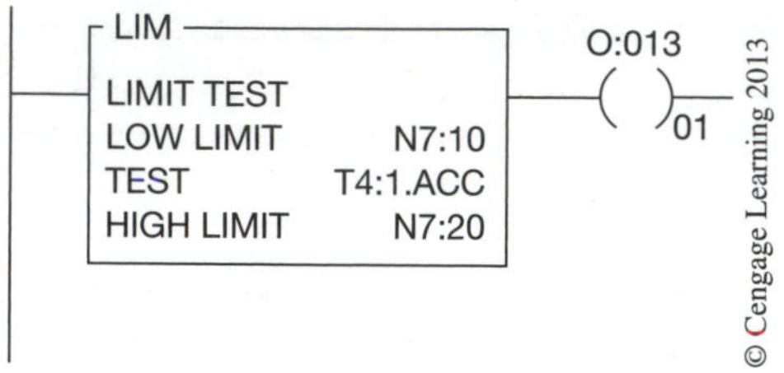

LIM
LIMIT TEST
LOW LIMIT N7:10
TEST T4:1.ACC
HIGH LIMIT N7:20
O:013
01

Figure 13–22 PLC-5 Limit Test Instruction

If the lower limit is set to 100 (value stored in N7:10) and the upper limit is set for 300 (value stored in N7:20), the instruction will be false as long as the accumulated value of T4:1 is less than 100 or greater than 300. When the accumulated value of T4:1 reaches 100 and becomes equal to or greater than the low limit, the instruction becomes true and the indicator lamp (output O:013/01) is set to 1, or turned *ON*. If the accumulated value of T4:1 becomes greater than 300, the instruction again goes false and the indicator lamp (O:013/01) is turned *OFF*. The instruction remains *OFF* as long as the accumulated value of T4:1 is outside the limit test range (100–300) that was established for the LIM instruction.

The values that are used for the low limit and high limit can be entered as numeric values when the instruction is being programmed. Instead of referencing N7:20 (which held a value of 300), a value of 300 could have been entered at the low limit prompt.

Another example of the limit test instruction (LIM) is shown in Figure 13–23.

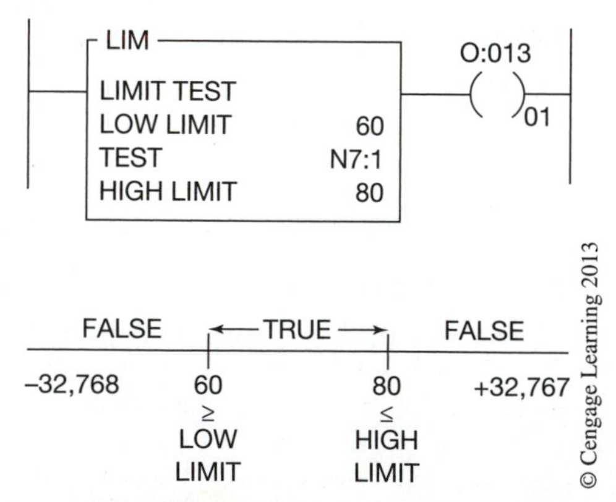

Figure 13–23 LIM Instruction with the Low Limit Value Set at 60

The low limit is a constant with a value of 60 whereas the high limit has a constant value of 80. The test address is N7:1. This could be the numeric value from an analog input device such as a thermocouple or resistive temperature device (RTD). Programmed this way, the instruction is true, and output device O:013/01 is *ON,* as long as the test value is equal to or greater than the low limit of 60 and less than or equal to the high limit of 80. Any time the value stored in N7:1 is below 60 or greater than 80, the instruction is false and output device O:013/01 is *OFF*. The output device could be an indicator lamp that is lit as long as the temperature value measured by the thermocouple is within the specified range of 60 to 80 degrees.

If an LIM instruction is programmed with the low limit value higher than the high limit value, as shown in Figure 13–24, the logic of the instruction is reversed.

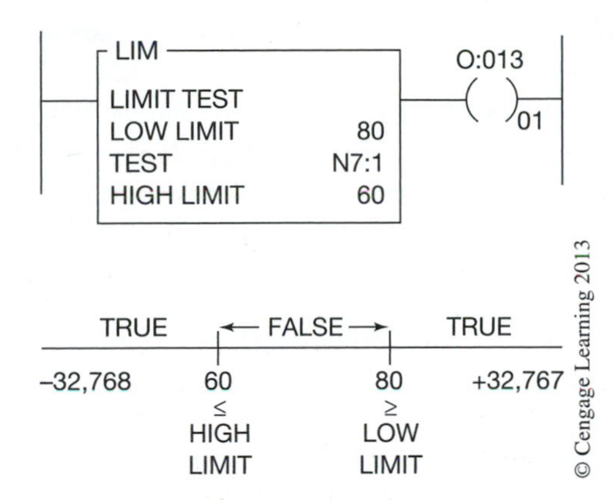

Figure 13–24 LIM Instruction with the Low Limit Value Higher Than the High Limit Value

In this illustration, the low limit has been set at 80, and the high limit has been set at 60. The test value is still the value stored in integer file N7 word 1. Programmed this way, the logic of the LIM instruction is true when the value stored in N7:1 is equal to or greater than 80 (low limit) and also true when the value is less than or equal to 60 (high limit). Figure 13–24 illustrates the logic for the instruction. Output device O:013/01 will be *ON* whenever the value stored in N7:1 is equal to or less than 60 and also will be *ON* whenever the value in N7:1 is equal to or greater than 80.

ALLEN-BRADLEY LOGIX5000 DATA MANIPULATION

The only difference between the data manipulation instructions for the Logix5000 controllers and those just discussed is that the Logix5000 controllers use tag-based addressing. Otherwise, the same data transfer and compare instructions are found in both. It is worth noting that the optimal data types for Logix5000 data manipulation instructions are DINT and REAL.

Once the electrician or technician becomes familiar with a specific PLC, the many applications and advantages of using the various data compare instructions become evident.

Chapter Summary

Although formats and instructions vary with each PLC manufacturer, the concepts of data manipulation are the same. Data manipulation enables an operator to transfer data from one word location to another, whereas data comparison allows a value in one word to be compared to another word or a constant value.

Both data transfer and data comparison instructions give new dimension and flexibility to motor-control circuits, and the application of either is only limited by programmer imagination.

Review Questions

1. Define the term *data transfer*.
2. When numerical information replaces data that already exists in a memory location, it is referred to as:
   a. exchanging info (data)
   b. replacement programming
   c. blanket move
   d. writing over
3. Match the symbols to their correct definitions. ***Note:*** *Not all nine definitions are used.*

   a. > ______ 1. less than 1
   b. < ______ 2. less than
   c. = ______ 3. less than or equal to
   d. ≥ ______ 4. greater than
   e. ≠ ______ 5. greater than or equal to
   f. <> ______ 6. equal to
   g. >= ______ 7. not equal to
   h. <= ______ 8. not equal to 1
   i. ≤ ______ 9. greater than 1
4. Define the term *data compare*.
5. Write a program that compares the accumulated value of T4:0 to a constant of 250. The instruction is to be true when the accumulated value of T4:0 is greater than 250.
6. Define the term *mask*.
7. Give an example of how an Allen-Bradley LIM instruction is used.

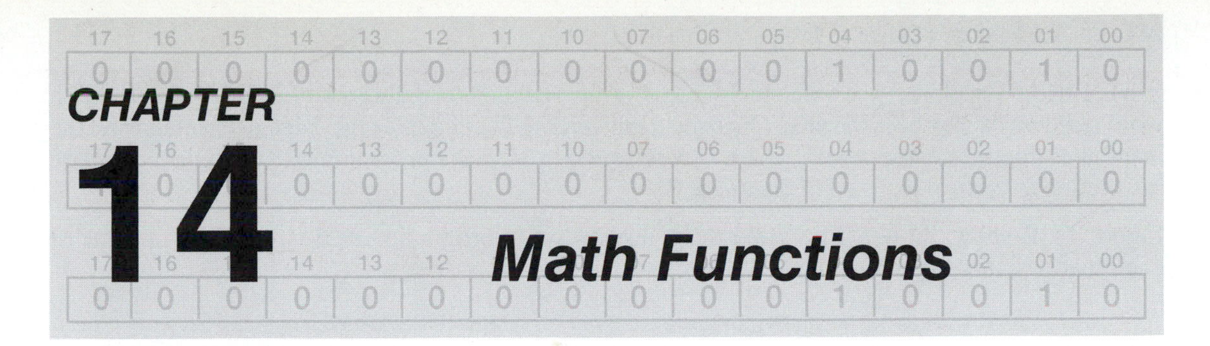

CHAPTER 14 Math Functions

Objectives

After completing this chapter, you should have the knowledge to:

- List the four standard math functions available with most PLCs.
- Discuss the math functions and give examples of how they are used.
- Program the square root function.

USING MATH FUNCTIONS

The four basic math, or arithmetic, functions are addition (+), subtraction (–), multiplication (×), and division (÷); they can be used with constant values or values stored in a storage register, holding register, input/output registers, or any other accessible word or tag locations.

A typical application of an arithmetic, or math, function may be a chemical batch plant where a given mix of two chemicals (A and B) is to have a 2:1 ratio. By using analog input devices (discussed in Chapter 2), the weight of the first chemical, A, can be converted to a binary equivalent and stored. For a 2:1 mix, only half as much of chemical B (by weight) is needed. The binary value of the weight that is stored can be divided by 2 to determine the amount of chemical B that is needed for a proper 2:1 mix.

ALLEN-BRADLEY PLC-5, SLC 500, AND MICROLOGIX MATH INSTRUCTIONS

Figure 14–1 illustrates a divide (DIV) instruction showing how the value of chemical A (Source A-N7:10) can be divided by 2 (Source B), and the result placed into destination N7:3. A data comparison is then made to the data stored in integer file N7:3 to limit the amount of chemical B to one-half the amount of chemical A.

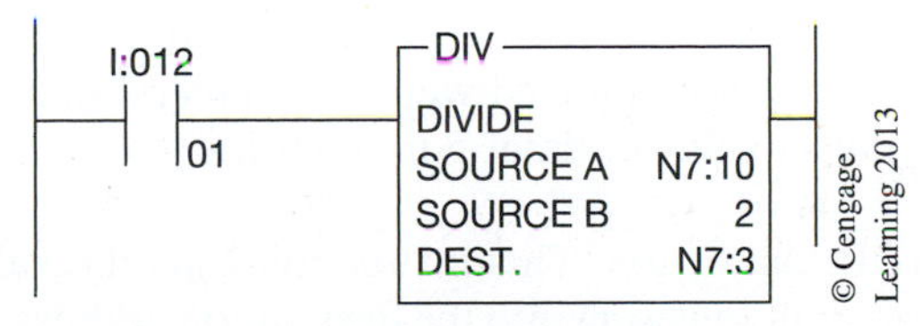

Figure 14–1 Divide (DIV) Instruction

When I:012/01 is closed, the value in Source A is divided by 2, and the result is stored in destination word N7:3 on each processor scan. Once the total weight of chemical A has been determined and the amount divided by 2, the result (which is now stored in N7:3) is programmed with a data compare instruction to control the feed of chemical B.

Figure 14–2 shows how a LES (less than) instruction is programmed to control the amount of chemical B that is to be mixed with chemical A.

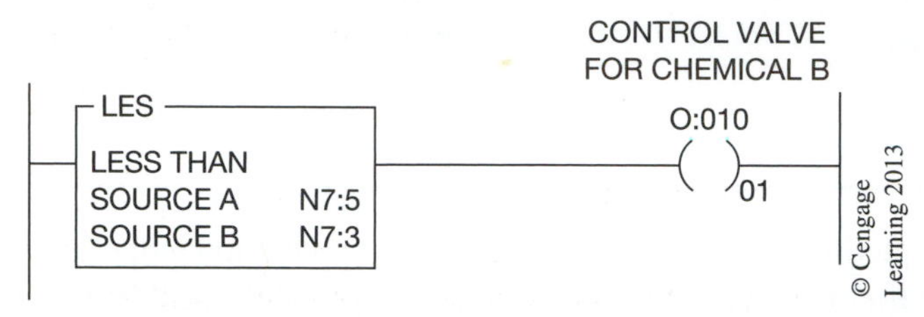

Figure 14–2 Data Compare Instruction LES to Control Amount of Chemical B

As shown in the figure, output O:010/01, the control valve for chemical B, is true, or *ON*, as long as the value in source A is less than Source B. Source A, N7:5, is the word that stores the weight of chemical B. When the weight of chemical B is equal to or greater than one-half the weight of chemical A (the value stored in N7:3, Source B), the instruction goes false and the valve to chemical B is closed.

The math function might also be used with timer and/or counter values, accumulated or preset, to change a given process machine operation under varying conditions. The DIV instruction divides the value in source A by the value of Source B and stores the answer in the destination word. Figure 14–3 shows the DIV instruction.

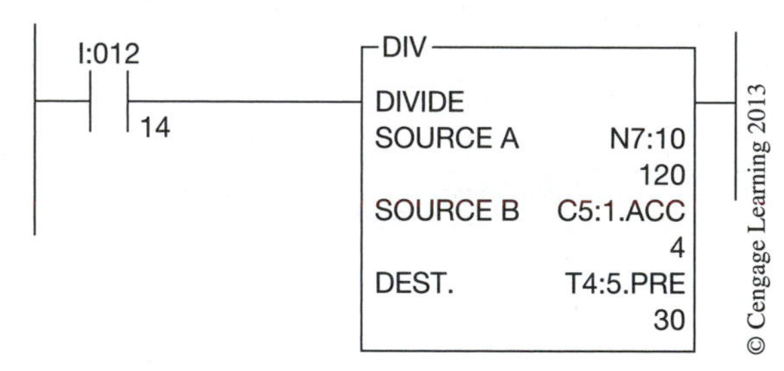

Figure 14–3 Divide (DIV) Instruction Used to Enter Preset Value in a Timer

When input device I:012/14 is true, the value of Source A, address N7:10 (120), is divided by the value of source B, 4, which is the accumulated value of counter 1 in counter file 5. The result of the division, 30, becomes the preset value of timer 5 in timer file 4.

Figure 14–4 shows the ADD instruction. This instruction adds the value in Source A to the value in Source B and places the result (answer) into the destination address. In the example, Source A is shown as N7:1, which indicates that this is word 1 of File 7, the integer file.

Source B is word 2 of the integer file (file 7), and the destination is shown as word 10 of File 7. When input address I:012/10 is true, the value in Source A (20) is added to the value in Source B (40) and the results (60) are placed in the destination word.

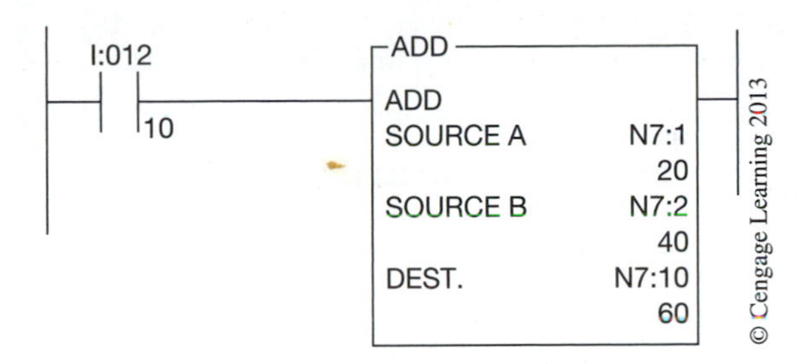

Figure 14–4 Addition (ADD) Instruction

Source A and B can each be the value of a word address or can be a constant. If the result of the addition is greater than +32,767 or less than –32,768, an overflow bit (bit 1) in word 0 of the status file, File S, will be set to 1. Additional status bits for word 0 are bit 0, which is set if a carry is generated; bit 2, which is set if the result of the math is 0 after the instruction has been implemented; or bit 3, which indicates a negative value after the math instruction has been executed. Figure 14–5 shows the arithmetic status bits.

STATUS BIT	CONTROLLER ACTION
S:0/0 Carry (C)	set to 1 if a carry is generated
S:0/1 Overflow (V)	set to 1 when the results of the math will not fit the destination address
S:0/2 Zero (Z)	set to 1 if the math operation results in an answer of 0
S:0/3 Sign (S)	set to 1 when the result of the math instruction is negative (less than 0)

Figure 14–5 Arithmetic Status Bits

With the MicroLogix and the SLC 5/02, 5/03, and 5/04 processors, addition and subtraction using 32-bit words is possible. This allows for addition and subtraction beyond the normal limits of +32,767 and –32,768. For instructions on how to add or subtract 32-bit words, refer to the programming manual for the processor being used.

Figure 14–6 shows the subtract (SUB) instruction. With this instruction, the value in Source B is subtracted from the value in Source A and the results are placed in the destination word (DEST).

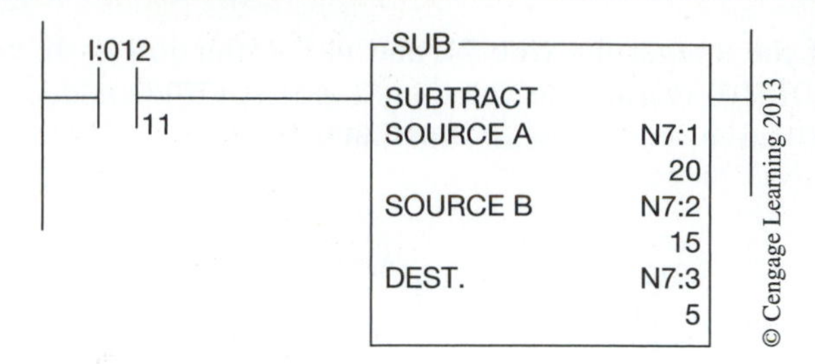

Figure 14–6 Subtract (SUB) Instruction

When input device I:012/11 goes true, the value in Source B, address N7:2 (15), is subtracted from the value in Source A, address N7:1 (20), and the result, 5, is stored in destination address N7:3 (5).

The multiply (MUL) instruction used for the PLC-5, SLC 500, and MicroLogix is shown in Figure 14–7.

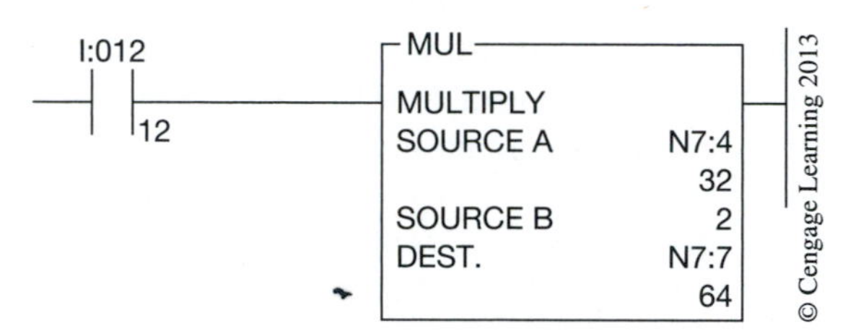

Figure 14–7 Multiply (MUL) Instruction

Like previous instructions, the MUL instruction multiplies the value in Source A times the value in Source B and places the answer into the destination address. In this example, when address I:012/12 goes true, the value stored in Source A, address N7:4 (32), is multiplied by a constant value of 2 and the answer, 64, is placed into destination word N7:7. As with previous math instructions, Sources A and B can be values (constants) or addresses that contain values.

For the Allen-Bradley PLC-5 and the SLC 500 5/03, 5/04, and 5/05 processors, a more powerful math instruction called the compute (CPT) is available. The CPT instruction is an output instruction that performs the operations defined in the expression and then writes the results to the destination address. Figure 14–8 shows the CPT instruction format.

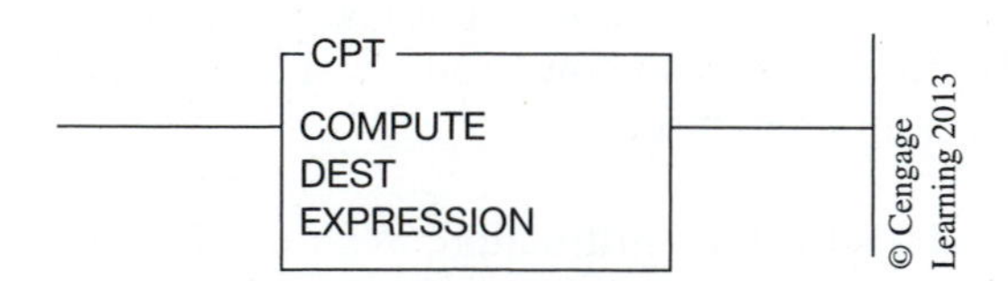

Figure 14–8 PLC-5 Compute (CPT) Instruction Format

The typical math functions that can be performed with the CPT instruction are add (+), subtract (–), multiply (*), divide (|), negate (–), square root (SQR), and exponential (**). The table in Figure 14–9 shows the functions and an example of each operation. The complexity of the math that can be performed will vary with the PLC model.

Note: *Different symbols are used for some operations due to limited symbols on the keyboard.*

OPERATOR	DESCRIPTION	EXAMPLE
+	ADD	2 + 3 + 7
–	SUBTRACT	12 – 5
*	MULTIPLY	6 * (5 * 2)
÷	DIVIDE	24 ÷ 4
–	NEGATE (NEGATIVE)	–N7:0
SQR	SQUARE ROOT	SQR N7:1
**	EXPONENTIAL	10 ** 3 OR 10^3

Figure 14–9 Compute (CPT) Operators (Symbols)

Figure 14–10 shows how a CPT instruction is typically programmed. In some PLC-5 models, but not all, any combination of operators may be used with various addresses and/or constants. The expression can be up to 80 characters in length.

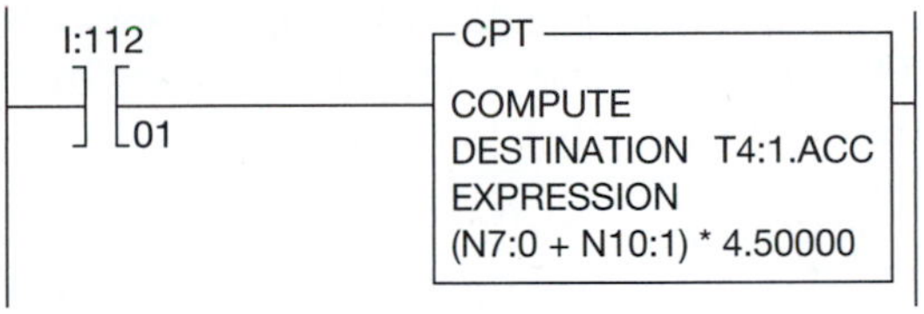

Figure 14–10 Programmed PLC-5 Compute (CPT) Instruction

When the instruction first appears on the screen during programming, the destination must be specified first. The destination address in the example is T4:1.ACC. Then the expression or math formula is entered. The expression states that the value in N7:0 is to be added to the value found in N10:1. The sum is then multiplied by 4.5. The result of the computation is then sent to the designated destination, in this case, the timer file, file 4, timer 1, accumulated value (T4:1.ACC).

Additional math functions available for the PLC-5, SLC 500, and MicroLogix are the clear (CLR) and square root (SQR) instructions.

The clear instruction (CLR), as the name implies, clears the value in a word to zero. Figure 14–11 shows the CLR instruction.

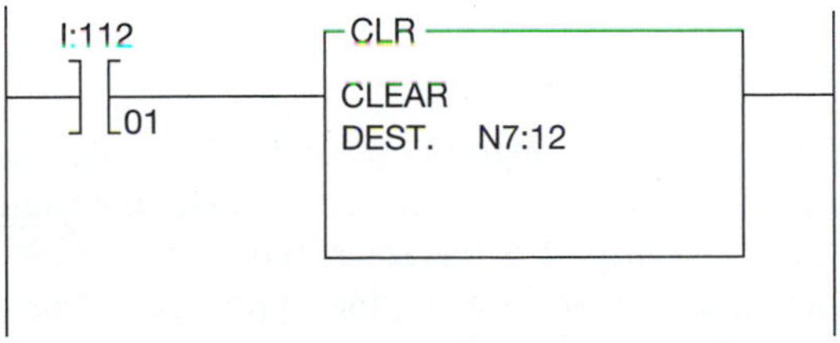

Figure 14–11 Example of the Clear (CLR) Instruction

When the input address is true, the value in the destination word is set to zero. That means that all 16 bits of word 12, in integer file 7, are cleared or set to zero.

The square root instruction (SQR) is shown in Figure 14–12.

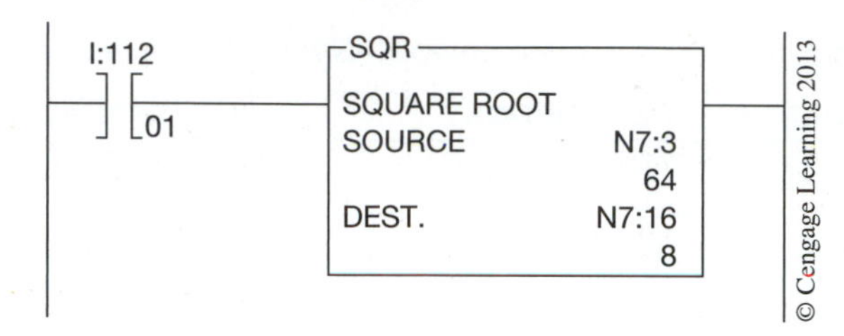

Figure 14–12 Example of the Square Root (SQR) Instruction

When the logic that precedes the SQR instruction is true, the square root of the source value is transferred into the destination word. In the example in Figure 14–12, the square root of the source address, N7:3 (64), which is 8, is placed into destination word 16 of the N7 file.

Some PLC-5 processors and the SLC-5/02, 03, 04, and 05 processors also can perform the trigonometric functions sine (SIN), cosine (COS), tangent (TAN), arc sine (ASN), arc cosine (ARS), and arc tangent (ATN). These PLCs can also raise a number to a power by using the X to the Power of Y (XPY) instruction. This instruction is shown in Figure 14–13.

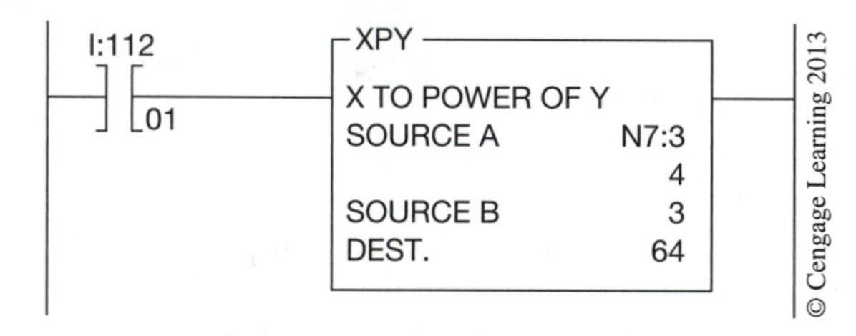

Figure 14–13 Example of the XPY Instruction

If we wanted to raise the value of a word by a power of 3, which is the same as saying X^3 or $X \bullet X \bullet X$, we could use the XPY instruction. If the value of Source A, word N7:3, is 4, and this value is to be raised by a power of 3, Source B, the destination would hold the result of raising 4 by a power of 3, which is 64. Raising 4 by a power of 3 could be written as 4^3 or could be written as $4 \bullet 4 \bullet 4$. In either case, the answer is 64 ($4 \bullet 4 \bullet 4 = 64$).

COMBINING MATH FUNCTIONS

An example of using more than one math instruction can be illustrated using the Pythagorean theorem, which is used to find the hypotenuse of a right triangle. The theorem states that $C^2 = A^2 + B^2$. This formula, or theorem, can be transposed and rewritten $C = \sqrt{a^2 + b^2}$. Figure 14–14 shows a right triangle. Side A is 3″ and side B is 4″. Side C, the hypotenuse, has no measurement and is the unknown that the theorem will solve.

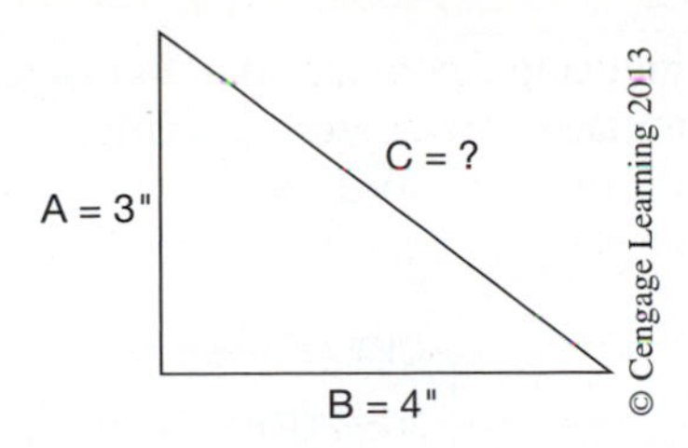

Figure 14–14 Solving the Hypotenuse of a Right Triangle Using the Pythagorean Theorem

Figure 14–15a shows how two XPY instructions, an ADD instruction, and an SQR instruction are used to find the dimension for side C of the right triangle shown in Figure 14–14. When the rung is true, the value in Source A (which is 3 from side A of the right triangle) of the first XPY instruction is increased by a factor of 2 (3^2) and the value of 9 is stored in word 3 of file N7. The next XPY instruction takes the dimension of side B, which is 4, and raises it by a factor of 2 (4^2), then stores the answer (16) in word 4 of file N7. The ADD instruction now takes the values in Sources A and B and adds them together to get the answer of 25, as shown in word 5 of file N7. The last instruction, SQR, finds the square root of the value that is stored in word 5 of file N7. The value is 25, and the square root of 25 is 5. And, as we all remember from our high-school geometry class, if side A is equal to 3, and side B is equal to 4, then side C must be equal to 5. This is often referred to as a 3, 4, 5 triangle.

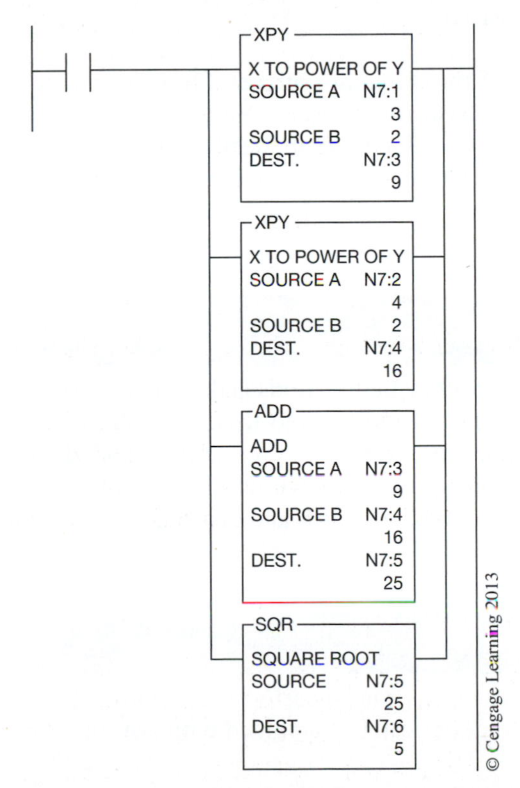

Figure 14–15a Combining Math Instructions

We could have solved the problem using a compute (CPT) instruction, as shown in Figure 14–15b. With the CPT instruction, a formula or expression is written that will solve the problem and the solution is then displayed in the destination word (N7:6).

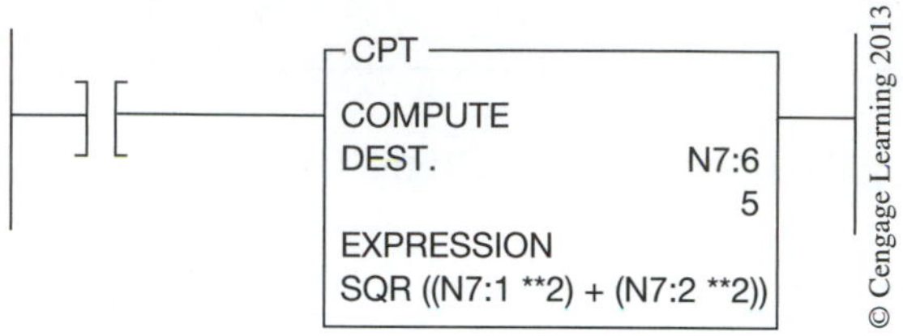

CPT
COMPUTE
DEST. N7:6
5
EXPRESSION
SQR ((N7:1 **2) + (N7:2 **2))

Figure 14–15b Using the CPT Instruction to Solve the Hypotenuse of a Right Triangle

ALLEN-BRADLEY LOGIX5000 BASIC MATH FUNCTIONS

The only difference between the basic math instructions for the Logix5000 controllers and those just discussed is the Logix5000 controllers use tag based addressing. Otherwise, the same basic math instructions are found in both. It is worth noting that the optimal data types for Logix5000 math instructions are DINT and REAL.

The limits and limitations of the different arithmetic functions vary from manufacturer to manufacturer, but the concepts are basically the same. Values or contents of storage registers, data table words, tags, or constants are combined arithmetically, and the results are stored in registers, data table words, or tags. The only way to learn how to apply the arithmetic functions on a given PLC is to read the manufacturer's literature and *work* with *that* PLC.

Chapter Summary

As with most other features, arithmetic functions and formats vary with each manufacturer. Arithmetic functions of add, subtract, multiply, and divide can be combined with data manipulation instructions (data transfer and data compare) to provide expanded control and information for and from process or driven equipment. Memory words such as holding, storage, and data can be used with the arithmetic functions as well as words and constants, or just constants.

Review Questions

1. List the four math functions that can be performed by most PLCs.
2. Data manipulation instructions can be combined with arithmetic (math) instructions.
   T F
3. Define the term *double-integer.*

4. Give an example of how a math function is used in a PLC program.
5. Using Allen-Bradley format, describe the following status bits:
   S:0/0
   S:0/2
   S:0/3
6. When using the Allen-Bradley CPT instruction, what operator is used for:
   a. Multiply ________________
   b. Divide ________________
   c. Exponential ________________
7. What is the function of the clear (CLR) instruction?
8. What does the XPY instruction do?
9. List the trigonometric instructions discussed in this chapter.

CHAPTER 15 Word and File Moves

Objectives

After completing this chapter, you should have the knowledge to:

- Describe the function of a synchronous shift register.
- Explain the function of *word-to-file, file-to-word,* and *file-to-file* instructions.
- Explain the difference between an *asynchronous shift register (FIFO)* and a *word-to-file* move.

Before word and file moves are discussed, the electrician and technician should understand the definition of both words and files.

Words, or registers as they are often called, are locations in memory that can be used to store different kinds of information. Typically, a word or register can store the status of inputs and outputs, hold numerical values used for math functions and other numerical data used for timers and counters, etc. Most words consist of 16 bits, although on newer PLCs, a 32-bit word is sometimes used.

A **file** is a group of consecutive memory words used to store information. Words 1 through 5 would make up a consecutive 5-word file. Words 1, 2, 3, 6, and 7 are not used as a 5-word file because the numbers are not consecutive. A file is also referred to as a table by some PLC manufacturers.

WORDS

Information stored in a word can be shifted within the word, or from one word to another. Information stored in a word may also be moved into a file, or the information stored in a file can be transferred into a word. All of these different possibilities are discussed later in this chapter.

SYNCHRONOUS SHIFT REGISTER

When information is shifted—one bit at a time—within a word, or from one word to another, it is called a synchronous shift register. The bits may be shifted forward (left) or reverse (right).

Note: *The synchronous shift register may also be referred to as a serial shift register or bit shift.*

Figure 15–1 shows a 16–bit word used as a forward synchronous shift register.

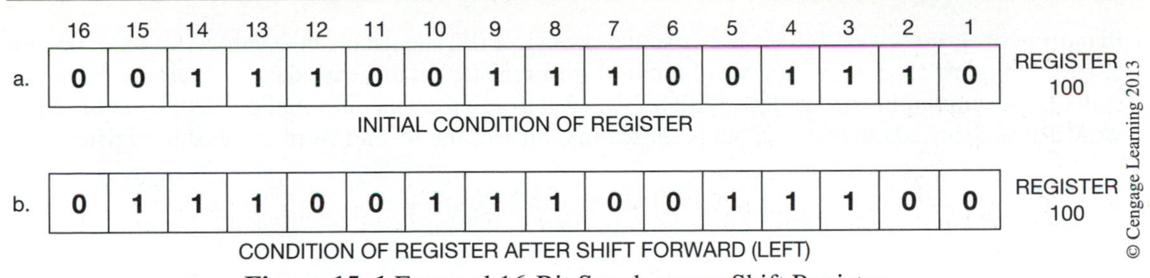

Figure 15–1 Forward 16-Bit Synchronous Shift Register

Figure 15–1(a) shows the bit status of register word 100 prior to the forward shift, while 15–1(b) shows how the register looks after the bits have been shifted one place to the left, or forward.

Notice that when the register is shifted, the information (1 or 0) in bit 16 is shifted out, and is lost. If the register is continually shifted with a zero (0) in bit location 1, all of the 1s are shifted left (forward) until only 0s remain (Figure 15–2).

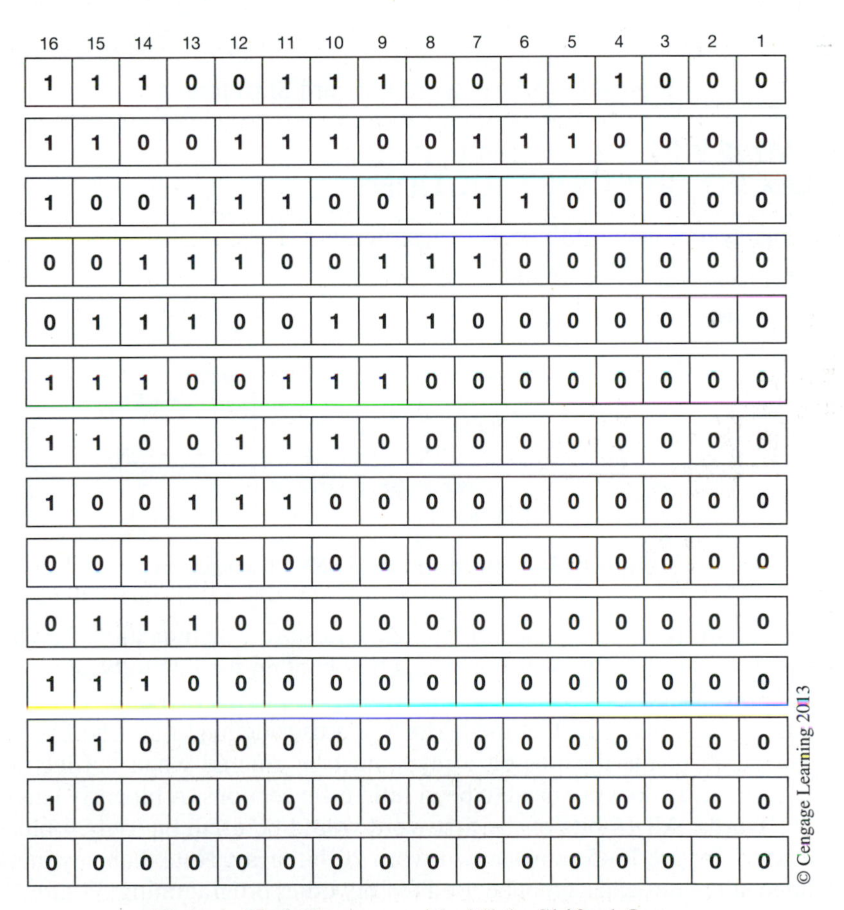

Figure 15–2 Register with All 1s Shifted Out

In a forward shift register, bit 1 is used to enter data (1 or 0). The data is then shifted forward, one bit at a time. Figure 15–3 shows a 2-word forward shift register. In this case data is entered at bit 1 and shifted one bit at a time to the left. With a 2-word shift register, the information (1 or 0) in bit 16 of word 1 is not shifted out and lost, but is shifted into bit 1 of the second word of the shift register.

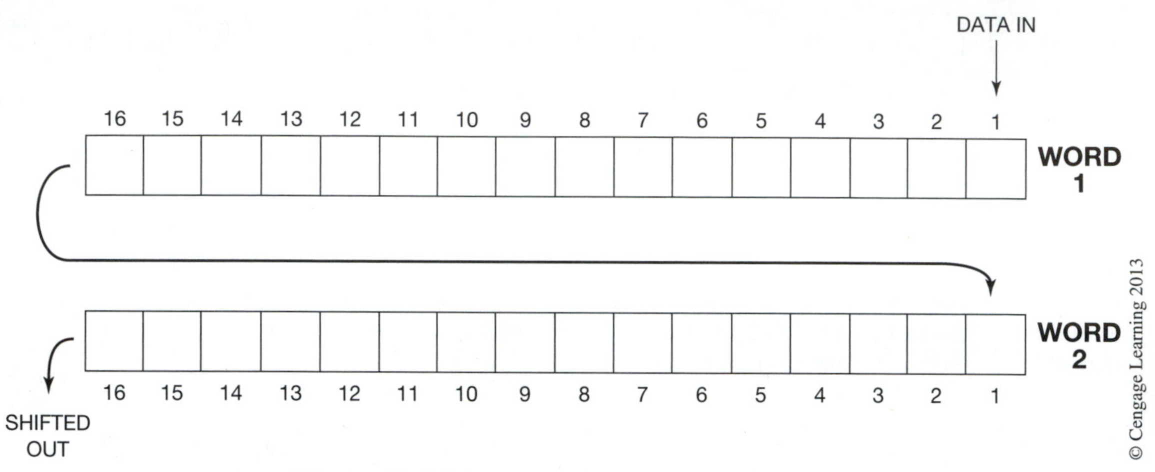

Figure 15–3 Two-Word Forward Shift Register

ALLEN-BRADLEY PLC-5, SLC 500, AND MICROLOGIX BIT SHIFT INSTRUCTIONS

Allen-Bradley has synchronous shift register instructions. They are Bit Shift Left (BSL) and Bit Shift Right (BSR). Figure 15–4 shows a BSL instruction.

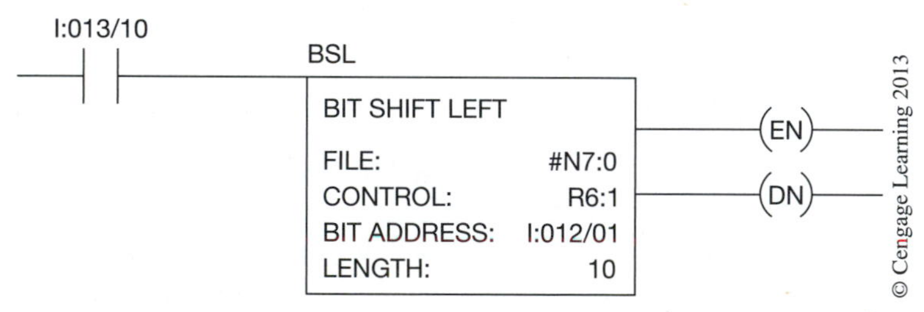

Figure 15–4 Bit Shift Left (BSL) Instruction

The BSL instruction shifts all bits to the left with each false-to-true transition. The file information that is entered is the address of the word or words that contain the bits to be shifted. This address must be a 16-bit word address and the address must be preceded with the file indicator symbol (#). The file indicator symbol indicates that the address is a user-defined file. The file is referred to as a bit array. The length of the file—or bit array—must be entered when the instruction is being programmed. The bit array does not need to be in full 16-bit sections. A bit array length of 20 would use two memory words. All 16 bits of the first word would be used, but only 4 bits of the second word would be used. Figure 15–5 shows a two-word 20-bit array. Note than any unused bits in the file are invalidated addresses and cannot be used for any other programming. Figure 15–6 shows the allowable bit array length for each type of Allen-Bradley PLC.

15	14	13	12	11	10	9	8	7	6	5	4	3	2	1	0	WORD 1
–	–	–	I	N	V	A	L	I	D	–	–	19	18	17	16	WORD 2

Figure 15–5 Two-Word 20-Bit Array

ALLEN-BRADLEY PLC	MAXIMUM BIT ARRAY
PLC-5	15,999
SLC 500	2,048
MICROLOGIX 1000	1,680

Figure 15–6 Allowable Length of the Bit Array

The Control shown on the instruction in Figure 15–4 is the element that stores the status of the BSL instruction and the size of the bit array. Figure 15–7 shows the three-word *control* element that is used for the BSL instruction.

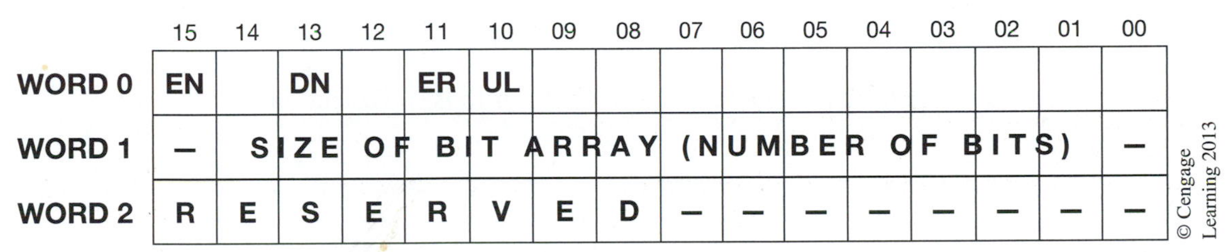

	15	14	13	12	11	10	09	08	07	06	05	04	03	02	01	00
WORD 0	EN		DN		ER	UL										
WORD 1	–	SIZE OF BIT ARRAY (NUMBER OF BITS)														–
WORD 2	R	E	S	E	R	V	E	D	–	–	–	–	–	–	–	–

Figure 15–7 BSL Three-Word Control Element

Word 0 of the control element holds the status bits for the instruction.

Unload Bit—UL (bit 10) stores the status of the last bit in the array that is shifted out, or exits, from the array on each false-to-true transition.

Error Bit—ER (bit 11) is set to 1 when an error in programming the BSL instruction is detected. This bit would be set if a negative number is entered for the length or position of the array.

Done Bit—DN (bit 13) is set to 1 each time the bit array is shifted to the left.

Enable Bit—EN (bit 15) is set to 1 on each false-to-true transition and indicates that the instruction is enabled.

When the array is shifted and the rung condition goes false, the enable, done, and error bits are reset to zero.

The Bit Address portion of the BSL instruction is the address of the bit whose information will be shifted into the bit array. The information will be entered into the lowest bit position of the array.

The Length portion of the instruction determines the length of the bit array.

Figure 15–8a shows a one-word (B3:0), 16-bit array, with the input bit address of I:012/00. The status of bit 00 of input word 012 will be inserted into the first bit of the array, which is bit 00, when the bit array is shifted left. The information in bit 15 (position 16 of the array) will be shifted out and into the unload bit (UL) of the control element. Figure 15–8b shows the 16-bit array after the BSL instruction has made a false-to-true transition and the information has been shifted left.

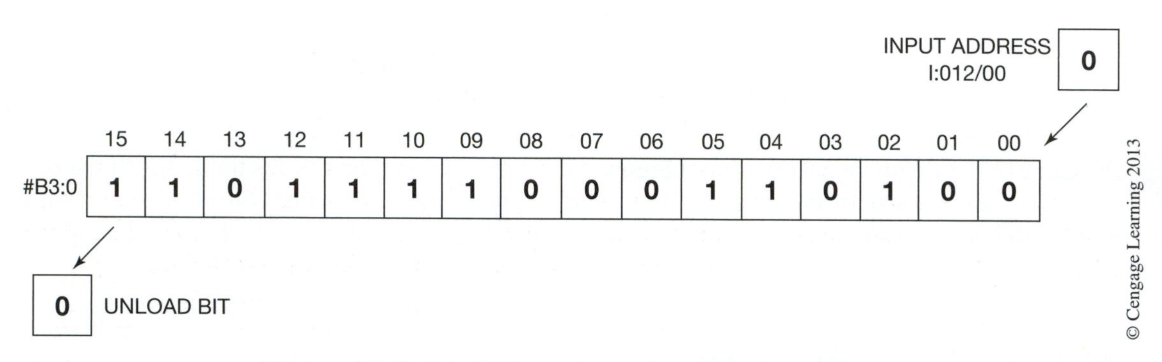

Figure 15–8a Bit-Shift Array Prior to BSL Execution

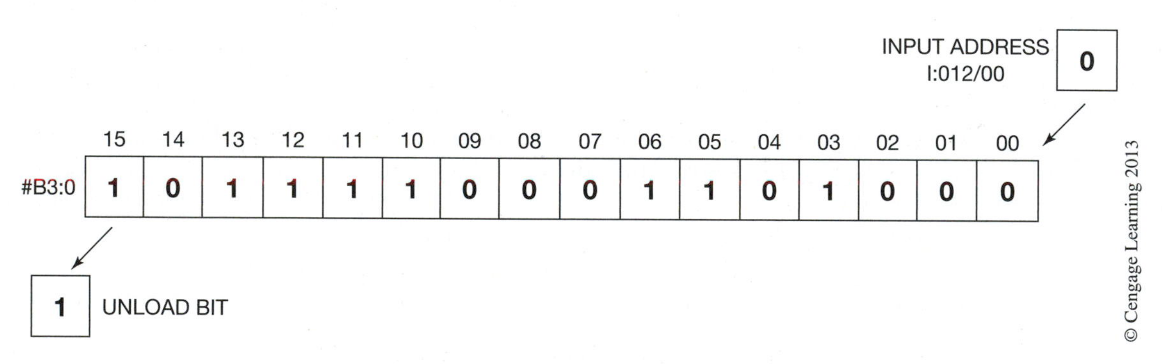

Figure 15–8b Bit-Shift Array After BSL Execution

An example of a practical application for a shift register is the overhead parts conveyor in Figure 15–9, which is used to transport parts into a paint booth for painting. If a part is on the hook as it enters the paint booth, limit switch 1 (LS-1) is activated. Limit switch 2 (LS-2) is activated each time a hook on the conveyor passes, even if no part is present.

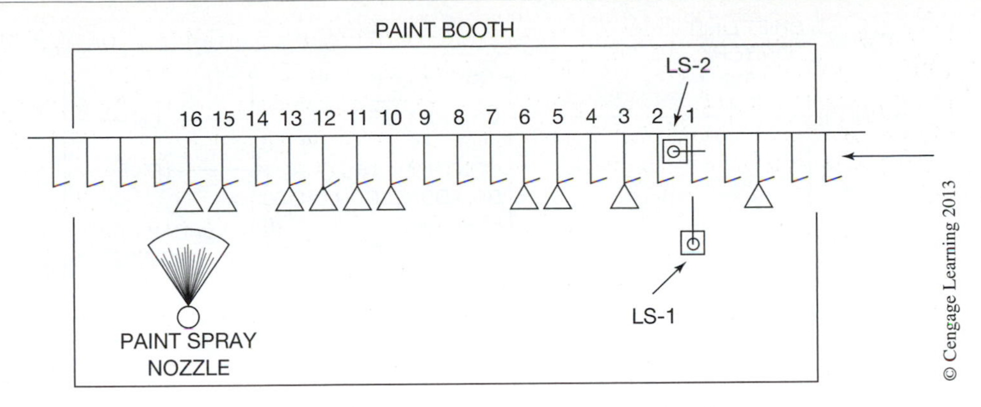

Figure 15–9 Applying a Forward Shift Register

Both limit switches (LS-1 and LS-2) are wired to an input module of a PLC, and the solenoid that operates the paint spray nozzle is wired to an output module, as shown in Figure 15–10.

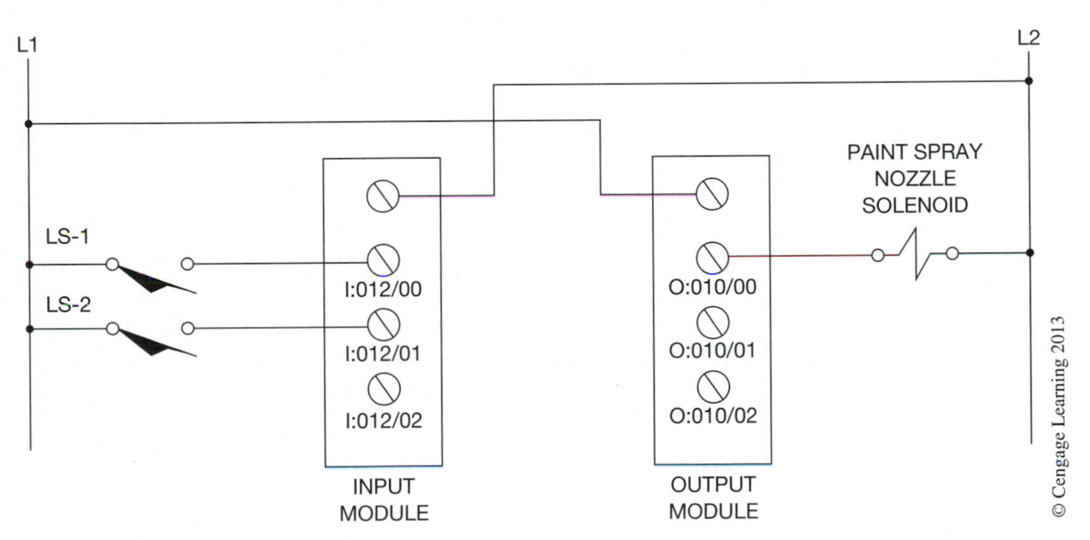

Figure 15–10 Input and Output Devices Wired to I/O Modules

The input addresses of the limit switches and the output address of the paint spray nozzle are now programmed with a forward shift register (Figure 15–11).

When a part activates LS-1 (address I:012/00), a 1 is placed in bit 00 of input word 012. As the part moves toward the spray nozzle, LS-2 is activated by the hook that closes LS-2. The closing of the

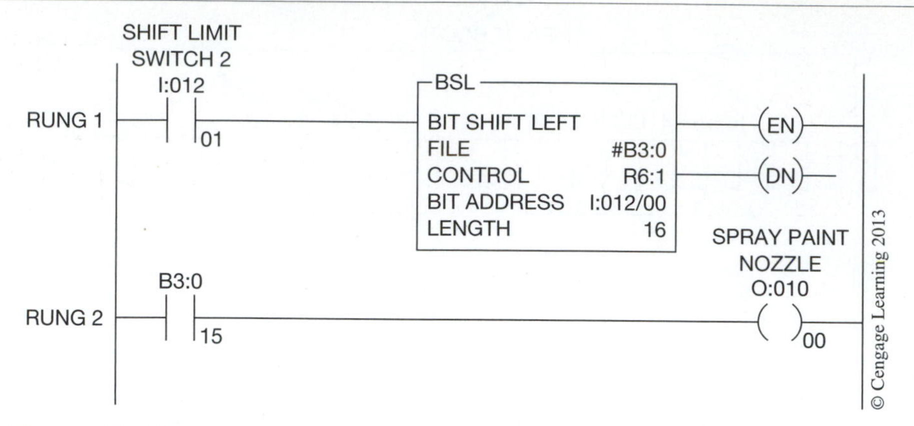

Figure 15–11 Programming a Forward Shift Register Using BSL Instruction

LS-2 contacts (address I:012/01) gives a false-to-true transition for the BSL instruction, and the instruction loads the status (1 or 0) of bit 00 of input word 012 into bit 00 of the bit array. The other information in the bit array is shifted left. The information in bit 15 of the array, the last bit of the 16-bit array, is shifted out and into the UL bit of the control element. As the conveyor continues to run, a 1 or 0 is entered into bit 00 of input word 012, depending on whether a part is present or not. The data is then shifted as LS-2 is activated and deactivated by the moving hooks, which causes a succession of false-to-true transitions of the BSL instruction. As the data shifts to the left, the paint spray nozzle solenoid (O:010/00) in Rung 2 is activated each time a 1 is shifted into bit 15 of the bit array. Bit 15, which is the 16th bit of the array, is equivalent to location 16 in the spray paint booth.

A reverse shift register is programmed using a Bit Shift Right (BSR) instruction. This instruction is the same as the BSL instruction except that instead of entering data at the lowest numbered bit in the bit array and unloading data from the highest numbered bit, this instruction enters (loads) data into the highest numbered bit and unloads at the lowest numbered bit. Figures 15–12a and 15–12b show the load and unload order for a BSR instruction.

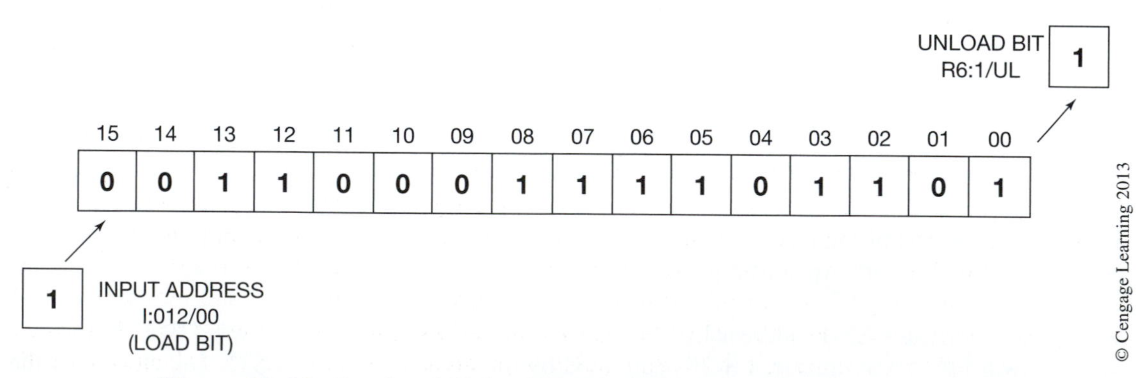

Figure 15–12a Load and Unload Bits for BSR Instruction Prior to Shift

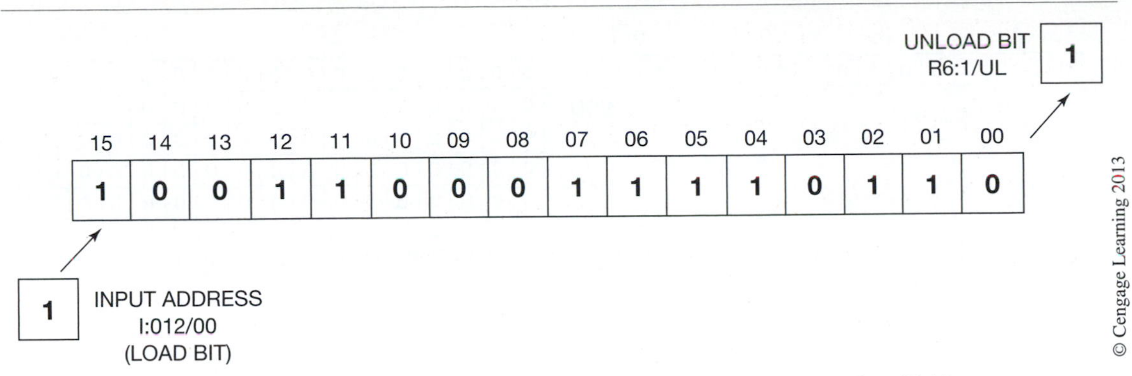

Figure 15–12b Load and Unload Bits for BSR Instruction After Shift

Similar to other functions, shift register formats vary from manufacturer to manufacturer, but the basic function of the synchronous shift register is the same.

FILE MOVES

As indicated earlier, a file, or table, is a group of *consecutive* words used to store or hold information. A file can consist of just a few words or can be several hundred words in length, depending on the PLC program. Figure 15–13 shows a five-word file using consecutive memory words 50 through 54.

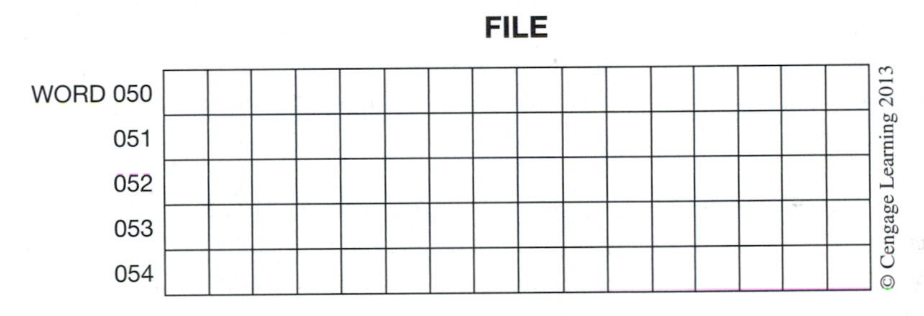

Figure 15–13 Five-Word File

Information (data) may be transferred into or out of a file by using data transfer instructions. The three most common data transfer instructions are word-to-file, file-to-word, and file-to-file.

WORD-TO-FILE INSTRUCTION

The word-to-file instruction is used to transfer data from a word into a file. For example, word 110 stores the temperature of the die for a plastic injection molding machine. A thermocouple is attached to the heated die and then connected to a thermocouple input module. Depending on the module, the temperature of the die is then stored in an input word in either binary or BCD format. By using a word-to-file data transfer instruction, the data (temperature) in word 110 can be transferred into a file. Once the word-to-file instruction has been programmed, the information stored in word 110 is transferred into a file each time the instruction is implemented. Figure 15–14a shows a 5-word file prior to a word-to-file instruction being implemented, and Figure 15–14b shows the file after the data transfer instruction is implemented.

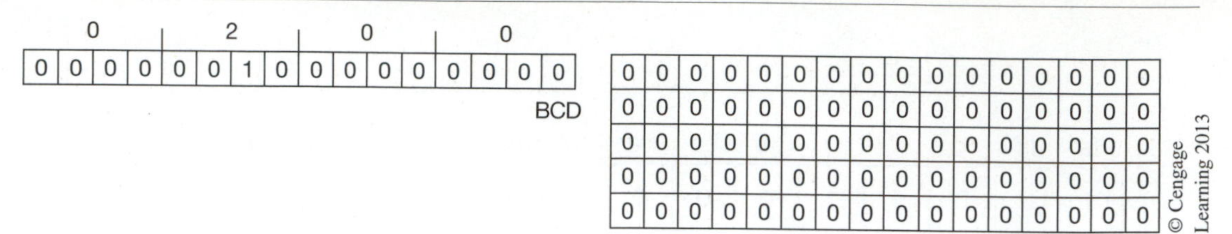

Figure 15–14a File Prior to Word-to-File Data Instruction Implementation

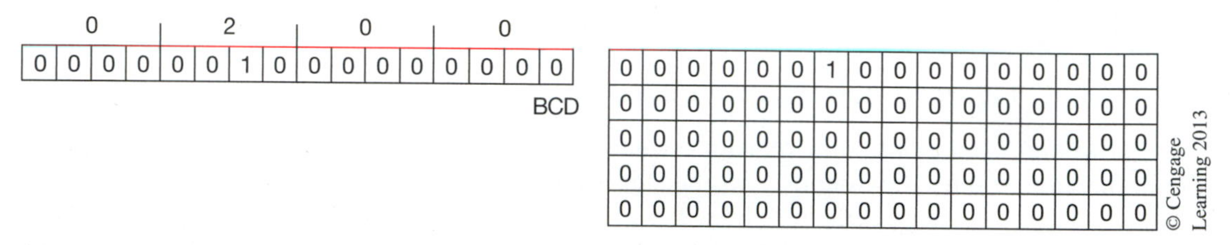

Figure 15–14b File Content After First Word-to-File Instruction Is Implemented

The next time the word-to-file instruction is implemented (indexed), the current value in word 110 is transferred to the file. At the next word location in the file, Figure 15–15 illustrates this point.

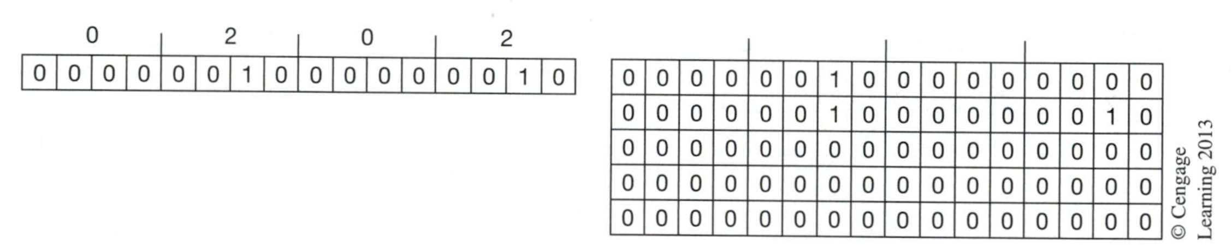

Figure 15–15 File After Second Word-to-File Instruction Is Implemented

By using a timer, the word-to-file instruction could be implemented every 15 minutes. By increasing the size of the file, a record of the die temperature for an 8-hour shift can be stored, the data (temperature) from the file could be printed out, and the temperature of the die compared to quality-control records. The application of this instruction, like all other instructions, is limited only by imagination.

FILE-TO-WORD INSTRUCTION

The file-to-word instruction transfers data from a file into a word.

Using the previous example, the temperature of the injection molding machine die can be transferred to an output word (011) that controls an LED display. By incrementing or indexing the file-to-word

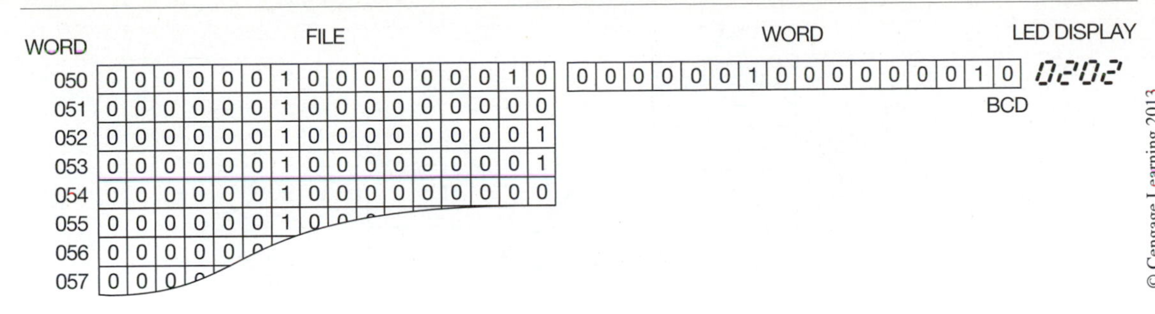

Figure 15–16a File-to-Word Instruction at First Word of File (050)

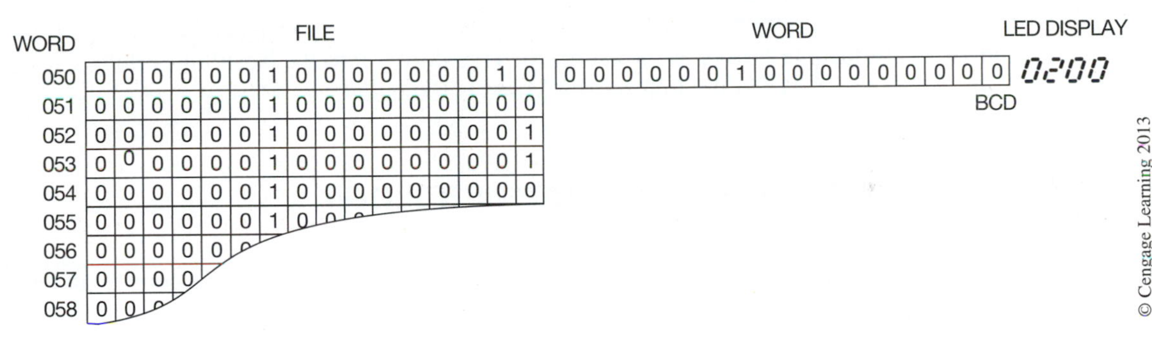

Figure 15–16b File-to-Word Instruction at Second Word of File (051)

instruction, the temperature of the die in 15-minute intervals is displayed. Figures 15–16a and 15–16b illustrate how a file-to-word instruction functions.

FILE-TO-FILE INSTRUCTION

This instruction moves data from one file to another. The data from the source file may be moved to the destination file one word at a time, or the entire contents of the file can be moved in one move, depending on the PLC.

An example of using the file-to-file move might be a chemical batch plant where different amounts and types of chemicals are mixed for a variety of products. The different mix ratios (recipes) are stored in different files, and could be transferred to a file that controls machine and/or plant operation for a given product.

Figure 15–17 shows how a file-to-file move works.

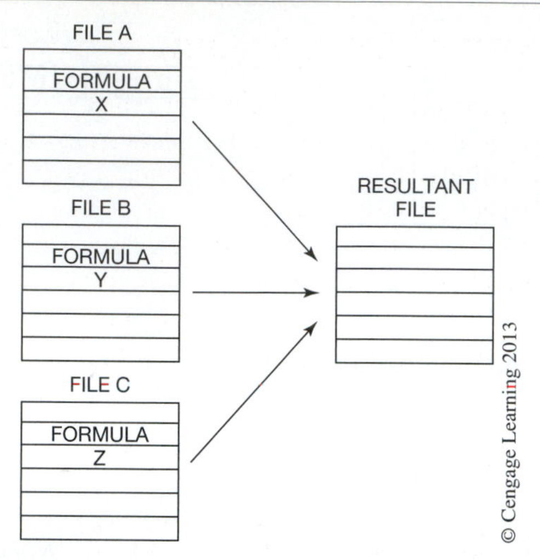

Figure 15–17 Data in File A, B, or C Can Be Transferred into the Resultant File When the File-to-File Move Is Executed

ALLEN-BRADLEY PLC-5, SLC 500, AND MICROLOGIX FILE COPY INSTRUCTION

Allen-Bradley uses a file copy (COP) instruction for making file-to-file moves. File copy is an output instruction, and is programmed as shown in Figure 15–18.

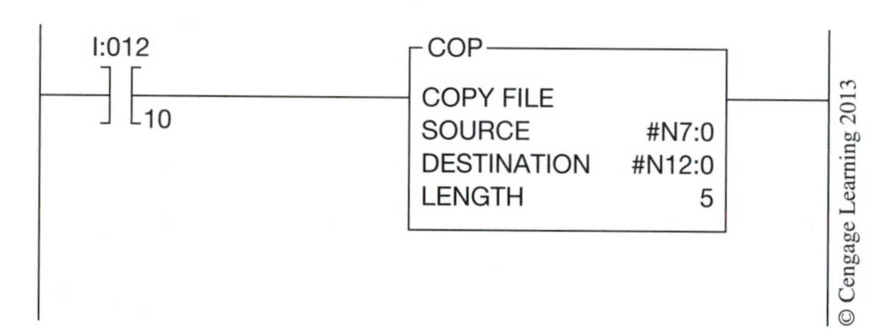

Figure 15–18 PLC-5 File Copy (COP) Instruction

The address of the first word in the source file is specified as well as the first address of the destination file. As both of these files are to be user-defined, the file numbers must start with the file indicator symbol (#). The length of the file is then specified. In this example, the source file starts with N7:0, the destination file is N12:0, and the file length is 5 words. When the input device is closed, the COP instruction is enabled, and the instruction copies data from the 5-word file starting at N7:0 into the 5 words of the destination file starting at N12:0.

ASYNCHRONOUS SHIFT REGISTER (FIFO)

The asynchronous shift register, instead of shifting bits of information within a word, or words, like the synchronous shift register, shifts the data from a complete word into a file, or stack. Although this appears to be just another name for a word-to-file instruction, it is not. There are similarities between the two, but there is also one major difference. In the asynchronous shift register, the information from a word is shifted into the top of the file and moved down through the file with each implementation, or indexing, of the instruction. The word-to-file move, however, allows the information transferred from the word to go to the last *unused* word of the file. This difference is why the asynchronous shift register is often referred to as a FIFO stack (first-in first-out). Figure 15–19 compares the asynchronous shift register to the word-to-file instruction to demonstrate the difference.

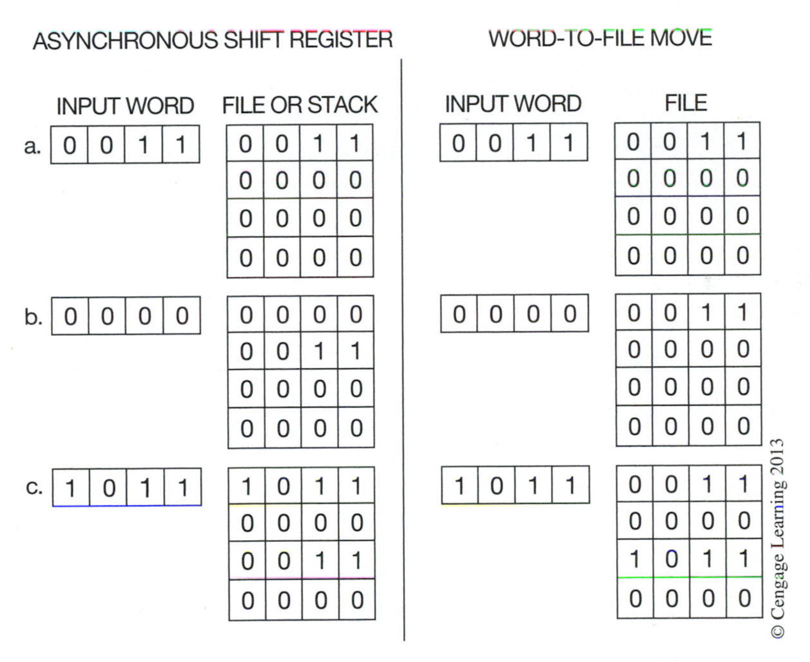

Figure 15–19 Comparison of Asynchronous Shift Register (FIFO) to Word-to-File Move

When data is transferred at position (a), the asynchronous shift register places the data into the first (top) word of the file or stack. In the word-to-file move, data is also moved into the first (top) word of the file, just like the asynchronous shift register.

At the next implementation at position (b), no data is entered at the top of the stack and all previous data is shifted *down* in the asynchronous shift register, whereas the data of the input word is transferred to the next available word in the file in the word-to-file instruction.

When the instructions are indexed again at position (c), all previous data is shifted down in the asynchronous shift register and new data is entered at the top of the stack, whereas the word-to-file instruction transfers the new data into the next available location in the file.

ALLEN-BRADLEY PLC-5, SLC 500, AND MICROLOGIX FIFO INSTRUCTION

Allen-Bradley uses a pair of output instructions to store and retrieve data in a prescribed order, or FIFO (First-In First-Out). Words are loaded into a file and unloaded in the same order in which they were loaded. Figure 15–20 shows the First-In First-Out Load (FFL) and First-In First-Out Unload (FFU) Instructions.

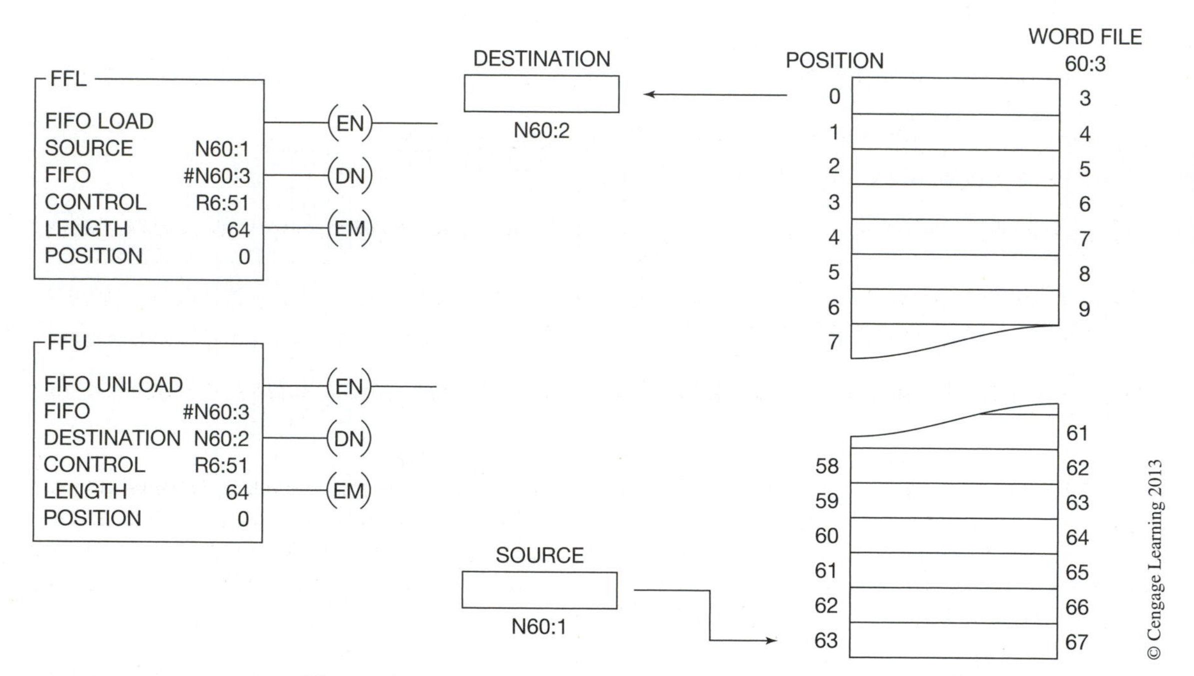

Figure 15–20 FIFO Load and Unload Instructions

The FIFO Load instruction loads the information stored in one word into a file, or stack. The FIFO Unload instruction retrieves information stored in the file, or stack, and places it into a destination word. The instruction components are as follows:

The Source is the address that stores the "data" that will be loaded into the file, or stack.

FIFO is the address of the first word in the file, or stack.

Control is the address of the control structure (48 bits—three 16-bit words) in the control area (R) of memory. The control structure stores the instruction's status bits, stack length, and next available position (pointer) in the stack, as shown in Figure 15–21.

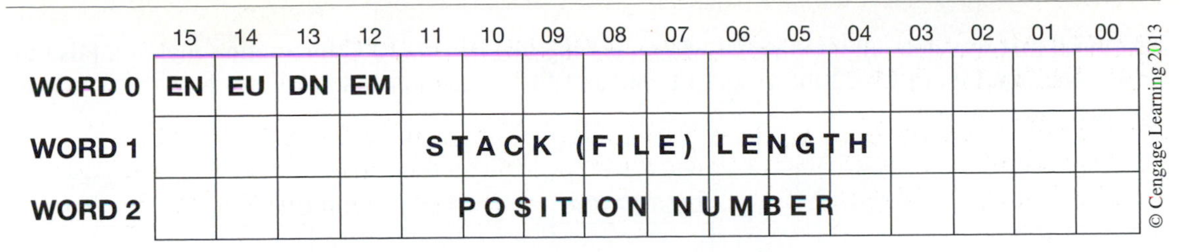

Figure 15–21 FFL Three-Word Control Element

Empty Bit—EM (bit 12) is set to 1 when the stack, or file, is empty.
Done Bit—DN (bit 13) is set to 1 when the stack is full. This prevents any additional data from being loaded into the stack.
FFU/LFU Enable Bit—EU (bit 14) is set to 1 on a false-to-true transition and is reset to 0 on a true-to-false transition.
FFL/LFL Enable Bit—EN (bit 15) is set to 1 on a false-to-true transition and is reset to 0 on a true-to-false transition.

The Length portion of the instruction defines the length, or number of words, that make up the stack, or file.

The Position portion of the instruction indicates the position in the file, or stack, in which information from the source word will be entered.

Destination (FIFO Unload) is the address that stores the data that exits, or is removed from the file, or stack.

In Figure 15–20, note that the FIFO address is #N60:3 for both the FIFO Load and FIFO Unload instructions. Address #N60:3 is the start of the 64-word file, and address N60:67 is the last address for the file. Position 0 is the first word of the file, or stack, and Position 63 would be the last word of the 64-word file. With each false-to-true transition of the FFL instruction, data from the Source is loaded into the file (FIFO) and the position indicator will advance, or increment, by one.

When the rung that contains the FFL instruction goes true, the processor sets the EN bit (bit 15) *ON*, and loads the source data (N60:1) into the next available position in the stack. The processor loads data from the source into the stack with each false-to-true transition. When the stack is full, the processor sets the DN bit (bit 13) to 1. The program should be programmed so that when the stack is full, no additional data can be loaded from the source.

When the rung that contains the FFU instruction goes from false to true, the processor sets the EU (enable unload bit 14) to 1 and unloads data from the first element of the stack into the destination word N60:2. As the data is shifted out, the processor shifts the remaining data in the stack *up* one position toward the first word. The processor continues to unload the stack each time the rung goes from false to true until it empties the FIFO stack.

LAST-IN FIRST-OUT (LIFO) INSTRUCTIONS

The Last-In First-Out Load (LFL) instruction loads words into a file. Then, using an LFU (LIFO Unload) instruction, it retrieves them in inverse order. In other words, the instruction removes the last word that was entered into the file when the rung that has the LFU instruction makes a false-to-true transition. Figure 15–22 shows an LFL and an LFU instruction and illustrates how the words are entered and retrieved.

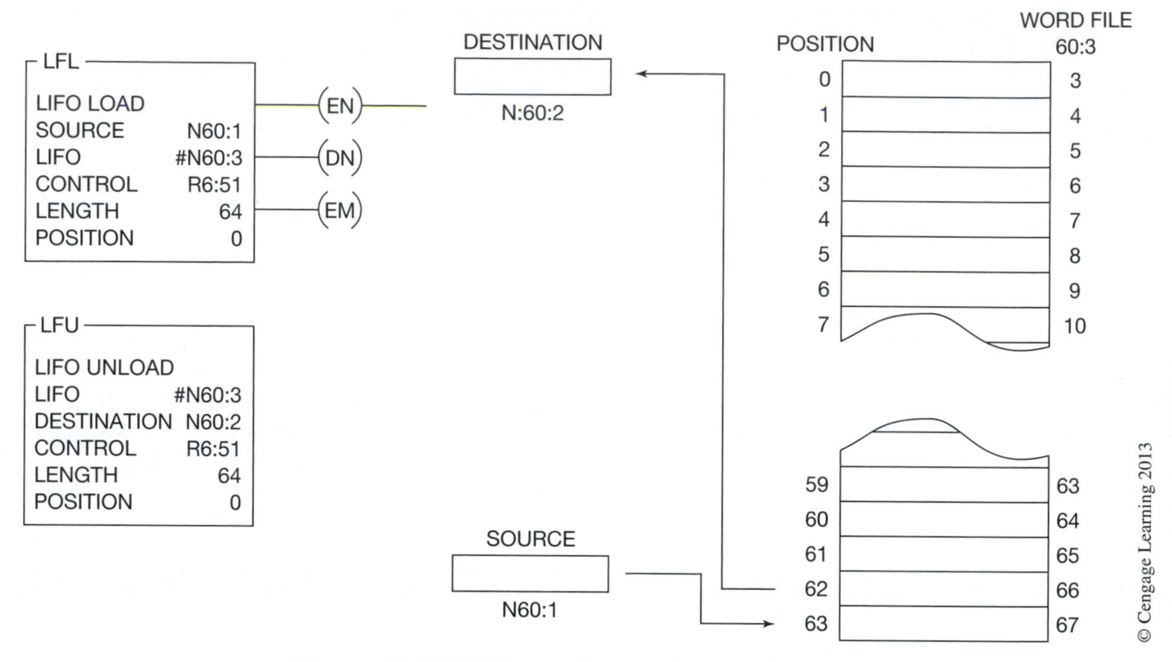

Figure 15–22 LIFO Load and Unload Instructions

When the rung condition makes a false-to-true transition, the LFL EN (enable bit) will be set to 1, the LFL instruction will load the contents (data) from the source word N60:1 into the stack, and the position value will be incremented by 1. The LFL instruction will continue to load information into the file, or stack, on each false-to-true transition until the stack is full (64 words). When the stack is full, the processor will set the DN bit (bit 13) to 1. This prevents any additional information from being loaded into the file.

When the LFU rung makes a false-to-true transition, the LFU EU (unload bit) is set to 1 and the LFU instruction will unload the last word that was loaded into the file and place it into the destination word (N60:2). The file words will continue to be unloaded from the file each time the rung makes a false-to-true transition. Each time a word is removed from the file, or stack, the position indicator will decrement (decrease in value) by 1.

ALLEN-BRADLEY LOGIX5000 FILE (ARRAY) INSTRUCTIONS

The Logix5000 controllers have the same basic shift, word, and file move instructions as those just discussed for the PLC-5, SLC 500, and MicroLogix PLCs. In addition, the Logix5000 controllers have the following array instructions:

FAL—Performs arithmetic, logic, shift, and function operations on values in arrays.

FSC—Searches for and compares values in arrays.
CPS—Copies the contents of one array into another array without interruption.
AVE—Calculates the average of an array of values.
SRT—Sorts one dimension of array data into ascending order.
STD—Calculates the standard deviation of an array of values.

As you recall, Logix5000 controllers use tag based memory and therefore do not have data files like the other Allen-Bradley PLCs. So, the only difference between the basic instructions just covered and the corresponding Logix5000 instructions is that the "*File*" parameter is replaced with "*Array*".

Logix5000 controllers use arrays to organize data and are a key element when programming file type instructions. If you recall from Chapter 4, an array is a tag that contains a block of multiple pieces of data and can be one-, two-, or three-dimensional. An array is similar to a file that contains individual pieces of data called *elements*. Each element uses the same data type, such as DINT. The elements of an array tag are arranged in a contiguous block of memory in the controller with each element in sequence. An example of a single-dimension array called "Parts" is shown in Figure 15–23. In this example, the array has been configured for six DINT elements. A subscript identifies each individual element within the array. A subscript starts at 0 and extends to the number of elements in the array. In our array example, we have six elements identified as Parts[0]...Parts[5].

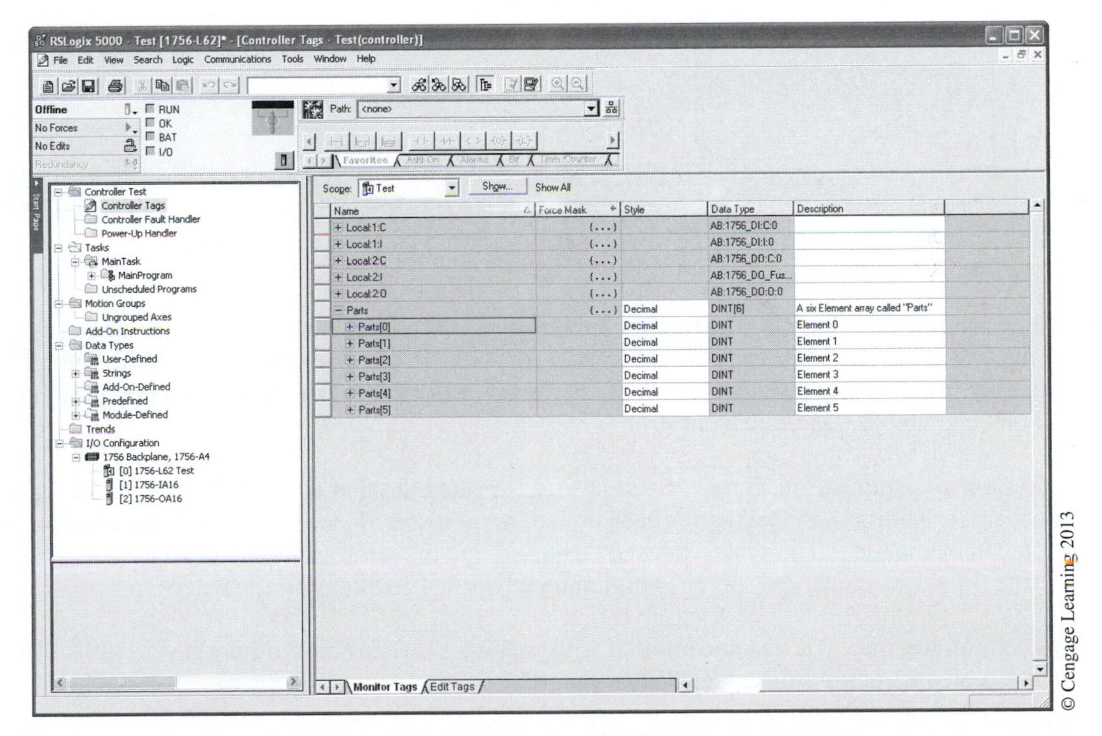

Figure 15–23 Example Array Tag

To better understand and see how the array tag is used, the same Bit Shift Left (BSL) instruction that was used earlier in the paint booth example (Figure 15–11) is shown programmed in the Logix5000 format. See Figure 15–24.

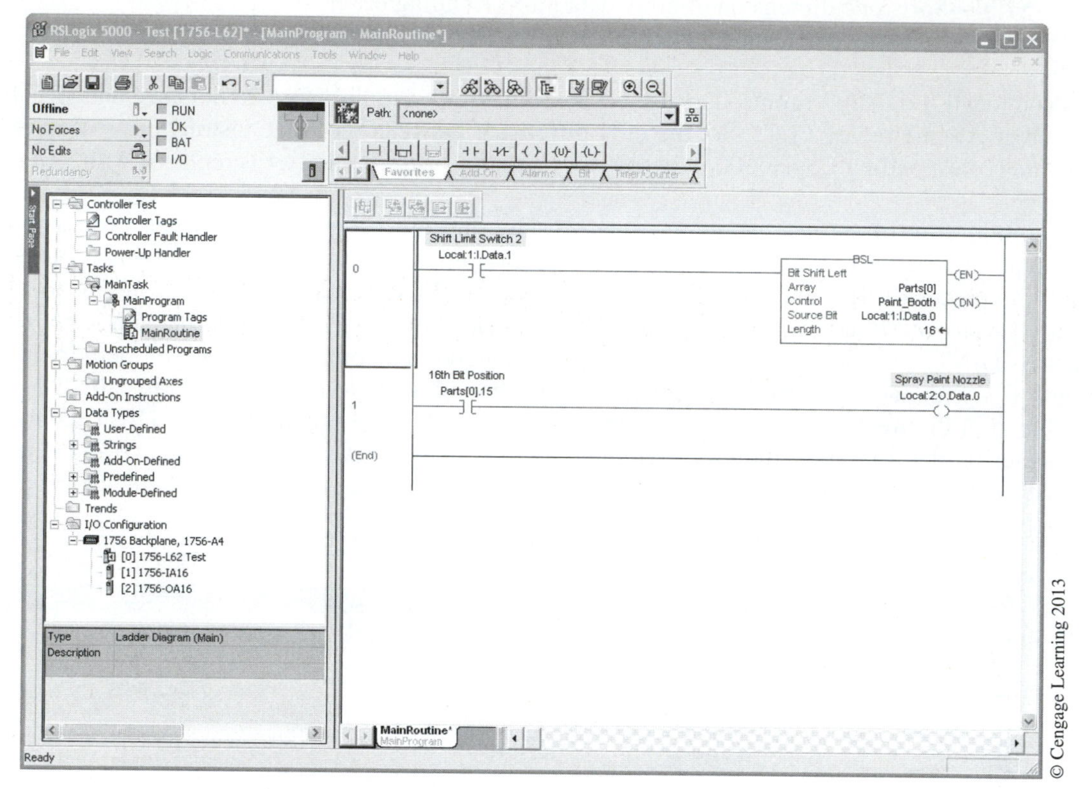

Figure 15–24 Logix5000 Bit Shift Left (BSL) Instruction

When programming the BSL instruction in the Logix5000 controller, the *Array* parameter is the tag of the first element of the group of elements in the array. In our example we specified element [0] in the array called "Parts". With a length of 16, the bits will shift to the left starting at bit 0 of element [0] and end with bit 15 of element [0].

The *Control* parameter in Figure 15–24 is a user-created tag that is defined as a control type. You must create a unique control tag for each type of array instruction you program. The *Source* bit is a bit tag in memory containing the status to be moved into the array each time the rung goes from false to true. In our example, the source is digital input Local:1:I.Data.0 (limit switch).

As you can see, once you become familiar with creating and using arrays there is very little difference in the basic instructions discussed in this chapter.

Note: *For more information on using the available array instructions found in the Logix5000 controllers, refer to the reference manual "Logix5000™ Controllers General Instructions."*

Chapter Summary

Although the keystrokes and instructions vary with each PLC manufacturer, the principles of word and file moves are the same. The synchronous shift register shifts bits of information left or right (forward and reverse) within a word or words, while file moves transfer data from words to files, files to words, or files to files. The asynchronous shift register is referred to as FIFO, or first-in first-out, as data transfers or falls to the bottom of the stack and uses the last unused word. The data is retrieved in the order it enters the stack (first-in first-out). The file can also be programmed to retrieve the data that was last in to be the first data out. This convention is referred to as a LIFO stack.

Review Questions

1. Define the term *word* as used in this chapter.
2. Define the term *file* as used in this chapter.
3. The synchronous shift register shifts data in a forward direction only.
   T F
4. In a 1-word shift register, the data is entered at bit:
   a. 1
   b. 2
   c. 4
   d. 8
   e. 16
   f. none of the above
5. In a 1-word shift register, the data is shifted out at bit:
   a. 1
   b. 2
   c. 4
   d. 8
   e. 16
   f. none of the above
6. Define the term *FIFO*.
7. Define the term *LIFO*.
8. Briefly describe the difference between *synchronous* and *asynchronous shift registers*.
9. When using the PLC-5 FFL instruction, which bit is set to 1 when the stack is full?
10. When using the PLC-5 FFU instruction, which bit is set to 1 when the stack is empty?
11. List two other terms that could be used to refer to a file.
12. When using a AB PLC-5 PLC, what is the maximum length of a FIFO stack?
13. What is the purpose of the PLC-5 *COP* instruction?
14. Which of the following group of words could *not* be a file?
    a. 50, 51, 52
    b. 50, 51, 52, 53
    c. 100, 101, 102, 103
    d. 100, 101, 102, 103, 105

15. Briefly describe the function of a word-to-file move.
16. Briefly describe the function of a file-to-word move.
17. Briefly describe the function of a file-to-file move.
18. Which instruction is also known as a FIFO?
    a. synchronous shift register
    b. word-to-file move
    c. file-to-word move
    d. asynchronous shift register
    e. file-to-file move
19. When data is transferred into a file using a word-to-file move, the data is entered at the last unused word of the file.
    T F
20. The control element for a BSL instruction requires how many words? What information does each word hold?
21. File numbers must start with what file indicator symbol when programming an Allen-Bradley PLC-5, SLC 500, or MicroLogix PLC?

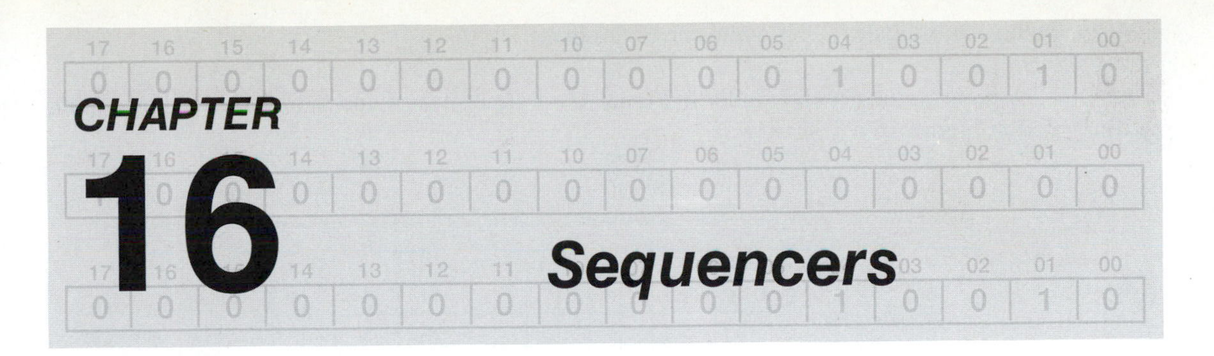

CHAPTER 16 Sequencers

Objectives

After completing this chapter, you should have the knowledge to:

- Describe what a sequencer instruction does.
- Understand the basics of sequencer operation.
- Define the term *mask*.

The sequencer instruction transfers information from memory words into output words. Sequencer instructions are typically used to control automatic assembly machines that have consistent and repeatable operations.

A programmed **sequencer** replaces the mechanical drum sequencer that was used in the past. On the mechanical sequencer, when the drum cylinder rotated, contacts opened and closed mechanically to control output devices. Figure 16–1 shows a mechanical drum cylinder with pegs placed at varying horizontal positions for step 1 of the sequence. When the cylinder rotated, contacts that aligned with the pegs closed, and contacts where no pegs existed remained open. In this example, the presence of a peg should be thought of as a 1, or *ON,* and the absence of a peg as a 0, or *OFF*.

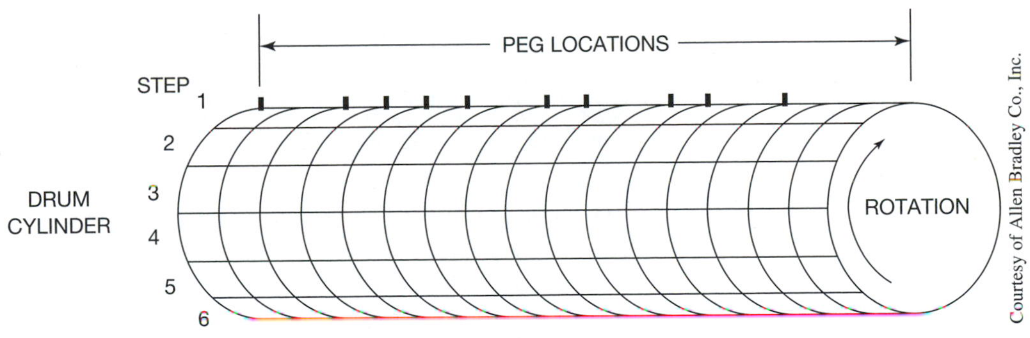

Figure 16–1 Drum Cylinder

To program a sequencer, binary information is entered into a series of consecutive memory words. These consecutive memory words are referred to as a file. Information from the words in the file is transferred sequentially to the output word to control the outputs.

If the first six steps on the drum cylinder in Figure 16–1 are removed and flattened out, they appear as illustrated in Figure 16–2.

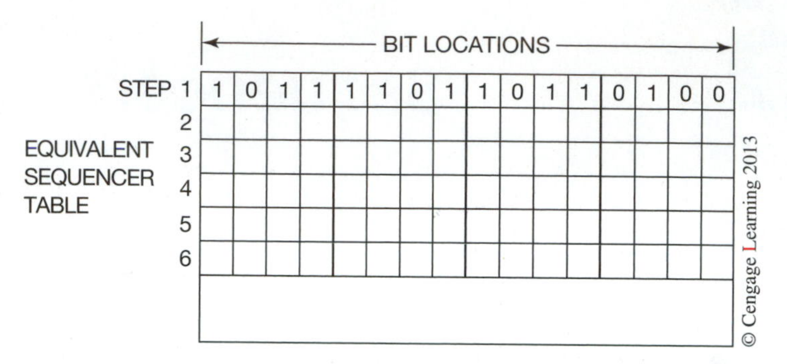

Figure 16–2 Sequencer Table

For step 1, each horizontal location where a peg was located is now represented by a 1 (*ON*), and the positions where there were no pegs are represented by a 0 (*OFF*).

The six steps could also be viewed as a 6-word file with each 16-bit word representing a sequencer step. If one enters different binary information (1s and 0s) into each word of the file, the file replaces the rotating drum cylinder.

To illustrate how this works, 16 lamps are used for outputs (as shown in Figure 16–3).

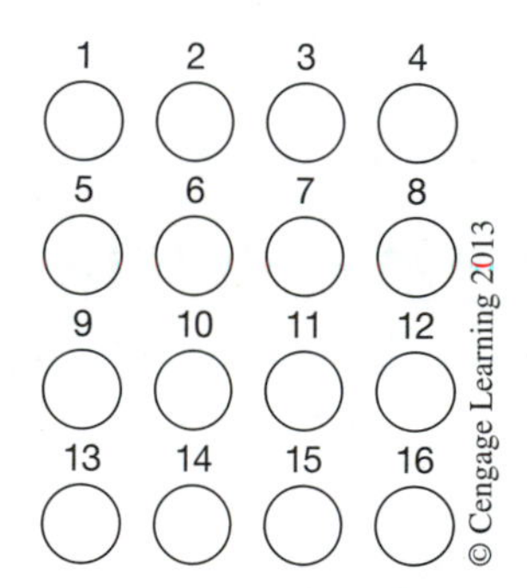

Figure 16–3 Output Lamps

Each lamp represents one bit address (1 through 16) of output word 25.

Assume, for the sake of discussion, that the operator wants to light the lamps in the 4-step sequence shown in Figures 16–4a, 16–4b, 16–4c, and 16–4d.

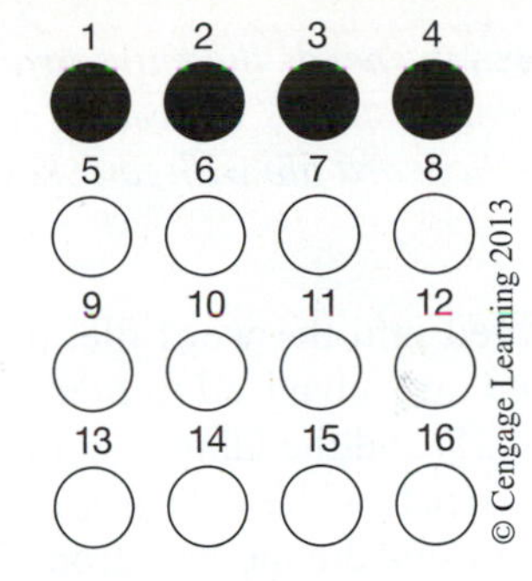

Figure 16–4a Sequence 1

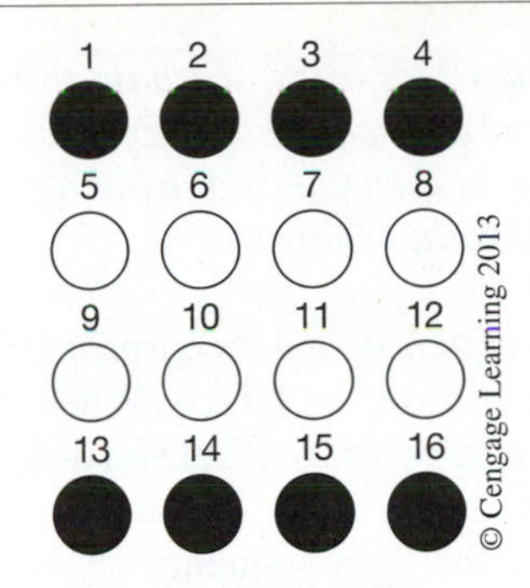

Figure 16–4b Sequence 2

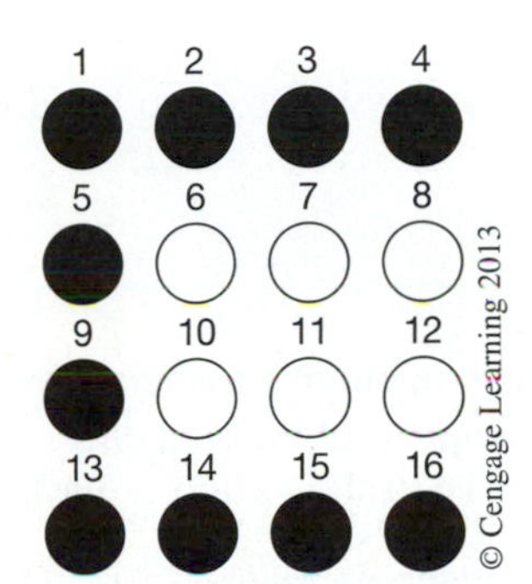

Figure 16–4c Sequence 3

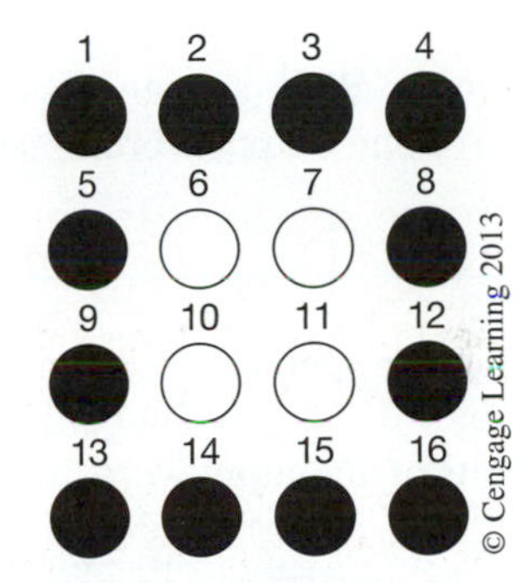

Figure 16–4d Sequence 4

The bit addresses of the lamps that are to be lit in each step of the sequence are written down to assist in entering data into a word file.

Step 1. 1, 2, 3, 4
Step 2. 1, 2, 3, 4—13, 14, 15, 16
Step 3. 1, 2, 3, 4—5, 9—13, 14, 15, 16
Step 4. 1, 2, 3, 4—5, 8, 9, 12—13, 14, 15, 16

The next step is to define a word file to store the binary data required for each step of the sequencer. Words 30, 31, 32, and 33 are used for the 4-word file. By using the programmer, one enters binary information (1s and 0s) into each word of the file to reflect the desired lamp sequence (Figure 16–5).

	16	15	14	13	12	11	10	9	8	7	6	5	4	3	2	1	
WORD 25	0	0	0	0	0	0	0	0	0	0	0	0	0	0	0	0	OUTPUT
WORD 30	0	0	0	0	0	0	0	0	0	0	0	0	1	1	1	1	STEP #1
WORD 31	1	1	1	1	0	0	0	0	0	0	0	0	1	1	1	1	#2
WORD 32	1	1	1	1	0	0	0	1	0	0	0	1	1	1	1	1	#3
WORD 33	1	1	1	1	1	0	0	1	1	0	0	1	1	1	1	1	#4

Figure 16–5 Binary Information for Each Sequencer Step

Note: *Some PLCs allow the data to be entered using BCD, which speeds the entry process. To use this feature, the required binary information for each sequencer step is converted to BCD. The information is then entered using a programming device into the word file with four key strokes for each word, rather than 16.*

Once the sequencer has been programmed and the data entered into the word file, the sequencer is ready to control the lamps. When the sequencer is activated and advanced to Step 1, the binary information in word 30 (Figure 16–5) is transferred into word 25, and the lamps light in the pattern shown in Figure 16–4a. Advancing the sequencer to Step 2 transfers the data from word 31 into word 25 for the light sequence shown in Figure 16–4b. Step 3 transfers the data from file word 32 into word 25, and Step 4 transfers information from word 33 into word 25. When the last step is reached, the sequencer can be reset and sequenced again.

Depending on the PLC, sequencers can be programmed from a few steps up to hundreds of steps, and can control one output word or several.

MASKS

When a sequencer operates on an entire output word, there may be outputs associated with the word that the operator does not want controlled by the sequencer. To prevent the sequencer from controlling certain bits of an output word, a **mask** word is used. Figure 16–6 shows how a mask word works.

	16	15	14	13	12	11	10	9	8	7	6	5	4	3	2	1	
WORD 025	1	1	1	1	1	0	0	1	1	0	0	1	1	1	1	1	OUTPUT
WORD 20	1	1	1	1	1	0	0	1	1	0	0	1	1	1	1	1	MASK
WORD 30	0	0	0	0	0	0	0	0	0	0	0	0	1	1	1	1	FILE
WORD 31	1	1	1	1	0	0	0	0	0	0	0	0	1	1	1	1	
WORD 32	1	1	1	1	0	0	0	1	0	0	0	1	1	1	1	1	
WORD 33	1	1	1	1	1	0	0	1	1	0	0	1	1	1	1	1	
WORD 34	1	1	1	1	1	1	1	1	1	1	1	1	1	1	1	1	

Figure 16–6 Using a Mask Word

The mask word is a means of selectively screening out data from the sequencer word file to the output word. For each bit of output word 025 that the operator wants the sequencer to control, the corresponding bit of mask word 20 must be set to 1.

In Figures 16–4a, 16–4b, 16–4c, and 16–4d, bits 6, 7, 10, and 11 are not used. *Not* setting bits 6, 7, 10, and 11 of the mask word to 1 means that these bits can be used independently of the sequencer.

In Figure 16–6, a fifth step is added to the sequencer. File word 34 and bits 6, 7, 10, and 11 are set to 1. With bits 6, 7, 10, and 11 of mask word 20 set to 0, the data in file word 34 is screened out and prevented from being transferred into output word 025.

The sequencer works much like the file-to-word move discussed in Chapter 15. For programmable controllers that don't have a dedicated sequencer instruction, a file-to-word move instruction can be used.

ALLEN-BRADLEY PLC-5, SLC-500, AND MICROLOGIX 1000 SEQUENCER INSTRUCTION

The Allen-Bradley Sequencer Output Instruction (SQO) creates a sequencer file of information that is used to control various output devices. When the rung that contains the SQO instruction makes a false-to-true transition, the instruction increments to the next word in the sequencer file and loads the information into the destination word. Figure 16–7 shows an SQO instruction.

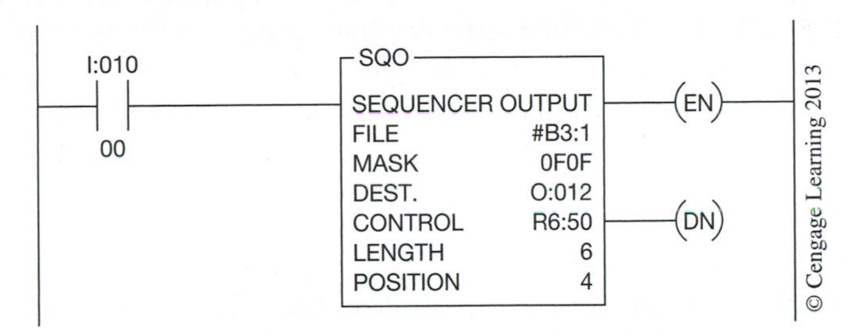

Figure 16–7 Sequencer Output Instruction (SQO)

The File portion of the instruction is the address of the sequencer file. You must use the file indicator symbol (#) for this address. In this illustration, the file has been addressed #B3:1. This address indicates that a file has been created in the Bit File portion of the processor memory starting with word 1.

The Mask, as explained earlier, is a filter through which all data from the sequencer file must pass before being placed into the destination, or output word. Allen-Bradley uses a hexadecimal number that represents the bit pattern that is desired by the mask for screening information, or data, from the sequencer file. A 1 must be placed in the mask bit location for information to be passed through.

The Destination is the address of the output word that is to be controlled by the sequencer instruction. In Figure 16–7, the destination address is O:012, or word 12 of the output image table.

The Control is the three-word element that stores the status bits for the sequencer, as well as the length of the sequencer file and the position of the sequencer. Figure 16–8 shows the three-word element for the SQO instruction.

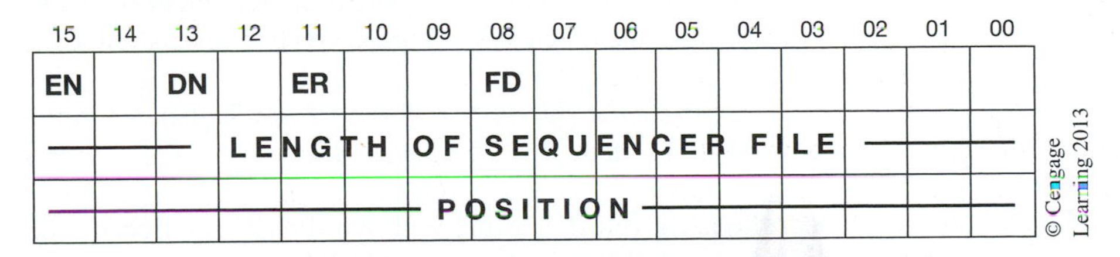

Figure 16–8 Three-Word Element for an SQO Instruction

The status bits shown in word 1 include:

Error Bit—ER (bit 11) is set to 1 if the processor detects a negative *position* value or a negative or zero *length* value.

Done Bit—DN (bit 13) is set to 1 when the last word of the sequencer file has been transferred into the *destination* word.
Enable Bit—EN (bit 15) is set to 1, or is true, when the SQO instruction is enabled.
Found Bit—FD (bit 08)—SQC instruction only. When the status of all nonmasked bits in the source address matches the reference word bits, bit 08 is set to 1.

The Length value in the instruction is the number of steps in the sequencer file starting with Step 1. Step 0 is the startup, or default value of the sequencer on startup. On the next false-to-true transition, after the last step of the sequencer has been loaded into the destination word, the sequencer will be reset to Step 1. The maximum length of a sequencer file is as follows:

PLC-5	1–1000
SLC 500	1–255
MicroLogix 1000	1–104

The Position indicates the step, or word, where the sequencer is currently positioned. The position number will increment with each false-to-true transition of the instruction.

Figure 16–9 shows a programmed SQO instruction with the sequencer file, mask word, destination word, and the status of the external output devices when the sequencer instruction is on Step 4.

As programmed in Figure 16–9, the sequencer instruction transfers data from the sequencer file each time that input device I:010/00 closes. In Figure 16–9, the sequencer is shown on Step 4. Even though the data in word B3:5 (Step 4) has a 1 set for bits 0, 1, 4, 6, 8, 9, 11, 12, and 14, the external output devices associated with output word O:012 only show that outputs 0, 1, 8, and 11 are *ON*. The reason outputs 4, 6, 12, and 14 are not *ON* is that these bits are masked.

When a sequencer instruction has a mask address, data will transfer only from the sequencer file to the bits of the destination word that have a 1 in the mask address. The mask address in Figure 16–7 was hexadecimal number 0F0F. This address is the binary equivalent of 0000 1111 0000 1111.

Another Allen-Bradley sequencer instruction is the Sequencer Compare Instruction (SQC). This instruction compares all of the masked bits of the source word to the current step of the sequencer file. If all of the bits *match,* bit 08 of the control word is set to 1. Bit 8 is the Found Bit (FD). This instruction can be used to compare the status of machine/equipment input devices with what is normal operation. This is a great way to do machine diagnostics. Figure 16–10 shows an SQC instruction.

Note: *The PLC-5 family of processors calls this instruction SQI for Sequencer Input. While the initials are different, the operation is the same.*

As programmed, the sequencer file is Bit File 3, starting with word 8 (B3:8). Because this is a user-defined file, the address is preceded with the # symbol. The Mask has been set to all 1s by entering the hexadecimal number FFFF. The Source is listed as I:010, or word 10 of the input image table. This is the word that holds the status of the input devices for the process equipment. The Control is a three-word element in the R (control) file that holds the status bits, the length of the sequencer file, and the current position (step) of the sequencer. The Length of the file is shown as 2, and current Position (step) is shown as 0.

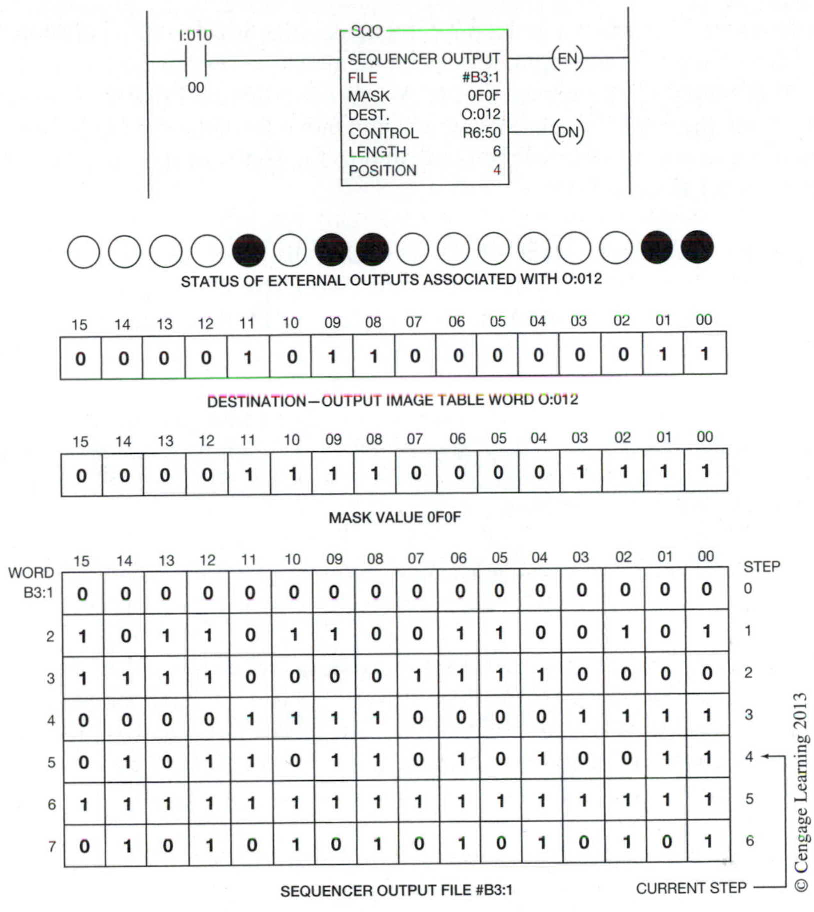

WORD	15	14	13	12	11	10	09	08	07	06	05	04	03	02	01	00	STEP
B3:1	0	0	0	0	0	0	0	0	0	0	0	0	0	0	0	0	0
2	1	0	1	1	0	1	1	0	0	1	1	0	0	1	0	1	1
3	1	1	1	1	0	0	0	0	1	1	1	1	0	0	0	0	2
4	0	0	0	0	1	1	1	1	0	0	0	0	1	1	1	1	3
5	0	1	0	1	1	0	1	1	0	1	0	1	0	0	1	1	4
6	1	1	1	1	1	1	1	1	1	1	1	1	1	1	1	1	5
7	0	1	0	1	0	1	0	1	0	1	0	1	0	1	0	1	6

Figure 16–9 Sequencer (SQO) Instruction

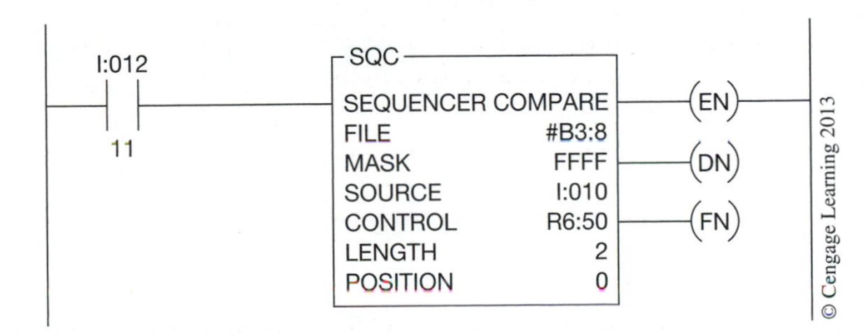

Figure 16–10 Sequencer Compare Instruction

The status bits for this instruction are the same as the SQO instruction previously discussed (Figure 16–8), with the addition of the Found Bit (FD) (bit 8). This bit will be set to 1 whenever the status of all of the masked bits in the source word matches the status of the reference word in the sequencer file.

As an example, let us say that when a given production machine is operating normally and ready to produce Product A, the status of the input devices on the machine (*ON* or *OFF*) should be indicated by the data stored in Step 1 of the sequencer file. When input device I:012/11 is closed, making a false-to-true transition, the sequencer will increment and move from the startup position of 0 to Step 1. The bit status of input word I:010 will be compared to the status of the 16-bit word at Step 1 of the sequencer file. Step 1 is word B3:9.

As shown in Figure 16–11, if the bit status of I:010 matches the bit status of B3:9, the processor sets bit 08 (FD) of the control word to 1. If bit 08 of control word R6:50 was programmed to an output device, like the indicator lamp (O:014/00) shown in Figure 16–11, the output device would turn *ON* when bit 08 was set to 1. This light indicates that the input devices of the process machine are operating correctly and the machinery is ready to start producing Product A. If the data of the two words does not match, bit 08 will not be set to 1. The processor will continue to compare the status of input word I:010 with the data stored in Step 1 of the sequencer on each program scan. When the input

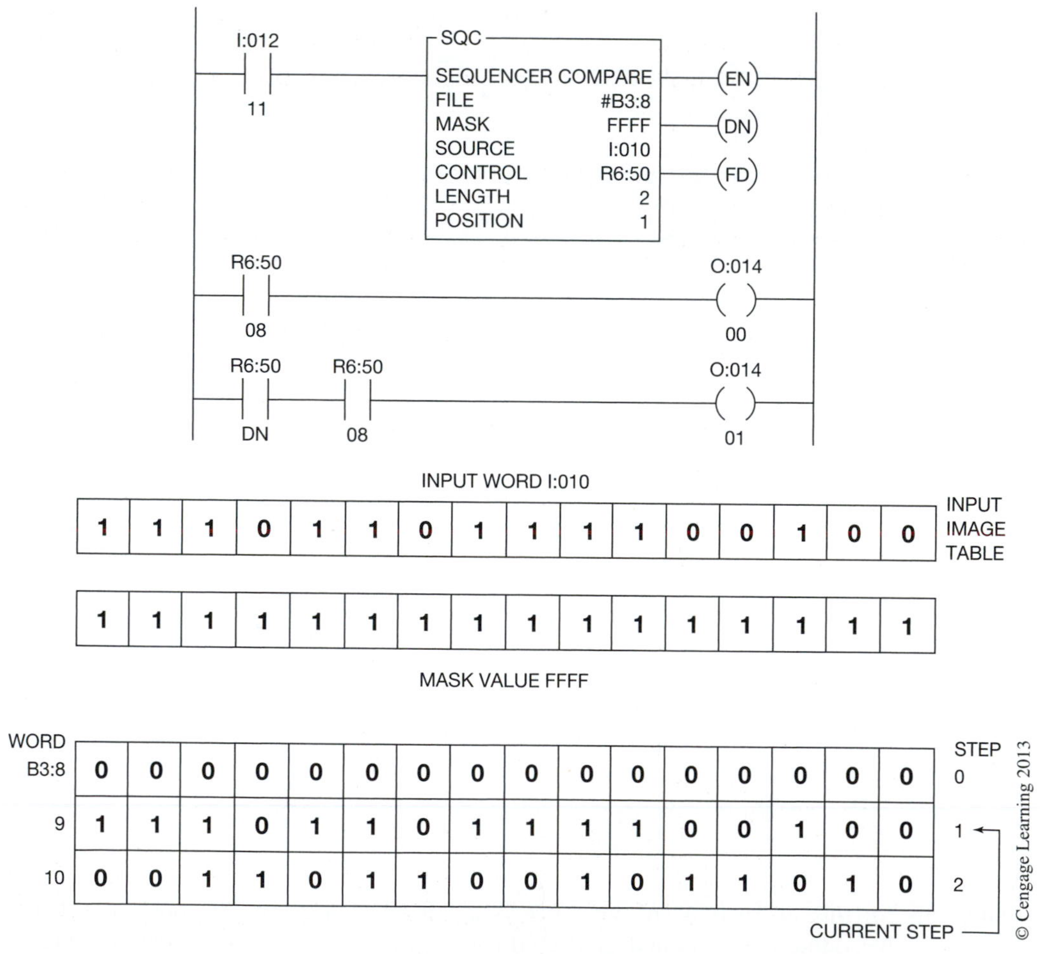

Figure 16–11 SQC Instruction with File Compare

devices are set (*ON* or *OFF*) correctly and match sequencer Step 1, bit 08 will be set to 1, and output indicator lamp O:014/00 will turn *ON*.

Step 2 of the sequencer could be loaded with the bit status of the machine when it is correctly set up to produce Product B. By opening and closing input device I:012/11, the sequencer is activated and moves to Step 2. The bit status of Step 2 mirrors the status of the input devices on the machine when it is ready to produce Product B. If output light O:014/01 comes *ON*, the operator knows the input devices are operating correctly and he or she can start the process of producing Product B. As programmed, output O:014/01 can only be turned *ON* when the DN bit (bit 13) of the sequencer is set to 1, indicating that the sequencer is on the last step and the FD bit (bit 8) has been set to 1. The program could have been written by entering FD instead of 08, and 13 instead of DN. Either way, output light O:014/01 can only come *ON* when the sequencer is on the last step, Step 2, and the data in sequencer Step 2 matches the status of input word I:010. If the light fails to come *ON*, the operator knows that one or more of the input devices are not operating correctly. Using the video display on the programming device, it could be determined which device(s) are not set properly.

This instruction can be used as a powerful diagnostic tool to determine correct machine operation. This instruction can be used to compare input words as well as output words that represent input and output devices used for the manufacturing process.

Another sequencer instruction is the Sequencer Load (SQL) instruction. This instruction allows data to be loaded into a sequencer file from a source address on each false-to-true transition. Figure 16–12 shows the Sequencer Load instruction, the Source word, and the five-word sequencer file.

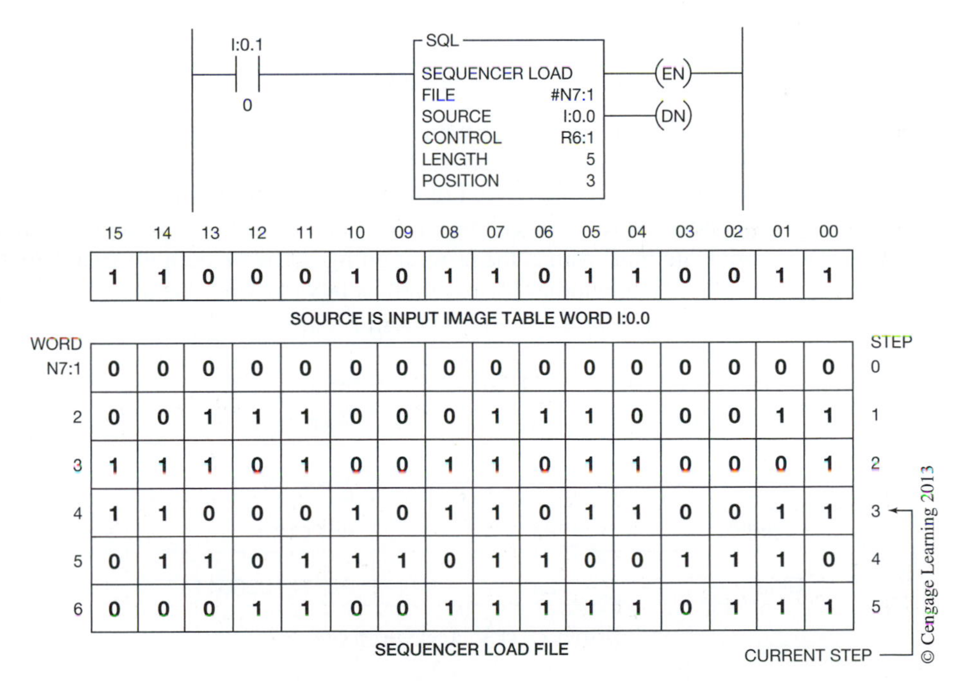

Figure 16–12 Sequencer Load (SQL) Instruction

As with the other sequencer instructions, the file used for the SQL instruction is a user-defined file and must be preceded by the # symbol. The source is shown as I:0.0. This address tells us that this instruction has been programmed using either an SLC 500 or a MicroLogix PLC with fixed I/O. The source word address is Input Word 0 in slot 0.

The input that controls the SQL instruction has an address of I:0.1/0, indicating that this is word 1, slot 0, bit 0. If you refer back to Chapter 4, Figure 4–8, you will see that bit 0 of word 1 of the input image table is input device number 16.

On each false-to-true transition of input device I:0.1/0, the current status (1 or 0) of each input device represented by input image word 0 will be transferred into the sequencer file. Figure 16–12 shows the SQL instruction in Step 3. The information shown in the Source word I:0.0 has been written into Step 3 of the sequencer file. On each false-to-true transition, the instruction will be incremented by 1 and the current status of input devices in input image table word 0.0 will be written into the file N7:1. When the instruction has reached Step 5, the DN bit (bit 13) will be set to 1. On the next false-to-true transition of the instruction, the Sequencer Load instruction will recycle to Step 1 and the status of word 0.0 will be loaded into the file, overwriting previous information that had been loaded into Step 1 of the file.

In this SQL example, an input word was used for the source. The source can also be a file address or a constant (−32,768 to +32,767).

The SQL instruction is like a word-to-file move instruction and could be used to store numeric data from RTDs, thermocouples, and the like.

ALLEN-BRADLEY LOGIX5000 SEQUENCERS

The only difference between the sequencer instructions for the Logix5000 controllers and those just discussed is that the Logix5000 controllers use array tags instead of files. Otherwise, the same sequencer instructions (SQI, SQO, SQL) are found in both. It is worth noting that the *Mask* parameter in the Logic5000 sequencer instructions can be entered as a tag address or an immediate mask value. If entered as an immediate mask value, the programming software defaults to decimal values. If you want to enter a mask by using another format such as hexadecimal, octal, or binary, precede the value with the correct prefix. The prefixes are shown in Figure 16–13.

Data Type	Prefix	Example
Hexadecimal	16#	16#0F0F
Octal	8#	8#7417
Binary	2#	2#111100001111

Figure 16–13 Mask Prefixes

Chapter Summary

Although sequencers, like other data manipulation and arithmetic instructions, are programmed differently with each PLC, the concepts are the same. Data are entered into a word file for each sequencer step, and, as the sequencer advances through the steps, binary information is transferred sequentially from the word file to the output word(s). Output word bits can be masked so they can operate independently of the sequencer.

Review Questions

1. Briefly describe a *sequencer*.
2. A series of consecutive words is referred to as a:
   a. deck
   b. group
   c. file
   d. chain
3. What is the purpose of a *mask word* in a sequencer?
4. What device is commonly replaced by a sequencer instruction?
5. Set up the file in the following figure so the sequencer will operate the motors as shown in steps 1, 2, 3, and 4. Program the circuit so motors 01015, 01016, and 01017 cannot be energized.

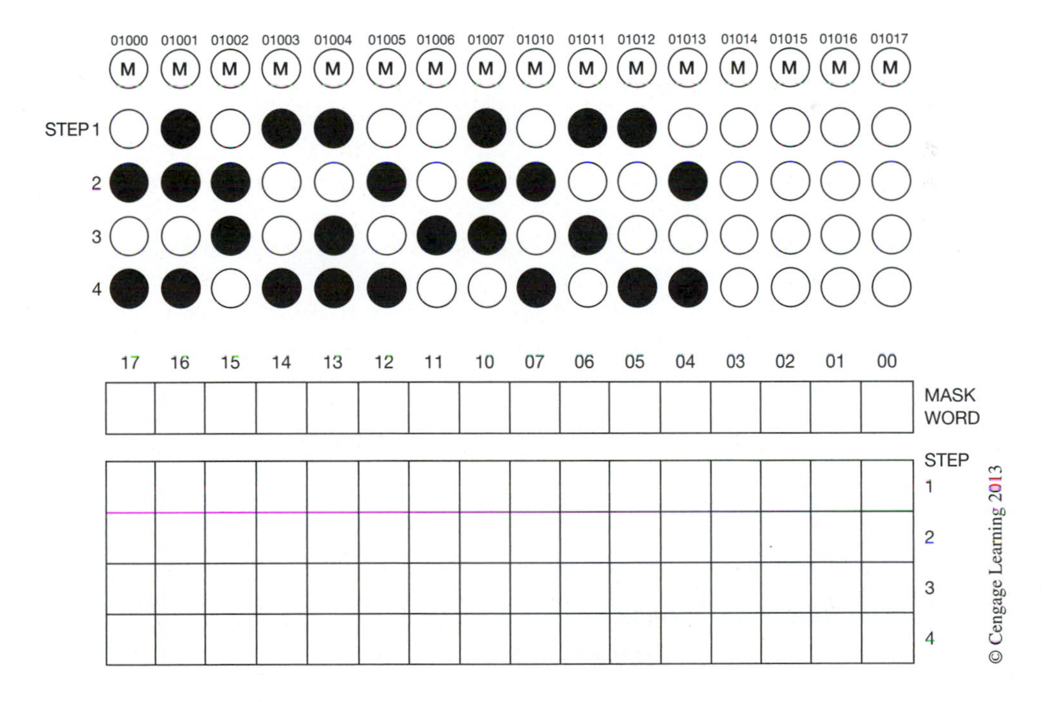

6. What is the maximum length of a sequencer file when using the SLC 500?
7. Which Allen-Bradley sequencer instruction is like a *file-to-word* move?
8. Which Allen-Bradley sequencer instruction is like a *word-to-file* move?
9. The maximum length of an SLC 500 sequencer file is ______________.
10. What condition would cause bit 08 (FD) of an SQC instruction control word to be set to 1?
11. What would be the mask word bit status if the mask was set at FFFF?
12. What would be one use for an SQC instruction?

CHAPTER 17 Process Control Signals, Scaling, and PID Instructions

Objectives

After completing this chapter, you should have the knowledge to:

- Understand process control signals and scaling.
- Apply linear scaling equations.
- Program the Allen-Bradley SCL and SCP instructions.
- Understand process controllers.
- Understand and program the Allen-Bradley PID instruction.
- Program a basic process control loop using the Allen-Bradley SLC 500.

This chapter covers basic process control signals, linear scaling of analog process signals, and the Allen-Bradley SLC 500 Scale (SCL), Scale with Parameters (SCP), and Proportional Integral Derivative (PID) instructions. Most of what you will learn in this chapter can be applied to any PLC on the market today, provided that it has a basic set of math instructions and a PID instruction. Because PID instructions and the processes that they control can be extremely complex, this chapter will cover only the basics elements of the PID instruction, its application, and tuning.

PROCESS CONTROL SIGNALS AND SCALING

The ability to monitor and/or control a process depends on having accurate and meaningful information about the process. This information can include such things as temperature, pressure, level, flow, weight, etc., and is commonly referred to as the **process variable (PV).** Electrical and pneumatic signals are used to represent the process variables and are typically generated by field-mounted devices such as transducers and transmitters. These process signals are sent to PLCs, loop controllers, displays, and other devices as one of the following standard analog type signals:

Low DC electrical current (4–20 mA, 0–20 mA, etc.)
Low DC electrical voltage (0–10 V DC, 1–5 V DC, +/−10 V DC, etc.)
Low air pressure signal (3–15 psig)

A standard analog signal has a value that uniquely corresponds to the process variable being measured. That is, as the measured process variable changes, so does the analog signal representing it (refer to Figure 17–1).

Figure 17–1 shows a graph of an electrical analog signal compared to the process variable being measured. The process variable is shown along the vertical or *y*-axis and is in degrees Fahrenheit.

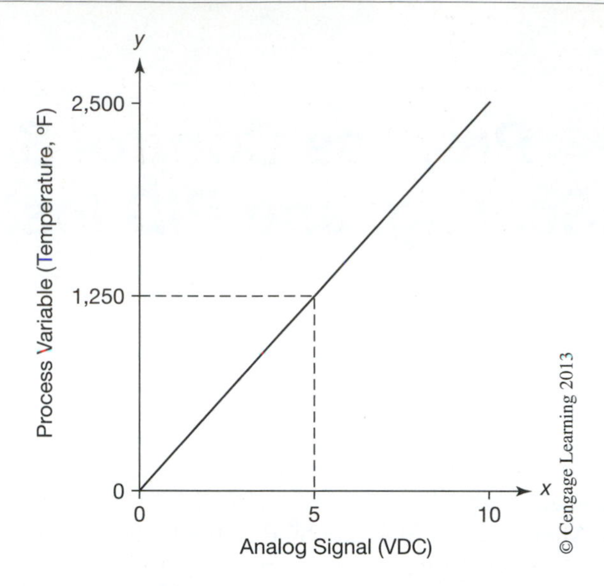

Figure 17–1 Relationship Between Analog Signal and Process Measurement

The electrical analog signal is shown along the horizontal or *x*-axis and is represented as a low-level DC voltage. This low-level DC electrical voltage is the analog signal that would be sent to the PLC or other device representing the measurement of the process variable, temperature in this example.

As you study the graph in Figure 17–1, you will notice that the process variable has a range of 0 to 2500 °F and the corresponding electrical analog signal representing that process variable has a range of 0 to 10 V DC. There is a linear relationship that exists between these two signals, meaning that when the temperature is at 50% of its range (1250 °F), the electrical analog signal will be at 50% of its range (5 V DC). Simply put, for every change in the process variable there is an equally proportional change in the electrical analog signal representing that process variable.

Before an electrical analog signal can be used by the PLC or other digital device, it must first be converted into a corresponding digital value. This conversion from analog to digital is accomplished by an Analog-to-Digital or A/D converter that is part of the I/O hardware of the device. The digital range that the electrical analog signal is converted into depends on the electrical range and digital resolution of the A/D converter. No matter what that range is, there is again a linear relationship between the electrical analog signal and the corresponding digital value, just as there was with the process variable and the electrical analog signal described above. Figure 17–2 shows a graph of an electrical analog signal and the corresponding digital range of the A/D converter.

In Figure 17–2 the electrical analog signal is shown on the horizontal or *x*-axis and has a range of 0 to 10 V DC. The corresponding digital range is shown on the vertical axis or *y*-axis and has a range of 0 to 32,767. When the analog signal is at 50% of its range (5 V DC), the corresponding digital value will be at 16,384 or 50% of the digital range.

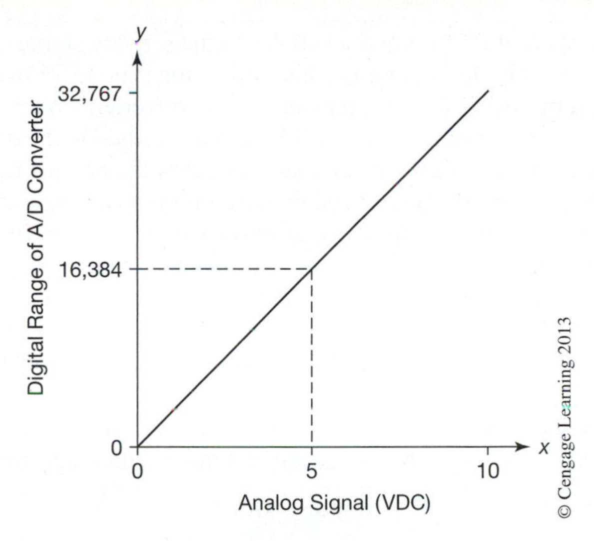

Figure 17–2 Relationship Between Analog Signal and Digital Range

In order for process information to be transmitted and made available for display and/or control, there will always be some type of signal conversion that will take place, and in most cases, more than one conversion. Shown in Figure 17–3, is an example of a complete process variable signal from the output of the transmitter in the field to a PLC memory location that stores the current process variable measurement in engineering units.

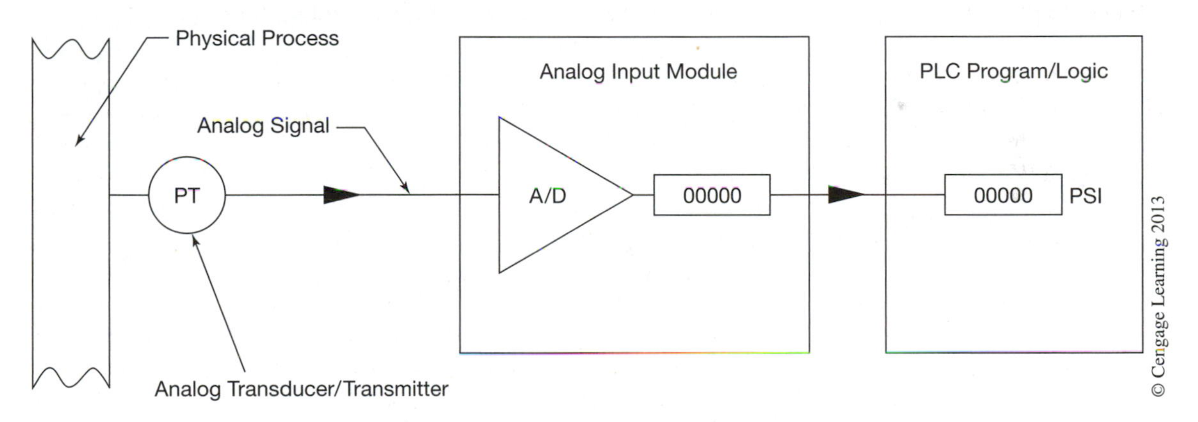

Figure 17–3 Process Variable Signal (Process to PLC)

As you can see from Figure 17–3, there are three signal conversions required in order to transmit and store the process variable measurement, in engineering units, into a PLC memory location. The first conversion converts the physical measurement into an electrical analog signal by the transducer/transmitter. The second conversion converts the electrical analog signal into a digital value by the A/D converter. The third conversion is required to convert the digital output of the A/D converter into a corresponding scaled value representative of the engineering units being measured, pressure in this example.

To maintain measurement accuracy of process variable signals, every signal conversion required must be done correctly and accurately. In most cases, this conversion is done for us by solid-state and digital solid-state devices like transducers, transmitters, and A/D converters. With these devices we merely need to make sure they are programmed and/or calibrated correctly. On the other hand, converting the digital output of an A/D converter into a value that represents the engineering units being measured will require that we calculate the linear conversion between the two ranges using linear interpolation or scaling formulas and program the required math instructions into the PLC. Some PLCs, like the Allen-Bradley SLC 500s, have special scaling instructions that can be programmed to perform this linear conversion for us. The Allen-Bradley SLC 500 scaling instructions will be covered later in this chapter. Let us begin by calculating the linear relationship between the output of an A/D converter and the corresponding scaled value in engineering units using linear interpolation or scaling formulas.

The following mathematical equations can be used to express the linear relationship between an input value and the resulting scaled value if the minimum and maximum ranges of both values are known:

Scaled Value = (input value × slope) + offset

The *slope* (sometimes called *rate*) and *offset* values in the above equation can be calculated from the following:

Slope = (scaled maximum − scaled minimum)/(input maximum − input minimum)
Offset = scaled minimum − (input minimum × slope)

Note: *The offset value will always be equal to what the scaled value would be when the input is at zero.*

EXAMPLE: The following example will use the graph data shown in Figure 17–4 to calculate a scaled value in psi. The digital output range of our A/D converter is shown on the horizontal or *x*-axis and has a range of 0 to 32,767. The corresponding scaled range for the process variable is shown on the vertical or *y*-axis and has a range of 0 to 250 psi.

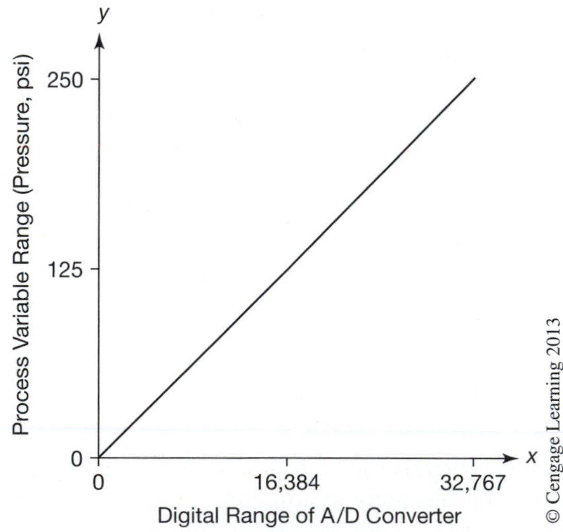

Figure 17–4 Relationship Between Process Variable and Digital Range

Let us assume in our example that the current digital output value from the A/D converter is 18,325. What would be the corresponding scaled value in psi?

SOLUTION: First we must calculate the *slope* (*rate*) using the following equation:

Slope = (scaled maximum − scaled minimum)/(input maximum − input minimum)
Slope = (250 − 0) / (32,767 − 0)
Slope = **0.0076296**

Next we must determine what the offset value should be. The offset can be calculated from the following equation:

Offset = scaled minimum − (input minimum × slope)
Offset = 0 − (0 × 0.0076296)
Offset = **0**

Remember that the offset value is always equal to what the scaled value would be when the input is at zero. In this case, we know that when the input is zero the scaled value will also be zero. Since there is no offset in our example, we could choose to omit the offset in the scaling equation.

The last step is to calculate the scaled value for the digital input value given (18,325) using the scaling equation:

Scaled Value = (input value × slope) + offset
Scaled Value = (18,325 × 0.0076296) + 0
Scaled Value = **139.8 psi**

Figure 17–5 shows the same graph as in Figure 17–4, but this time with dashed lines showing the intercept of the two ranges for the above example.

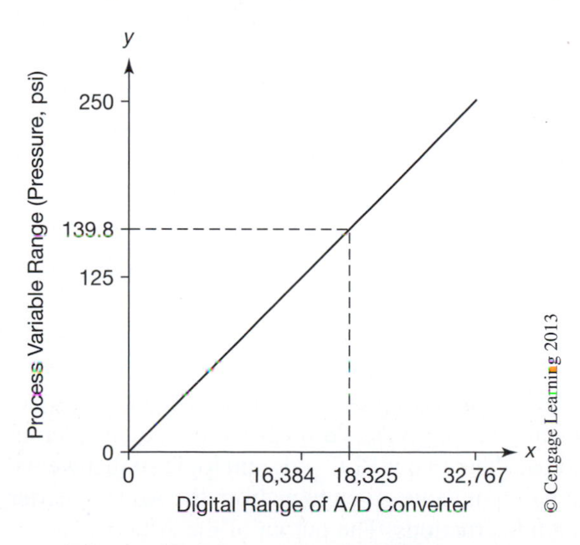

Figure 17–5 Intercept of Two Ranges

EXAMPLE: This example will use the range data shown in Figure 17–6 to calculate a scaled value. The digital input range is again shown on the horizontal or *x*-axis and this time has a range of 3,277 to 16,384. The corresponding scaled range is shown on the vertical or *y*-axis and has a range of 0 to 1500 °F.

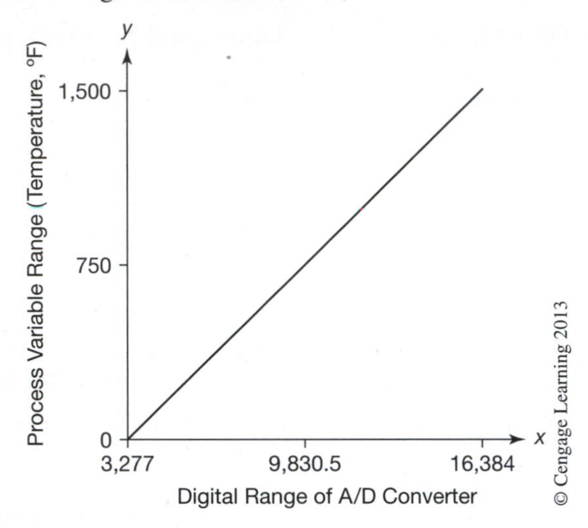

Figure 17–6 Relationship Between Process Variable and Digital Range

Let us assume that the current digital output value from the A/D converter is 10121. What would be the corresponding scaled value in °F for this value?

SOLUTION:

Slope = (scaled maximum − scaled minimum) / (input maximum − input minimum)
Slope = (1,500 − 0) / (16,384 − 3,277)
Slope = **0.1144426**

Offset = scaled minimum − (input minimum × slope)
Offset = 0 − (3,277 × 0.1144426)
Offset = **2375.0284**

Scaled Value = (input value × slope) + offset
Scaled Value = (10,121 × 0.1144426) + −375.0284
Scaled Value = **783.25 °F**

After working through the previous two examples you can see that once the *slope* (*rate*) and *offset* values have been determined based on the digital input and scaled ranges, we can calculate a corresponding scaled value for any input value given. To further illustrate this, let us take the slope and offset values just calculated and use them in our PLC program logic so that we will have a scaled value representing °F for any digital output value given to us from the A/D converter. Figure 17–7a shows our PLC logic using basic math instructions. The output of the A/D converter is stored in PLC memory address I:1.0 and the corresponding scaled value will be stored in floating point element address F8:0. Figure 17–7b shows the same math operation using the Allen-Bradley Compute instruction.

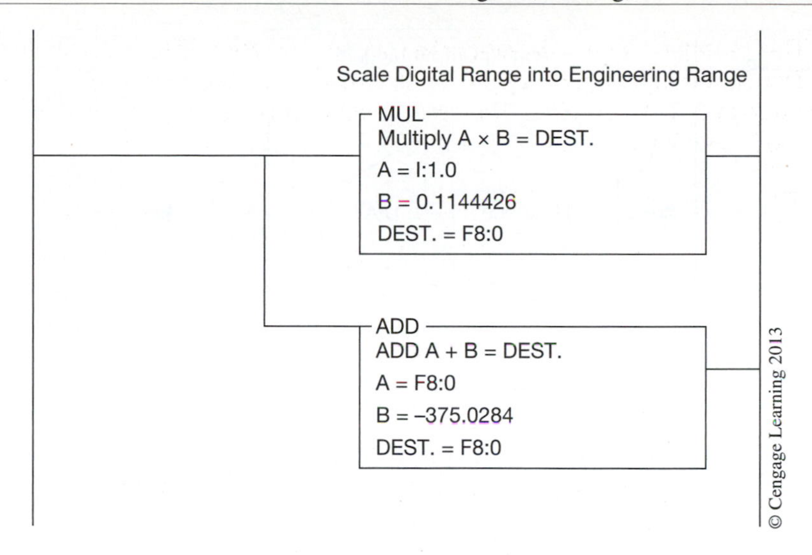

Figure 17–7a Analog Scaling Using Basic Math Instructions

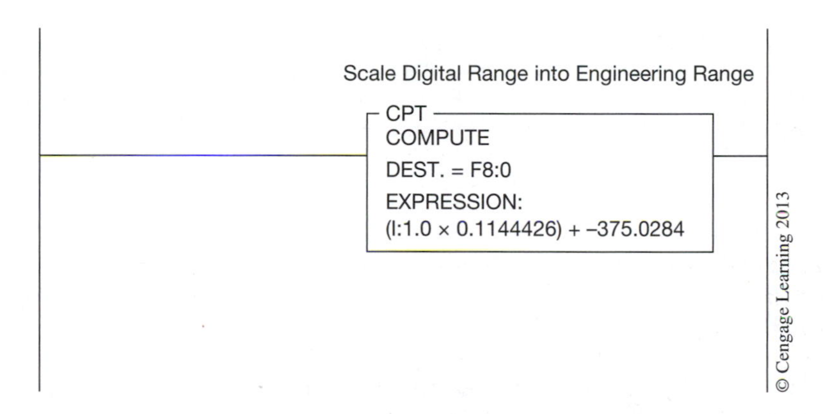

Figure 17–7b Analog Scaling Using Compute Instruction

Analog signals from process variables are not the only type of analog signals found in industry. There are analog signals that are used to control devices such as valves, motors, and pumps. An analog signal that is used to control a device that has a direct influence on the process being controlled is called the **control variable (CV),** or final control element. The control variable is also referred to as an analog output signal, and more correctly so if it does not have a direct influence on the process being controlled, as is the case with an analog display.

The same analog signal conversions described previously for process variable signals also apply to control variable signals, but only in reverse. This time a digital value is converted to a standard electrical analog signal by means of a Digital-to-Analog or D/A converter. The electrical analog signal

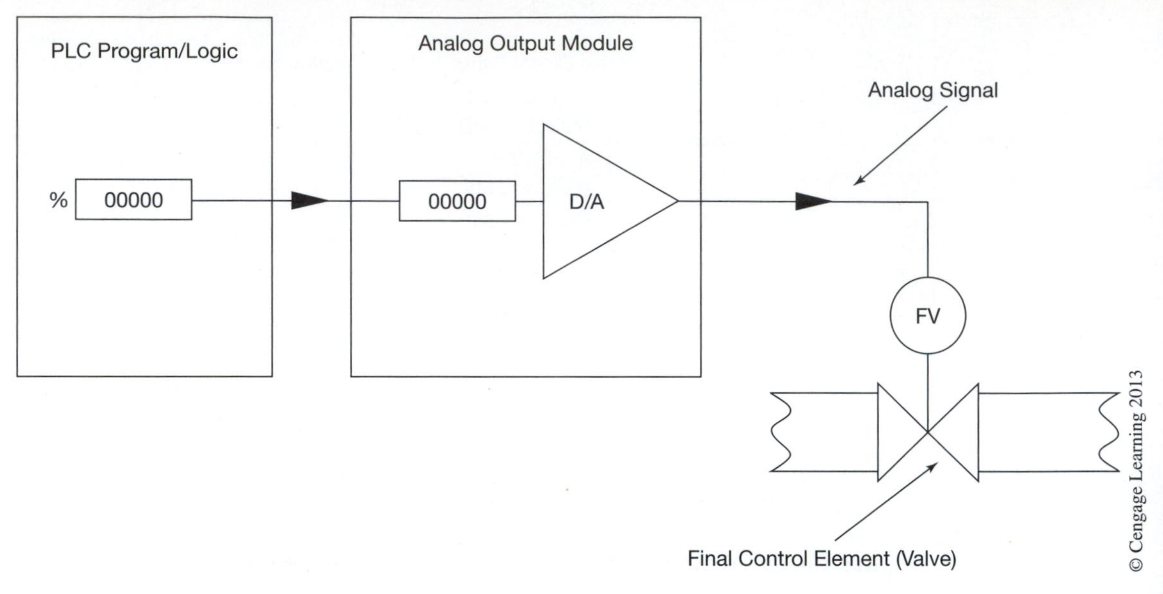

Figure 17–8 Control Variable Signal (PLC to Process)

is then converted into physical motion by the final control element device, or in the case of a digital display, back into a digital value. Shown in Figure 17–8 is an example of a complete control variable signal from a PLC memory location to the final control element that is controlling the process.

Many times when working with analog output signals in a PLC, you will be required to convert a decimal value in engineering units, such as valve position in % or speed in ft/min, into a digital value required by the D/A converter to output the proper electrical analog signal. The same linear interpolation or scaling equation and math instructions used previously can also be used here. The following example will help to illustrate this.

EXAMPLE: In this example an internal PLC memory word contains the position of a process control valve (0 to 100%) that we desire to control with an electrical analog output signal. We have determined that the required electrical analog signal to the valve is a 4 to 20 mA current signal and the corresponding digital range required by the D/A converter to produce such a signal is 6,242 to 31,208. We will again use a graph to help illustrate the two data ranges we will be working with (see Figure 17–9). The scaled range of the valve is shown on the horizontal or x-axis and has a range of 0 to 100%. The digital input range to the D/A converter to produce the 4 to 20 mA signal is shown on the vertical or y-axis and has a range of 6,242 to 31,208. This will also be our scaled range in the scaling formulas.

If we assume that the current position the valve needs to be at is 32%, what would be the corresponding digital value required by the D/A converter to produce the required analog current signal?

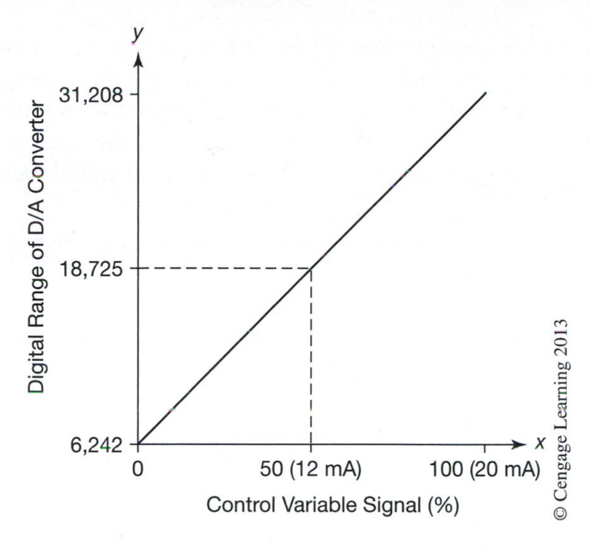

Figure 17–9 Relationship Between Scaled Range and Digital Range

SOLUTION:

Slope = (scaled maximum − scaled minimum) / (input maximum − input minimum)
Slope = (31,208 − 6,242) / (100 − 0)
Slope = **249.66**

Offset = scaled minimum − (input minimum × slope)
Offset = 6,242 − (0 × 249.66)
Offset = **6,242**

Scaled Value = (input value × slope) + offset
Scaled Value = (32 × 249.66) + 6,242
Scaled Value = **14,231.12**

The value of 14,231 would be moved into the location in PLC memory that is used by the D/A converter to output the desired analog signal.

In the above example what would be the analog current signal, in milliamps, to the control valve?

SOLUTION:

Milliamps (mA) = (((20 − 4) / (31,208 − 6,242)) * (14,231 − 6,242)) + 4
milliamps = **9.12 mA**

or

milliamps (mA) = ((20 − 4) * .32) + 4
milliamps = **9.12 mA**

You will find that when working with analog signals it is quite common to be required to convert analog signals from one form or value into another. Understanding the scaling formulas and examples presented here will allow you to make such conversions. You will also find that the preceding equations will be helpful when you want to find, for example, what the analog signal reading would or should be on a digital multi-meter for a given process or control signal value.

ALLEN-BRADLEY SLC 500 SCALE (SCL) INSTRUCTION

The Scale (SCL) instruction is an output-type instruction used to produce a scaled value that has a linear relationship to that of an input value. The instruction takes a source value and then, based on the *slope* (*rate*) and *offset* values you enter into the instruction, makes a linear conversion and stores the result into a destination address. This instruction is often used with analog input and output signals to convert the signals for use within your PLC program. Figure 17–10 shows the Scale (SCL) instruction.

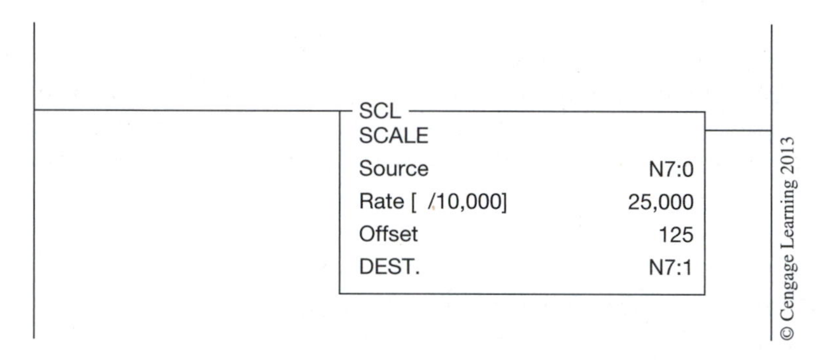

Figure 17–10 Allen-Bradley Scale Instruction

Whenever the instruction is *true,* the value at the source address is multiplied by the *slope* (*rate*) value. The result is then added to the offset value and placed into the destination address specified. The four parameters that must be entered into the instruction are outlined below:

Source—A word address containing the value to be scaled. Example: integer word N7:0, analog input word I:1.0, etc.

Rate—The *slope* (*rate*) can be a program constant or a word address. Most often this is a program constant that you have calculated. The rate value you enter, either positive or negative, is divided by 10,000 prior to being applied to the source. The rate value is limited to a range of −32,768/10,000 to 32,767/10,000 or more correctly a rate of −3.2768 to 3.2767. For example, if your calculated *slope* (*rate*) is 1.5, then enter 15,000 into the Rate parameter.

Offset—The *offset* can be a program constant or a word address. Like the rate, this is also most often a program constant that you have calculated. Valid range for the offset is−32,768 to 32,767.

Destination—A word address that the scaled value will be placed in. Example: integer word N7:1, analog output word O:2.0, etc.

All parameters, including the Source and Destination, are limited in value to an integer range between −32,768 and 32,767.

Note: *Any time an underflow or overflow occurs in the destination word, minor error bit S:5/0 will be set. At the end of the program scan, the processor checks the minor error bit and if set, will cause a major error to be declared and the processor will fault. It must be noted that this overflow can occur before the offset is added. To prevent this problem, check to ensure that the minimum and maximum input values expected will not cause an underflow or overflow condition.*

EXAMPLE: This example illustrates the use of the Scale (SCL) instruction to scale the output of a PID instruction into the appropriate digital range needed by an analog output module's D/A converter.

In this example, the PID instruction can output a digital value in the range of 0 to 16,383, which is equal to 0 to 100%. The analog output to be controlled by the PID instruction is a 4 to 20 mA signal. In order to produce a 4 to 20 mA analog signal, the analog output module's D/A converter must receive a digital value in the range of 6,242 to 31,208.

The Scale (SCL) instruction can be used in this example to convert the 0 to 16,383 output of the PID instruction into a corresponding range of 6,242 to 31,208 needed by the analog output module's D/A converter, which in turn will produce the required 4 to 20 mA analog current output signal. Before we can program the Scale (SCL) instruction, we must first calculate what the *slope* (*rate*) and *offset* values need to be.

SOLUTION:

Slope = (scaled maximum − scaled minimum) / (input maximum − input minimum)
Slope = (31,208 − 6,242) / (16,383 − 0)
Slope = **1.5239**

Offset = scaled minimum − (input minimum × slope)
Offset = 6,242 − (0 × 1.5239)
Offset = **6,242**

Figure 17–11 shows the Scale (SCL) instruction programmed with the *slope* (*rate*) and *offset* values just calculated. The PLC output address of the PID instruction is N7:1 and the PLC address of the analog output module's D/A converter is O:2.0.

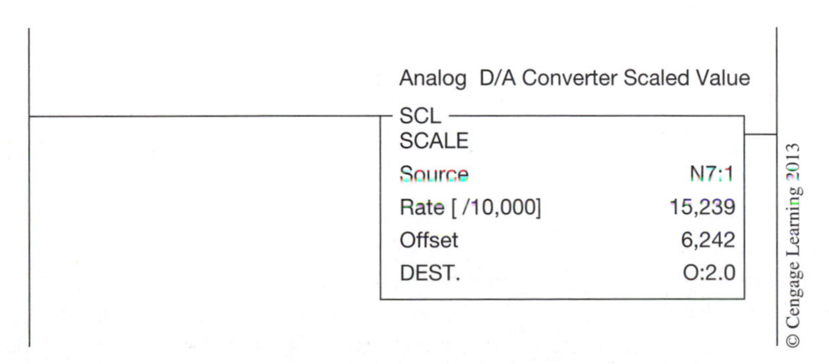

Figure 17–11 Programmed Allen-Bradley Scale Instruction

The Scale (SCL) instruction has limited applications when scaling operations are being performed. This is due in part to the range limits of the *slope* (*rate*) and *offset* values, integer-only values, and the potential math overflow problem. It is the author's belief that the Scale (SCL) instruction was intended to provide an easy means of scaling the various analog input and output ranges so that they would be compatible with the input and output range requirements of the PID instruction. For this reason the author recommends that the Scale (SCL) instruction be used in limited applications.

ALLEN-BRADLEY SLC 500 SCALE WITH PARAMETERS (SCP) INSTRUCTION

The Scale with Parameters (SCP) instruction is an output-type instruction used to produce a scaled value that has a linear relationship between an input value and the scaling parameters entered into the instruction. The SCP instruction is a true linear scaling instruction and does not share the same limitations as that of the Scale (SCL) instruction. The instruction takes a source value and, based on the minimum and maximum ranges entered into the instruction, makes a linear conversion and stores the result into a destination address. If you recall, the Scale (SCL) instruction required that we calculate the *slope* (*rate*) and *offset* values, whereas the SCP takes the minimum and maximum ranges you enter into the instruction and performs all required calculations to produce a scaled output value. You simply enter into the instruction the minimum and maximum ranges of both the input and output, and the instruction does the rest. This instruction supports both integer and floating point values. Figure 17–12 shows the SCP instruction.

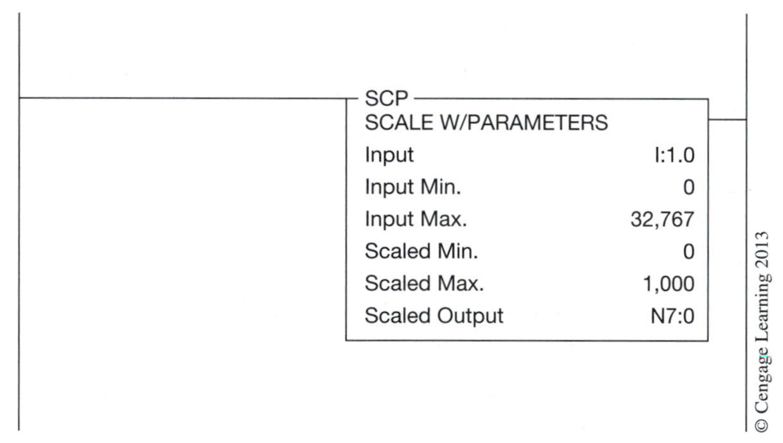

Figure 17–12 Allen-Bradley Scale with Parameters Instruction

When one is programming the SCP instruction, six parameters are required to be entered into the instruction. The following is a brief description of these six parameters.

Input—A word address containing the value to be scaled. Example: integer word N7:0, floating point element F8:0, analog input word I:1.0, etc.

Input Minimum and Input Maximum—These two values determine the range of data that can appear in the Input parameter. The values entered into these two parameters can be a word address, floating point element, or a constant value, either integer or floating point. Most often the Input Minimum and Input Maximum parameters are entered as constants.

Scaled Minimum and Scaled Maximum—These two values determine the range of data that appears in the Scaled Output parameter. The values entered into these two parameters can be a word address, floating point element, or a constant value, either integer or floating point. Most often the Scaled Minimum and Scaled Maximum parameters are entered as constants.

Scaled Output—A word address or floating point data element containing the scaled value. Example: integer word N7:0, floating point element F8:0, analog output word O:1.0, etc.

EXAMPLE: This example will illustrate the use of the SCP instruction to scale the digital input of an electrical analog input signal associated with a temperature transmitter into engineering units representative of the process being measured, in this case temperature in degrees Fahrenheit.

Figure 17–13 shows the temperature transmitter, analog input module, and A/D conversion of the analog signal. After studying Figure 17–13, you can see that the temperature transmitter has an effective measurement range of 200 to 2500 °F for a corresponding 4 to 20 mA analog signal. The analog input module converts the electrical signal into an equivalent digital value through the onboard A/D converter. The digital range equivalent to that of the 4 to 20 mA analog signal in our example is 3,277 to 16,384.

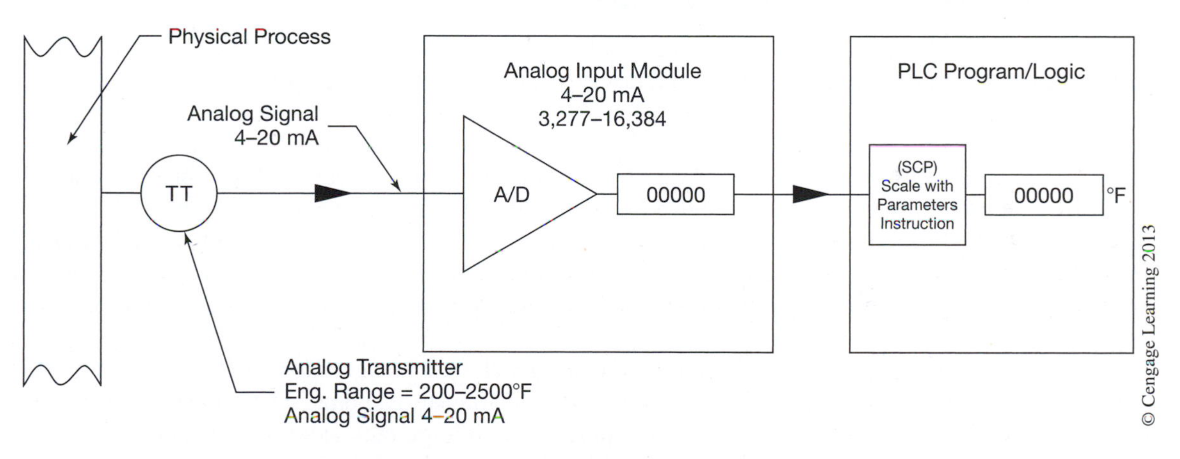

Figure 17–13 Process Variable Signal (Temperature)

With the information from Figure 17–13, we are able to enter into the SCP instruction the required values for the minimum and maximum ranges for both the output of the A/D converter and the measurement range of the temperature transmitter, as shown in Figure 17–14. The output from the A/D converter is located in I:1.0 and we have chosen to store the scaled value, in degrees Fahrenheit, in floating point element address F8:0.

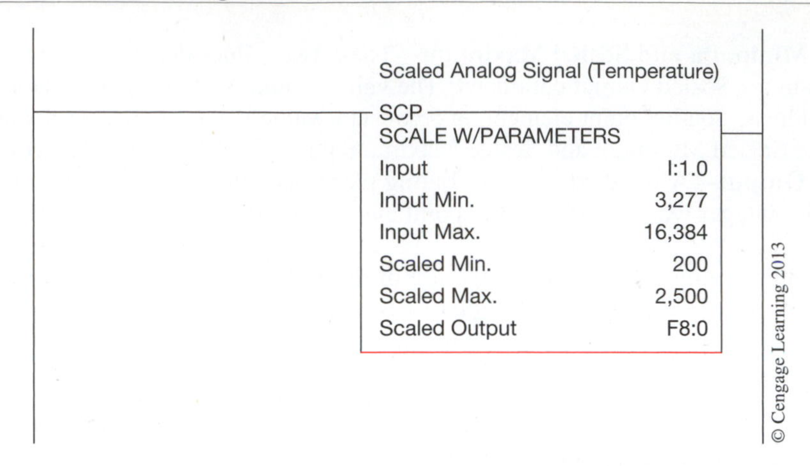

Figure 17–14 Programming for Scale with Parameters Instruction

In summary, the Scale with Parameters (SCP) instruction can be used any time you have a need to perform a linear conversion between two values.

ALLEN-BRADLEY SLC 500 PID INSTRUCTION

Before we cover the PID instruction, let us first look at what the elements of a process control system are, what a process control loop is, and how we might use the PID instruction.

Process control can be defined as a means by which we regulate a process. The heating/cooling system that maintains or regulates your home's temperature can be thought of as a process control system. The internal temperature of your home is the process under control and the means to regulate or control that temperature is your home's furnace or air conditioner. The hot water heater in your home is also a process system. Almost every day of our lives we are affected by some means of process control. Figure 17–15 illustrates the basic elements of a process control system.

In the system shown in Figure 17–15, a level transmitter (LT), a level controller (LC), and a control valve (LV) are all used to control the level of the liquid in the storage tank. The process control system is designed to maintain the height of the liquid in the storage tank at some predetermined level from the bottom of the tank. This predetermined level is called the process setpoint or **setpoint (SP).** It will be assumed that the rate at which the liquid enters the tank varies. The level transmitter is the measurement device that converts the physical level in the tank into a standard analog signal, also referred to as the **process variable (PV).** The process controller reads the level measurement and compares it against the setpoint or desired level. If the level in the tank is not at the desired level, then the process controller produces a series of corrective actions that are sent to the control valve in the outlet pipe in an attempt to bring the process under control. The control valve is referred to as the **control variable (CV)** or final control element. The control variable or final control element is a device that when operated exerts a direct influence on the process under control. Devices such as control valves, motors, fans, and pumps are all examples of final control elements.

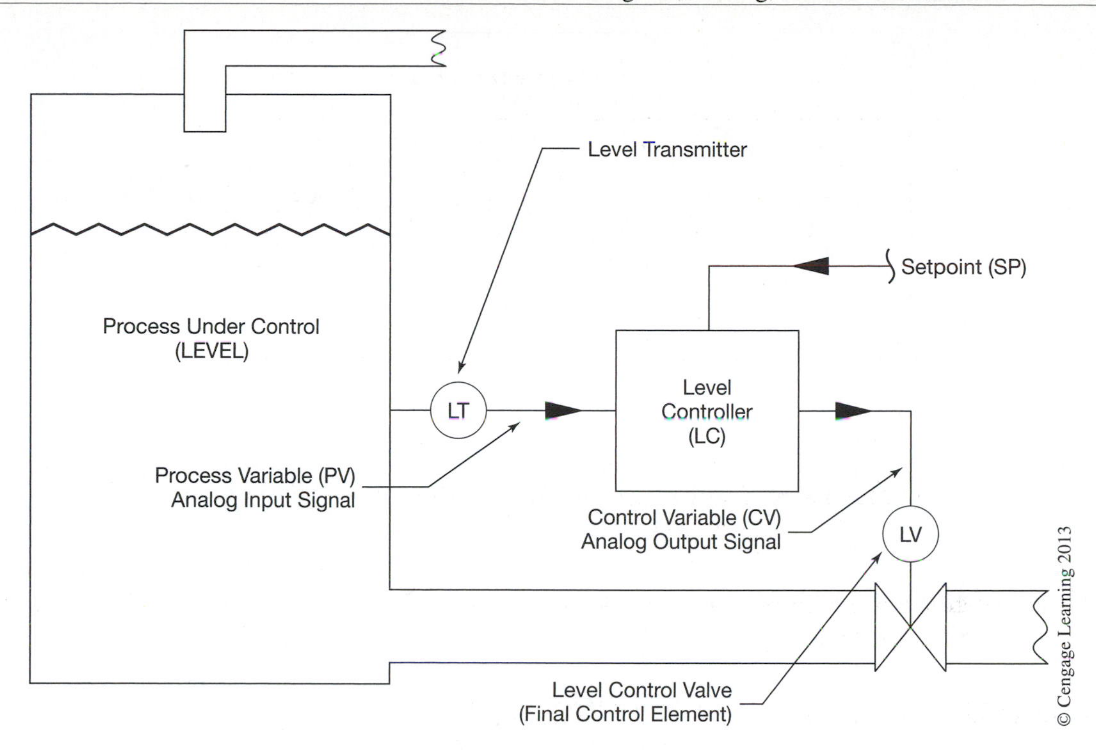

Figure 17–15 Elements of a Process Control Loop (System)

As you can see from the above example, a process control system consists of four key elements: the process, the measurement or process variable (PV), the process controller, and the final control element or control variable (CV). When all four elements are present, it is considered a *closed process control loop*. Figure 17–16 shows a simple block diagram of a closed process control loop.

The term *closed loop control* comes from the fact that you are continuously measuring the process under control to determine if corrective action is required to maintain the process within the desired limits (setpoint), whereas with open loop control there is no continuous measurement and corrective action taking place to maintain the process within limits. For example, in Figure 17–15, if you were to manually adjust the valve position based on current conditions, there would be no guarantee that the level would remain the same in the tank because of possible changing conditions. It is like operating blind.

The process controller can be a stand-alone device or a software instruction that is part of a larger control system. A process controller has analog inputs and outputs that are used to monitor and control the process. In addition to the process connections, the process controller will also have a means to enter the desired setpoint, as well as the ability to make adjustments to various parameters that are designed to tune the process controller for a given process. Some process controllers that are used for temperature control may have the analog output converted to a time proportioning on/off output for driving a heater or cooling unit.

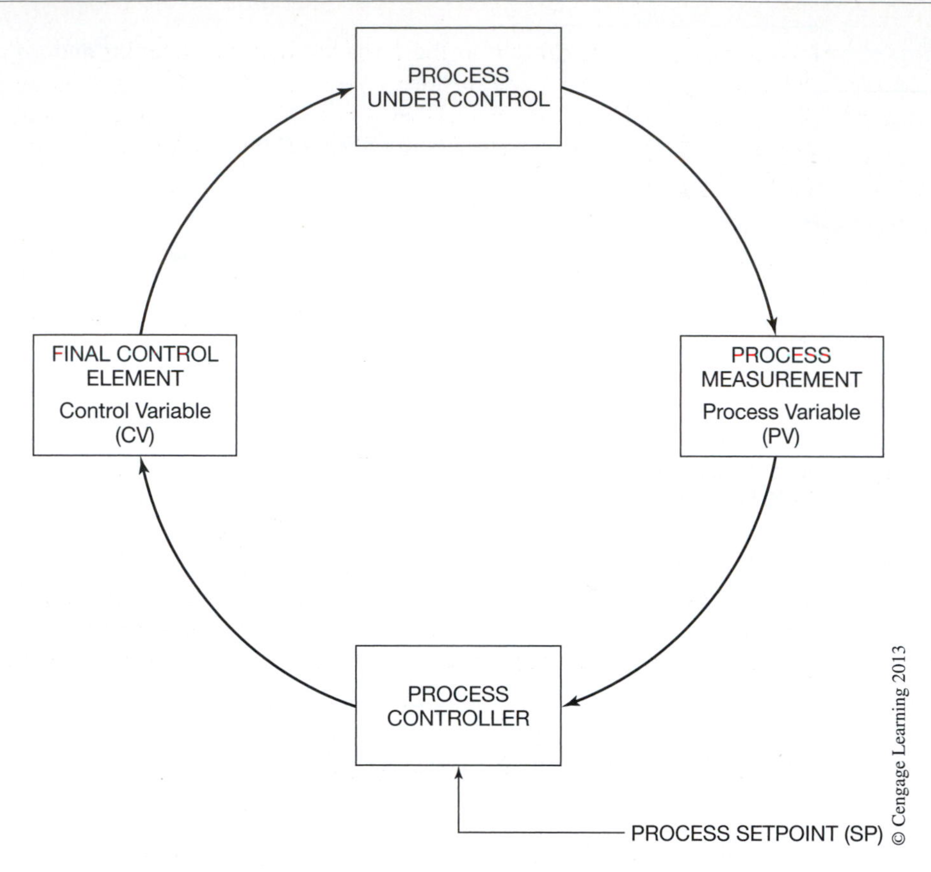

Figure 17–16 Block Diagram of Process Control Loop

The remainder of this chapter is devoted to describing the Allen-Bradley SLC 500 Proportional Integral Derivative (PID) instruction. The PID instruction is a software type of closed loop process controller, like that described above.

The Allen-Bradley SLC 500 PID instruction is an output instruction that can be operated in the **timed mode** or **selectable timed interrupt (STI) mode**. In the timed mode, the instruction updates its output at a rate selected by the user. In the STI mode, the instruction is placed in an STI subroutine program file and its output is updated each time that the instruction is scanned by the processor. The timed mode is more commonly used than the STI mode. In the timed mode, you are not required to create a separate STI subroutine program file for your PID instructions. More important, the instruction's update time can be easily changed without further impacting other PID instructions as would be the case in the STI mode.

Note: *When used in the STI mode, the STI time interval and the PID loop update rate parameter must be the same, in order for the PID instruction to operate properly.*

The PID instruction controls its output based on the error between the setpoint and process variable using a mathematical equation. The greater the error between the setpoint and process variable input, the greater the output signal will change, and vice versa. The PID instruction will automatically control the output signal, up or down, until the error between the setpoint and process variable is nearly gone.

The Allen-Bradley SLC 500 PID instruction uses the standard PID equation with dependent gains. This is the standard PID equation used in almost all process controllers. The equation is shown below:

$$\text{output (CV)} = Kc\,[(E) + 1/Ti \int (E)\,dt + Td \cdot D(PV)/dt] + \text{bias}$$

where
Kc = Proportional Gain Constant (Unitless)
1/Ti = Reset Gain (Repeats/Minute) or Integral
Td = Rate Gain (Minutes) or Derivative
E = Error (SP − PV or PV − SP)

The Proportional, Integral, and Derivative gains are constants entered into the instruction at the time that the instruction is implemented and are dependent upon the nature of the process and system components under control. We will cover these constants in greater detail a little later in this chapter. But first, let us take a look at the instruction. Shown in Figure 17–17 is the Allen-Bradley PID instruction.

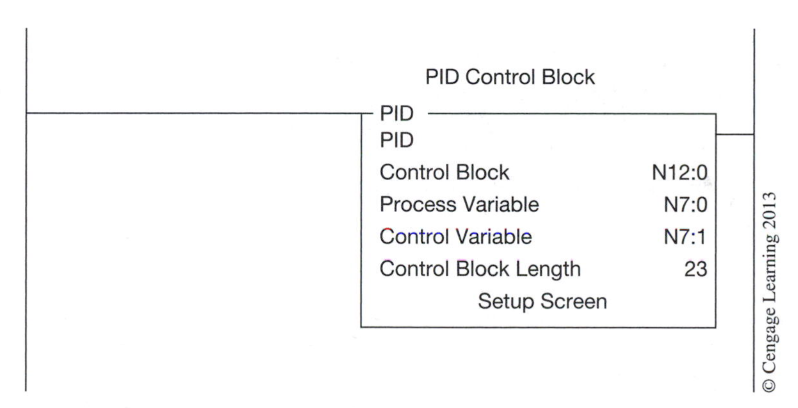

Figure 17–17 Allen-Bradley SLC 500 PID Instruction

Normally, you program the PID instruction as shown in Figure 17–17, without conditional logic. If the rung is false, the output will remain in its last state and the integral term will reset. You will notice that the PID instruction has three parameters that must be entered into the instruction at the time of programming. These three parameters are described as follows:

Control Block—The control block is a file that stores the required data to operate the instruction. The file length is fixed at 23 words and is an integer file address. The address entered in the control block should be the starting word of the integer file that will make up the control block. For example, if you enter N12:0 then the PID instruction will allocate and use elements N12:0 through N12:22 for control of the instruction. These words should not be used for any other purpose in your program. A good programming practice would be to create an integer file dedicated to your PID control blocks, such as N12. Figure 17–18 shows the 23-word control block layout.

WORD	
WORD 1	PID Status Bits
2	PID Sub Error Code
3	Setpoint (SP)
4	Gain Kc
5	Reset Ti
6	Rate Td
7	Feed Forward Bias
8	Setpoint (SP) Max.
9	Setpoint (SP) Min.
10	Deadband
11	Internal Use Only
12	Output Max.
13	Output Min.
14	Loop Update
15	Scaled Process Variable
16	Scaled Error
17	Output CV%
18	MSW Integral Sum
19	LSW Integral Sum
20	Internal Use Only
21	Internal Use Only
22	Internal Use Only
23	Internal Use Only

Figure 17–18 Allen-Bradley SLC 500 PID Instruction Control Block Layout

Process Variable (PV)—The process variable parameter is an integer word address that contains the measurement of the process under control. This word address could be the location in PLC memory where the analog input module stores the value from the A/D converter. The numerical scale for the process variable parameter is 0 to 16,383.

Control Variable (CV)—The control variable parameter is an integer word address that stores the output of the PID instruction. The output of the PID instruction has a numerical range from 0 to 16,383, where 16,383 equals 100%. The PID output value would most likely be scaled to a value required by the D/A converter of an analog output module in order to produce the required analog signal to the final control element.

The numerical scale for both the process variable and control variable is 0 to 16,383 for the PID instruction. You will most likely be required to scale your analog I/O ranges within the above numerical scale required by the PID instruction. To do this, use the Scale (SCL) instruction. The following two examples will help illustrate this:

EXAMPLE: If a 4 to 20 mA analog input signal has a digital range of 3,277 to 16,384, then it must first be scaled into the range of 0 to 16,383 to be used as the process variable in the PID instruction.

SOLUTION: Calculate the *slope* (*rate*) and *offset* values for the analog input range of 3,277 to 1,63,684 and the PID process variable range of 0 to 16,383.

Slope = (scaled maximum − scaled minimum) / (input maximum − input minimum)
Slope = (16,383 − 0)/(16,384 − 3,277)
Slope = **1.2499**

Offset = scaled minimum − (input minimum × slope)
Offset = 0 − (3,277 × 1.2499)
Offset =**24,096**

Enter the *slope* (*rate*) and *offset* values into the Scale (SCL) instruction as shown in Figure 17–19.

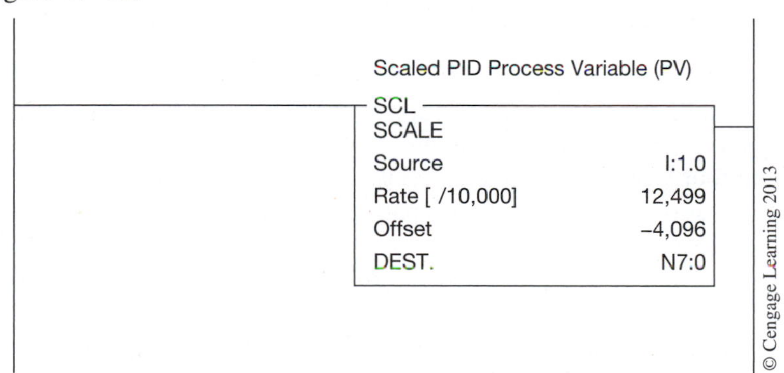

Figure 17–19 Scaling PID Process Variable with Scale Instruction

EXAMPLE: If the analog output of a D/A converter is 4 to 20 mA and requires a digital input in the range of 6,242 to 31,208 to produce it, then the 0 to 16,383 control variable output of the PID instruction must be scaled across this range of 6,242 to 31,208.

SOLUTION: Calculate the *slope* (*rate*) and *offset* values for the analog output range of 6,242 to 31,208 and the PID control variable range of 0 to 16,383.

Slope = (scaled maximum – scaled minimum) / (input maximum – input minimum)
Slope = (31,208 − 6,242) / (16,383 − 0)
Slope = **1.5239**

Offset = scaled minimum − (input minimum × slope)
Offset = 6,242 − (0 × 1.5239)
Offset = **6,242**

Enter the *slope* (*rate*) and *offset* values into the Scale (SCL) instruction as shown in Figure 17–20.

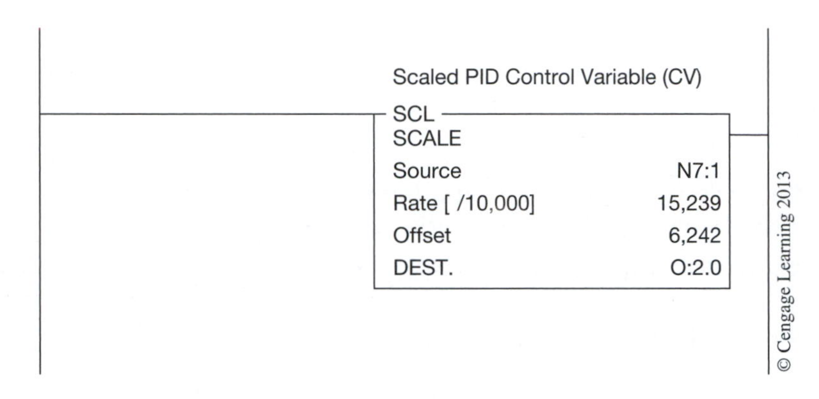

Figure 17–20 Scaling PID Control Variable with Scale Instruction

Note: *The PID instruction is an integer-only instruction and will not allow you to enter floating point values for any of its parameters. If you attempt to move a floating point value into one of the parameters using ladder logic, then a floating point-to-integer conversion occurs.*

After you have entered the Control Block, Process Variable, and Control Variable parameters into the PID instruction, a PID Setup Screen will appear that allows you to enter additional parameters to complete the configuration of the instruction. The PID Setup Screen is shown in Figure 17–21 and is divided into four groups or parameter types, Tuning, Inputs, Output, and Flags. The Tuning, Inputs, and Output parameters are additional parameters that you must configure in order for the PID instruction to properly control your process. The Flags section, located on the far right of the Setup Screen, displays the various status and control flags associated with the PID instruction.

The following is a brief description of each of the additional configuration parameters and status/control flags found on the PID Setup Screen:

PID Tuning Parameters

The PID Tuning area of the Setup Screen allows you to configure the PID instruction for the mode of operation and how the PID controller reacts and regulates your process loop.

Controller Gain Kc—Controller Gain is the proportional gain of the controller, ranging from 0 to 32,767 (0 to 3276.7). If the RG status bit is set to 1, then the valid range is 0 to 327.67 and is only valid on the 5/03 and higher processors.

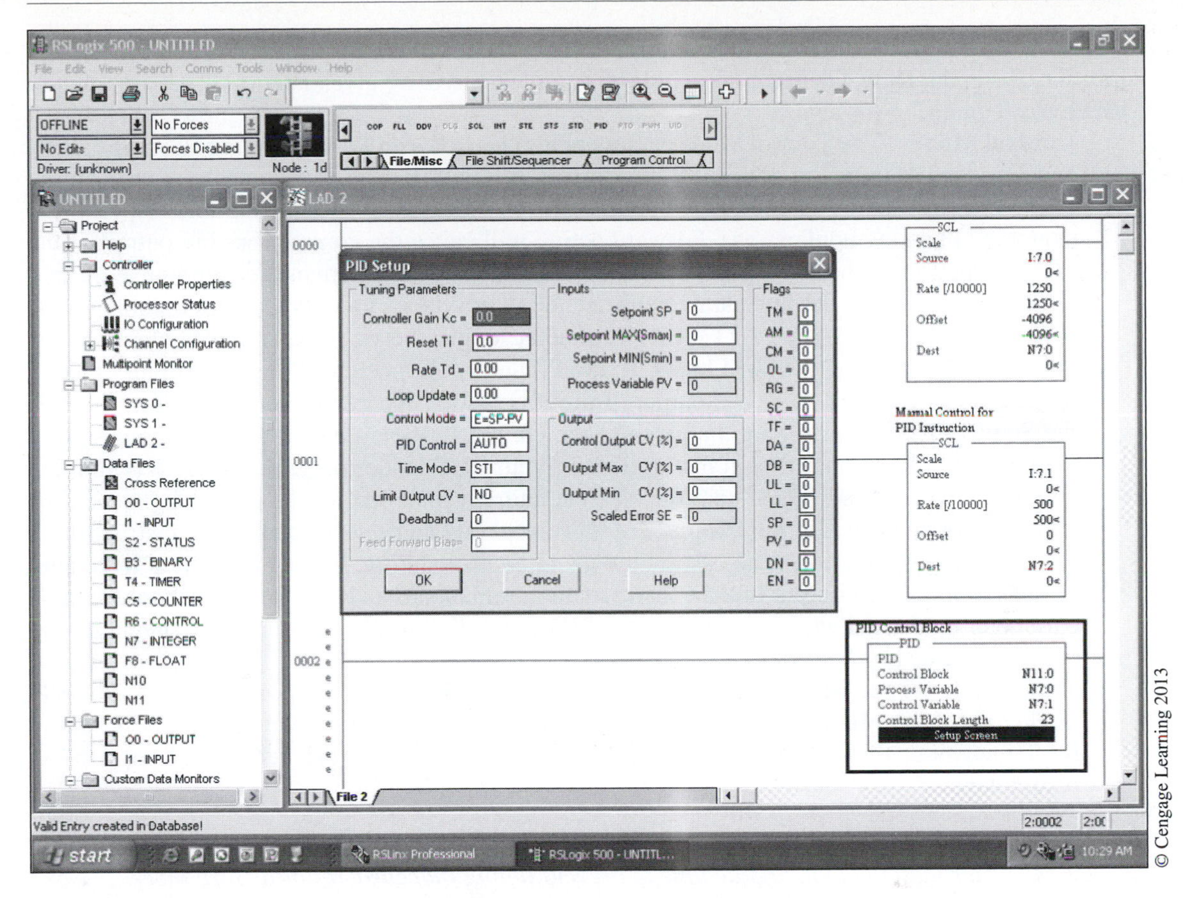

Figure 17–21 PID Instruction Setup Screen

Reset Ti—Reset is the integral action, ranging from 0 to 32,767 (0 to 3276.7) minutes per repeat. If the RG status bit is set to 1, then the valid range is 0 to 327.67 and is only valid on the 5/03 and higher processors.

Note: *The RG status bit will be covered later under the status/control flags section.*

Rate Td—Rate is the derivative action, ranging from 0 to 32,767 (0 to 327.67) minutes. This word is not affected by the reset and gain range (RG) bit.

Loop Update—Loop Update time is the time interval between PID calculations. The valid range is 1 to 1024 (0.01 to 10.24) second intervals. You should enter a loop update time that is five to ten times faster than the natural period of the loop. The faster the process responds to changes, the faster the loop update time needs to be. When in the STI mode, this value needs to be set equal to the STI interval time.

A more detailed description, along with several methods to determine the settings for the **Controller Gain Kc, Reset Ti, Rate Td,** and **Loop Update** parameters, will be given later in this chapter under "PID Loop Tuning."

Control Mode—The Control Mode parameter toggles between *Reverse Acting* (E = SP − PV) and *Forward Acting* (E = PV − SP). Reversing acting will cause the control variable output of the PID instruction to decrease when the process variable is greater than the setpoint (for example, a heating application). Forward acting will cause the control variable output of the PID instruction to increase when the process variable is greater than the setpoint (e.g., in a cooling application). Set this parameter to match your type of process loop application.

PID Control (AM)—The PID Control parameter toggles between Auto and Manual modes. When auto mode is active (AM bit = 0), the PID equation is controlling the output. Once in manual mode (AM bit = 1), the output of the PID instruction can be set by the user through the Setup Screen by changing the value of the Output (CV)% parameter; or, in the case of a manual control station, through instructions in your PLC ladder program by directly moving a value between 0 and 16,383 into the element address specified for the output *Control Variable* (*CV*) parameter. The *auto/manual* (*AM*) control bit can be set or cleared by instruction in your ladder program, as in the case of a manual control station. When you switch from auto to manual mode, the output of the PID instruction remains in its last state. You should set this parameter to Manual when programing the PID instruction until you are ready to begin tuning the PID instruction.

Time Mode—Time Mode toggles between Timed and STI mode of operation. Select the STI mode only if the PID instruction is programmed in an STI interrupt subroutine file. When in the Timed mode the PID instruction updates its output at the rate specified in the Loop Update parameter.

Note: *When the timed mode is selected, your processor scan time should be at least five to ten times faster than the loop update time to prevent timing inaccuracies or disturbances.*

Limit Output CV—The Limit Output CV parameter toggles between Yes and No. If Yes is selected, then the output of the PID instruction is limited to the minimum and maximum ranges as determined by the Output Min CV% and Output Max CV% parameters.

Deadband—The Deadband parameter is a nonnegative value that, when entered, extends a deadband above and below the setpoint by the value you enter. The deadband only takes effect after the process variable has entered the deadband and passed through the setpoint. This deadband sets a range above and below the setpoint where the PID output does not change as long as the error remains within this range. The deadband helps to reduce the effects of "hunting" in many applications. The valid range is based on whether scaling has been implemented or not. If scaling has been implemented, then the range is 0 to Setpoint MAX Smax parameter; otherwise the range is 0 to 16,383.

PID Input Parameters

The PID Input parameters area of the Setup Screen allows you to configure the PID instruction for the process setpoint and its scaled range.

Setpoint SP—The Setpoint SP parameter is the desired control point of the process variable. Enter a value between 0 and 16,383 or within the valid scaled range as defined by the Setpoint

MIN Smin and Setpoint MAX Smax parameters below. This parameter can be changed through your ladder program by moving a value into the third word of the control block; see Figure 17–18.

Setpoint MAX Smax—If the Setpoint SP parameter is to be entered into and displayed in engineering units, then the value entered into this parameter should correspond to the value, in engineering units, when the process variable is at maximum (16,383). For example, if the process variable has a full scale range of −200 (0) to 1500 °F (16,383), then enter 1,500 into this parameter. The valid range is +/−32,767 (if you are using the SLC 5/02 the valid range is +/−16,383).

Setpoint MIN Smin—If the Setpoint SP parameter is to be entered into and displayed in engineering units, then the value entered into this parameter should correspond to the value, in engineering units, when the process variable is at minimum (0). For example, if the process variable has a full scale range of −200 (0) to 1500 °F (16,383), then enter −200 into this parameter. The valid range is +/−32,767 (if you are using the SLC 5/02 the valid range is −16383 to +16382).

Note: *Entering scaling values in the* Setpoint MAX Smax *and* Setpoint MIN Smin *parameters allows you to enter the setpoint value in engineering units. In addition, the Deadband, Error, and Process Variable PV parameters will also be displayed in the same engineering units. The process variable input still must be in the range of 0 to 16,383.*

Process Variable PV—The Scaled Process Variable parameter is for display only. This parameter will display the scaled value of the process variable input. If scaling is not implemented, then the range is 0 to 16,383.

PID Output Parameters

The PID Output parameters area of the Setup Screen allows you to configure the PID instruction output limits and manual control.

Control Output CV%—The Control Output CV% parameter displays the actual 0 to 16,383 control variable output in terms of percent (0 to 100%). When the PID instruction is in auto mode, this parameter is for display only. When the PID instruction is in manual mode, this parameter will allow you to change the % output of the control variable of the PID instruction through the PID Setup Screen only. If you are using a manual control station, then change the output of the PID instruction through the manual control station and not through this parameter. This parameter is widely used during the tuning of the PID instruction and control loop.

Output Max CV%—When the Limit Output CV parameter is set to *Yes,* the value you enter into this parameter will determine the maximum output percent that the PID control variable will obtain. If the control variable should exceed this value, then the control variable will be clamped to the value you entered into this parameter and the Output Alarm, Upper Limit status bit will be set to 1.

If the Limit Output CV parameter is set to *No,* then the value you enter into this parameter will only determine when the Output Alarm, Upper Limit status bit will be set to 1.

Output Min CV%—When the Limit Output CV parameter is set to *Yes,* the value you enter into this parameter will determine the minimum output percent that the PID control variable will obtain. If the control variable should drop below this value, the control variable will be clamped to the value you entered into this parameter and the Output Alarm, Lower Limit status bit will be set to 1.

If the Limit Output CV parameter is set to *No,* the value you enter into this parameter will only determine when the Output Alarm, Lower Limit status bit will be set to 1.

Scaled Error SE—The Scaled Error parameter is for display only. This parameter will display the calculated error based on the mode selected (E = PV − SP or E = SP − PV). If scaling is implemented, then the value displayed will equal the scaled range of the process variable input; otherwise the range is 0 to 16,383.

PID Status & Control Flags

When one is viewing the PID Setup Screen, shown in Figure 17–21, the far right column or group displays the various status and control flag indicators associated with the PID instruction. The status and control flags are bit-level addresses that are located in the first word of the PID control block; see Figure 17–18. The following is a brief description of each of the status and control flags:

Time Mode Bit TM (bit 0)—This bit specifies the PID mode, Timed or STI. When set, the Timed mode is in effect. The status of this bit was determined at the time that you set up the PID instruction as described above.

Auto/Manual Bit AM (bit 1)—This bit specifies whether the PID instruction is in automatic or manual control. When set, the PID instruction is selected for manual operation. This bit can be set or cleared by instructions programmed in your ladder logic. See PID Control parameter above for further information on the uses of this bit.

Control Mode Bit CM (bit 2)—This bit specifies the PID control mode, *Forward* or *Reverse* acting. When set, forward acting mode is in effect. The condition of this bit was determined at the time that you set up the PID instruction as described above.

Output Limiting Enabled Bit OL (bit 3)—This bit specifies whether you have selected to limit the control variable output of the PID instruction. When set, output limiting is in effect. The status of this bit was determined at the time that you set up the PID instruction as described above.

Reset and Gain Range Enhancement Bit RG (bit 4)—This bit is specific to the SLC 5/03, and higher processors only. When this bit is set, it will cause the Reset Minute/Repeat value and the gain multiplier to be enhanced by a factor of 10 (reset multiplier of 0.01 and gain multiplier of 0.01). See Controller Gain Kc and Reset Ti parameters above for additional information on the effects of this bit.

Scale Setpoint Flag SC (bit 5)—This status bit is set to one (1) when setpoint scaling to engineering units is **not** being performed. The status of this bit was determined at the time that you set up the PID instruction scaling option under the PID Input Parameters section above.

Loop Update Time Too Fast TF (bit 6)—This bit is set by the PID algorithm if the loop update time you have entered cannot be achieved. This condition is due to scan time limitations. If this bit is set, try to increase your Loop Update parameter; this will slow down the interval time between calculations. If you are not able to update the PID loop at a slower rate, then move the PID instruction into an STI interrupt routine.

Derivative Action Bit DA (bit 7)—When the Derivative Action Bit is set to one (1), then the Derivative (Rate) calculation will be based on the error instead of the process variable. This bit is specific only to the SLC 5/03 and higher processors.

DB, Set When Error Is in DB (bit 8)—This status bit is set to 1 when the process variable is within the zero crossing deadband range of the Setpoint.

Output Alarm, Upper Limit UL (bit 9)—This status bit is set to 1 whenever the PID control variable output exceeds the upper limit as set by the Output (CV) % Max parameter.

Output Alarm, Lower Limit LL (bit 10)—This status bit is set to 1 whenever the PID control variable output is less than the lower limit as set by the Output (CV) % Min parameter.

Setpoint Out of Range SP (bit 11)—This status bit is set to 1 whenever the value entered into the Setpoint SP parameter is outside the range of the minimum and maximum scaled values. If scaling is not used, then the range is 0 to 16,383.

Process Variable Out of Range PV (Bit 12)—This status bit is set to 1 whenever the process variable input is outside the range of 0 to 16,383.

PID Done DN (bit 13)—The PID Done bit is set *On* during the scans in which the PID equation is calculated. The PID equation is calculated at the loop update time.

PID Enable EN (bit 15)—The PID Enable bit is set *On* whenever the rung of the PID instruction is true or enabled.

After studying the list of configuration parameters and status/control flags associated with the PID instruction, you should begin to have a better understanding of the instruction and its configuration. Of all the PLC instructions, the PID instruction is the most complex and requires the greatest knowledge to configure and implement properly.

PID Program Example

The following example should help you gain a better understanding of the application and configuration of the PID instruction. Figure 17–22 shows a basic process control loop for which we will program the Allen-Bradley SLC 500 PID instruction.

The process under control in Figure 17–22 is the outlet flow rate of a liquid storage tank. The flow rate is controlled by a flow control valve located in the outlet pipe. One of the first steps involved in programming any PID type controller is to identify the system components that will be used to monitor and control the process. In our example, we will begin by first identifying the process variable, then the control variable, and last we will identify any auxiliary control requirements (manual, etc.).

Process Variable (PV) The process variable under control in our example is the liquid flow, as measured in gallons per minute (gpm) by the flow transmitter located in the outlet pipe. The flow transmitter is designed to continuously measure the flow rate and output a corresponding electrical analog signal that is representative of that flow rate. The analog signal is wired to our PLC analog input module. Notice in Figure 17–22 that the flow transmitter has a range of 0 to 200 gpm and converts that range into a 4 to 20 mA electrical analog signal. The PLC analog input module is designed to convert the 4 to 20 mA analog signal, by way of the A/D converter, into a digital value with a range of 3,277 to 16,384.

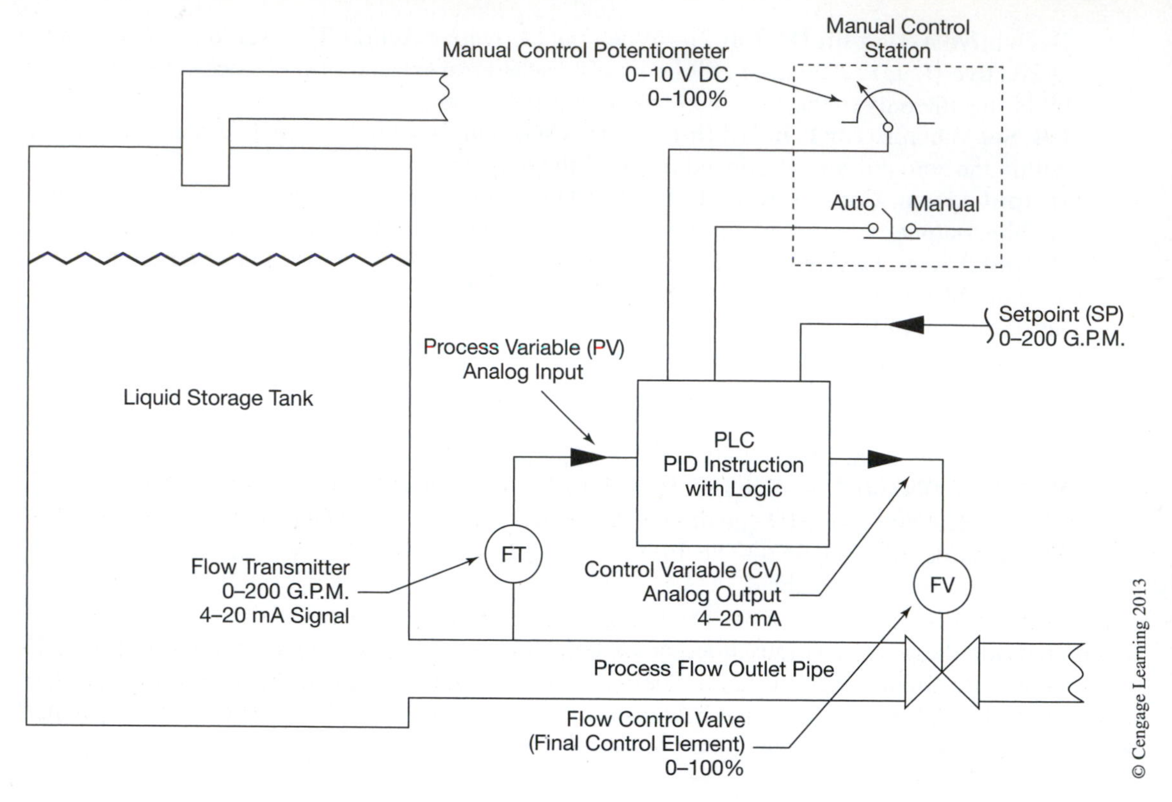

Figure 17–22 PID Program Example Control Loop

Remember from our study of the PID instruction, the range of the process variable input is from 0 to 16,383. In order to use the analog input value from the flow transmitter as the process variable input to the PID instruction, we must first scale the analog input into a range of 0 to 16,383 that is required by the PID instruction. We will use the Scale (SCL) instruction to scale the analog input signal. Before we can program the Scale (SCL) we must first calculate the *slope* (*rate*) and *offset* values, as follows:

Slope = (scaled maximum − scaled minimum) / (input maximum − input minimum)
Slope = (16,383 − 0) / (16,384 − 3,277)
Slope = **1.2499**
Offset = scaled minimum − (input minimum × slope)
Offset = 0 − (3,277 × 1.2499)
Offset = **24,096**

Enter the slope and offset values into the Scale (SCL) instruction as shown in Figure 17–23.

In the Scale (SCL) instruction in Figure 17–23, we are just assuming that the analog input value from the A/D converter is being placed into PLC Input word I:1.0. This address location would

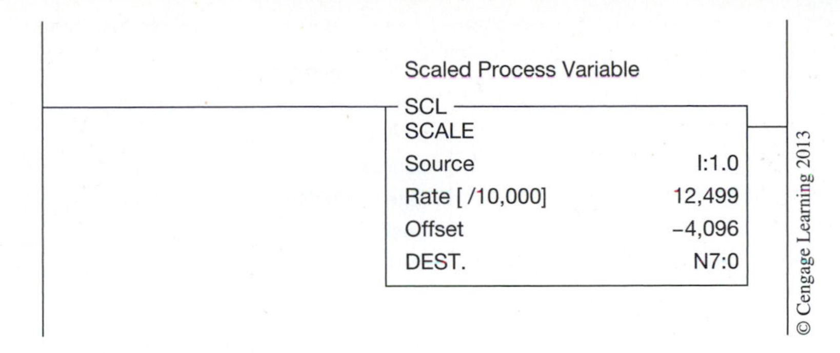

Figure 17–23 Scaling Process Variable for PID Instruction

depend upon the type of analog input module being used and its physical location in the PLC system. We have chosen to store the scaled value in integer element address N7:0. We can now use N7:0 as the process variable input parameter in our PID instruction.

Control Variable (CV) The control variable or final control element in our example is the flow control valve located in the outlet pipe; refer back to Figure 17–22. The flow of liquid in the outlet pipe is determined by the position or opening of the flow control valve. The position of the flow control valve in our example is determined by an analog current signal (4 to 20 mA) that is generated by the PLC.

The digital range required to produce a 4 to 20 mA analog signal by the D/A converter in the analog output module of our PLC is 6,242 to 31,208. Again, if you remember from our study of the PID instruction, the output range of the control variable of the PID instruction is 0 to 16,383. We must then scale the output of the PID instruction into a range that will produce the required 4 to 20 mA analog signal needed to control the valve. To scale the analog output signal of the PID instruction we will again use the Scale (SCL) instruction. Before we can program the Scale (SCL) we must first calculate the *slope* (*rate*) and *offset* values, as follows:

Slope = (scaled maximum − scaled minimum) / (input maximum − input minimum)
Slope = (31,208 − 6,242) / (16,383 − 0)
Slope = **1.5239**

Offset = scaled minimum − (input minimum × slope)
Offset = 6,242 − (0 × 1.5239)
Offset = **6,242**

Enter the *slope* (*rate*) and *offset* values into the Scale (SCL) instruction as shown in Figure 17–24.

In Figure 17–24, we are just assuming that the output address O:2.0 is the word location in PLC memory for the analog output D/A converter. This address location would depend upon the type of analog output module being used and its physical location in the PLC system. We have chosen to use the integer element address N7:1 as the control variable output of our PID instruction.

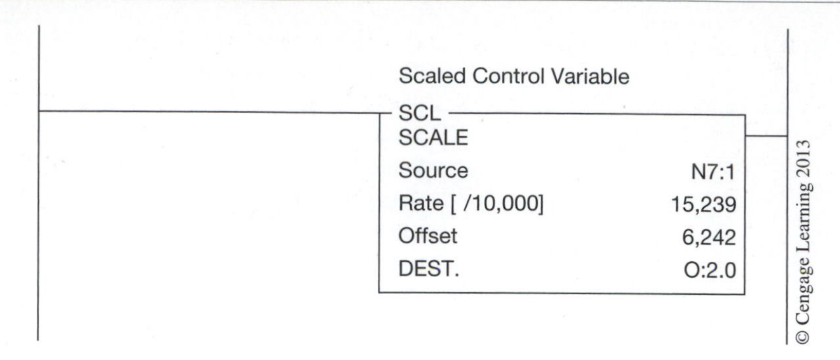

Figure 17–24 Scaling Control Variable Output of PID Instruction

As you can see, we were required to program two Scale (SCL) instructions into our PLC program in order for the analog input and output signals to be compatible with our PID instruction's process variable input and control variable output parameters.

Auxiliary Control Functions The last item to identify in setting up our PID instruction is to determine if there are any auxiliary control requirements required for our process loop. If you refer back to Figure 17–22, you will notice that there is a manual control station that can be used by an operator to manually control the flow control valve. The manual control station in our example consists of a two-position selector switch (Auto/Manual) and a 0 to 10 V DC potentiometer to control valve position.

The two-position selector switch is wired to our PLC as digital input (I:3/0). When the input is *true* or *on* the selector switch is in the Manual position.

The 0 to 10 V DC potentiometer is wired to our PLC as an analog input (input word address I:1.1). The 0 to 10 V DC output range of the potentiometer corresponds to a valve position of 0 to 100%. Before we can use the analog input from the potentiometer to manually control the valve through our PID instruction, we must first determine what the digital output range is for a 0 to 10 V DC analog signal and see if that range is compatible with the PID manual control feature. The digital output range for a 0 to 10 V DC analog input signal from the A/D converter is 0 to 32,767. If you recall from our study of the PID instruction, in order to control the PID instruction in manual requires that we set the Auto/Manual (AM) bit to 1 and write a value directly into the control variable output address of the PID instruction that corresponds to the desired output. The valid range that can be written to the control variable output address of the PID instruction is 0 to 16,383, which corresponds to 0 to 100% output.

As you can see, we will need to convert, or scale, the analog input from the potentiometer that has a range of 0 to 32,767 into a corresponding range of 0 to 16,383 in order for it to be compatible with the PID instruction's control variable output. To scale the analog input signal from the potentiometer, we will again use the Allen-Bradley Scale (SCL) instruction. Before we can program the Scale

(SCL), we must first calculate the *slope* (*rate*) and *offset* values, as follows:

Slope = (scaled maximum − scaled minimum) / (input maximum − input minimum)
Slope = (16,383 − 0) / (32,767 − 0)
Slope = **0.5**

Offset = scaled minimum − (input minimum × slope)
Offset = 0 − (0 × 0.5)
Offset = **0**

Enter the *slope* (*rate*) and *offset* values into the Scale (SCL) instruction as shown in Figure 17–25.

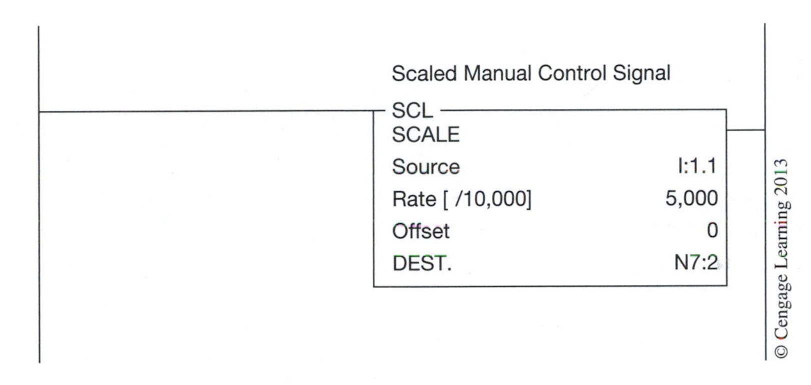

Figure 17–25 Scaling Manual Control Signal for Use with PID Instruction

In Figure 17–25, we have chosen the integer element address N7:2 to store the scaled value that represents the desired position of the flow control valve when in manual mode as determined by the remote potentiometer.

Now that we have scaled the process variable, control variable, and manual control analog signals, it is time to program and configure the PID instruction for our example application. The completed PLC program logic is shown in Figure 17–26.

In Rung 1 of Figure 17–26, we have programmed the Scale (SCL) instruction that is required to convert the analog input signal from the flow transmitter (I:1.0) into a digital value with a range that is compatible with the PID instruction's process variable input parameter (N7:0).

The Scale (SCL) instruction in Rung 2, as you recall, is required to scale the analog input signal from the manual potentiometer (I:1.1) into a digital range that is compatible with the PID instruction's control variable output parameter (N7:1). Once in manual (see Rung 3), the position of the flow control valve in the outlet pipe is determined by the position of the potentiometer, thus giving the operator manual control over the valve.

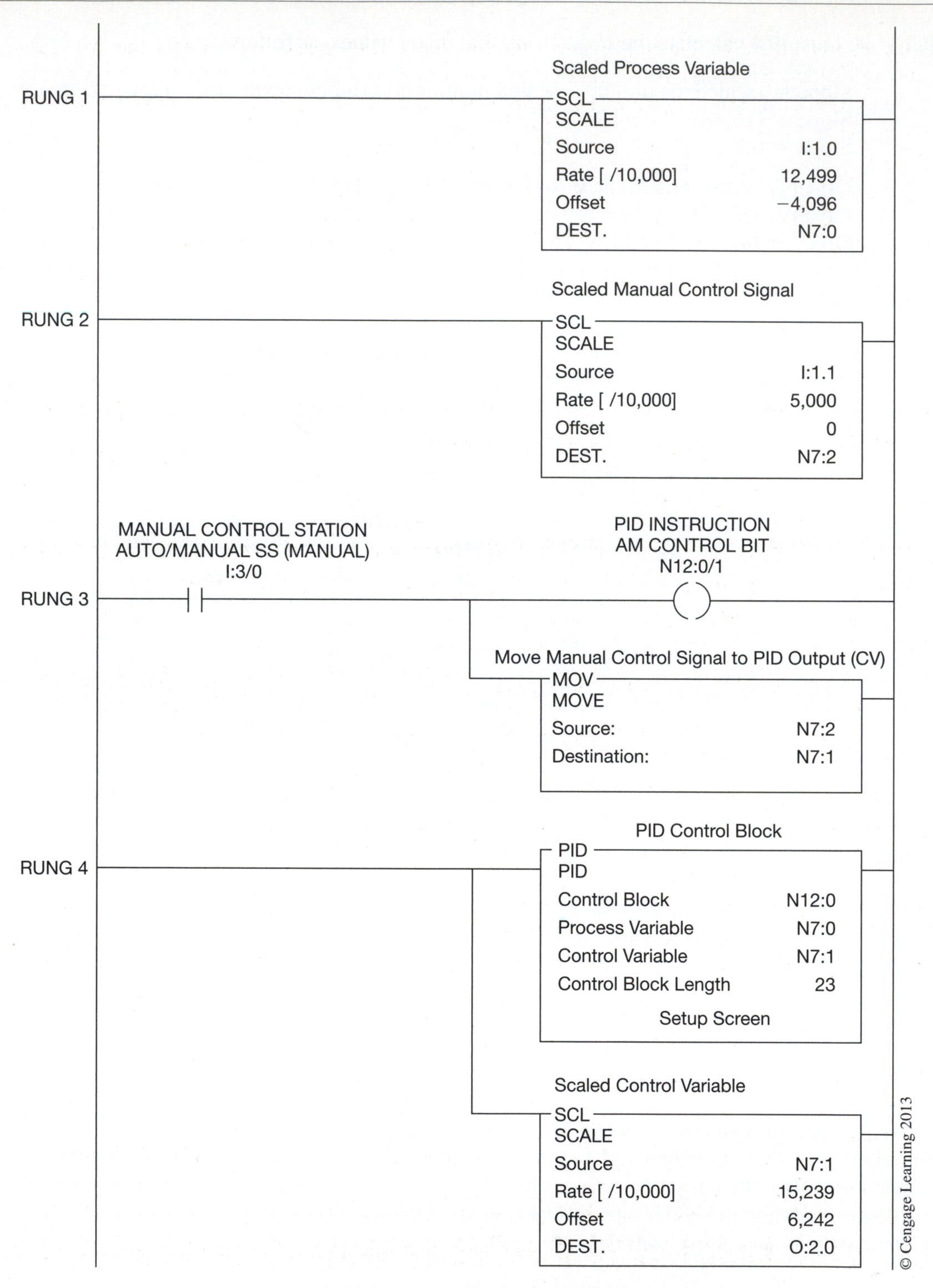

Figure 17–26 Program Logic for PID Example

When manual mode is selected by the operator, Input I:3/0 *true,* Rung 3 sets the Auto/Manual AM flag (N12:0/1) in the PID instruction to *true,* or *on,* and allows the output of the Scale (SCL) instruction (N7:2) located in Rung 2 to be moved into the PID instruction's control variable output parameter.

In Rung 4, we have programmed the PID instruction with a Control Block integer address of N12:0, process variable input as N7:0, and control variable output as N7:1. We have also programmed our third Scale (SCL) instruction that is required to convert the control variable output (N7:1) of the PID instruction into a digital range that will produce the 4 to 20 mA analog output signal (O:2.0) to the flow control valve.

After the PID instruction has been entered into Rung 3, the remainder of the PID instruction configuration is done through the PID instruction's Setup Screen. The PID Setup Screen in Figure 17–27 shows the parameter settings that we have entered for our example application. The following is a brief description of the parameter settings.

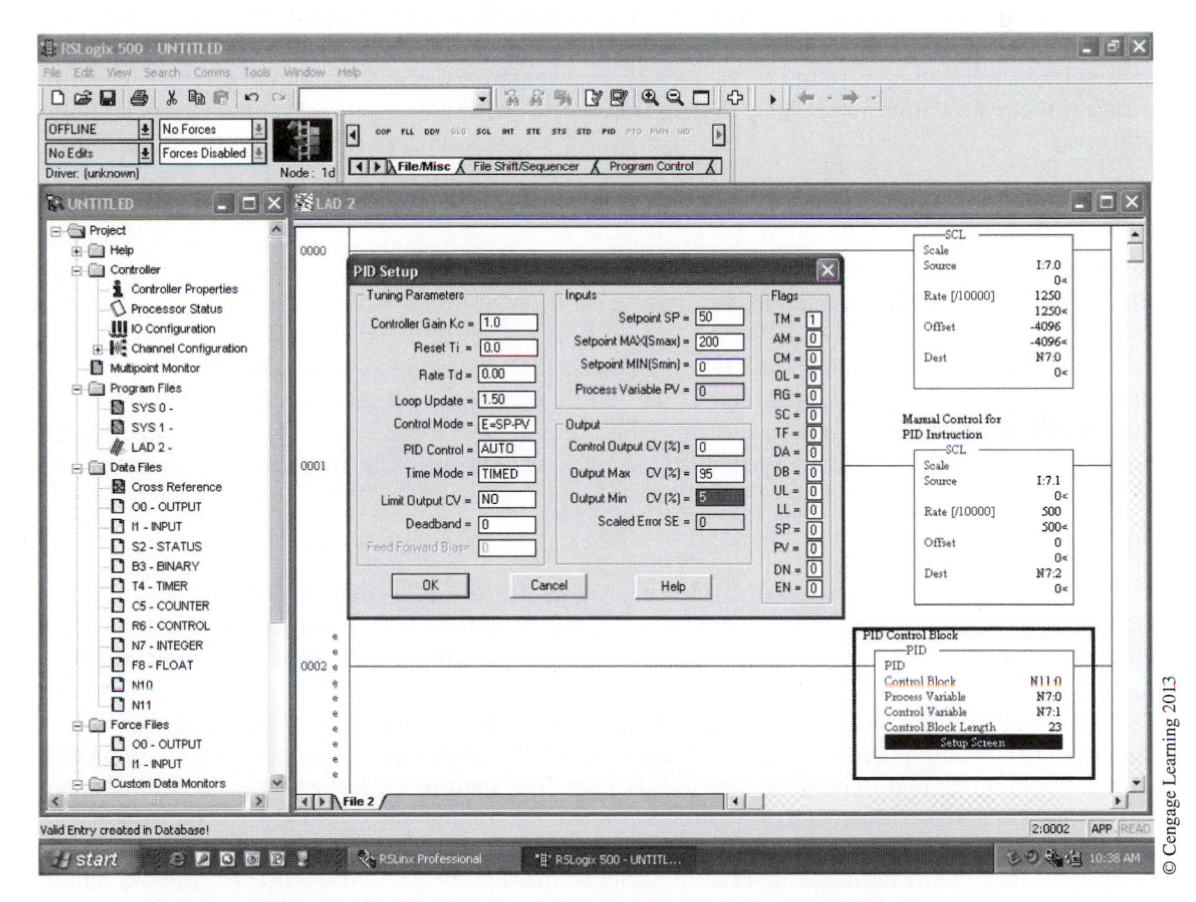

Figure 17–27 PID Configuration Parameters Setup Screen

The values entered into the Controller Gain Kc, Reset Term Ti, Rate Term Rd, and Loop Update Time parameters are just arbitrary values. The real values will be determined at the time the PID instruction and control loop is tuned. See the following PID Loop Tuning section for assistance in determining these values.

The Control Mode parameter is set to E = SP − PV (*reverse acting*). Whenever the flow rate (process variable) in the outlet pipe drops below the setpoint, we want the output of the PID instruction to increase, causing the control valve to open in an attempt to increase the flow (*reverse acting*). The PID Control (Auto/Manual) parameter is determined by the output (N12:0/1) in Rung 3 (see Figure 17–26). The Timed Mode parameter is set to "Timed" because the PID instruction is programmed in a standard PLC program file and not in an STI interrupt subroutine. We have chosen not to limit the output of the PID instruction, so we have set the Limit Output CV parameter to No. With the Limit Output CV parameter set to No, the output of the PID instruction will be allowed to operate from 0 to 100%. We have chosen no deadband around the setpoint, so the Deadband parameter was set to 0, or *no* deadband.

The value in the Setpoint SP parameter is just an arbitrary value that we have set at 50, or 50 gpm. The value you enter into this parameter is your desired flow rate in gpm. You can change this parameter through PLC program logic if desired. The Setpoint MAX Smax and Setpoint MIN Smin parameters are set to correspond to the engineering range of the process variable input, which in our case is the flow transmitter. The flow transmitter has an engineering measurement range of 0 to 200 gpm.

The Output Max CV(%) and Output Min CV(%) parameters have been set to 95 and 5 respectively. Anytime the PID output falls below 5% or above 95%, the corresponding output alarm flags will be set in the PID instruction status word. We can use these status flags as desired in our PLC program. They have no effect on the actual PID output.

Now that we have programmed and configured our PID instruction for our example application, the last and final step would be to adjust the tuning parameters in our PID instruction based on the characteristics that would be unique to our process control loop. The next section will cover in greater detail the loop tuning parameters and several procedures for determining their values.

PID LOOP TUNING

Tuning a PID loop is the process of selecting values for the PID tuning parameters (Controller Gain Kc, Reset Ti , Rate Td, and Loop Update Time) so that the PID instruction is able to quickly eliminate an error between the setpoint and process variable without causing excessive fluctuations in the process loop under control. PID loop tuning is probably one of the most difficult procedures to master and requires a knowledge of general process control, process controllers, and the process under control. Many people would say that tuning a PID loop is an art. Although not completely true, I would agree that to properly tune a PID loop does require the knowledge and skill that can only come from experience. If you thoroughly understand what each tuning parameter does, you are more likely to be able to tune a PID loop with confidence and success. Before we discuss the techniques that can be used to tune PID loops, let us first review each of the PID instruction tuning parameters.

Controller Gain Kc, or proportional gain, refers to the amount by which the error signal will influence controller output. In other words, the controller gain changes the output of the controller by an amount proportional to the error between the setpoint and the process variable. The higher the gain, the greater the output will change for a given error. Too much gain may cause the control loop to become unstable or oscillate. You should always try to start out with a small controller gain value and then gradually increase it while observing how the process loop responds to an upset.

The **Reset Ti,** or integral gain parameter, is used to change the output of the controller by a rate proportional to the error over time. The Reset Ti parameter is expressed in minutes per repeat. This means that as long as there is an error between the setpoint and the process variable, the integral action will add to the controller's proportional output by repeating the previous proportional action over the time, and at the frequency specified in the Reset Ti parameter. As long as there is an error, the integral action will continue to add to the output of the PID instruction until the process variable equals or nearly equals (with deadband) the setpoint. Some integral gain is required to eliminate the effects of offset, which can occur with proportional-only control. Keep in mind that the integral action is expressed in minutes per repeat, so the larger the value, the longer the time between repeats. See Figure 17–28 for an illustration of both proportional and integral response to a step change.

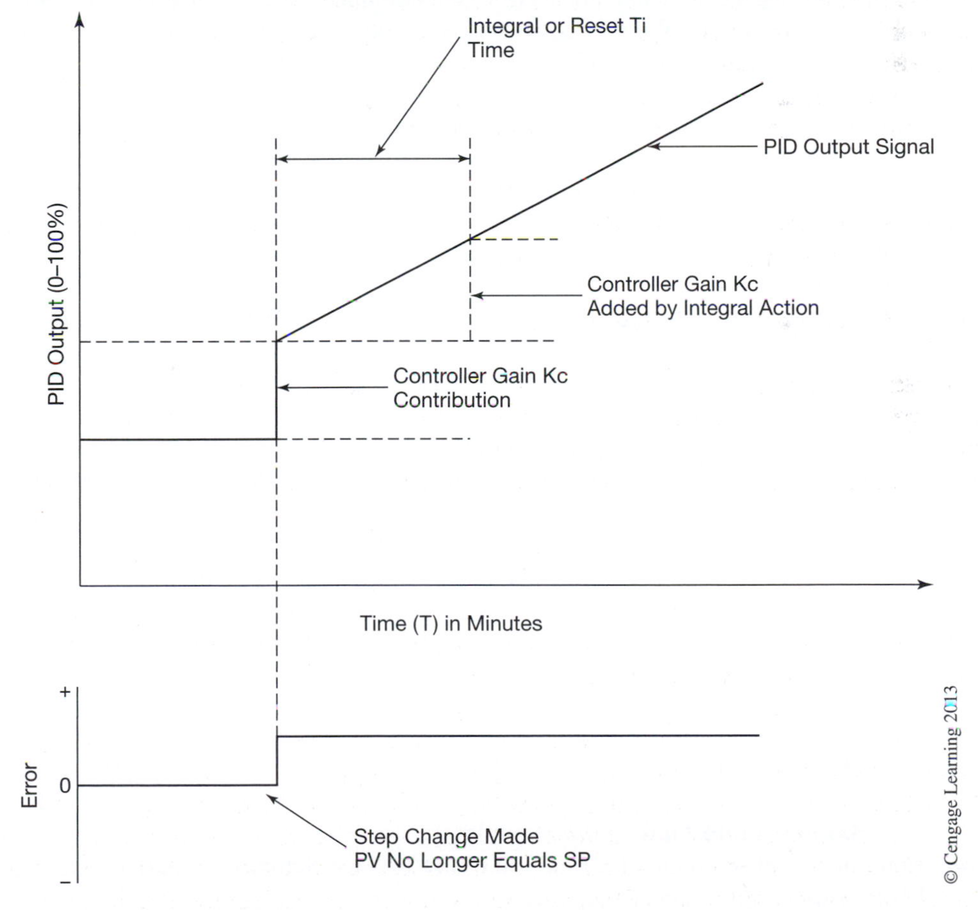

Figure 17–28 Proportional and Integral Response to Step Change

The **Rate Td,** or derivative action, is used as a method of changing the PID output in proportion to the rate of change of the process variable. The faster the process variable is changing, the greater the influence that the derivative action will have on the output. The derivative action acts as an anticipator, or "brake," in the control loop, helping to minimize the amount of overshoot and undershoot in the control loop's response. You typically see derivative action used when a process has considerable time constant lags, such as temperature loops. You should set the Rate Td parameter to 0, derivative term *off,* when controlling process loops that have a fast response time.

The **Loop Update Time** parameter sets the time between PID calculations and should be set to a time that is five to ten times faster than the natural period of the process under control. The faster that the process responds to a change in the PID output, the faster the loop update time needs to be. Determining the natural period of the process will be covered next under "Loop Tuning Techniques." If you are unsure, start with a loop update time of one-half second and adjust from there.

Loop Tuning Techniques

In 1942 J.G. Ziegler and N.B. Nichols published two loop tuning techniques that are still used today by many control engineers and technicians. The two techniques are called the *ultimate gain method* and the *reaction curve method.* We will cover both methods, but before we do we need to address some safety concerns related to loop tuning.

Caution: These tuning methods should only be done on noncritical applications in terms of personal safety and equipment damage. During loop tuning it is possible that the PID control variable output may become unstable and oscillate between 0 and 100%. For this reason, be sure that the process can safely tolerate such instability and that personal safety and equipment damage have also been considered before the tuning process begins.

Ziegler and Nichols Ultimate Gain Method (Closed Loop Technique) The ultimate gain method requires that you determine the ultimate gain and ultimate period of the process control loop. The ultimate gain is a controller gain value that will cause a sustained but stable oscillation in the process variable from the slightest error. The period of these sustained oscillations is called the ultimate period or the natural period of the control loop. This method is conducted with the PID instruction in automatic mode and with the integral action (Reset Ti) and derivative action (Rate Td) parameters turned *off.* Once you have determined what the ultimate gain (Gu) and ultimate period (Pu) of the control loop is, you can then use these two values to calculate the Controller Gain Kc, Reset Ti, Rate Td, and Loop Update Time parameters for the PID instruction.

To determine the ultimate gain and ultimate period of your control loop, perform the following steps.

Step 1. Enter the following values into the tuning parameters of your PID instruction through the PID Setup Screen:

- **Controller Gain Kc** = 1
- **Reset Ti** = 0
- **Rate Td** = 0
- **Loop Update Time** = 0.5 seconds

Step 2. Enter an initial setpoint value that you desire into the Setpoint SP parameter through the PID Setup Screen.

Step 3. You will need to observe the process variable as it varies with time and with respect to the setpoint value you entered in Step 2. A strip chart recorder or trending chart works well for making this observation, particularly when working with slow-responding process loops.

Step 4. Place the PID instruction in manual mode and adjust the output of the PID instruction until the process variable equals or nearly equals the setpoint. If you are using a manual control station, then use the manual control station for this adjustment; otherwise adjust the PID output by setting a value in the Control Output CV% parameter through the PID Setup Screen. Remember that the value you enter is a percent (0 to 100).

Step 5. After the process loop is under control manually, place the PID instruction in auto mode.

Step 6. While observing the relationship of the process variable to the setpoint over time, impose an upset (create an error) on the control loop and observe the response. The simplest and easiest way to impose an upset is to change the setpoint by a small amount.

Step 7. If the response curve of the process variable produced by Step 6 becomes unstable (see Figure 17–29a), the controller gain is too high. Reduce the Controller Gain Kc value and repeat Step 6 until you have obtained a sustained but stable oscillation of the process variable, as shown in Figure 17–29b.

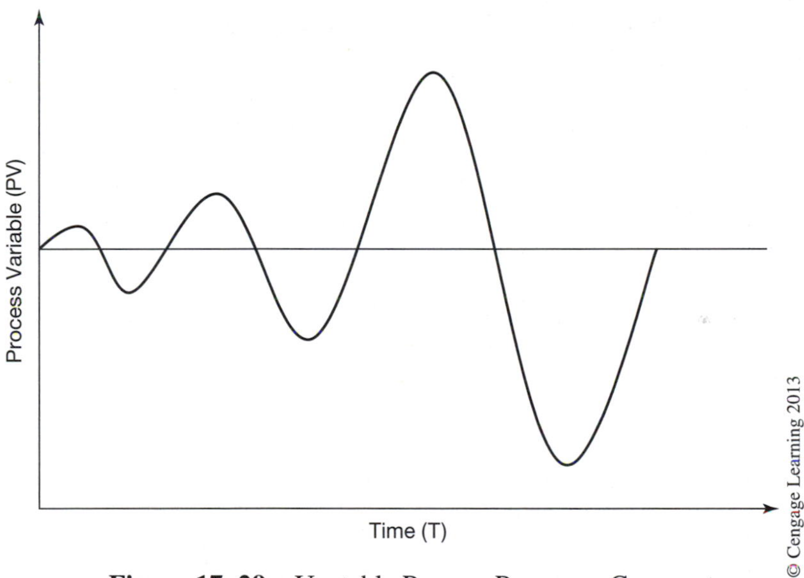

Figure 17–29a Unstable Process Response Curve

Step 8. If the response curve of the process variable produced by Step 6 dampens out over time, as shown in Figure 17–29c, then the controller gain is too low. Increase the Controller Gain Kc value and repeat Step 6 until you have obtained a sustained but stable oscillation, as was shown in Figure 17–29b.

Step 9. When you obtain a stable response in which the process variable is oscillating above and below the setpoint in an even manner, record the values of the ultimate gain (Gu) and ultimate period (Pu). The ultimate gain is the current value in the Controller Gain Kc parameter.

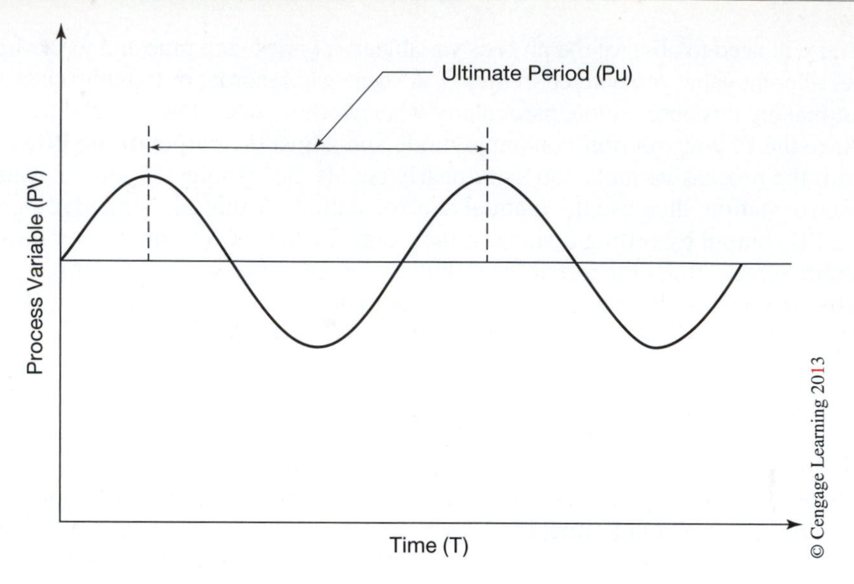

Figure 17–29b Stable Process Response Curve

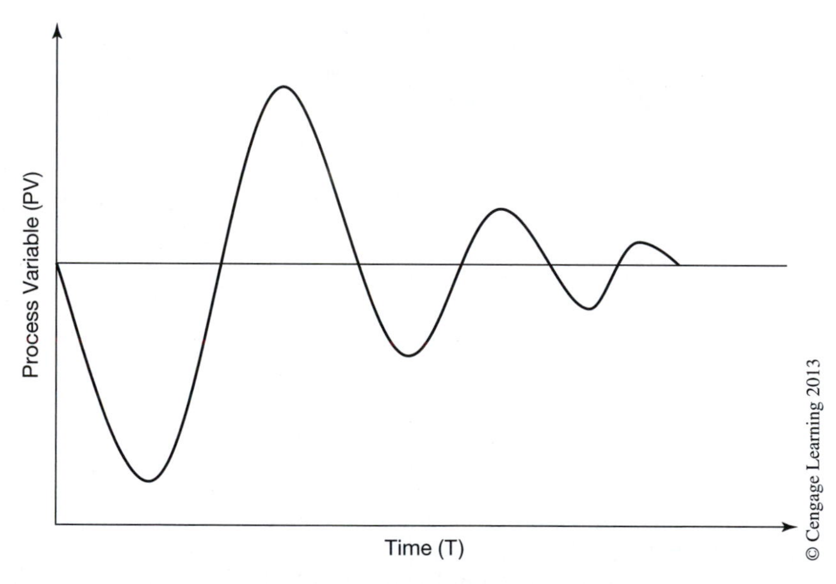

Figure 17–29c Damped Process Response Curve

The ultimate period is the time of one cycle, measured between successive peaks (see Figure 17–29b). The ultimate period is also referred to as the natural period of the process.

Step 10. Return the PID instruction to manual mode and gain control of the process. At this time you may stop controlling the process if desired.

Step 11. Set the Loop Update Time parameter to a value that is from 5 to 10 times faster than the ultimate (natural) period as observed in Step 9. If, for example, the ultimate period is 30 seconds and we desire a loop update time 10 times faster than the ultimate period, then the Loop Update Time parameter would be set to 3 seconds.

Step 12. Calculate the value of the Controller Gain Kc parameter by taking the ultimate gain (Gu) recorded in Step 9 and multiplying it by 0.45 (Kc = Gu · 0.45). For example, if the ultimate gain (Gu) recorded in Step 9 was 40, you would then set the Controller Gain Kc parameter to 18.

Step 13. Calculate the value of the Reset Ti parameter by taking the ultimate period (Pu) recorded in Step 9 and multiplying it by 0.5 (Ti = Pu · 0.5). For example, if the ultimate period (Pu) recorded in Step 9 was 30 seconds, you would then set the Reset Ti parameter to 2 (0.2 minutes approximates 15 seconds).

Step 14. Calculate the Rate Td parameter by taking the ultimate period (Pu) recorded in Step 9 and dividing it by 6.3 (Td=Pu/6.3). For example, if the ultimate period (Pu) recorded in Step 9 was 30 seconds, you would then set the Rate Td parameter to 8 (0.08 minutes approximates 4.8 seconds). Only set the Rate Td parameter on process loops that have large time constant lags.

Note: *If you do not add derivative action (Rate Td = 0) then change the Controller Gain Kc parameter calculation in Step 12 to Controller Gain Kc = Gu · 0.32 and change the Reset Ti parameter calculation in Step 13 to Reset Ti =Pu / 1.2.*

Step 15. Place the PID instruction in auto mode. If you have an ideal process, the PID controller should achieve a decay ratio of one-quarter wave when an error occurs, which Ziegler and Nichols defined as good control (see Figure 17–30). In reality the tuning parameters just calculated and entered into the PID instruction are more likely just a good starting point, and additional tweaking of the tuning parameters will be required. It will take experience and sometimes a little luck to achieve the right values for Kc, Ti, and Td. This is where the art of loop tuning comes into play.

Ziegler and Nichols Reaction Curve Method (Open LoopTechnique) The second method of tuning control loops developed by Ziegler and Nichols was based on data taken from a reaction curve during an open loop test. The open loop technique involves making a manual step change in the output of the controller and observing the reaction of the process variable over time. This observed reaction over time is called the reaction curve, and it indicates the reaction of the control system loop when a step change is made to the process. The reaction curve is obtained by using a strip chart recorder or trending chart and recording the process variable signal over time

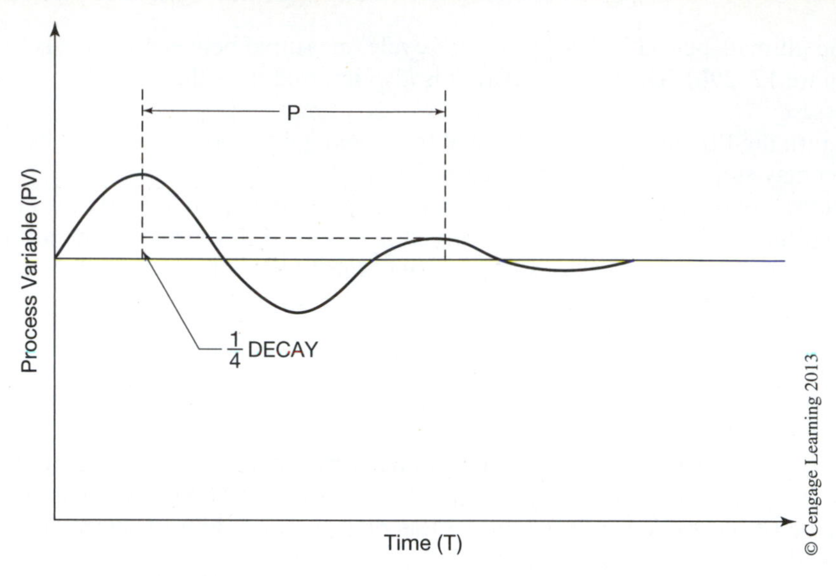

Figure 17–30 Decay Ratio of One-Quarter

after a step change has been made to the controller output. An example reaction curve is shown in Figure 17–31.

In Figure 17–31, the sloped line drawn tangent to the curve at its point of maximum slope is the process reaction rate and denotes how fast the process reacted to the step change made in the controller output. The inverse of this line's slope is a measure of the severity of the process lag as measured in time (T). The process deadtime (Dt) can be determined by measuring the time from when the step change was made to the intersection of the tangent line with the baseline. The process gain (K) is a measure of how much the process variable increased relative to the size of the step change.

Perform the following steps to calculate the Controller Gain Kc, Reset Ti, and Rate Td parameters for the PID instruction using the reaction curve method:

Step 1. Place the PID instruction in manual mode and adjust the output of the PID instruction until the process variable is at about 50% of its range. If you are using a manual control station then use the manual control station for this adjustment; otherwise adjust the PID output by setting a value in the Control Output CV% parameter in the PID Setup Screen.

Step 2. Turn on your strip chart recorder or trending chart and allow the system to stabilize. You should be trending the process variable at this time. Make sure that your strip chart recorder or trending chart is set fast enough to capture the reaction curve.

Step 3. Manually introduce a 10% step change (50% to 60%) in the output of the PID instruction. Record on the chart the time when you made the step change. Record the reaction curve of the process variable to the step change.

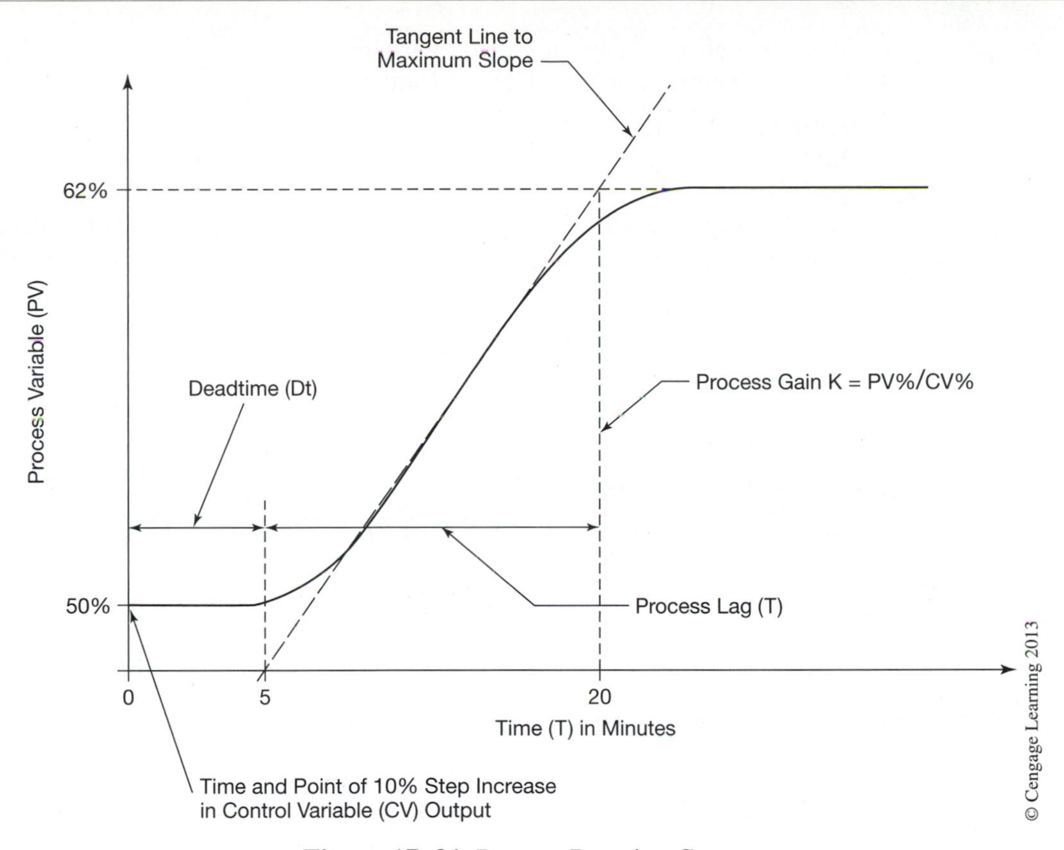

Figure 17–31 Process Reaction Curve

Step 4. Once you have recorded the process variable reaction curve, you may stop controlling the process at this time if desired.

Step 5. Using the process reaction curve recorded in Step 3, draw a line tangent to the reaction curve at its point of maximum slope. Next draw a line that is the inverse of the tangent line just drawn and measure the severity of the process lag in time (T). Determine the process deadtime (Dt) by measuring the time from when the step change was made to the intersection of the tangent line with the baseline. Finally, measure the process gain (K) by measuring how much the process variable increased relative to the size of the step change in the control variable.

Step 6. Compute the values for the Controller Gain Kc, Reset Ti, and Rate Td parameters from the following equations:

Controller Gain Kc = (1.2 · T) / (Dt · K)

K = ΔPV% / ΔCV%

$\Delta PV\%$ = Percent change in Process Variable
$\Delta CV\%$ = Percent change in Control Output

Reset Ti = $2.0 \cdot Dt$
Rate Td = $0.5 \cdot Dt$

For example, let's take the data from the reaction curve shown in Figure 17–31 to calculate what the tuning parameter values would need to be. We will assume the reaction curve was made when a 10% step change was made to the output of the PID controller. Studying the reaction curve of Figure 17–31, we can determine the following values:

Process Lag Time (T) = 15 minutes (20 − 5)
Process Deadtime (Dt) = 5 minutes
Process Gain (K) = percent change of PV / percent change of CV = 12% / 10%
Process Gain (K) = 1.2

Controller Gain Kc = $(1.2 \cdot 15) / (5 \cdot 1.2)$
Controller Gain Kc = 3

Reset Ti = $2.0 \cdot 5$
Reset Ti = 10 minutes

Rate Td = $0.5 \cdot 5$
Rate Td = 2.5 minutes

Note: *If you do not add derivative action (Rate Td = 0) then change the Controller Gain Kc parameter calculation in Step 6 to Controller Gain Kc =* $(0.9 \cdot T) / (Dt \cdot K)$ *and change the Reset Ti parameter calculation in Step 6 to Reset Ti =* $3.33 \cdot Dt$.

Step 7. After calculating the values for the Controller Gain Kc, Reset Ti, and Rate Td parameters, enter them into the PID instruction using the Setup Screen.

Step 8. The PID instruction is now ready to be placed in the auto mode. As in the previous method, if you have an ideal process the PID controller should achieve a decay ratio of one-quarter wave when an error occurs. See Figure 17–30 for an example of a decay ratio of one-quarter wave. As stated before, the tuning parameters just calculated and entered into the PID instruction are more likely just a good starting point, and additional tweaking of the tuning parameters may be required.

One of the major advantages of the reaction curve method over the ultimate gain method is that the tuning of the PID parameters can be done much quicker when tuning very slow process loops. Another advantage of the reaction curve method is that there is less process disturbance, which can be important when dealing with sensitive or critical processes.

There are many commercial PID tuning software products available that can be used to find the ideal tuning values. Some PLC manufacturers offer PID tuning as an addition to their PLC programming software.

Chapter Summary

Process control signals fall into two types, those that measure the process, and those that control the process, referred to as *process variables* (*PVs*) and *control variables* (*CVs*). These process control signals are typically transmitted as low-level electrical analog signals to and from devices like PLCs. In order for analog input signals to be used in the PLC program, they must first be converted into a corresponding digital value by an *Analog-to Digital Converter* (A/D converter), or in the case of an analog output signal a *Digital-to-Analog Converter* (D/A converter). When working with digital values in the PLC that represent analog signals, you may be required to convert, or scale, the digital values from one range to another. In order to make this linear conversion from one range to another, mathematical formulas can be employed, or in the case of the Allen-Bradley SLC 500, the Scale (SCL) and Scale with Parameters (SCP) instructions can be used.

Process control can be defined as a means by which you regulate a process. All process control systems consist of four key elements; the process, the measurement or process variable, the process controller, and the final control element or control variable. When all four elements are present, it is considered a *closed loop control* system. An *open loop control* system on the other hand does not have the process variable, process controller, and control variable providing automatic regulation of the process. A process controller is a device or software instruction that regulates a process by monitoring an analog input from a process variable and outputting an analog output to a control variable based on a programmed setpoint. The process controller is often referred to as a PID controller, as in the case of the Allen-Bradley SLC 500 PID instruction. PID stands for *proportional, integral,* and *derivative* control that uses a mathematical formula to regulate a process. Loop tuning is the process of finding the right values for the proportional, integral, and derivative constants in the PID equation so the PID controller is able keep the process under control.

Review Questions

1. If an analog input has a digital range of 0 to 32,767 and we wish to scale that range between 200 and 1500 °F, what would be the *slope* (*rate*) and *offset* values required for our scaling math formula?
2. If the current digital input value in Question 1 was 12,537, what would be the scaled value in °F?
3. A 4 to 20 mA analog input signal is converted into 3,277 to 16,384 by the PLC A/D converter. What would be the digital value that corresponds to 9.34 mA?
4. The transmitter producing the 4 to 20 mA signal in Question 3 is a pressure transmitter with a range of 0 to 250 psig. What is the pressure reading when the transmitter is putting out 17.42 mA?
5. The *slope* (*rate*) and *offset* values entered into the Scale (SCL) instruction must first be divided by 10,000.
   T F
6. The maximum value that can be entered into the Source parameter of a Scale (SCL) instruction is 32,767.
   T F

7. List the six parameters that must be entered into the Scale with Parameters (SCP) instruction.
8. What are the four key elements that make up a closed loop process control system?
9. The Allen-Bradley PID instruction has three integer address parameters that must be entered into the instruction at the time of programming. What are the names of those three parameters?
10. PID stands for what?
11. What are the digital input and output ranges of the Allen-Bradley PID instruction?
12. J.G. Ziegler and N.B. Nichols in 1942 published two loop tuning methods that are still used today. What are those two methods called?

CHAPTER 18 Function Block Diagram and Structured Text Programming

Objectives

After completing this chapter, you should have the knowledge to:

- Understand Function Block Diagrams.
- Use function block elements.
- Understand Structured Text programming.
- Apply structured text components.

FUNCTION BLOCK DIAGRAM PROGRAMMING

Function Block Diagram (FBD) programming is a method of programming that uses function blocks to make decisions or perform calculations. FBD programming is typically found in process control applications where there is more data handling and calculations, as compared to discrete machine control applications. There are many types of function blocks available. In fact, Allen-Bradley's Logix5000 controllers have over 80 different function blocks available to perform various tasks.

A function block takes one or more inputs, makes a decision or calculation, and then generates one or more outputs. An output of one function block can also be the input to other function blocks. In the case of the Allen-Bradley Logix5000 controllers, the user must create a function block routine to program and use FBD instructions. A Jump-to-Subroutine (JSR) instruction is typically used to run a function block routine from the main routine or another routine.

The following examples will illustrate the Function Block Diagram programming features of the Allen-Bradley Logix5000 controllers. The function block diagram is made up of function block elements (Figure 18–1). The elements consist of the function blocks themselves and the elements used to get information into and out of the function blocks.

The function block performs an operation on an input value or values and produces an output value or values. As you can see in Figure 18–1, there are four basic input and output elements used with the function blocks: Input Reference (IREF), Output Reference (OREF), Input Wire Connector (ICON), and Output Wire Connector (OCON).

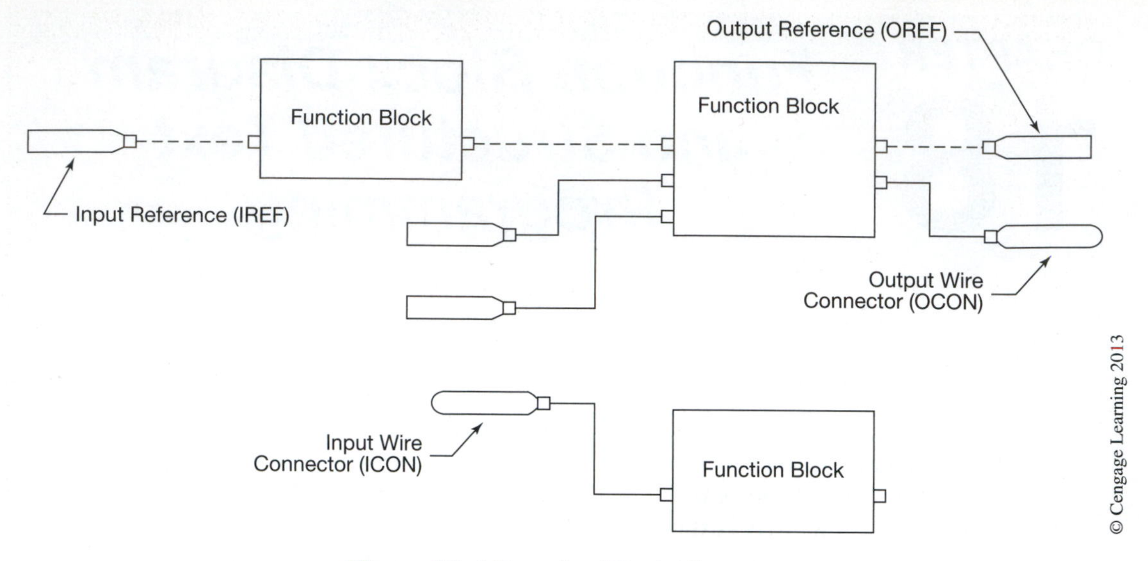

Figure 18–1 Function Block Elements

The Input Reference (IREF) supplies a value from an input device or tag to a function block input. The Output Reference (OREF) sends a value from the function block to an output device or tag. The Input Wire Connector (ICON) and Output Wire Connector (OCON) transfer data between function blocks when they are far apart or on different sheets, or to disperse data to several points in the routine.

Each function block uses a tag to store configuration and status information. The RSLogix 5000 software will automatically create a tag for the function block when it is created. This default tag can be used as is, or you can assign a different tag. For the IREF and OREF elements, you have to create a tag or assign an existing tag.

The order of execution (flow of data) is done by wiring elements together as shown in Figure 18–2. The actual location of the function blocks does not affect the order of execution. As you can see in Figure 18–2, there are two types of wire symbols: a solid line indicates a SINT, INT, DINT, or REAL value data path, and a dashed line indicates a BOOL value (0 or 1) data path.

This should begin to give the reader a basic understanding of the function block diagram (FBD) programming language. Additional information on using this type of programming language with the Logix5000 controllers can be found in their programming manual entitled "Logix5000 Controllers Function Block Diagram," publication 1756-PM009C-EN-P.

STRUCTURED TEXT PROGRAMMING

Structured text is a textual programming language similar to BASIC, C, or C++ that uses statements to define what to execute. It is best used for complex mathematical operations or specialized array/table loop processing.

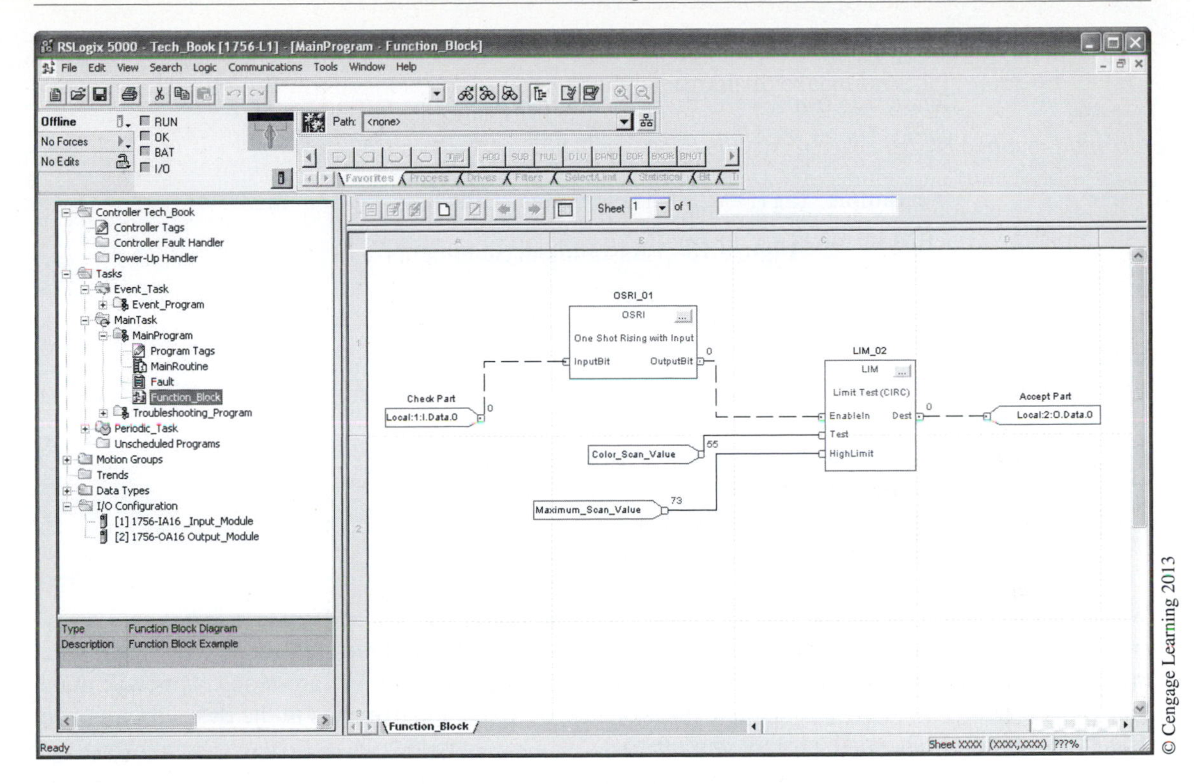

Figure 18–2 Function Block Example

Because structured text programming language is difficult to monitor and troubleshoot, it should be left to special applications that, once debugged, rarely require troubleshooting by the electrician/technician. Many computer programmers will feel right at home programming in structured text.

Structured text can contain the following components: assignments, expressions, instructions, constructs, and comments. Each will be discussed to an extent that the reader should have a basic understanding of their function. Before getting started, it is worth mentioning that structured text is not case sensitive. You also use tabs and carriage returns to make the structured text easier to read.

Assignments

You use an assignment statement to assign values to tags. The operator (symbol) that is used to indicate an assignment statement is ": = ". You terminate the assignment statement with a semicolon ";". Figure 18–3 shows the syntax for an assignment statement along with several examples.

The tag must be a BOOL, SINT, INT, DINT, or REAL data type. If you use a BOOL tag, then the expression must of the BOOL type; otherwise, use a numeric expression.

tag: = expression

Where tag is any valid memory tag and is the tag getting the new value. The expression contains the new value to assign to the tag.

Example: Tag1: = 250; Tag1: = Tag2; Tag1: = Tag2 + Tag3;

Figure 18–3 Assignment Statements

Expressions

An expression can be a tag name, equation, or comparison. When writing an expression, you can use any of the following elements in the expression:

- A memory tag that stores a value
- An immediate value or number
- Function elements such as: LOG, TAN
- Operator elements such as: + , − , <, >, And

When writing expressions, you can use any combination of uppercase and lowercase letters. For example, the AND operator can be entered as AND, And, or and. All three are acceptable. When writing a complex expression, use parentheses to group expressions within expressions. This will make the whole expression easier to read and ensure that the expression executes in the desired sequence. The example below shows the use of parentheses.

tag1: = (tag2 + tag3) * (tag4 − tag5);

Expressions can be either BOOL or numeric expressions. A BOOL expression produces a value of 1 (true) or 0 (false), whereas a numeric expression calculates an integer or floating-point value.

A BOOL expression uses BOOL tags and operators (relational and logical) to compare values or check if conditions are true or false. The example below uses a relational operator to check if the value stored in tag1 is greater than 150. If tag1 is greater than 150, the expression is true or 1, otherwise the expression is false or 0.

tag1 > 150

You typically use BOOL expressions to condition the execution of other logic. This will become more apparent later in this chapter.

A numeric expression uses arithmetic operators and functions to calculate a value. The example below adds 10 to the value stored in tag2 and puts the results in tag1.

tag1: = tag2 + 10;

A numeric expression can also use bitwise operators to return a value based on two values. In the example below, DINT tag1 and tag2 are logically ANDed together and the result is placed in DINT tag3.

tag3: = tag1 AND tag2;

Many times in structured text programming you nest a numeric expression within a BOOL expression. In the example below, 10 is added to the value stored in tag1 and the result is then compared to see if it is greater than 100.

(tag1+10)>100

To calculate arithmetic values, use one of the Arithmetic or Function Operators shown in Figure 18–4.

To/For	Operator/Function	Optimal Data Type
Add	+	DINT, REAL
Subtract	–	DINT, REAL
Multiply	*	DINT, REAL
Exponent(x to the power of y)	**	DINT, REAL
Divide	/	DINT, REAL
Module-divide	MOD	DINT, REAL
Absolute value	ABS (numeric_expression)	DINT, REAL
Arc cosine	ACOS (numeric_expression)	REAL
ARC SINE	ASIN (numeric_expression)	REAL
ARC tangent	ATAN (numeric_expression)	REAL
COSINE	COS (numeric_expression)	REAL
Radians to degrees	DEG (numeric_expression)	DINT, REAL
Natural log	LN (numeric_expression)	REAL
Log base 10	LOG (numeric_expression)	REAL
Degrees to radians	RAD (nemeric_expression)	DINT, REAL
SINE	SIN (nemeric_expression)	REAL
Square root	SQRT (nemeric_expression)	DINT, REAL
Tangent	TAIN (nemeric_expression)	REAL
Truncate	TRUNC (nemeric_expression)	DINT, REAL

Figure 18–4 Arithmetic and Function Operators

To compare two values, use one of the Relational Operators shown in Figure 18–5.

Comparison	Operator	Optimal Data Type
Equal	=	DINT, REAL
Less than	<	DINT, REAL
Less than or equal	<=	DINT, REAL
Greater than	>	DINT, REAL
Greater than or equal	>=	DINT, REAL
Not equal	<>	DINT, REAL

Figure 18–5 Relational Operators

To compare the bits within values, use one of the Bitwise Operators in Figure 18–6.

For	Operator	Data Type
Bitwise AND	&, AND	DINT
Bitwise OR	OR	DINT
Bitwise exclusive OR	XOR	DINT
Bitwise complement	NOT	DINT

Figure 18–6 Bitwise Operators

To check if conditions are true (1) or false (0), use one of the Logical Operators in Figure 18–7.

For	Operator	Data Type
Logical AND	&,AND	BOOL
Logical OR	OR	BOOL
Logical exclusive OR	XOR	BOOL
Logical complement	NOT	BOOL

Figure 18–7 Logical Operators

The operators you write into an expression are not necessarily performed from left to right, but rather in a prescribed order. Operations of equal order are performed from left to right. When expressions contain more than one operator or function, group them in parentheses "()" as this will ensure the correct order of execution and make it easier to read the expression. Figure 18–8 shows the order of execution for the various operations.

Order	Operation
1	0
2	Function (...)
3	**
4	−(negate)
5	NOT
6	*, /, MOD
7	+, − (subtract)
8	<, <=, >, >=
9	=,<>
10	&,AND
11	XOR
12	OR

Figure 18–8 Order of Execution

Instructions

Structured text statements can also be instructions. An instruction is a stand-alone statement that performs a given task such as copying the contents of one array into another array or jumping to another routine. A structured text instruction executes each time it is scanned unless it is used within a construct. Constructs will be covered next.

A structured text instruction uses parentheses to contain its operands, and depending on the instruction, there can be zero, one, or multiple operands. Always terminate an instruction with a semicolon ";". The example below shows the structured text format for a Copy File instruction. Where "tag1_array[0]" is the source array, "tag2_array[0]" is the destination array, and 10 is the number of elements to copy.

```
COP(tag1_array[0], tag2_array[0], 10);
```

Another example of a structured text instruction is the Jump-to-Subroutine instruction shown below.

```
JSR(routine_name, input_count, input_parameter, return_parameter);
```

Notice how the operands of the instruction are contained within the parentheses and separated by commas. Also notice that the instruction is terminated with a semicolon.

Even though they are similar, instructions differ from functions in that instructions cannot be used in expressions; only functions can be used in expressions.

Refer to the Logix5000 Controllers General Instructions manual (instruction locator section) for a complete listing of the available structured text instructions.

Constructs

Constructs are conditional statements used to trigger structured text code (statements). Constructs can be programmed singly or nested within other constructs. Always terminate a construct with a semicolon ";".

Constructs are very powerful statements and one of the key elements in structured text programming. The following constructs are available.

- IF…THEN
- CASE…OF
- FOR…DO
- WHILE…DO
- REPEAT…UNTIL

IF…THEN

Use the IF…THEN construct to do something if or when specific conditions occur. For example: *If* the motor faults *then* turn on the alarm light or *if* the water is greater than 100° *then* start the cooling pump.

The syntax for the IF…THEN construct is:

```
IF bool_expression THEN
    <statement>;
END_IF;
```

The operand can be either a BOOL tag or an expression that evaluates to a BOOL value. The example below shows a BOOL tag operand.

```
IF motor_fault THEN
    alarm_light : = 1;
END_IF;
```

where "motor_fault" is a BOOL tag that is either 1(true) or 0(false). If the BOOL value is 1 or true, then the statement is executed. In our example, if the "motor_fault" tag contained a value of 1(true), then the "alarm_light" tag would be set to a value of 1.

The example below shows an expression that evaluates to a BOOL value.

```
IF tank1_temperature > 100 THEN
    cooling_pump : = 1;
END_IF;
```

where "tank1_temperature" tag contains a numeric value that is compared to a constant. If the value stored in "tank1_temperature" is greater than 100, then the statement is true or 1 and the "cooling_pump" tag would be set to a value of 1 or on.

The IF...THEN construct also has two optional statements that can be used with it, ELSIF and ELSE. The ELSIF statement allows you to select from several possible groups of statements, where each ELSIF represents an alternative path. The ELSE statement does something when all of the IF or ELSIF conditions are false. The following examples will help to illustrate the use of the ELSIF and ELSE statements.

EXAMPLE 1: IF...THEN...ELSE

```
IF motor_fault THEN
        alarm_light : = 1;
ELSE
        alarm_light : = 0;
END_IF;
```

As you can see in this example the ELSE statement turns the alarm light off (0) if the condition of the IF statement is false.

EXAMPLE 2: IF...THEN...ELSIF

```
IF tank1_temperature > 100 THEN
        cooling_pump : = 1;

ELSIF tank1_temperature < 75 THEN
        cooling_pump : = 0;

END_IF;
```

In this example, the ELSIF provides an alternative path if the temperature is not greater than 100. In this case, if the temperature is below 75, then turn the cooling pump off. As you can clearly see, we have created an IF, THEN, and ELSIF statement that turns the cooling pump on anytime the temperature in the tank is above 100° and leaves the cooling pump on until the temperature drops below 75°.

The table in Figure 18–9 summarizes the various combinations of IF, THEN, ELSIF, and ELSE.

If you want to	And	Then use this construct
Do something if or when conditions are true	Do nothing if conditions are false	IF...THEN
	Do something else if conditions are false	IF...THEN...ELSE
Choose from alternative statements based on input conditions	Do nothing if conditions are false	IF...THEN...ELSIF
	Assign default statements if all conditions are false	IF...THEN...ELSIF...ELSE

Figure 18–9 IF, THEN, ELSIF, and ELSE

CASE...OF

Use the CASE...OF construct to select what to do based on a numerical value. For example: A recipe number determines which ingredients to add in a batching process.

The syntax for the CASE...OF construct is:

```
CASE numeric_expression OF
        selector1:      <statement>;
        selector2:      <statement>;
                "          "
                "          "
        selectorN:      <statement>;
ELSE
        <statement>;
END_CASE;
```

The *numeric_expression* operand can be either a tag or expression that evaluates to a number. The *selector* operand is an immediate number of the same type as the *numeric_expression* operand. The "N" represents the last *selector* number. The *statement* is the statement to execute when the *selector* is true. The ELSE is optional and will only execute the statements if the *numeric_expression* does not equal any of the *selectors.*

The *selector* values can be a single value, multiple distinct values, a range of values, or a combination of distinct and range of values. The syntax for the *selector* values is shown in the example below.

```
value: <statement>;
value1, value2, value3, valueN: <statement>;
value1..valueN: <statement>;
value1, value2, value3..valueN: <statement>;
```

Use a comma (,) to separate each value and two periods (..) to identify a range.

The following example will help to illustrate the CASE…OF construct.

```
CASE batch_recipe_number OF
    1:              ingredient_valve1 : = 1;
                    ingredient_valve2 : = 1;
                    ingredient_valve3 : = 0;
                    ingredient_valve4 : = 0;
    2,3,4:          ingredient_valve1 : = 0;
                    ingredient_valve2 : = 1;
                    ingredient_valve3 : = 1;
                    ingredient_valve4 : = 1;
    5..7:           ingredient_valve1 : = 1;
                    ingredient_valve2 : = 1;
                    ingredient_valve3 : = 1;
                    ingredient_valve4 : = 1;
    8,9..12:        ingredient_valve1 : = 1;
                    ingredient_valve2 : = 1;
                    ingredient_valve3 : = 0;
                    ingredient_valve4 : = 0;
ELSE
    ingredient_valve1 : = 0;
    ingredient_valve2 : = 0;
    ingredient_valve3 : = 0;
    ingredient_valve4 : = 0;
END_CASE;
```

In the example above, the value stored in the batch_recipe_number tag is compared to the selector values for a match. If a match is found, then the ingredient valves are either opened or closed based on the selector statements. If no match is found then the ingredient valves are closed.

FOR…DO

Use the FOR…DO construct to do something a specific number of times before doing anything else. For example: Clear an array of bits or check an array of part numbers for a match.

The syntax for the FOR…DO construct is:

```
        FOR count : = initial_value TO final_value BY increment
DO
        <statement>;
END_FOR;
```

The *count* operand is a tag that stores the count position as the construct executes and assigns a unique tag to this operand. The *initial_value* operand specifies the initial value for the count. This can be a tag, expression, or immediate number. The *final_value* operand specifies the final value for the count, which determines when to exit the loop. This can be a tag, expression, or immediate number. The *increment* operand is optional and is the amount to increment the count by each time through the loop. If you don't specify an *increment,* the count increments by 1.

Caution: The controller does not execute any other statements in the routine until it completes the loop. This could lead to a processor fault if the loop time is greater than the watchdog timer for the task.

The following examples will illustrates the FOR…DO construct.

EXAMPLE 1:

```
FOR count_tag : = 0 TO 20 DO
      bit_array1[count_tag] : = 0;
END_FOR;
```

EXAMPLE 2:

```
FOR count_tag : = 0 TO 100 DO

IF part_number = part_number_array[count_tag] THEN
      found_tag : = 1;
      EXIT;
ELSE
      found_tag : = 0;
END_IF;
END_FOR;
```

In the first example, bits 0–20 are cleared in the BOOL array called "bit-array1". Notice how the *count_tag* is used as the subscript in the array tag. Since the *increment* operand was omitted, the count increment defaulted to 1. In the second example, the value in the tag called "part_number" was compared against an array of numbers (part_number_array) for a match. If a match was found, the BOOL tag "found_tag" was set to 1 and the loop was stopped by using the EXIT statement. If a match was not found then the "found_tag" was set to 0.

WHILE…DO

Use the WHILE…DO construct to keep doing something as long as certain conditions are true.

The syntax for the WHILE…DO construct is:

```
WHILE bool_expression DO
      <statement>;
END_WHILE;
```

The *bool_expression* operand is a BOOL tag or expression that returns a value of 1(true) or 0(false). The *statement* is what is executed as long as the *bool_expression* returns a 1(true).

Caution: The controller does not execute any other statements in the routine until it completes the WHILE...DO loop. This could lead to a processor fault if the loop time is greater than the watchdog timer for the task.

REPEAT...UNTIL

Use the REPEAT...UNTIL construct to keep doing something until certain conditions are true. The REPEAT...UNTIL construct executes the statements in the construct first before checking if conditions are true. If conditions are not true, then the controller executes the statements within the loop again.

The syntax for the REPEAT...UNTIL construct is:

```
REPEAT
        <statement>;
UNTIL bool_expression

END_REPEAT;
```

The *bool_expression* operand is a BOOL tag or expression that returns a value of 1(true) or 0(false). The *statement* is what is executed as long as the *bool_expression* returns a 0(false).

Caution: The controller does not execute any other statements in the routine until it completes the REPEAT...UNTIL. This could lead to a processor fault if the loop time is greater than the watchdog timer for the task.

Comments

You can add comments to your structured text program to help describe how your program works. This will make your program much easier to interpret by someone else or if you must work with it at a later date. Comments will not affect the execution of the program and are downloaded to controller memory.

You can add comments on a single line, at the end or within a line of structured text. The following syntax is used for entering comments.

```
//comment – single line only
(*comment*)
/*comment*/
```

EXAMPLE COMMENTS:

```
//Check for high temperature conditions
        IF water_temp > 100 THEN....
        IF water_temp (*sludge tank 1 water temperature*) > 100 THEN....
        pressure_valve1 : = 1;/*open the pressure valve on tank 1*/
```

Structured Text Example Program

The following structured text program loads the correct temperature set points into a working temperature array based on which recipe the operator has selected. It also monitors the temperatures for abnormally high conditions.

```
//Check for operator input and then select the correct temperatures to load

IF operator_select THEN

CASE recipe_number OF
    1:      tank_temp[0] : = 150;
            tank_temp[1] : = 175;
            tank_temp[2] : = 200;
            tank_temp[3] : = 100;
            batch_start : = 1;
    2:      tank_temp[0] : = 125;
            tank_temp[1] : = 150;
            tank_temp[2] : = 175;
            tank_temp[3] : = 75;
            batch_start : = 1;
    3:      tank_temp[0] : = 200;
            tank_temp[1] : = 250;
            tank_temp[2] : = 300;
            tank_temp[3] : = 150;
            batch_start : = 1;
    4..5:   tank_temp[0] : = 225;
            tank_temp[1] : = 300;
            tank_temp[2] : = 325;
            tank_temp[3] : = 180;
            batch_start : = 1;

ELSE        tank_temp[0] : = 100;
            tank_temp[1] : = 100;
            tank_temp[2] : = 100;
            tank_temp[3] : = 75;
            batch_start : = 0;

END_CASE;

END_IF
```

```
//Monitor batch temperatures during batching operation

IF batch_start THEN
      FOR count : = 0 TO 3 DO
      IF temp[count] => tank_temp[count] + 20 THEN
      batch_start : = 0;
      over_temp_alarm : = 1;

END_FOR;

END_IF

(*End of Program*)
```

Chapter Summary

Function Block Diagram (FBD) programming uses function blocks to make decisions or perform calculations and is typically found in process control applications. A function block takes one or more inputs, makes a decision or calculation, and then generates one or more outputs.

Structured text programming language uses statements to define what to execute and is similar to BASIC. It is best used for complex mathematical operations or specialized array/table loop processing. Structured text programming contains assignments, expressions, instructions, constructs, and comments.

Review Questions

1. A function block diagram is made up of function block______________.
   a) references
   b) assignments
   c) elements
   d) operators
2. What is the function of the Output Wire Connector (OCON)?
3. “OREF” stands for what?
4. Structured Text programming is similar to what type of programming language?
   a) Ladder Logic
   b) SFC
   c) Word
   d) BASIC

5. What are constructs?
6. The operator (symbol) that is used to indicate an assignment statement is______________.
   a) ";"
   b) "*"
   c) "^"
   d) ": = "
7. List any two constructs discussed in this chapter.
8. A dashed line indicates what type of data path in a function block diagram?
9. What type of expression is "tag1>150"?
10. A structured text instruction executes once unless it is used within a construct.
    a) True
    b) False

CHAPTER 19 Sequential Function Chart Programming

Objectives

After completing this chapter, you should have the knowledge to:

- Understand the building blocks used to create a sequential function chart program.
- Understand steps and transitions.
- Develop a basic sequential function chart program.

Sequential function chart (SFC) programming is a method of programming that is similar to designing a flowchart of your process. You program steps and transitions that are arranged like a flowchart and, when executed, control your system in a prescribed order. Some advantages of using SFC programming to specify your process are:

- Easier to organize and read (graphical representation)
- Faster execution of your logic
- Faster and easier troubleshooting
- Faster to design and debug

Sequential function chart programming is most often used with an application that is sequential in nature and has definable steps. For example, a burner control application would have definable steps that would typically be executed in sequential order. Figure 19–1 shows an example of what our burner control application might look like in an SFC format.

An SFC program is made up of steps, actions, and transitions. The "steps" are the major building blocks of any SFC program. They perform timing and counting functions as well as executing any actions associated with them. Actions are used for turning tags on or off and performing other functions. Transitions are conditions that must be met before moving on to the next step.

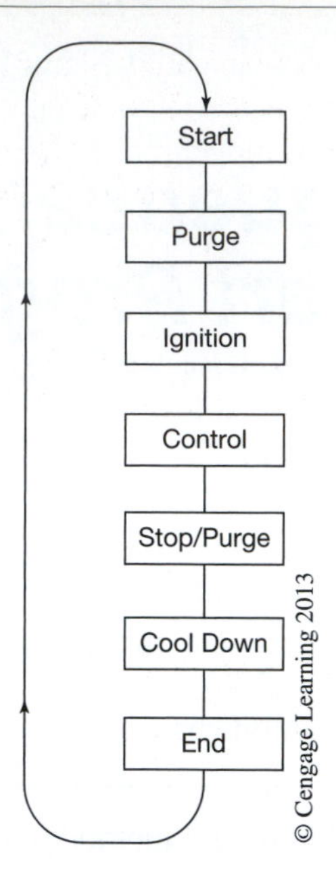

Figure 19–1 Burner Control Application (purge, ignition, control, shutdown, purge)

The key elements of an SFC program are shown in Figure 19–2. Notice how an SFC program can have simultaneous branches that execute more than one step at a time.

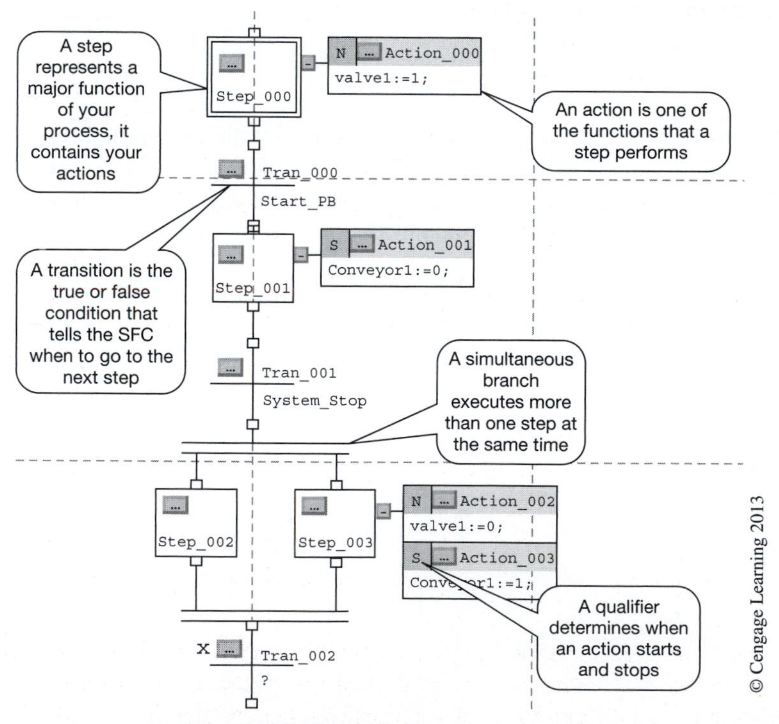

Figure 19–2 SFC Program Key Elements

In Figure 19–2, Step_000 is the first step to be executed and its actions performed until transition Tran_000 is true, at which time execution is moved to Step_001. The default names for each step are shown, but other names can be given to a step to help define its purpose such as "Purge" or "Ignition." The first two steps are linear steps, meaning they execute one right after the other, whereas the next two are concurrent. Concurrent steps are steps that are executed simultaneously.

When developing an SFC program, you must organize the execution of your steps to match the process. You can use linear, concurrent, and branch sequences to control the execution of your steps. Figure 19–3 shows an example of each type.

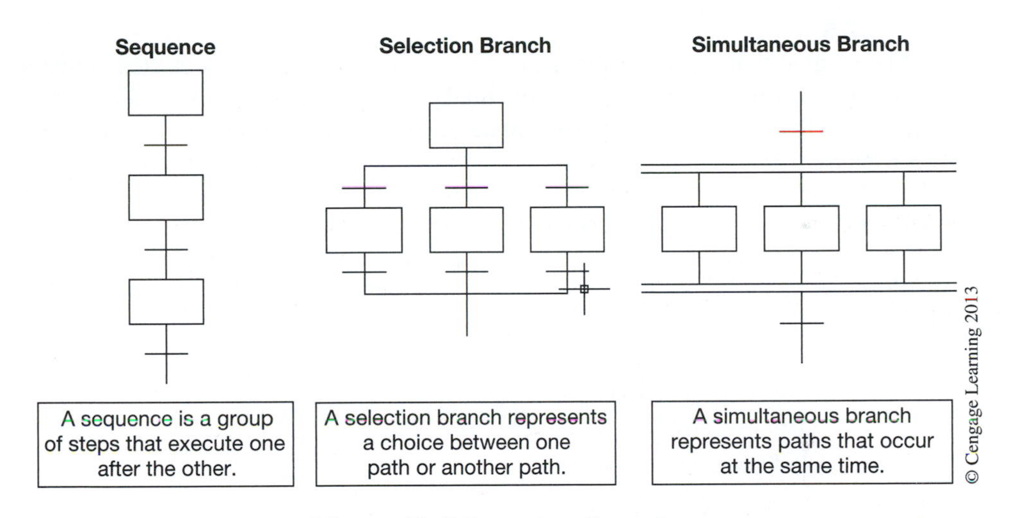

Figure 19–3 Execution Control

A linear sequence is used to execute one step after another. A concurrent or simultaneous branch is used to execute two or more steps at the same time, and a selection branch is used to select between steps depending on logic (transition) conditions.

Note: *A simultaneous branch is indicated by parallel horizontal lines at the top and bottom of the branch, whereas a selection branch uses a single horizontal line.*

In a selection branch, the SFC program checks each path from left to right; the first path with a true transition is executed and all others are ignored. If all path transitions are false, the program starts over with the first path. A path can have only a transition and no step if desired. This is useful if you want to bypass the entire branch group. See Figure 19–4.

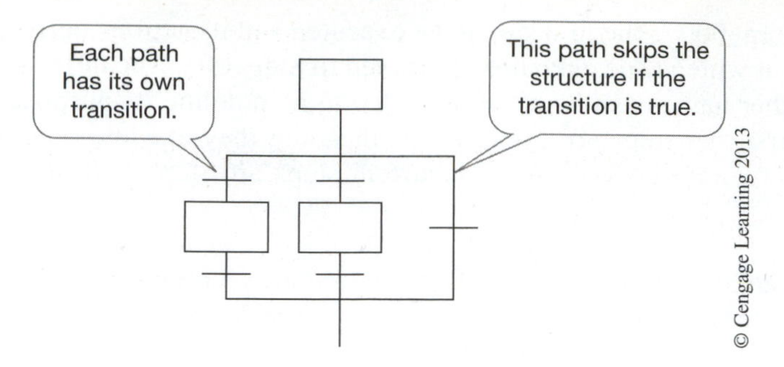

Figure 19–4 Selection Branch with Bypass

Creating Steps

Steps should be created that represent the major functions of your process. A step contains the actions to be performed during that part of the sequential process. When creating a step, a tag is created that provides information about the step. The tag name, by default, follows the step number created, such as Step_000, Step_001,…Step_005, etc. You can change the tag name to something that is more meaningful, such as Initialize, Purge, Ignite, etc. The information contained in the tag structure for an SFC step is shown in Figure 19–5. The members of the tag provide a substantial amount of information about the step such as, how long has the step been active, is this the first scan or the last scan, etc. All of this information can be used as needed when programming your SFC program. Refer to the detailed SFC programming manual for a complete explanation of each tag member.

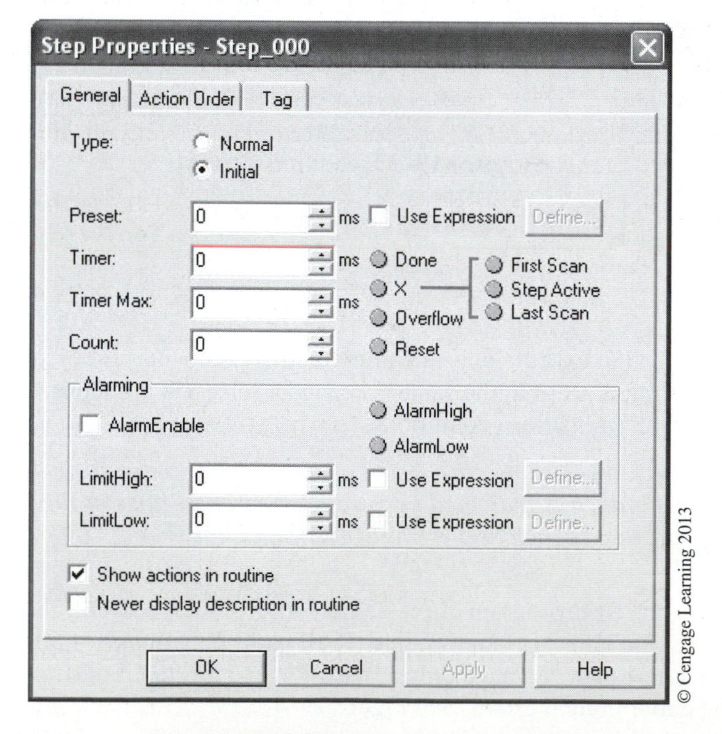

Figure 19–5 SFC Step Information

You add actions to steps to perform functions such as opening a valve, starting a motor, or initializing a mode. To add an action to a step, right-click on the step and then choose *Add Action*. See Figure 19–6. A step can have multiple actions associated with it.

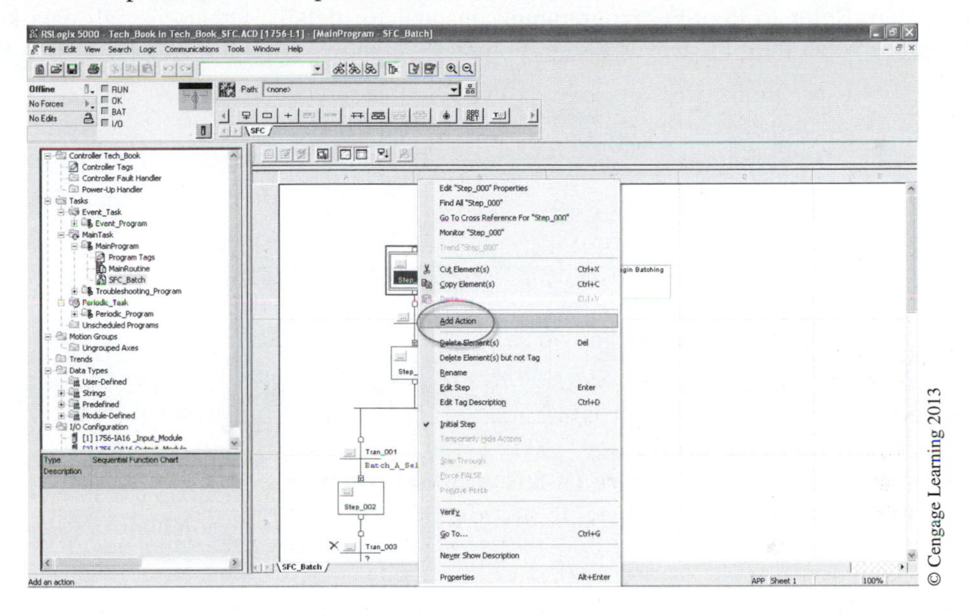

Figure 19–6 Adding Actions

When creating an action, you create a tag that provides information about the action. The tag name, by default, follows the action number created, such as Action_001, Action_002,…Action_005, etc. You can change the tag name to something that is more meaningful, such as Valve Open, Motor1 Start, etc. The information contained in the tag structure for an action is shown in Figure 19–7. This should sound familiar as this is the same as a step tag, which was discussed earlier.

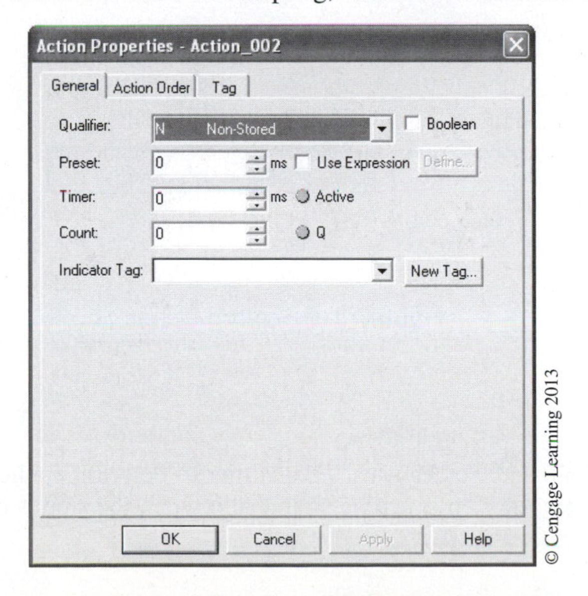

Figure 19–7 Action Properties

Actions can be either non-Boolean or Boolean. Non-Boolean actions use Structured Text programming to execute assignments and instructions (logic) or call a subroutine. When programming non-Boolean actions you have the option to automatically reset the assignments and instructions before leaving a step. Refer to the SFC programming manual for details. Otherwise, all data keeps its current values when the SFC leaves a step. Figure 19–8 shows several examples of non-Boolean actions.

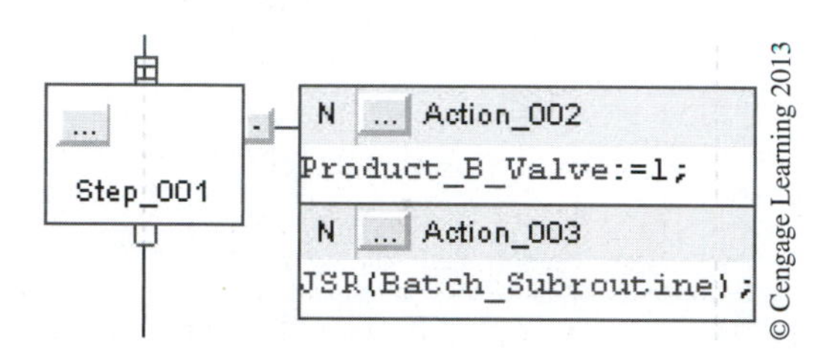

Figure 19–8 Non-Boolean Actions

A Boolean action on the other hand contains no logic. It simply sets a bit in its tag structure that can be monitored by other logic located elsewhere, such as in a ladder logic routine. This method allows you to reuse a Boolean action multiple times within the same SFC program. Figure 19–9 shows an example of a Boolean action.

Figure 19–9 Boolean Action

Each action (non-Boolean and Boolean) uses a qualifier to determine when it starts and stops. By default, actions start when the step is activated and stop when the step is deactivated. A list of the qualifiers is shown in Figure 19–10.

If you want the action to	And	Then this qualifier
Start when the step is activated	Stop when the step is deactivated	N
	Execute only once	PI
	Stop before the step is deactivated or when the step is deactivated	L
	Stay active until a Reset action turns off this section	S
	Stay active until a Reset action turns off OR a specific time expires, even if the step is deactivated	SL
Start a specific time after the step is activated and the step is still active	Stop when the step is deactivated	D
	Stay active until a Reset action turns off this action	DS
Start a specific time after the step is activated, even if the step is deactivated before this time	Stay active until a Reset action turns off this action	SD
Execute once when the step is activated	Execute once when the step is deactivated	P
Start when the step is deactivated	Execute only once	PO
Turn off (reset) a stored action		R

Figure 19–10 Action Qualifiers

When a step contains multiple actions, the order of execution is from top to bottom as shown in the "Action Order" window of the step's properties dialog box. See Figure 19–11.

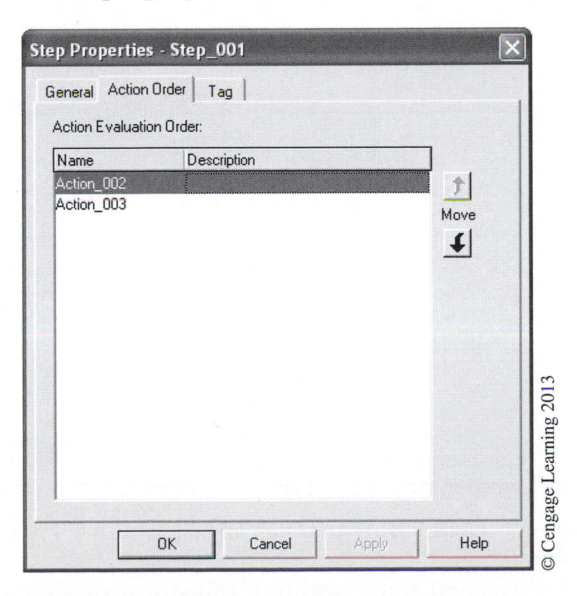

Figure 19–11 Action Order Window

Creating Transitions

Transitions control when the SFC program moves from step to step. If a transition is true, the SFC program goes to the next step; otherwise it repeats the step above it. Transitions occur between steps in a sequence, after a simultaneous branch, and in each branch of a selection branch. See Figure 19–12.

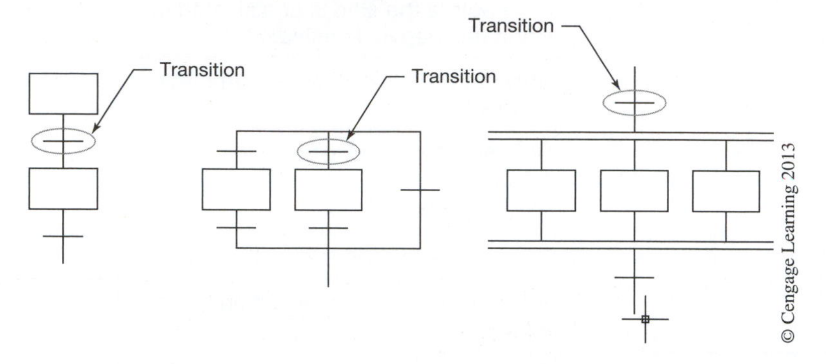

Figure 19–12 Transitions

Each transition uses a BOOL tag to represent the true or false state that is evaluated by the SFC program. When programming the transition, you enter the condition as a BOOL expression that uses BOOL tags, relational operators, and logical operators to compare values or check if conditions are true or false. The BOOL expression is entered in Structured Text format. You can also call a Subroutine that contains an End Of Transition (EOT) instruction that returns the state of the conditions to the transition. Figure 19–13 shows examples of both a BOOL expression and Subroutine transitions.

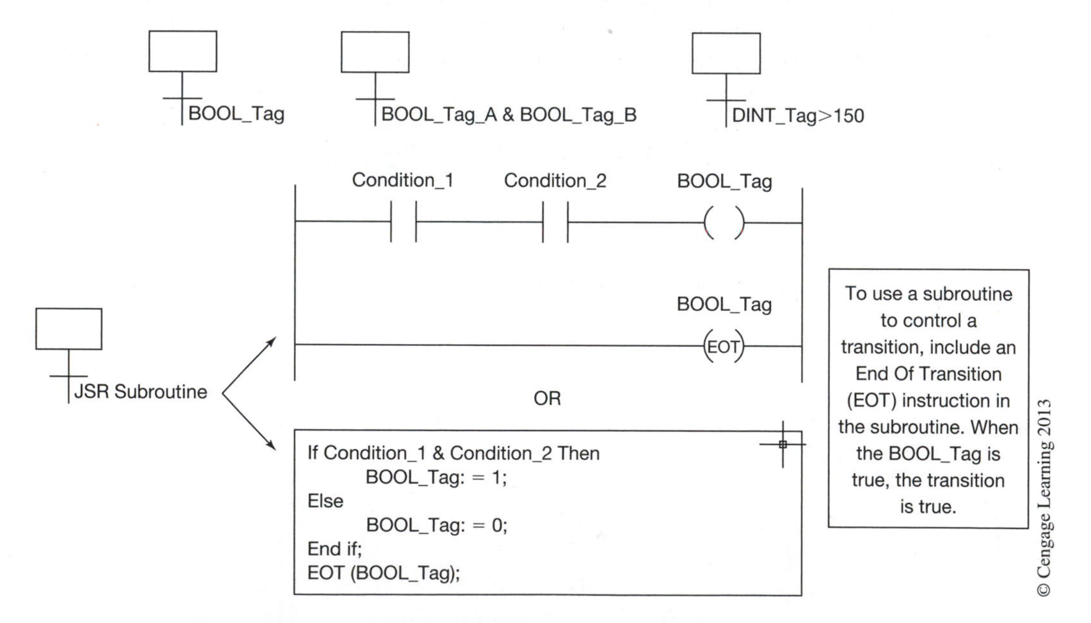

Figure 19–13 BOOL Expression and Subroutine Transitions

You can use the step millisecond timer within a transition to signal when the step has run the required time or too long and the SFC program should go to the next step. This is a very good use of the millisecond timer contained in the step tag. An example of this is shown in Figure 19–14.

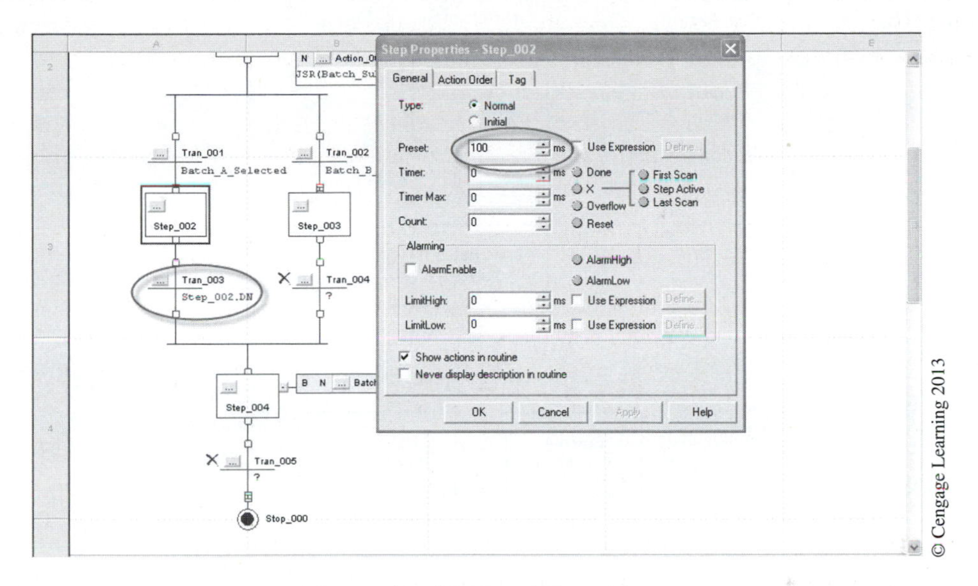

Figure 19–14 Timed Transition

When an SFC completes its last step, you must either wire the last transition to the top of the first step or use an SFC Stop element. When you use a Stop element, the SFC stops execution for part or the entire SFC program based on the location of the Stop element. If the Stop element is located after a sequence or selection branch then the entire SFC program stops. If the Stop element is located within a path of a simultaneous branch, then only that path is stopped and the rest of the SFC continues to execute. Once an SFC Stop element has been activated, you can restart the SFC by using either an SFC Reset (SFR) instruction or logic to clear the status bit of the Stop element. The following example illustrates the method of ending an SFC program. See Figure 19–15.

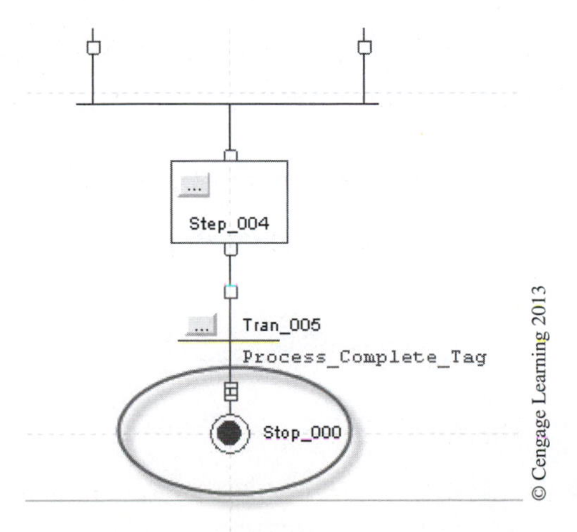

Figure 19–15 Ending SFC Program

SFC Programming Example

In this example, we will program a simple SFC program to control a batch operation as shown in Figure 19–16.

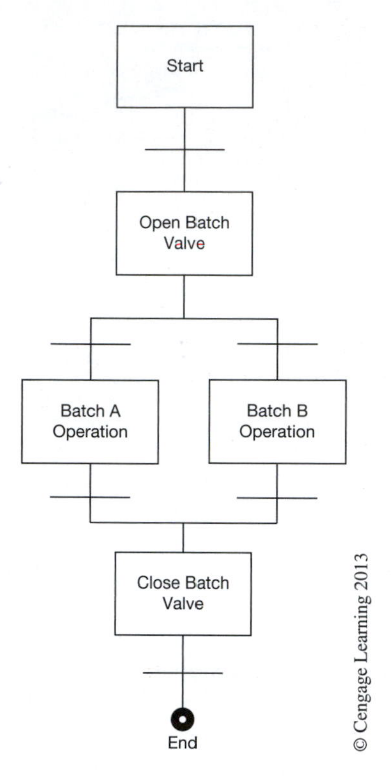

Figure 19–16 Batch Operation Example

The first step is to create the SFC routine by right-clicking on the Main Program and adding a new routine as shown in Figure 19–17. The new SFC routine was given the name SFC_Batch.

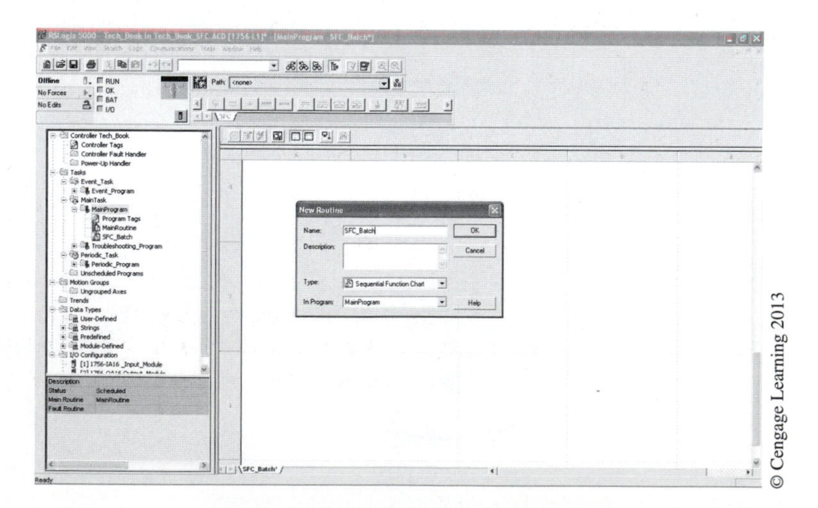

Figure 19–17 New SFC Routine

Double-click on the new routine (SFC_Batch) and the SFC program will appear in the right pane. See Figure 19–18. Note the SFC toolbar at the top of the window. The toolbar is used to select the SFC element to be added to the program.

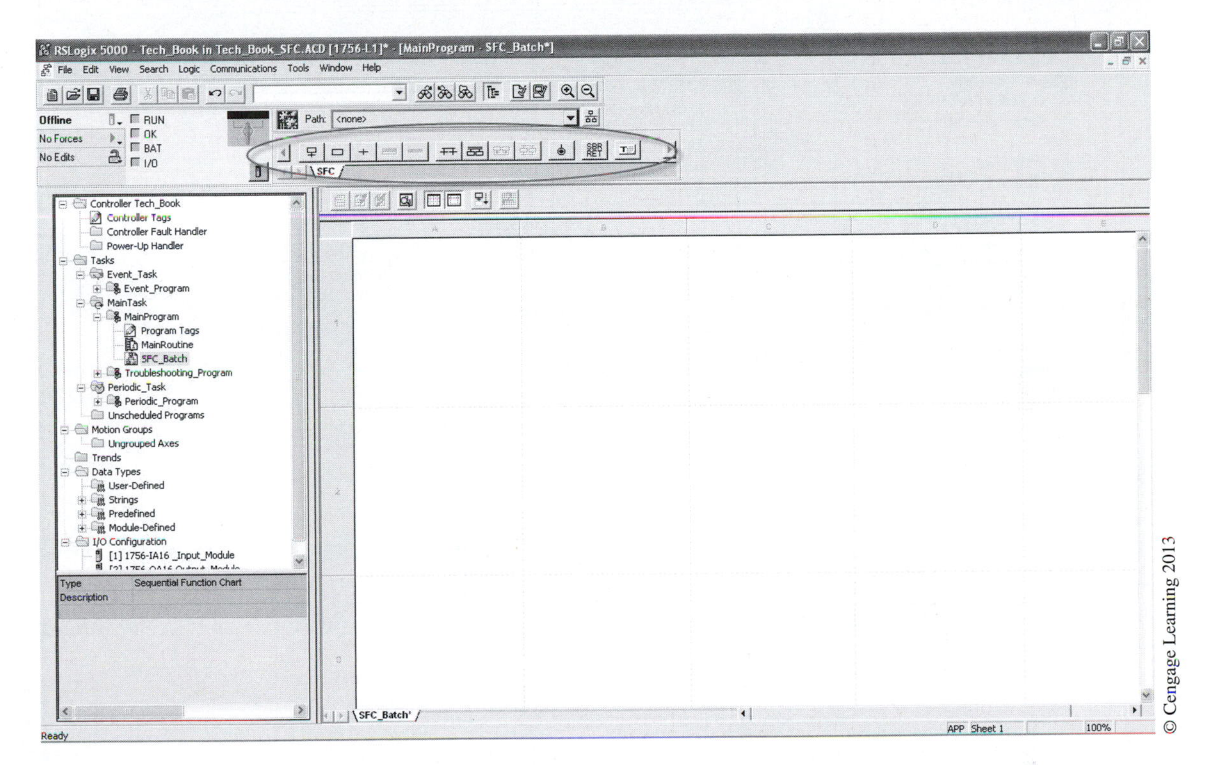

Figure 19–18 SFC Programming Window

Programming consists of dragging and dropping program elements from the toolbar onto the SFC program window replicating your SFC steps and transitions. After your steps and transitions have been placed, the last step is to configure your elements and add your actions. In Figure 19–19, we have added all of our steps and transitions for the batch operation example shown in Figure 19–16. We are now ready to configure our steps and transitions.

To configure each step, right-click on the step and select properties or click on the ellipses button located on the step. The properties screen will appear for that step as shown in

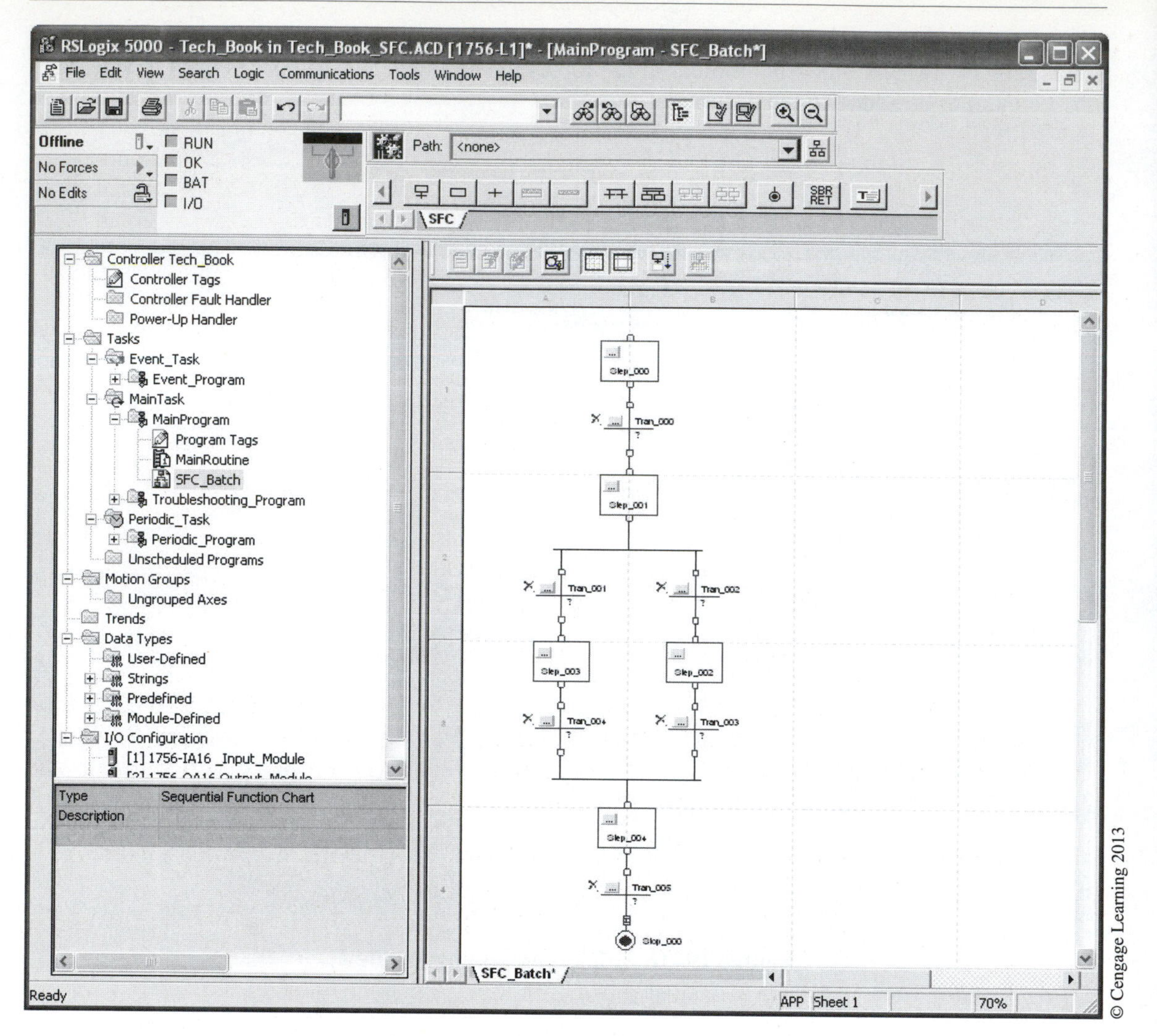

Figure 19–19 SFC Elements Added for Batch Example

Figure 19–20. If this is the first step of the SFC program, then you must choose the Initial checkbox; otherwise select Normal. The properties screen allows you to configure the step for such things as preset time, alarming, action order, etc.

Actions are added to a step by right-clicking on the step and then selecting Add Action. To configure an Action, right-click on the action and select properties or click on the ellipses button located

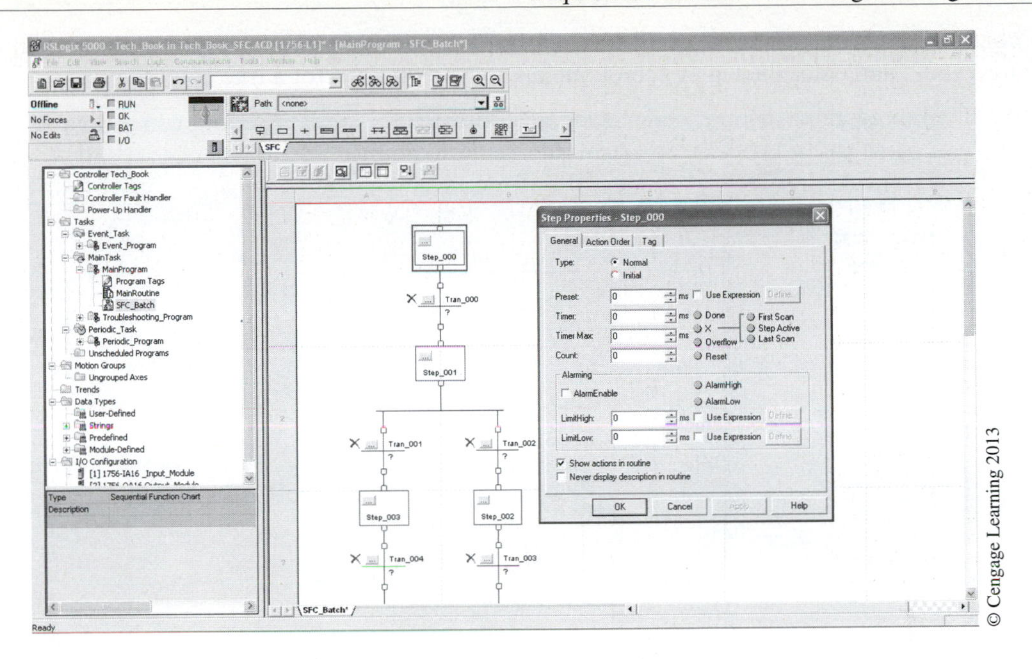

Figure 19–20 Step Properties Window

on the action. The properties screen will appear for that action as shown in Figure 19–21. Select the appropriate properties for that action. After selecting the properties for the action, then enter the action by double-clicking on the action window of the action box if a Non_Boolean action was selected. Continue to add actions to your steps as required for your SFC program.

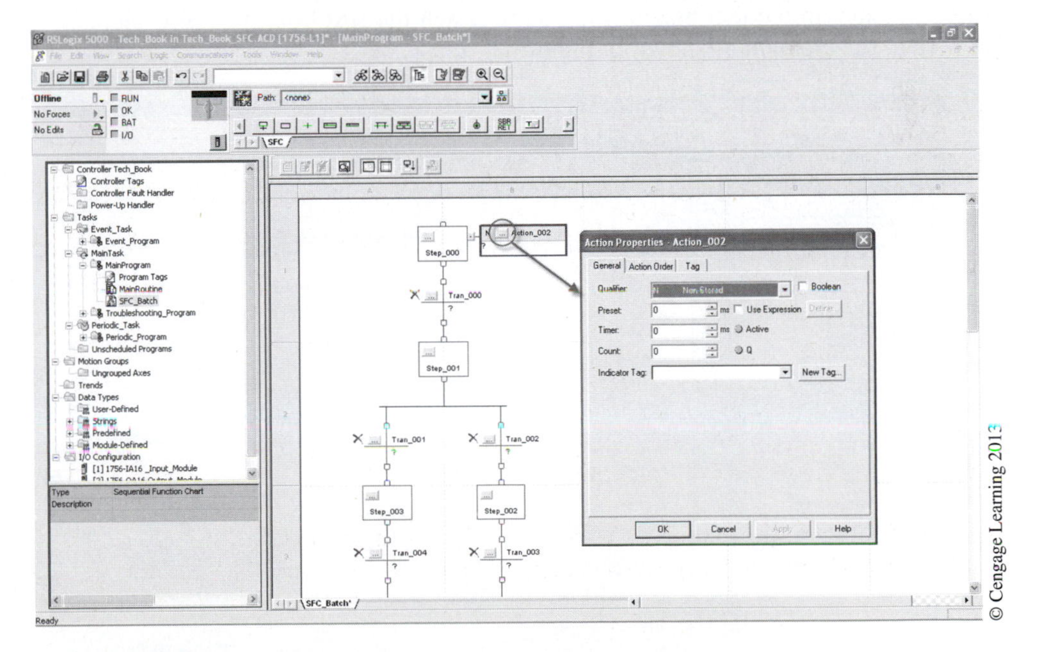

Figure 19–21 Action Properties Window

To configure each transition, double-click on the "?" just below the transition name as shown in Figure 19–22, and enter the tag or subroutine tag to be evaluated for a true condition.

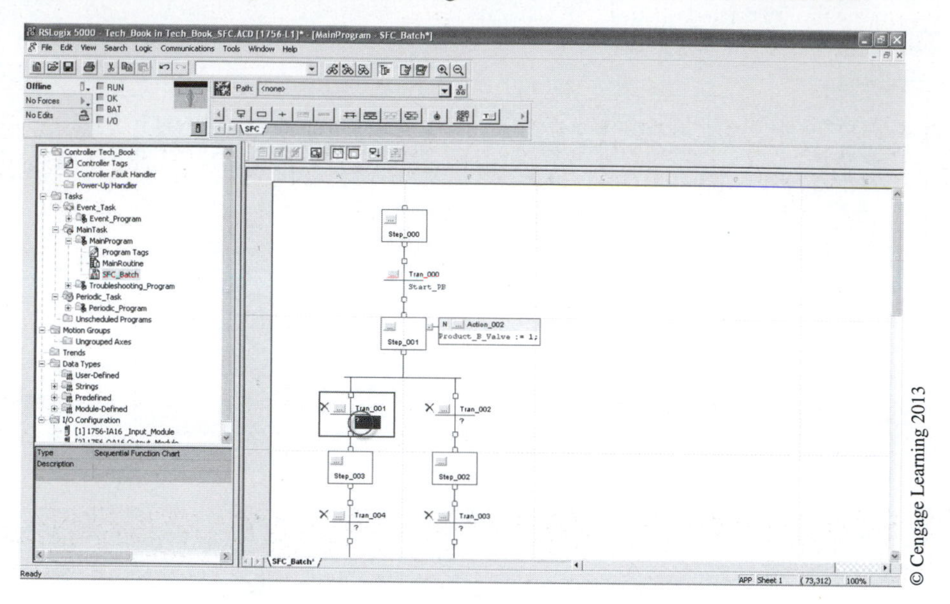

Figure 19–22 Transition Configuration

Continue entering your action and transition conditions until your program is complete. You can add text boxes to your SFC program by selecting the text box icon on the toolbar. After you select the text box on the toolbar, a text box will appear on your SFC window that you can move around. The pushpin symbol in the text box allows you to attach the text box to an SFC element by dragging the wire to the element as shown in Figure 19–23.

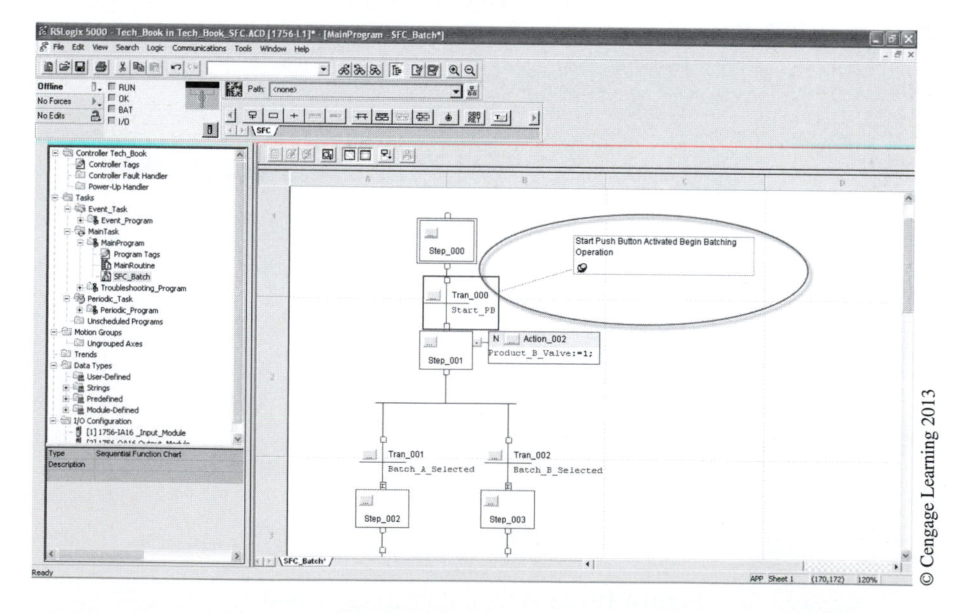

Figure 19–23 Text Box

Only a general overview of SFC programming has been presented here. For a complete description on how to program the Logix5000 controllers using SFC programming language, refer to the programming manual entitled "Logic5000 Controllers Sequential Function Charts," Publication 1756-PM006D-EN-P.

Chapter Summary

Sequential function chart (SFC) programming is similar to designing a flowchart of your process using steps and transitions. Steps contain actions that are used for turning tags on or off and performing other functions. Transitions are the conditions that must be met before moving on to the next step(s). Steps can be configured as linear steps (one step after another), concurrent steps (executed simultaneously), or selected branch (selected between steps).

SFC programming is easy to organize, read, design, and troubleshoot. It is most often used with an application that is sequential in nature and has definable steps.

Review Questions

1. List the key elements of an SFC program.
2. You can use ______________, ______________, and __________ to control the execution of your SFC steps.
3. Each step and transition is assigned a tag, which provides information about the step or transition.
   a) True
   b) False
4. Actions can be one of two types. List the two types and their differences.
5. A transition can use a subroutine to determine the state of the transition. What instruction is used in the subroutine for this?
   a) END
   b) OTE
   c) SFC
   d) EOT

CHAPTER

20 Understanding Communication Networks

Objectives

After completing this chapter, you should have the knowledge to:

- Understand networking principles.
- Describe the different network categories.
- Explain the different network topologies.
- Understand the media used to construct a network.
- Explain the different methods used to access a network.
- Understand what a network protocol is and how packets work.
- Understand the different industrial communication networks and protocols.

With PLCs used in greater numbers and distributive control being the norm, the control systems of today are becoming more complex, requiring various communication schemes to tie these system components together. Passing, or exchanging, information is not only a desire but a requirement in many control systems today. A communication scheme can be as simple as having two PLCs ten feet apart passing information between them, to as complex as a plant-wide control and information network in which PLCs, **human machine interfaces** (HMIs), intelligent I/O devices, and information technology (IT) systems are able to exchange information. This chapter on networking is not intended to provide the reader with a comprehensive knowledge of communication networks or schemes, but rather to give a general overview of the control and information networks that are being used today in many industrial plants and factories.

HISTORY

Data communication has become an integral part of modern control systems, and continues to evolve as technology advances at an ever-increasing rate. One of the first methods used to communicate between PLCs was a pair of wires used to connect the output card of one PLC to the input card of a second PLC. As simple as this method may seem, it did satisfy some of the criteria that we look for in many modern industrial control and information networks today, such as:

- real-time or nearly real-time control
- high data integrity
- high noise immunity
- reliability in harsh industrial environments

As industrial plants began to implement PLCs in a distributive or modular approach to controlling equipment, it was necessary to develop fast, secure, and reliable communications schemes to tie the various PLCs and their systems together. PLC manufacturers began developing their own control networks to meet this growing demand. These were considered proprietary networks since they were only compatible with their own equipment. These proprietary networks are often called highways, data highways, or control networks. Many of these proprietary networks were installed and continue to function today in many plants and factories across the world.

With the advancements in digital communications technology and the increased number of PLC manufacturers, it became clear that the proprietary networks of the past were limiting the ability for modern control systems to work together. This also became apparent to plant engineers who were under pressure to find ways to integrate the various PLCs, intelligent I/O devices, information, and HMI systems. Today most manufacturers of PLC and control equipment support open communication networks. Open networks are based on international standards developed through industry associations. This has opened the door for control equipment manufactured by different vendors to communicate across a common network.

NETWORKING PRINCIPLES

A communication network exists when two or more devices are connected together by some type of media for the sole purpose of exchanging information.

When devices are connected to network media, they are called **nodes.** Nodes can be PLCs, personal computers, HMIs, intelligent I/O devices, routers, and bridges, to name just a few examples. Some of these devices may also be called by other names like "stations," "network devices," or just plain "devices." Nodes are typically divided into two classes: devices that produce and consume data (i.e., computers, PLCs, HMIs) and devices that only receive and forward the data (i.e., repeaters, switches, routers, bridges). See Figure 20–1.

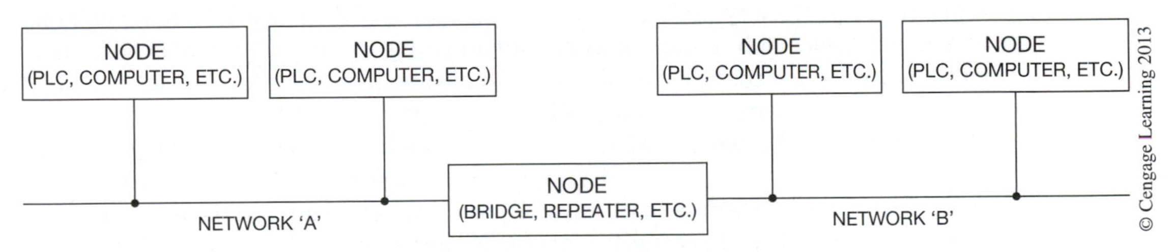

Figure 20–1 Nodes and Network Media

Modern data communications uses a digital method of sending and receiving binary information by way of low-level DC digital pulses. These digital pulses represent binary digits or bits that are sent as a bit stream called a **packet** over a common media format (i.e., wires, cables, optical fibers) to the multiple devices connected to the network media. A packet is a unit of data represented by a series of digital signals (being either *on* or *off*) that represents the message to be transmitted. Think of a packet as a letter that you would mail. The length of the bit stream or packet size is dependent on many factors such as data length, format, and protocol. These digital signals that make up the messages are used to communicate information between such devices as PLCs, computers, and intelligent I/O devices. The remote I/O bus network covered earlier is an example of a digital communication network that allows PLCs to communicate with remote I/O devices over a common media. Messages, packets, and protocols will be covered in more detail later in this chapter.

You will often hear the term **bandwidth** when working with communications networks. Bandwidth is the speed at which information can be transferred. Network bandwidth can be thought of as the ability of a network to pump data through the communication media. In other words, it's like the pipe that brings water to your house: the bigger the pipe, the greater the volume of water that can be delivered to your house in a given amount of time. Network bandwidth is usually expressed in *bits per second* transmitted, abbreviated bps. Most networks operate in the millions of bits per second range, so the abbreviation Mbps is typically used. Theoretically a 10 Mbps network can be transmitting data at the rate of 10 million bits per second, but this is not always the case. Some networks, such as Ethernet, may really only be operating at 40 to 50 percent of the rated bandwidth, depending on the mode of operation and hardware devices used.

NETWORK CATEGORIES

The geographic area that they encompass typically categorizes communication networks. Two of these categories, **Local Area Networks** (LANs) and **Radio Area Networks** (RANs), can be found in the control and information networks of many factories and cities. The following is a description of the four network categories:

- **Local Area Networks** (LANs) are typically high-speed, low-error data networks covering a relatively small geographic area (up to several thousand meters). A LAN connects such devices as PLCs, computers, HMIs, printers, servers, and other devices in a single building or geographically limited area (see Figure 20–2). All industrial control and most information networks fall in this category. Not all LANs are the same. LANs have standards or protocols that specify the cabling and signaling methods used. Some of these standards or protocols are proprietary and others are not.
- **Metropolitan Area Networks** (MANs) are data communications networks that generally span a metropolitan area. A MAN can be found, for example, connecting the departments of a city government if those departments are scattered throughout the city. A MAN network will typically use transmission devices provided by a local common carrier such as the telephone or cable TV provider. A MAN is larger than a LAN but smaller than a WAN. This category is used primarily for information networks like those found in large offices and buildings.
- **Wide Area Networks** (WANs) are data communications networks that encompass a broad geographic area, such as between cities or countries. A WAN will typically use transmission devices provided by common carriers such as telephone and cable TV providers. These transmission devices can include copper cable, optical fiber,

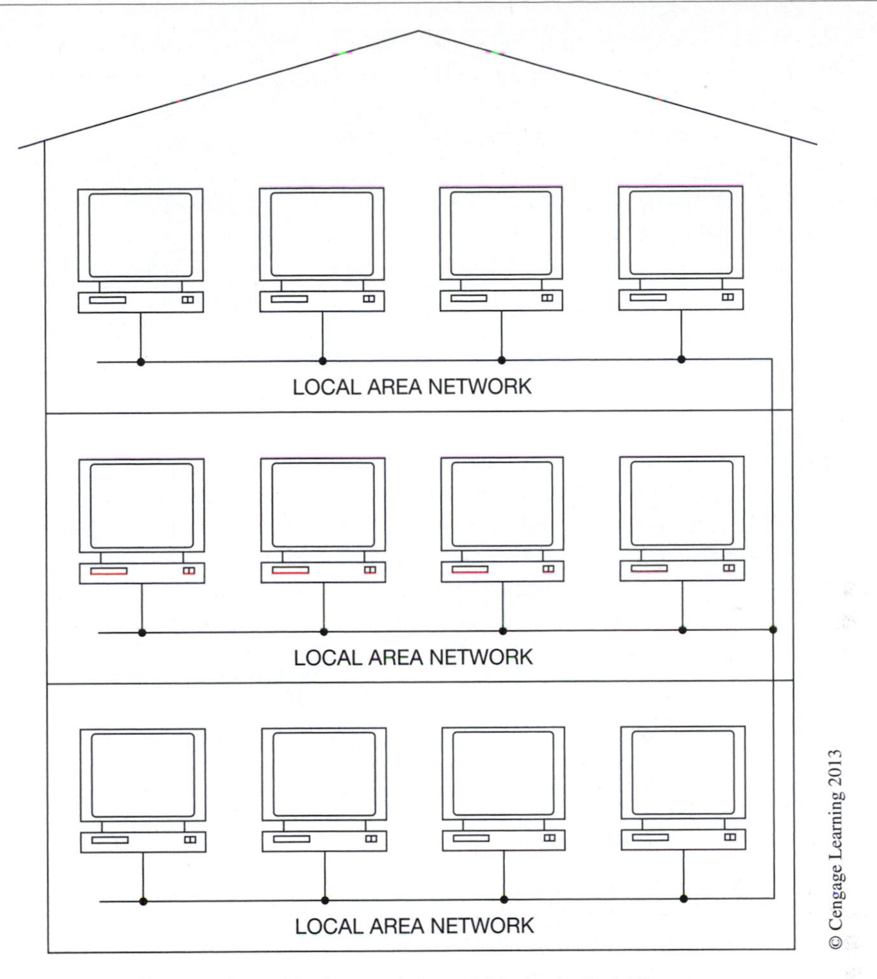

Figure 20–2 Local Area Network (LAN)

microwave, and satellite communications devices. A WAN is a network that spans the largest geographical area. When you connect to the Internet you are connecting to a WAN.

- **Radio Area Networks** (RANs) are data communications networks that use radio or microwave signals as the network media to communicate between devices. This category of network is typically found in supervisory control and data acquisition (SCADA) systems. You may find this type of network used by a city municipal water department, for example, to monitor and/or control the city's water reservoirs and pumping stations that may be located throughout the city (see Figure 20–3). This type of network can sometimes provide a cost savings over traditional methods like leasing phone lines from a local common carrier. In some cases in which no physical media are available or practical this may be the only option.

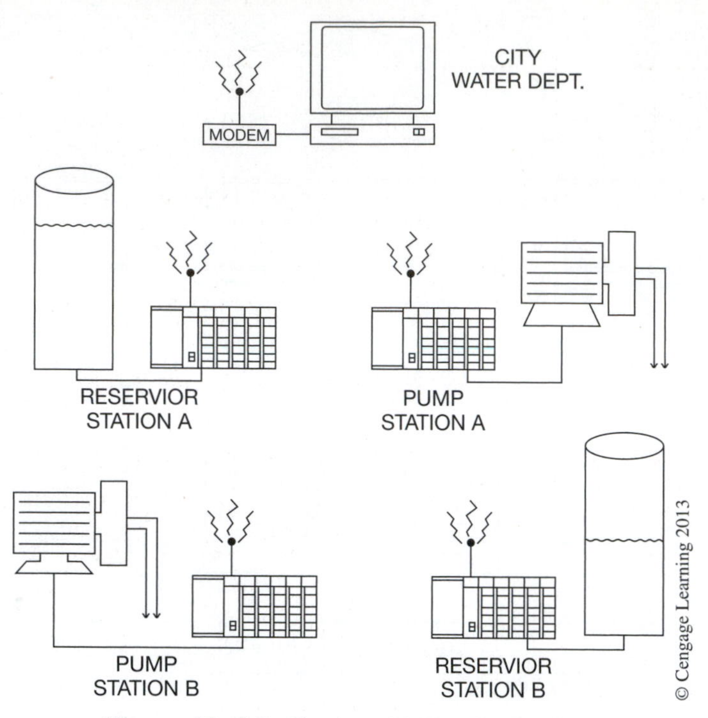

Figure 20–3 Radio Area Network (RAN)

NETWORK CONFIGURATIONS

Communication networks can be physically constructed in several different configurations. These configurations are called network **topologies.** A network topology can be in the form of a *bus, star,* or *ring* configuration. Each topology serves different network requirements, and in many cases more than one topology might be used to construct the network. The following is a description of the three network topologies.

- **Bus Topology**—A bus network is a topology in which nodes are connected to a common bus in a daisy chain or multidrop fashion, as shown in Figure 20–4. In the common bus topology, each device is capable of receiving all information packets on the bus, and network communication can occur between any two devices without having to pass the information through a central network controller (as is the case with a star topology). Common bus topologies are well suited to industrial control applications since each device has equal priority and can exchange information at any time. Most industrial I/O networks are of the bus topology. When adding or removing devices from this type of topology very little, if any, reconfiguration is required. The main disadvantage with this topology is that all devices share a common bus and a break in the bus can affect many devices.
- **Ring Topology**—A ring network is a topology in which nodes are connected in series to form a ring (see Figure 20–5). This type of network requires each node to participate in the distribution of the information along the network. Data packets are transmitted

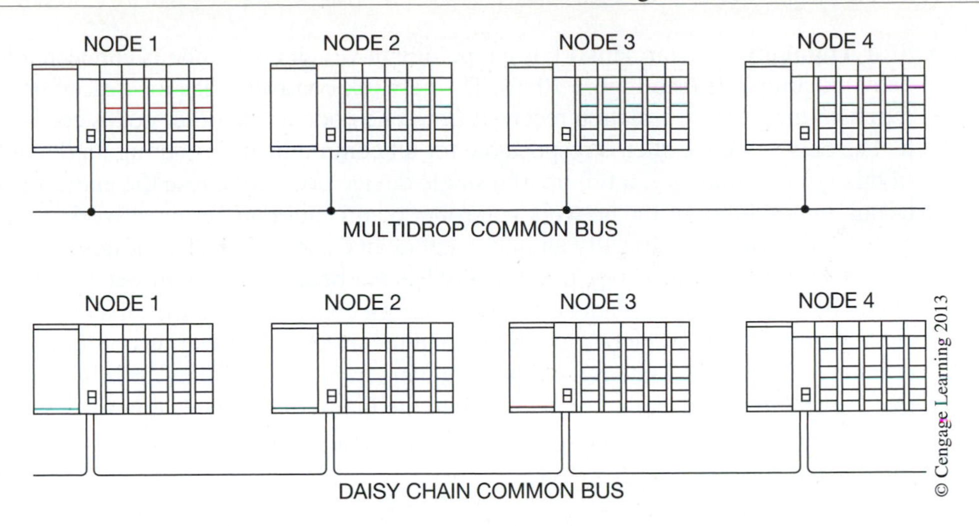

Figure 20–4 Bus Topology

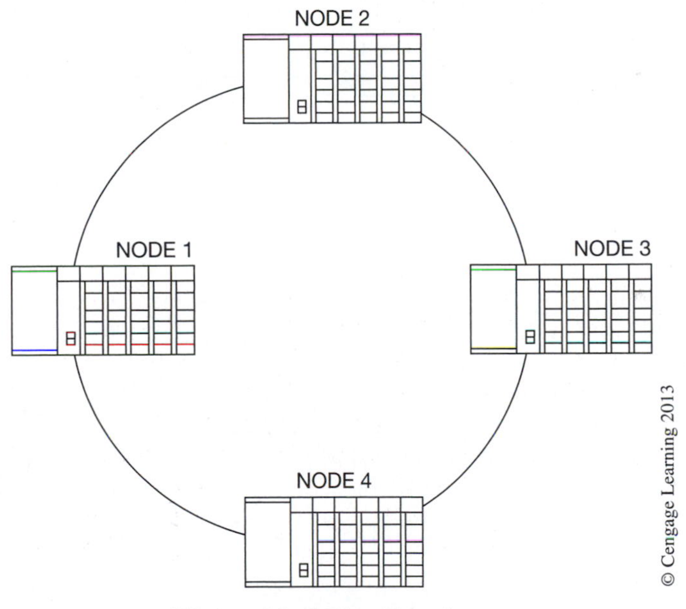

Figure 20–5 Ring Topology

from node to node along the network with each node checking the address attached to the packet. If the address attached to the packet matches the node address, then the node processes, or consumes, the information. If it does not match, then the node retransmits the packet to the next node in the network. The ring topology is not typically used in industrial control networks because of the dependency on each node to participate in the transfer of data.

- **Star Topology**—A star network is a topology in which each node is connected to a network controller (see Figure 20–6). The network controller has the task of passing or retransmitting the information received from one node to the other nodes connected to it. The network controller is responsible for all communication routing on the network. In this type of topology, a failure of a single device does not cause the entire network to fail, but a failure of the network controller could cause an entire network to fail. The network controller is typically an intelligent device such as a network repeater (hub) or switch. This type of topology traditionally has not been used in industrial control networks because of the dependence on the network controller, the installation cost (more cable required than in the bus topology), and the unreliable data transfer time. The star topology is often used in networking computers in business offices. It is worth noting that the star topology is finding its way into the industrial controls environment due to advances in technology and requirements to integrate information networks with control system networks.

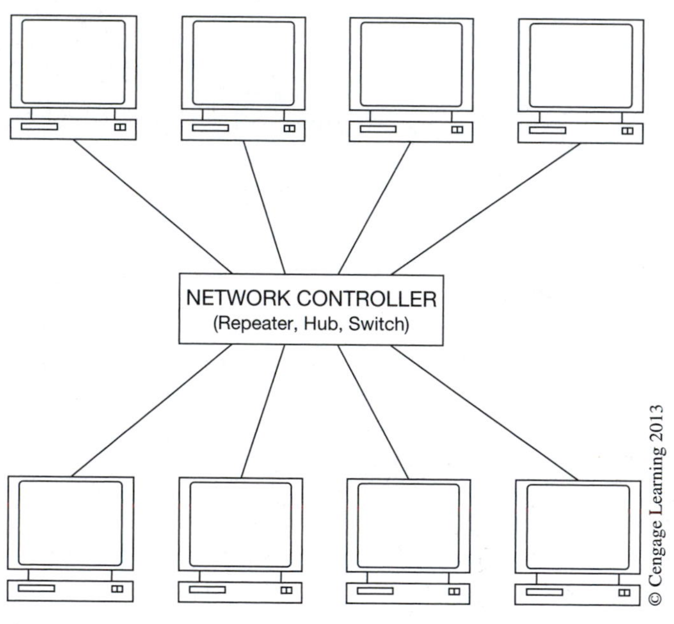

Figure 20–6 Star Topology

NETWORK MEDIA

The *network media* are the wire, cable, and optical fibers used to physically connect the various communication devices together within a network topology. The most commonly used media include *twisted-pair, coaxial,* and *optical fiber* cables. To implement a communications network you must have a good understanding of the media and hardware used to construct a network.

Twisted-Pair Cable

Twisted-pair is the basic networking cable used today in most offices, buildings, and industrial plants. This type of cable is relatively inexpensive and has a fair degree of noise immunity. The name comes from the way that each pair of wires are twisted around one another, as shown in Figure 20–7.

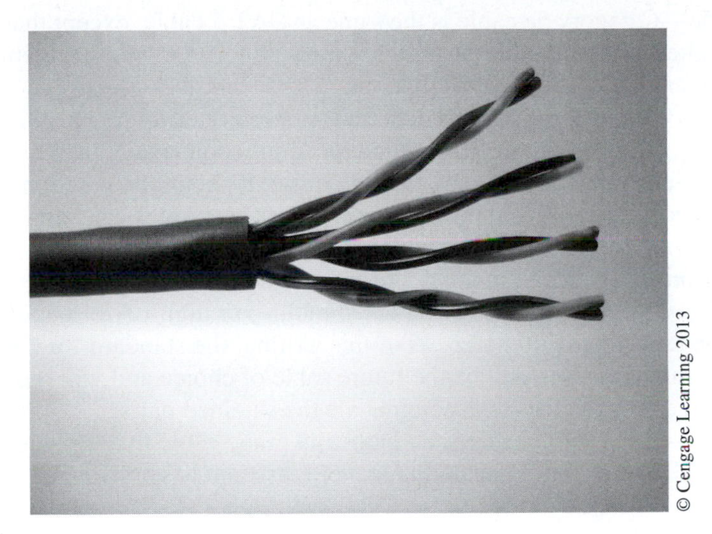

Figure 20–7 Twisted-pair Cable

Twisted-pair cables help reduce crosstalk and noise susceptibility by reducing magnetic coupling between pairs. Some networks using twisted-pair wiring, such as Ethernet, use balanced transmission. In balanced transmission, two wires are used to carry the electrical signal, and each carries a signal (and noise) that is of equal potential with the other conductor, but has opposite polarity. The receiving device only measures the difference between the two conductors. This has the effect of canceling the noise.

Twisted-pair cable can be unshielded or shielded. Unshielded Twisted-Pair cable (UTP) is the most common cable used in offices, buildings, and areas that have low levels of electrical noise. Shielded Twisted-Pair cable (STP), on the other hand, greatly improves the electrical noise immunity of the cable and is normally the type used in industrial network applications. It must be noted that when using STP cable, maximum shielding can be achieved only if the shield is properly terminated. Remember that the purpose of the shield is to pick up the noise and conduct it to ground. The cable, connectors, and equipment must provide a continuous path to ground for the noise, or shielded cable is no better than unshielded cable and in most cases worse.

When twisted-pair cables are used in industrial control networks, the manufacturers of the equipment will typically specify the type and quality of the cable to be used for optimum performance. A shielded twisted-pair cable with an outer jacket that can withstand high temperatures, dirt, oil, solvents, and abrasions is typically specified. Recommendations given by the manufacturers should be followed to minimize network problems.

The twisted-pair cables used in most offices, buildings, and factories for Ethernet communication networks are rated by a category (CAT) numbering system representing the quality and data rate of the cable. The following are the most common of the CAT cable ratings being used today.

- CAT 5—Category 5 cable is a multi-pair (usually 4) cable used primarily for data transmission at rates up to 100 Mbps.
- CAT 5e—Category 5e cable is the same as CAT 5 cable, except that it is manufactured to a higher standard with transmission rates up to 1000 Mbps (gigabit). CAT 5e cable is recommended for all new installations.
- CAT 6—Category 6 cable has data transmission rates up to 400 MHz (10 gigabit). Category 6 cable is the fastest communications cable available that is unshielded. This is a high-end cable and the emerging favorite. It should be noted that Category 6 cable is not recommended for industrial applications because it can carry higher frequencies better than Category 5e cable, which makes it more susceptible to the higher-frequency noise from arc welders, motors, VFDs, and other noise sources.
- CAT 7—Category 7 cable is a shielded multi-pair high-performance cable with transmission rates up to 700 MHz. As of this writing, the standard for CAT 7 cable had not been completed. This will be the future cable of choice and will require new connectors and connecting devices to accommodate the shielded pairs.
- CAT 5i—Category 5i cable is an industrial grade cable that has a higher temperature range and an outer jacket that is more resistant to abrasion, oils, chemicals, and solvents.

All communication cables must be connected at each end. In some cases the means of connection is as simple as stripping the cable back and connecting the wires on the proper terminals, and at other times a connector is used. Most twisted-pair cables used for Ethernet networks use an RJ-45 type connector, as shown in Figure 20–8.

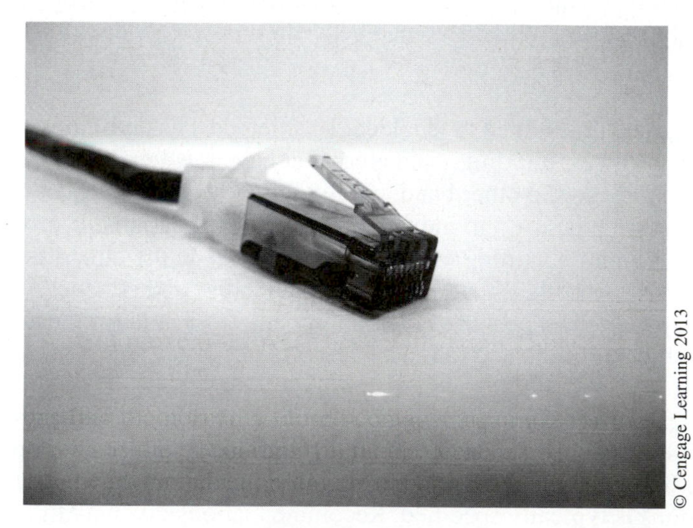

Figure 20–8 RJ-45 Connector

There are industrial RJ-45 type connectors available that are equipped with threaded boots that provide protection from moisture, liquids, and vibration. The equipment that the cable is to be connected to determines the type of connection. Cable terminations are a common source of network problems, due to such factors as improper terminations, miswiring, and mechanical or environmental damage. Care should be taken when terminating communication cables to minimize network problems.

Coaxial Cable

Coaxial cable was, and still is, widely used in industrial control networks because of its excellent immunity to electrical noise generated by factory equipment; it offers the best performance of all copper cables. On the other hand, coaxial cable is hard to work with and costs more than twisted-pair cable. The basic structure of a coaxial cable is a center conductor surrounded by a dielectric substance, shield, and outer jacket (see Figure 20–9).

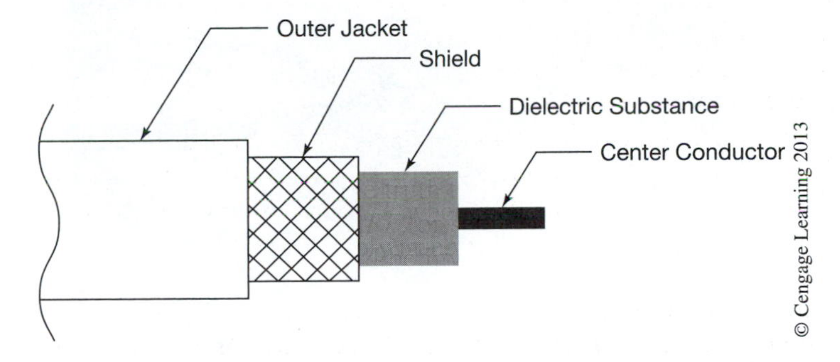

Figure 20–9 Basic Structure of a Coaxial Cable

Coaxial cables have characteristic impedances (50 ohms, 75 ohms, etc.), and must be terminated at their ends for proper operation. For example, if coaxial cable is used in a bus topology in which the ends are not connected to a device that terminates the cable, a special connector called a terminator must be placed on the ends, as shown in Figure 20–10.

Figure 20–10 Coaxial Cable Terminator

The terminator absorbs all energy reaching it and this prevents reflections that can cause network problems. Terminators must have the proper resistance value (50 ohms for 50-ohm coaxial cable, etc.) and be installed for proper network operation. The connector type typically used with coaxial cable is the BNC connector, as shown in Figures 20–11a and 20–11b.

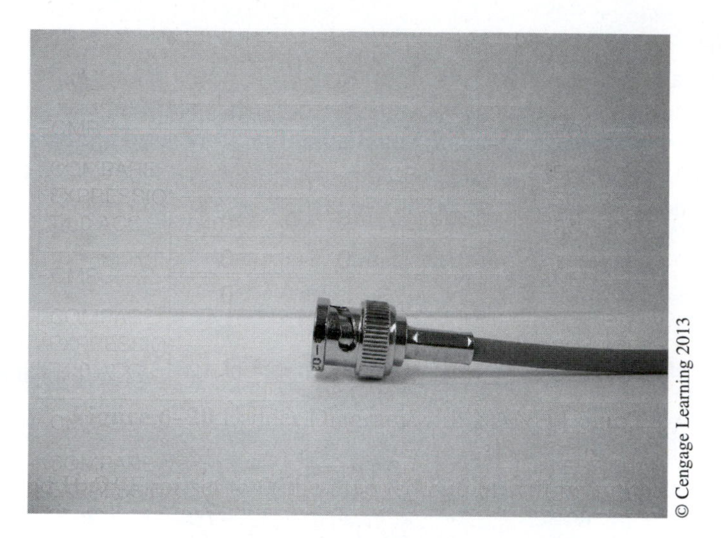

Figure 20–11a BNC Connector

Figure 20–11b T-tap and BNC Connectors

Fiber Optic Cable

Fiber optic cables use thin glass strands called optical fibers to carry data signals by means of light. The process is simple. Electrical signals are converted to light signals and transmitted down a glass fiber. At the other end, the light signals are converted back to electrical signals. A single optical fiber strand consists of two parts, the inner core made of glass and the outer cover or cladding. The cladding helps to confine the light inside the core by acting as a mirror to reflect the light down the core (see Figure 20–12).

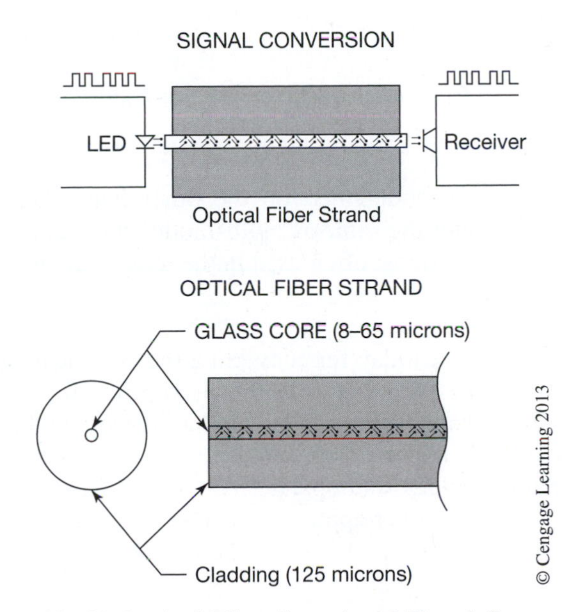

Figure 20–12 Optical Fiber Strand and Signal Conversion

The optical fibers themselves are enclosed within a protective structure. This structure typically consists of a fiber coating, buffer, strength member, and outer jacket, as shown in Figure 20–13. Most fiber cables consist of multiple optical fibers. Of the three network media types, fiber optic is considered to be the best because of its high data rates over long distances, its immunity to electrical noise, and the fact that it carries no current, so that it is not a spark or fire hazard.

Optical fibers come in two types: *multi-mode* and *single-mode*. Multi-mode fiber has a core diameter of around 50–62.5 microns (1 micron = 1 millionth of a meter) with a cladding diameter of about 125 microns, and can have transmission distances up to 12 Km without the use of repeaters. Multi-mode fiber has traditionally been the optical fiber of choice for most industrial communications networks because of its compatibility, ease of coupling, cost, and sufficient bandwidth. Not all multi-mode fibers are the same, however; there are several grades of fiber that determine such things as bandwidth and *attenuation*. Attenuation is a decrease in optical power from one point to another and is measured in decibels (dB). You could compare attenuation in optical fibers to the

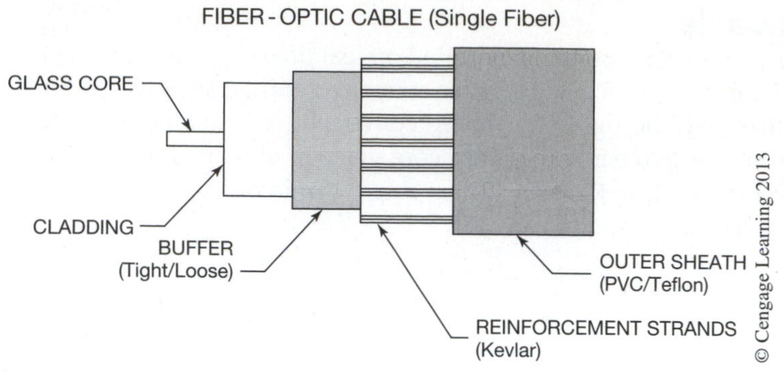

Figure 20–13 Optical Fiber Cable

I^2R loss in copper wires. Single-mode fiber, on the other hand, has a core diameter around 8 microns with a cladding diameter the same as multi-mode, and can have transmission distances up to 100 Km. Single-mode fiber is most often used in the telecommunication and cable television industries.

There are two optical devices in use today for converting the electrical signals into light. They are the LED and the laser. The LED is inexpensive and is used primarily with multi-mode fibers. The LED emits a wide band of light that lowers the bandwidth of the signal that the fiber can carry, making it impractical for single-mode fibers. Lasers, on the other hand, have a very narrow and intense beam, are high-speed capable devices, and operate with a very narrow spectrum of light, making them ideally suited for single-mode fiber applications. The wavelength of the light employed with most optical fiber systems is in the infrared region. The human eye cannot see light in the infrared region of the light spectrum.

Caution: Never look directly into the end of an optical fiber that is connected to a power source or permanent eye damage could result.

Fiber optic connectors provide an easy means to connect the optical fiber to the equipment with as little loss of power as possible. With optical fiber connectors, the challenge becomes one of alignment of the light-carrying optical cores to the optical cores on the equipment connectors. Any misalignment of the fiber cores means a loss of optical power at the connection. Remember that the optical core diameter of a multi-mode fiber is only about 50 microns—about the width of a human hair, if not smaller. Fiber connectors use what are called **ferrules** to hold the fiber cores. A ferrule has a precision hole that the fiber core is inserted into that provides for accurate positioning of the fiber. Ferrules are typically made of ceramic, plastic, or stainless steel. Epoxy has traditionally been used to secure the fiber in the ferrule, but recent advancements in manufacturing have developed alternate methods that are fast, efficient, and do not require the equipment and skill level of the traditional epoxy methods. Fiber optic connectors come in different types. Two of the most common used in industrial applications are the ST and SC connectors, as shown in Figures 20–14a and 20–14b. Some newer fiber optic connectors to hit the market are the LC and MT-RJ connectors, which are about half the size of the ST and SC connectors.

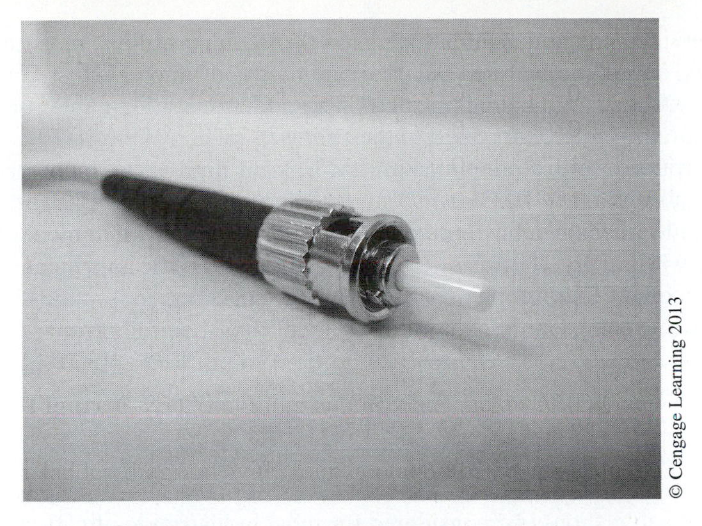

Figure 20–14a “ST” Fiber Connector

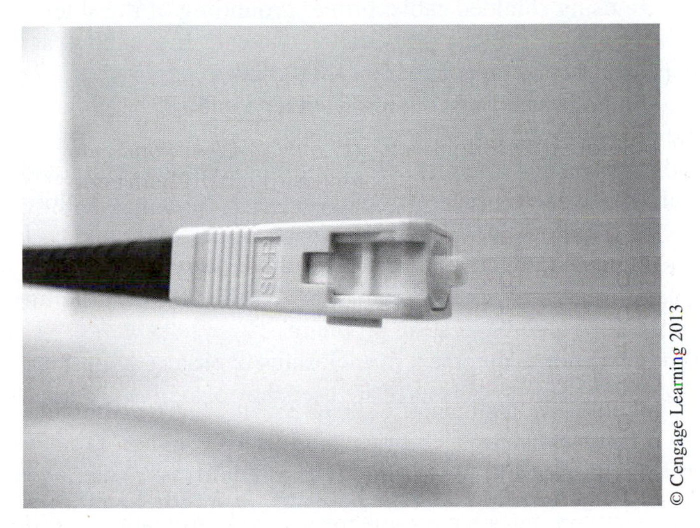

Figure 20–14b “SC” Connector

Reducing Electromagnetic Interference

Industrial environments are notorious for producing electromagnetic interference (EMI or electrical noise) in communications and other low-voltage cables. In order to minimize EMI problems, proper selection and installation of communications cables cannot be overemphasized. The following is a checklist that you can use to help minimize EMI problems.

- ☐ Survey the environment the cable will be located in. What kinds of sources of EMI may be present, such as power cables, high-voltage sources, motors, transformers, welders, variable-frequency drives, generators, etc.? This would be a good time to also consider other environmental concerns such as temperature extremes, moisture, oils, chemicals, solvents, abrasion, vibration, shock, etc.

- ☐ To the extent that would be practical, plan communication cable runs to avoid potential EMI sources that are identified in the survey.
- ☐ Do not route copper communication cables near high-voltage power sources and/or power cables.
- ☐ To provide maximum EMI and physical protection to communications cables that are located within industrial plants, consider running the cable in steel conduit. This provides excellent EMI protection. Never run communications cables in the same conduit or raceway with power wires or cables.

Note: *Some manufacturers recommend against running unshielded cable in metal conduits, as it may affect the electrical performance of the cable. Always check with the manufacturer.*

- ☐ Shielded cable should be considered for most industrial applications. For maximum noise immunity or in areas of high EMI sources, fiber optic cable should be considered.
- ☐ When you are using shielded cable, proper grounding of the shield is important. Shielded cable should only be grounded at one end. If grounded at both ends, noise-inducing ground loops can occur. You should always follow the equipment manufacturer's recommendations on grounding shielded cables.
- ☐ Make sure all cables are properly terminated.
- ☐ Other items to consider when installing communications cables and equipment:
  - Avoid sharp bends in communication cables. A good rule to follow is that the radius of the bend should be four (4) times the diameter of the cable and no less than 1 inch.
  - Connectors should be properly selected for the environment and network type.
  - Good cable termination practices cannot be overemphasized. Failure in this regard is the source of most communication problems.
  - Install communications equipment in enclosures to protect against moisture, dirt, and other contaminants. When steel enclosures are used they also offer a degree of EMI protection.

NETWORK ADDRESSING

Devices connected to a network must have some means of identification in order for messages to be received only by the intended device or devices. A network address, also called a **station** or **node address,** is like your home address. In order for the mail carrier to deliver a letter intended for you, he or she must identify your house by its address. In communication networks the same basic principle applies. Each node connected to the network is assigned a unique address. When a data packet (message) is transmitted across network media, each node connected to the network checks the destination address attached to the packet. If the address matches the node's address, then the packet is accepted; if not, the packet is discarded. The format and assignment of an address for a node on a network depends on the type of network and the protocols in effect for that network.

On most networks the address assigned to a device is a number such as 1, 10, 77, 115, etc. On some networks, it can be a series of numbers such as 256.256.168.23 separated by a character(s). The number(s) can be set, or assigned, to a device by setting switches physically located on the device or through software. On many industrial control networks the address is assigned to a device by switches, such as dual in-line package (DIP) switches or rotary switches, as shown in Figures 20–15a and 20–15b.

Figure 20–15a Dual In-line Package (DIP) Switches

Figure 20–15b Rotary Switches

On some networks, the assignment of node addresses is done automatically by a network controller across the network media by first scanning the network and then assigning each node a unique address. On other devices, the network address is programmed into the device using software and stored in memory on the device.

NETWORK ACCESS METHODS (ACCESS PROTOCOLS)

PLCs, computers, and other devices must employ a common network access-control method. An *access-control* method defines how and when devices can access and communicate information across the network. There are many methods used but the most common ones are *polling* or *master/slave, token passing,* and *carrier sensing multiple access with collision detection* (CSMA/CD). The following is a description of how each of these network access-control methods works.

- **Polling or Master/Slave Access Method**—In a polling or master/slave access-control method, a single PLC or computer is designated as the master and all other devices or nodes connected to the network are designated as slaves. In polling, the master is programmed to interrogate, or poll, each slave device in sequence to see if it has data to transmit. The master will send an inquiry to a slave and then wait a predetermined time for the slave to respond. If the slave does not respond within the allotted time, the master will assume that the slave is dead, or inactive, and continue to poll the next slave in sequence. When the master has polled the last device in the sequence, the master then repeats the process. In a master/slave access-control method, the slaves can only respond to the master, not to each other. If a slave wishes to send data to another slave device, the master must act as a mediator, receiving the data from one slave and then sending the data on to the second slave. Many of the bus networks used in industrial communications use this method of access-control or provide the capability.
- **Token Passing Access Method**—In a token passing access method all devices or nodes connected to the network have equal access to the network. There is no master; each is a peer to the other. Another common name for this type of network access method is *peer-to-peer*. In a token passing method, each node on the network is allowed to send data packets directly to other nodes and each node has a scheduled turn and allotted time to send data. In order for a node on the network to send data, it must first possess what is called the token, and only one node can possess the token at any time. The token is a special packet that, when received by a node, gives that node the exclusive but temporary right to transmit an allotted amount of data. If a node has more than the allotted data to transmit, it will send additional data packets each time it receives the token until it has sent the entire message. When a node that possesses the token has completed transmitting a data packet or if it has no message to transmit, it immediately passes the token to the next node in ascending order of the node addresses. The node with the highest address passes the token to the node with the lowest address, so the token continues to circulate in a loop throughout the network.

  Each node has a defined token holding time. If for any reason a node does not pass the token within the allotted time given, the originating node then begins polling node addresses in ascending order until it finds a node that will accept the token. It must be noted that in a token passing network method there is a maximum number of node addresses available, limiting the number of devices that can be connected to the network.

  The size of the data packets is defined as well as the token hold time, so it is possible to calculate the network access time. For this reason the token passing access method is the most common network access method used in industrial control networks for passing time-critical control information between PLCs and other controllers. A second advantage is that a failure of any one node does not cause a failure of communications between the other nodes on the network.

- **CSMA/CD Access Method**—The previous two access methods (master/slave and token passing) stipulate that a node on the network can only transmit when it has permission, either from the master or by possessing the token. In the CSMA/CD access method each node has equal right to attempt to transmit, without waiting for permission. Whenever a device is ready to transmit data, it checks the network for the presence of traffic on the network. If the network is clear, the device then transmits its data. If the network is busy, the device waits until the network is clear. A "collision" occurs when two or more devices attempt to transmit at the same time. When a collision occurs, a collision detection signal is sent to all devices on the network. Each of the colliding devices must then back off and wait a brief but random time before beginning the process again.

  A collision is not an event to be avoided, but simply a method used to arbitrate access to the network. The resolution of a collision occurs very quickly. The two devices transmitting almost immediately abort their transmission and wait a random amount of time before reattempting the transmission. The number of attempts with collision and a random number determine the *back off time*. The back off time is typically in the microsecond range but can be in the millisecond range if there are significant collisions occurring.

  Because it is impossible to predict the amount of time required for all colliding devices to successfully complete their transmission, the CSMA/CD access method may not be the best choice for time-critical control networks. It is worth noting that data transmission updates are typically processed in a fast (millisecond) time frame. Advances in network bandwidth and hardware (active switches) have made this access method more widely accepted in the industrial controls industry.

NETWORK PROTOCOLS

Network communication protocols are sets of formal rules describing how to transmit and share data across a network. Without these formal rules, devices on a network would not be able to understand each other (we must speak the same language). Low-level protocols define the electrical and physical standards to be observed, bit and byte ordering, transmission execution, error detection, and correction of the bit stream. High-level protocols deal with the data formatting, including the syntax of messages, the computer-to-computer dialogue, character sets, sequencing of messages, etc. The Open Systems Interconnect (OSI) reference model defines many of the protocols used. The OSI is a reference model developed by the ISO (International Organization for Standardization) in 1978 as a framework for international standards in network architecture. The OSI model is split into seven layers, from lowest to highest, as shown in Figure 20–16.

Each layer uses the layer immediately below it and provides a service to the layer above. In some implementations a layer may itself be composed of sublayers. It must be noted that a network requires only layers 1, 2, and 7 of the OSI model to operate. Individually, each layer of the OSI model is responsible for a specific task.

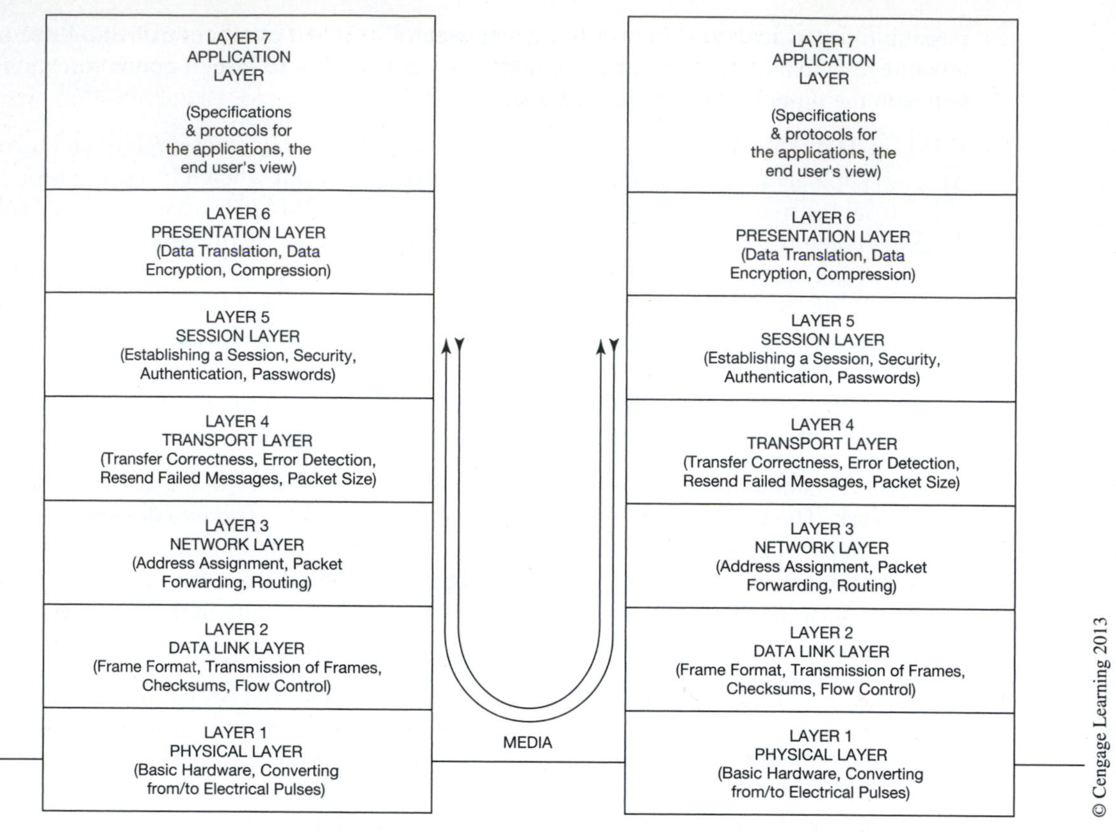

Figure 20–16 Two OSI Models and Common Media

Most communication networks today contain all or most of the OSI layers to allow other networks and devices to share information. When a device is sending data over the network, the data starts at the application layer and works its way down to the physical layer, where it's placed onto the network media. When a device receives a message, it is received by the physical layer and works its way up to the application layer. The final result of this effort is to ensure that the data sent from one device is the same exact data received by another device on the network (refer back to Figure 20–16).

NETWORK MESSAGES

When it comes to network communications, the terms *message, packet,* and *frame* can be confusing for even the most experienced technician. It seems that many of these terms are used interchangeably. We will make an attempt to define and describe some of these terms:

- **Message**—*Message* is a common term used to describe information transmitted via a network. A message is the complete information to be conveyed regardless of size or protocols.

- **Packet or Data Packet**—*Packet* is a generic term used to describe a unit of data at any layer of the OSI model, but it is most correctly used to describe the application layer data units. An application layer data unit is a packet of data exchanged between two application programs across a network. This is the highest-level view of communication in the OSI seven-layer model. A single packet exchanged at this level may actually be transmitted as several smaller packets at a lower layer, as well as having extra information (headers) added for routing, etc.
- **Frame**—A *frame* is a data link layer packet that contains the header, data, and trailer information required by the physical media. That is, network layer packets are encapsulated to become frames. A typical frame consists of a header section, a data section or payload, and a trailer section. A frame refers to the structural container of a packet. In most cases, a frame as outlined above is the equivalent of a data packet (see Figure 20–17).

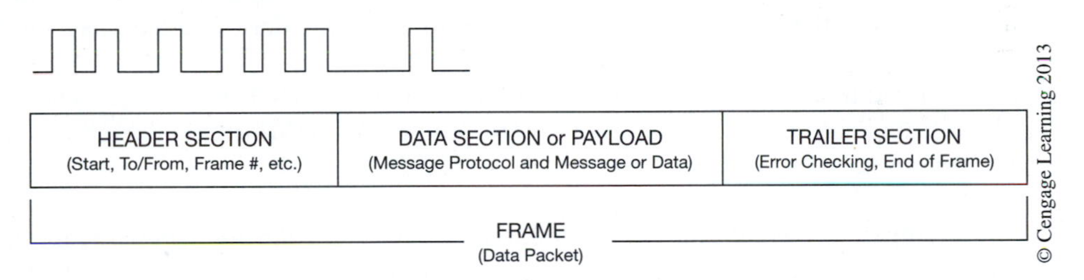

Figure 20–17 Basic Structure of a Data Frame

The following is a brief outline of the three sections of a data packet frame.

- **Header**—The header section contains instructions about the data carried by the data packet. These instructions may include: *length of data* (some networks have fixed-length data packets, while others rely on the header to contain this information); *synchronization* (a few bits that help the packet match up to the network); *packet number* (which packet this is in a sequence of packets); *protocol* (on networks that carry multiple types of information, the protocol defines what type of packet is being transmitted); *destination address* (where the packet is going); and *originating address* (where the packet came from).
- **Data**—The data section, also called the *payload,* is the actual data that the packet is delivering to the destination. If a packet is fixed-length, then the payload may be padded with blank information to make it the right size.
- **Trailer**—The trailer section, sometimes called the *footer,* typically contains a couple of bits that tell the receiving device that it has reached the end of the packet. It may also

have some type of error checking. The most common error checking used in packets is **Cyclic Redundancy Check** (CRC). CRC is pretty simple, but effective. Here is how it works in certain networks: It takes the sum of all the 1s in the payload and adds them together. The result is stored as a hexadecimal value in the trailer. The receiving device adds up the 1s in the payload and compares the result to the value stored in the trailer. If the values match, the packet is good. But if the values do not match, the receiving device discards the packet and sends a request to the originating device to resend the packet.

A message transmitted over a network may be composed of many data packets arranged in a numbered sequence. As mentioned above, most network protocols have fixed-length data packets. If a message is too large for one data packet, the message is broken up into several data packets, each given a sequence number before being sent on to the network media. When the data packets arrive at the destination node, they are reassembled according to their sequence number.

The subject of protocols, packets, frames, and messages may seem overwhelming to the reader at this point, so let's use the following example to illustrate the transfer of data across a network.

Let us assume that we have the message "Feed Boiler Number 2 High Temperature Warning" that needs to be sent from the PLC controller operating the boiler to the HMI (human machine interface) computer located in the main control room.

When the high temperature condition is detected, the boiler PLC controller prepares the message for transfer across the network by breaking the message into data packets (payloads) of the proper length. Each data packet is given a header section containing such things as a packet sequence number, destination address (node address of the HMI), source address (its node address), etc. Each data packet is also given a trailer section containing the error-checking value and end of data packet flag. Once the data packets are assembled, they are called frames or data packet frames. Each data packet frame is transmitted onto the network media as a bit stream message according to the network protocol, as seen in Figure 20–18.

The HMI computer checks the destination address on each packet being sent across the network. If the destination address matches, then the complete data packet frame is received. Once received, it is checked for errors. If an error is detected, the frame is discarded and a message is sent to the sending node to resend the data packet frame. If no errors are detected, the data packet frame is held until all the data packet frames have been received for that message. After all data packet frames have been received, the header and trailer sections are stripped from the frame and each payload data section is reassembled according to the packet sequence number. The message "Feed Boiler Number 2 High Temperature Warning" would then appear on the HMI computer screen alerting the operators to the problem.

The process just described is how most information is shared across network media. It must be made clear that the protocol for the type of network you are using will determine how the messages, packets, frames, and electrical or optical signals are constructed and used.

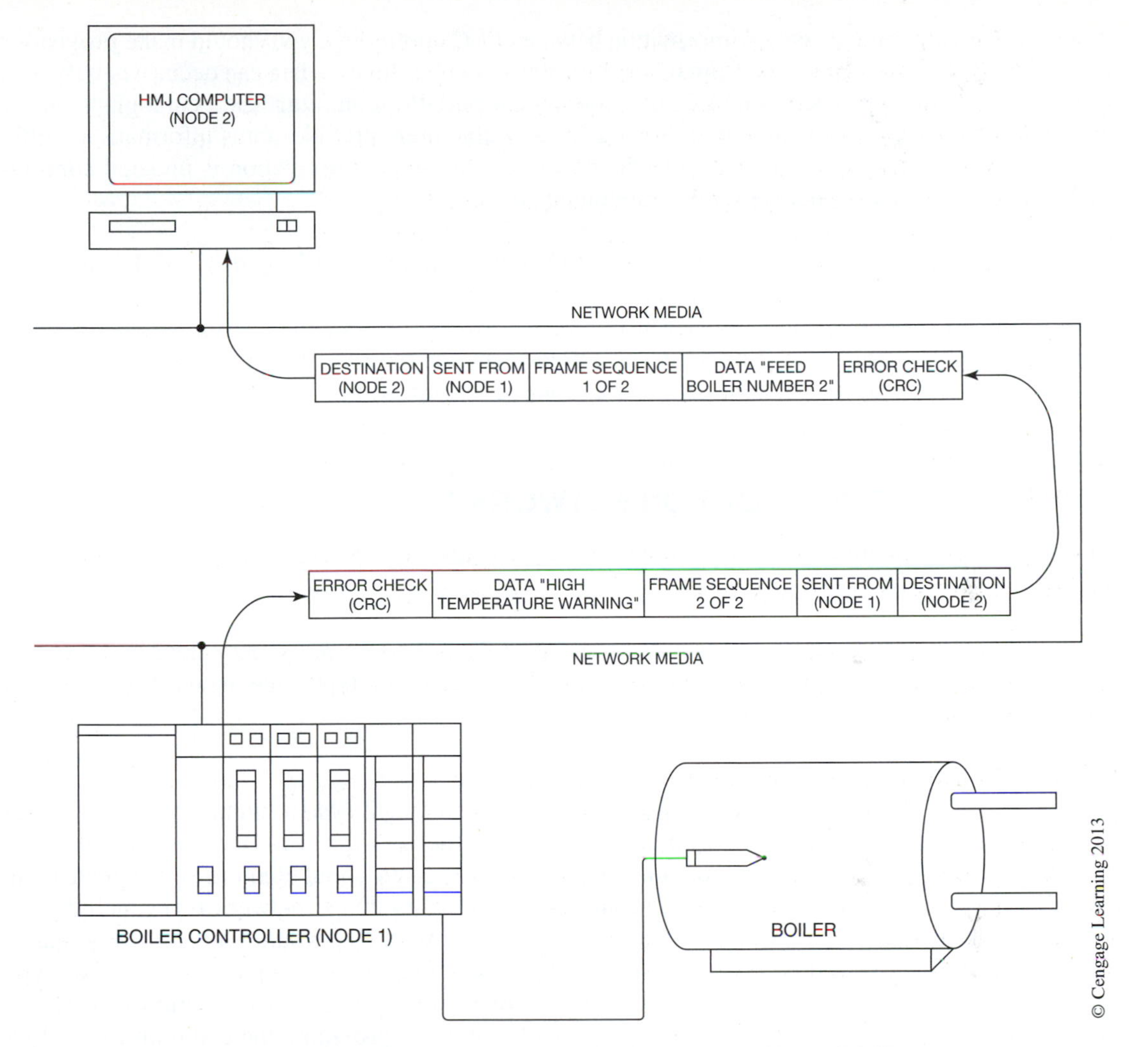

Figure 20–18 Data Packet Frames from Controller to HMI

NETWORK COMMUNICATION INSTRUCTIONS

Many PLC manufacturers provide network communication instructions that allow the transfer of information like the status of inputs, outputs, and registers between PLCs. These programmed instructions typically can be used to read and/or write multiple data registers between PLCs of the same manufacturer. Some PLC controllers provide for the transfer of information using a *producer/consumer* approach. In the producer/consumer approach, data memory locations are assigned as global producers or consumers during setup. This allows data to be shared between PLCs over a peer-to-peer network without programming special instructions into the controllers.

When transferring critical control information between PLC controllers, you should make provisions to ensure that the data being used from other controllers is valid. Invalid data can occur when there is a loss of communications between PLCs that goes undetected. Programming heartbeat logic in one or both PLC controllers may be required. Some PLC manufacturers provide status information within the PLC on all active nodes connected to the network. This status information is updated continuously and can be used to check for valid communications.

Most of the transfer of information between PLC controllers and non-PLC devices (HMIs, computers, information systems, etc.) does not require the programming of PLC instructions to carry out the transfer. The HMI and computer information systems are typically configured during setup to poll information from the various PLCs at a predetermined interval. The transfer of information is typically handled by an onboard communications processor within the PLC when a request for data is received.

INDUSTRIAL COMMUNICATION NETWORKS

Industrial communication networks can be divided into three types: *I/O and device, control,* and *information,* as shown in Figure 20–19.

Depending on the needs and requirements of the control systems on the factory floor and those of operators, supervisors, and managers, some or all of these network types may be used. Let's take a closer look at the three types of industrial networks.

I/O and Device Networks

I/O and Device Networks work at the lowest level of the control system architecture. These are networks that provide the communications link between the real-world devices and the machine or process controllers (PLCs). In previous chapters you learned about Remote I/O, which is an example of a communications network that falls in this category. Other network types include Device Bus and Process Bus networks. Some of the characteristics that would typically be found in this type of network would be relatively small data packet sizes, master/slave protocols, bus type topologies, high degree of reliability, fault tolerance, real-time operation, and deterministic operation. Remote I/O has long dominated this category of networks, providing the communication link between remote I/O devices and the PLC. Most remote I/O networks are proprietary, meaning that the hardware and communications protocols used have been developed by the manufacturer to support only their PLC systems.

With advancements in digital communications technology, distributive control, and increased desire for more information, two additional network types have been developed in recent years to join Remote I/O. In this category of industrial networks, they are *Device* and *Process* bus networks. These two network types are not unlike remote I/O in that they communicate I/O information to PLCs, and process control systems and other devices on the network. What sets these networks apart from remote I/O is the following:

- They allow you to connect devices such as photo sensors, proximity sensors, valve manifolds, smart motor controllers, pressure sensors, flow sensors, etc., directly to controllers and other control systems without hardwiring each device into an I/O module.

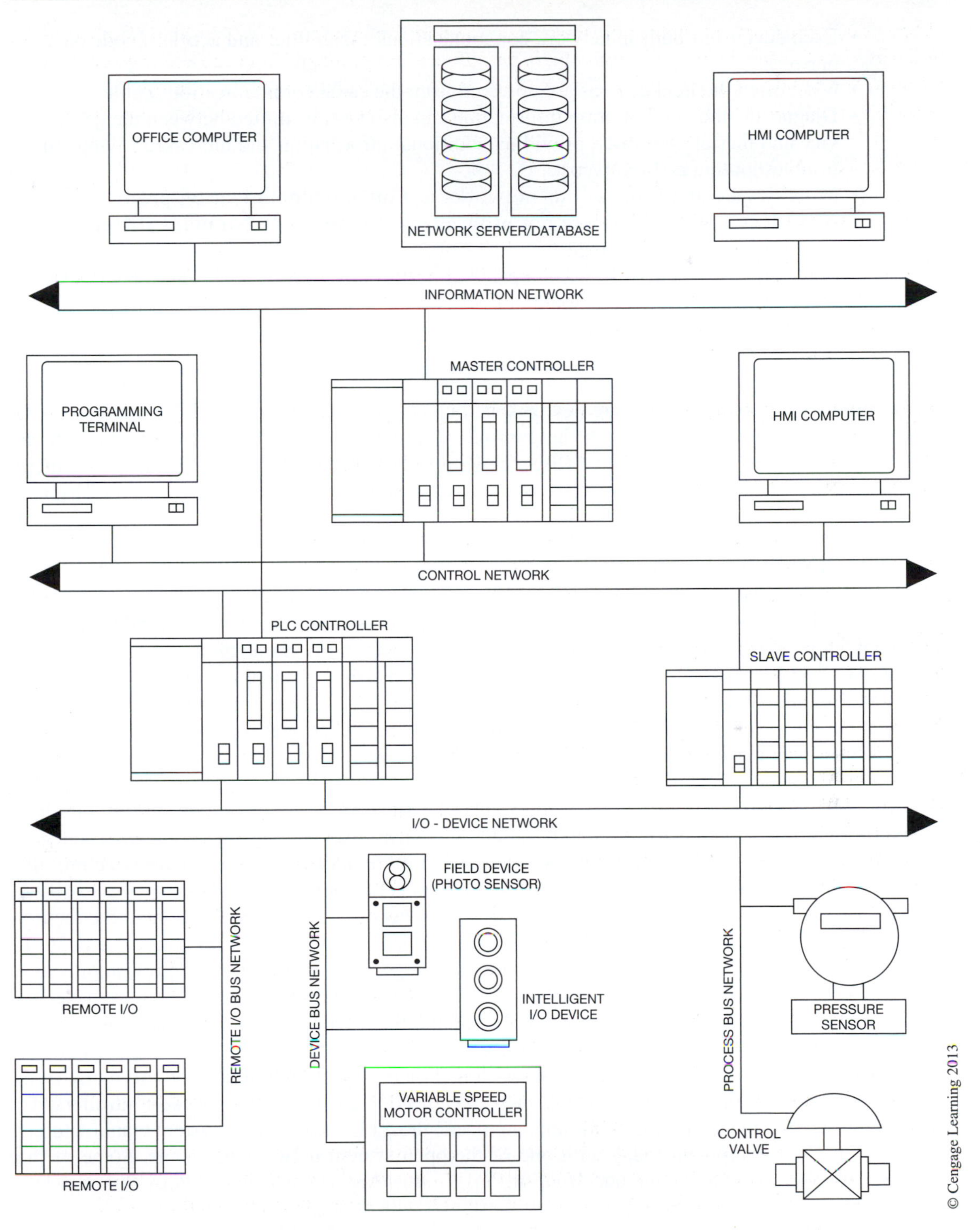

Figure 20–19 Industrial Communication Networks

- Each device has built-in network communications capabilities and acts as a node on a network.
- Most often the field devices are powered from the same communications cable.
- Diagnostic and configuration information can also be transmitted between the device and master controller, providing additional information and increased savings in troubleshooting and setup cost.
- Most device and process bus networks use open communication protocols, meaning that they are not proprietary, so devices from many different manufacturers can be used.
- Some process bus networks are capable of performing control functions at the device level using predefined function blocks.
- Scanner ports and modules are available from most PLC and controller manufacturers that support the open device and process bus protocols.

Device and process bus networks are device-specific networks and are intended for small amounts of data to be communicated between the controller and the devices. These networks provide an effective way to gain access to the intelligence within those types of devices. Some of the more common device and process bus protocols used today include *DeviceNet, Foundation Fieldbus,* and *Profibus-DP.*

Control Networks

Control Networks work at the middle level of the control system architecture. These are networks that provide the communications link between multiple process controllers (PLCs), HMIs, and other intelligent devices that are critical for machine or process operations. Some of the same characteristics that we found with I/O and device bus networks would also apply with control networks. High degrees of reliability, fault tolerance, high speed, bus topologies, and determinism would be equally important at this level of network communications. The data packet sizes would typically be larger, and a peer-to-peer (token passing) protocol is most often used.

Control networks provide for the communications of critical machine or process information between the PLC controllers in a system, programming devices, master controllers, and HMI stations. As control and plant engineers have worked towards a more *distributive* control architecture, the need for reliable communications between these various controllers and systems has also evolved. A distributive control architecture is one in which controllers are distributed throughout the factory floor, controlling single machines or processes, rather than a single or central controller controlling all the various machines or processes. The advantages of a distributive control architecture are increased processing speed with smaller controllers, no single source point of failure for complete shutdown of operations, decreased troubleshooting time, reduced expansion and reconfiguration limitations and problems, etc. Most of the early control networks were proprietary. As more and more controllers of different manufacturers found their way onto the factory floor, many from equipment suppliers, there became an ever-increasing need to easily integrate these controllers. To the benefit of many plant and control system engineers, most PLC and control system manufacturers are building their equipment today with one of the open system network protocols. Some of the control networks found today include *Modbus Plus, ControlNet, Data Highway Plus, Profibus-DP,* and others.

Information Networks

Information Networks work at the highest level of the control system architecture. These are networks that provide the communications link between process and system controllers (PLCs) on the factory floor and the business information systems used by supervisors, process engineers, managers, and inventory and production management systems. These networks also provide the backbone for plant-wide integration of control systems.

As PLCs multiplied on the factory floor, so did the need to retrieve the information that was in them to help increase production, correct process errors sooner, track products, and monitor production. With factory floor information in the hands of supervisors, managers, and process engineers, decisions can now be made more quickly, problems averted, and money saved. The days of the daily production reports are nearly gone, thanks to information networks. Some of the characteristics typically found in information networks include large data amounts, high network compatibility with existing business systems networks, and reliability. Topologies used with information networks include star and bus. Some common information network protocols include *Ethernet TCP/IP, Ethernet/IP, Modbus/IP, ProfiNet,* and others.

INDUSTRIAL PROTOCOLS

The following section provides a brief description of some of the more common industrial control and information protocols being used by PLC manufacturers to implement their industrial automation systems. The protocols mentioned in this section are but a few of the many being used today. Because the descriptions are brief, they are intended only to give you a general overview of the protocol.

DeviceNet

DeviceNet, originally developed by Allen-Bradley (Rockwell Automation), is an open system protocol based on the **C**ontroller **A**rea **N**etwork (CAN) technology. DeviceNet is a digital network that uses a trunk-line/drop-line topology (bus topology) in which node devices (i.e., sensors, motor controllers, small I/O blocks) can be connected either daisy chain or on short (20′ maximum) drop lines from the main trunk. Power and signal are provided on the same network cable using separate twisted-pair busses. DeviceNet supports both isolated and non-isolated physical layer design of devices. An opto-isolated option allows externally powered devices (i.e., smart motor controllers, solenoid valve manifolds) to share the same bus cable. The end-to-end network distance varies with data rate and cable thickness. A maximum end-to-end distance of 1,640′ can be achieved using thick cable and a communication rate of 125 kbs. Each DeviceNet network supports up to 64 nodes. DeviceNet systems can be configured to operate with a master-slave or peer-to-peer communication method. Figure 20–20 shows an example of a DeviceNet network.

Foundation Fieldbus

Foundation Fieldbus was established in 1994 to create a single international fieldbus standard for hazardous environments and is an open bus standard that enables devices of different manufacturers to be integrated into one system. Fieldbus technology was intended to replace the conventional 4 to 20 mA analog wiring methods found in process control applications with a digital communications

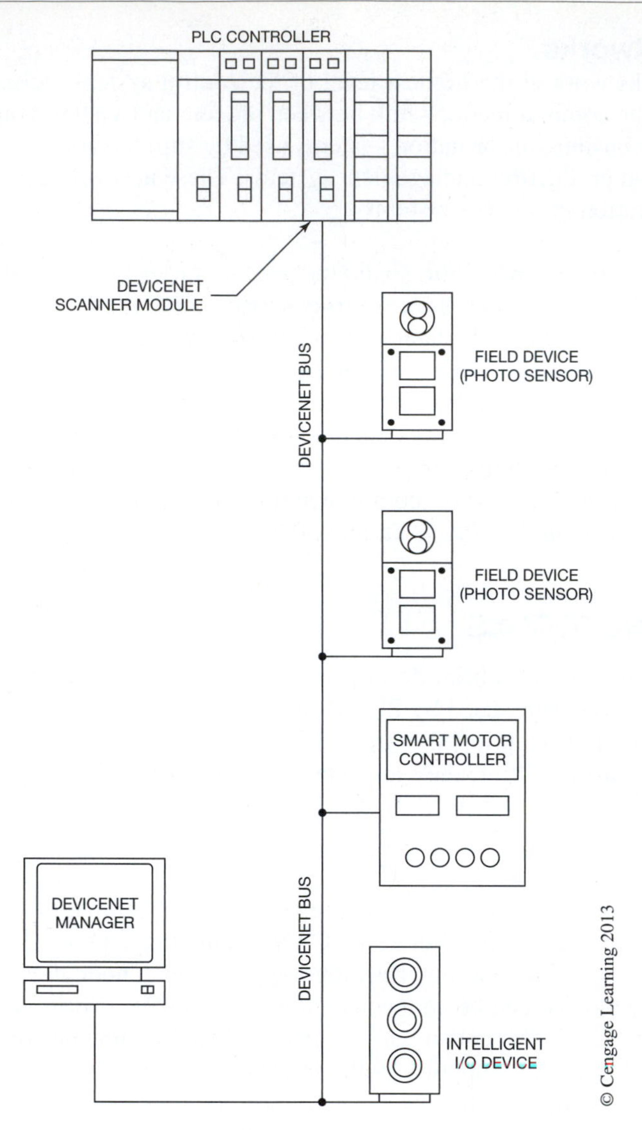

Figure 20–20 DeviceNet Network

network. Fieldbus technology supports bus-powered field devices such as process sensors, actuators, and I/O. Foundation fieldbus protocol allows for the capability to distribute the control functions to the devices using predefined function blocks. This allows field devices the capability of assuming process control functions, thus reducing the risk of system failure and the amount of I/O and control equipment needed. The Foundation Fieldbus protocol is based on the OSI reference model and operates only at layers 1, 2, and 7 of the model, as is the case with most open fieldbus systems. Foundation fieldbus uses two communications bus systems: the slow, intrinsically safe H1 bus and a higher-level, high speed Ethernet HSE bus. The H1 bus is used at the device level using shielded twisted-pair cable in a bus topology. Other topology combinations are possible when equipped with junction boxes. Field devices are connected to the bus using tee connectors and short drops called *spurs*. The maximum length of an H1 segment without repeaters is 6,000 feet including spurs, with a maximum number of devices limited to 32 per segment. Data transmission rates for the H1 bus are 31.25 Kbps. The H1 bus of the Foundation fieldbus uses a central communication control system called a Link Active Scheduler (LAS) that controls and schedules the communication on the bus. This device is called the Link Master. During network setup of the LAS, a transmission schedule is constructed. The schedule determines when devices process their function blocks and when it is time to transmit data. The LAS also allows for unscheduled transmissions for things like device setup and diagnostic data when needed. The LAS handles unscheduled transmissions using a token passing access method. The higher-level HSE bus is based on standard Ethernet technology and runs at 100 Mbps. (Ethernet will be covered later in this section.) The HSE bus allows for the high-speed integration of controllers (PLCs), workstations, HMIs, and H1 bus subsystems using a bridge device. Figure 20–21 shows a Foundation fieldbus architecture. For further information visit the Foundation fieldbus website at http://www.fieldbus.org.

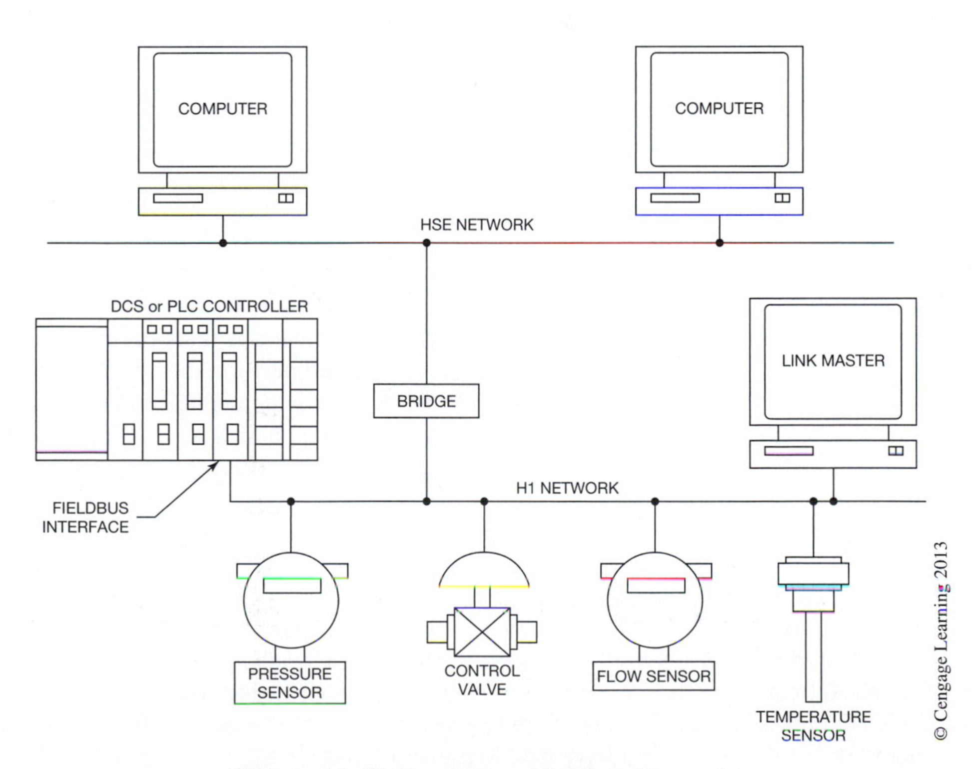

Figure 20–21 Foundation Fieldbus Network

Profibus

Profibus is a group of protocols originally designed by Siemens and adopted into the European standards in 1996. Profibus is an open bus system with protocols (Profibus-FMS, Profibus-DP, and Profibus-PA) that are compatible with each other. Profibus-PA is equivalent to the Foundation fieldbus just discussed. In fact the physical bus design is the same, and it exhibits many common features such as function block control at the device level, master/slave, and peer-to-peer communications. Profibus-DP is optimized especially for communication between automation systems and decentralized field devices. It is designed to handle time-critical communications. Profibus-FMS is optimized for communication between automation systems (i.e., PLCs, HMIs) as well as data exchange with field devices. For further information on Profibus and the available protocols visit the Profibus website at http://www.profibus.com.

Modbus Plus

Modbus Plus is a high-speed, peer-to-peer LAN network developed by Modicon (currently Groupe Schneider). The original Modbus network, one of the first, was introduced by Modicon in 1979 and was a master/slave network that permitted a host computer to communicate to one of several PLCs to perform programming, data transfer, upload/download, and other host operations. Modbus Plus is a local area network that allows host computers, PLCs, and other sources to communicate as peers using twisted-pair cable media. Modbus Plus networks can support up to 32 nodes at distances up to 1,500 feet at communications rates of 1 Mbps. Additional nodes and distances can be achieved by using repeaters. Modbus Plus networks use a bus topology in which nodes on the network function as peer members, gaining access to the network using a token passing access protocol.

Data Highway Plus

Data Highway Plus (DH+) is a peer-to-peer control network developed by Allen-Bradley (Rockwell Automation). This proprietary LAN network was Allen-Bradley's primary control network for many years to link their PLCs, programming, and HMI devices together. The DH+ network is considered obsolete and has been replaced by ControlNet as their network of choice, but many factories and industrial plants are still operating DH+ networks. A DH+ network allows computers, PLCs, and other devices to communicate using shielded twisted-pair cable media. DH+ networks can support up to 64 nodes at distances up to 10,000 feet at communications rates of 57.6 Kbps. Allen-Bradley recommends a maximum of 15 nodes per link for optimal performance. DH+ networks use a bus topology in a trunk-line/drop-line or daisy chain fashion, in which nodes on the network function as peer members, gaining access to the network using a token passing access protocol. Routing between DH+ networks, Ethernet/IP, ControlNet, and DeviceNet networks is possible using the ControlLogix Bridge modules.

ControlNet

ControlNet is an open systems (nonproprietary) control network originally developed by Allen-Bradley (Rockwell Automation). ControlNet uses the Common Industrial Protocol (CIP), which allows messages to be routed to other networks that support the CIP protocol, such as DeviceNet and Ethernet/IP. ControlNet can support control of I/O, peer-to-peer messaging between PLC controllers, and messaging of information on the same network media. ControlNet uses the Producer/Consumer model for network access. Nodes that have data to transmit produce a data packet with a data identifier and place it onto the network. All nodes connected to the network then consume the data packet and determine (based on the identifier) whether they are configured to further process the data. This allows a data packet from a single node to be consumed by multiple nodes without having to be retransmitted. ControlNet uses a bus topology with coaxial cable as the main trunk with passive taps to connect each node to the main trunk, as shown in Figure 20–22.

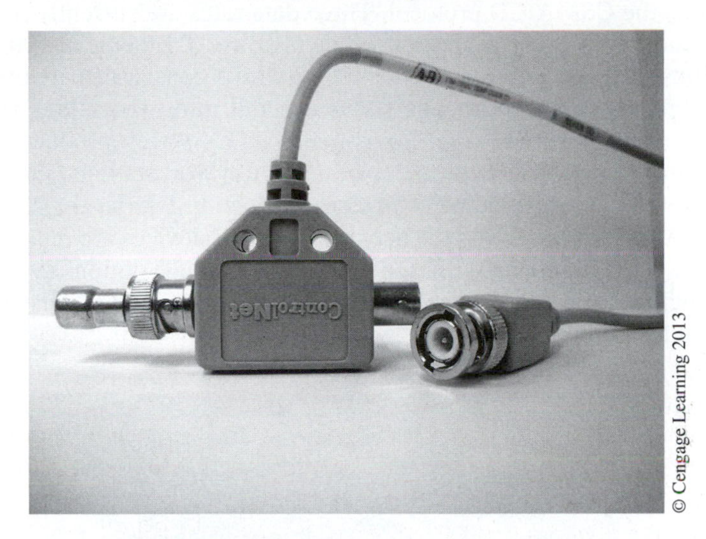

Figure 20–22 ControlNet Passive Tap

The maximum number of nodes for a single coaxial-only segment (i.e., no repeaters) is 48. With 48 nodes, the maximum segment length is 250 meters. The number of nodes per segment determines the maximum segment length. Additional segments and nodes can be added by using repeaters. The maximum number of nodes that can be connected to a single ControlNet network is 99. The data transmission rate for ControlNet is 5 Mbps.

Ethernet

The original Ethernet was developed as an experimental LAN in the 1970s by Xerox Corporation and operated on coaxial cable at a data rate of 3 Mbps using CSMA/CD protocol. The success of the original Ethernet led to the development of the 10 Mbps Ethernet through a joint effort by Digital Equipment Corporation, Intel Corporation, and Xerox Corporation in 1980. The American National Standards Institute (ANSI) and Institute of Electrical and Electronics Engineers (IEEE) published Ethernet as an official standard in 1985 as ANSI/IEEE std.802.3-1985. It has become known as IEEE 802.3 standard.

Ethernet is the major LAN technology used in the world today to connect personal computers (PCs) and workstations within offices, buildings, and across the Internet. Some of the reasons that the Ethernet protocol has monopolized the computer networking industry are that it is easy to understand, implement, manage, and maintain, not to mention the low cost of implementation. Not too many years ago, Ethernet was not a consideration for the factory floor because it was slow and response time varied greatly based on network traffic. With the development of high bandwidth and inexpensive Ethernet switching technology, Ethernet is rapidly emerging on the factory floor. It is worth the reader's time to take a few minutes and take a closer look at Ethernet because of its widespread use and future in factory automation and control.

The term Ethernet refers to LAN products covered by the IEEE 802.3 standard that defines what is commonly known as the CSMA/CD protocol. Three data rates are currently defined for operation over optical fiber and twisted-pair cables, 10 Mbps (10Base-T Ethernet), 100 Mbps (Fast Ethernet), and 1000 Mbps (Gigabit Ethernet). Ethernet networks consist of network nodes or devices that connect to interconnecting media. The nodes can fall into two classes of devices. The first class of devices is called *Data Terminal Equipment* (DTE), which produces and consumes the data packet frames on the network. Typical devices are PCs, workstations, servers, printers, PLCs, etc. The second class of devices is called *Data Communication Equipment* (DCE), which receives and forwards data packet frames across the network. Typical devices are repeaters (hubs), routers, switches, etc., as shown in Figures 20–23a and 20–23b. The media options available to connect the different devices include unshielded twisted-pair cable (UTP), shielded twisted pair-cable (STP), and optical fiber cable.

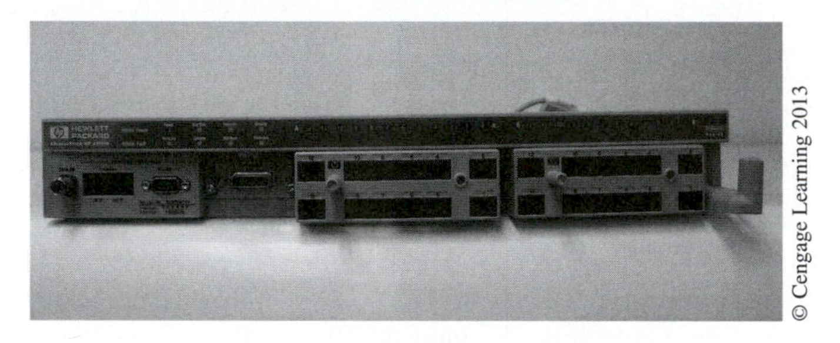

Figure 20–23a Ethernet Hub

Figure 20–23b Ethernet Switches

Ethernet LANs can have different network topology configurations including point-to-point, bus, and star. Since the early 1990s, the star-connected topology has been used as the standard topology. Data terminal equipment (DTE) is connected to a central repeater (known as a hub) or to a network switch that has multiple connection ports. Each DTE device is connected in a point-to-point configuration with the central repeater or switch. The maximum distance that a DTE device can be located from the central repeater device is typically 265 feet but can depend on the network data rate and media used. Most Ethernet LANs are made up of many small star topologies interconnected to form the network (see Figure 20–24).

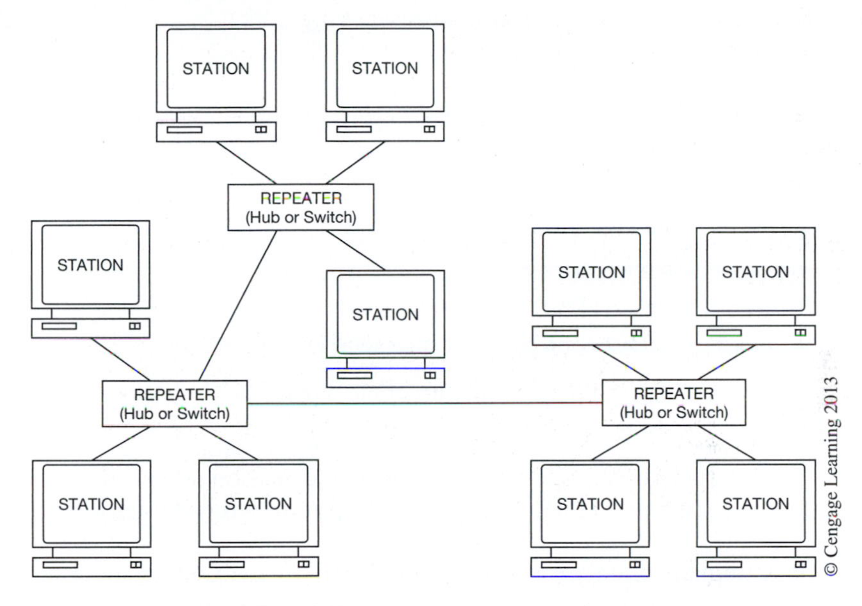

Figure 20–24 Typical Ethernet Topology

The Ethernet protocol itself operates at only layers 1 and 2 of the OSI reference model. The IEEE 802 standard divides the data link layer (layer 2) of the OSI model into two sublayers called the *Media Access Control* (MAC) sublayer and the MAC-client sublayer. The MAC-client sublayer provides the interface between the MAC sublayer and the upper layers of the OSI model. The MAC sublayer is responsible for the data frame assembly before transmission, frame disassembly and error-checking during and after receipt of data frames, media access control including initiation of frame transmission, and recovery from transmission failure. The format of an Ethernet data frame is shown in Figure 20–25. The physical layer (layer 1) specifies the transmission data rate (Mbps), signal encoding, and the type of media interconnecting the devices.

The IEEE 802.3 standard currently requires that all Ethernet MACs support half-duplex operation and the CSMA/CD access method. There are physical network size limits for half-duplex operation based on the network data rate. Full-duplex operation is an optional capability that allows simultaneous two-way transmission between two devices (point-to-point link). Full-duplex transmission is much simpler than half-duplex transmission, because full-duplex uses two pairs of network wires or links for sending and receiving data frames. Media contention and data collisions are all but eliminated with full-duplex transmission, leaving more time available for transmissions and in effect

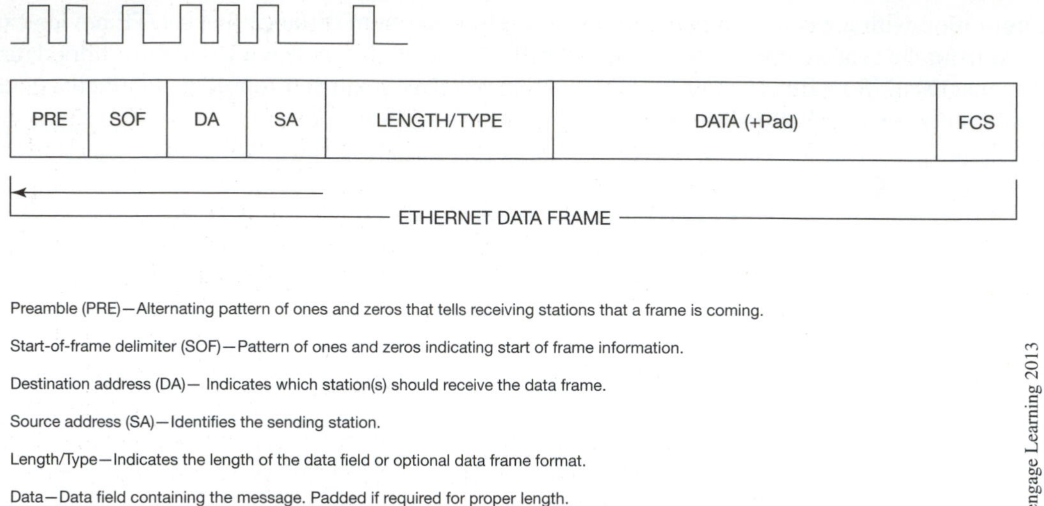

Figure 20–25 Structure of an Ethernet Data Frame

doubling the link bandwidth. In order for full-duplex transmission to take place, the physical layers (cables and network equipment) must be capable of supporting full-duplex operation.

Ethernet devices implement only the bottom two layers of the OSI protocol model, and for this reason they are typically implemented as *Network Interface Cards* (NICs), or as built-in Ethernet network interface ports. The NICs and interface ports are identified by a three-part naming convention based on the physical layer standards. The three parts of the naming convention are data transmission rate, transmission method, and the media type. For example: 10BaseT stands for 10 Mbps data transmission rate, baseband transmission method, and two twisted-pair cable as the media type.

Most Ethernet networks today are constructed using category 5 (CAT 5) or better cable and terminated at each end by an 8-pin RJ-45 connector. Cables lower than Category 5e should not be considered for new installations to allow for future data transmission rate needs.

As mentioned, DTE is connected to a central repeater (known as a hub) or network switch having multiple connection ports. Each DTE device is connected in a point-to-point configuration with the central repeater or switch. When central repeaters or hubs are used, they typically accept only one data packet frame at a time and then resend it to all active ports on the repeater. Switches, on the other hand, have ports with I/O frame buffers that isolate the port from traffic being sent at the same time on other ports. Multiple internal data paths allow data packet frames to be received on one port and then rapidly switched to the appropriate output port. Switches typically build and maintain internal tables that map Ethernet addresses to ports. The switch uses the addressing information within each Ethernet data packet frame to forward the frame only to the port connected to the destination device. This may not seem like that much of a difference between a switch and hub but it has a major effect on network operation. Because switches provide access to a high-speed network bridge (switch) that interconnects only two ports, the collision domain in the network is reduced to a series of small domains in which there are only two devices. This markedly increases network bandwidth and begins to increase the determinism of an Ethernet network, making Ethernet attractive for industrial control systems. For this reason network repeaters or hubs are all but obsolete for industrial and large networks.

In summary, Ethernet itself is simply a protocol that works at the bottom layers of the OSI stack (layers 1 and 2), and is a way to transport data between two devices; it does not guarantee that a device that receives the data will know how to interpret the data. To communicate and use information over an Ethernet network you only have to implement protocols for layers 1, 2, and 7 of the OSI model as a minimum. Remember that the applications that utilize this data operate at the highest level of the OSI model (layer 7), and if the applications are not using the same protocols at this level they will not be able to interpret the data received. Most industrial control manufacturers implementing Ethernet in their control equipment are implementing their own protocols. What this means is that devices connected to an Ethernet network from different manufacturers are not guaranteed to communicate with each other, even though they are all on the same network called "Ethernet."

Chapter Summary

Communication networks provide the means for devices like PLCs, computers, and intelligent I/O devices to share or pass information using common media. A communication network is typically called a Local Area Network (LAN), but other names like WAN, MAN, and RAN are also used depending on physical size. How the network cables are routed and the role that network devices play in the transmission of data is called the network **topology.** The three basic topologies are **bus, ring,** and **star.** The most common media types used in the construction of a network include **coaxial, twisted-pair,** and **fiber optic** cables. One or all three of these types could be used in the construction of a single network. In order for devices to share a common media, a method called **Network Access-Control** is used. Network access-control is like a traffic cop that determines when devices can access the network to transmit their information, and it also resolves any conflicts. The three most common network access-control methods include **master/slave, peer-to-peer,** and **CSMA/CD.** When devices share common media, they must speak the same language. This common language is called the **network protocol.** A network protocol is a set of rules that determines how a message is compiled into data packets, converted into electrical signals, transmitted onto the network media, received only by the intended device, and then converted back into the original message without any loss of information or translation.

Different types of networks are used in industrial automation and control based on the type of information to be communicated. An **I/O** or **device network** is used to transmit the information between field devices (i.e., sensors, actuators, motor controllers, push buttons) and the controllers (PLCs, computers) that are controlling the machines or process. **Control Networks** link the controllers, HMIs, and other control systems. **Information Networks** link the factory floor to the business system of the company and can also provide the backbone to link the smaller control networks together in a large plant.

Review Questions

1. When devices are connected to network media they are typically called what?
2. Network bandwidth is usually expressed in
   a. meters
   b. bytes per hour
   c. bits per second
   d. packets per second
3. What category of network would you typically find monitoring and/or controlling a city's water reservoirs and pump stations, which may be in located in remote areas?
4. Star topology is a network topology in which each device is connected:
   a. to the other devices in a daisy chain fashion
   b. to a PLC
   c. to a network controller
   d. to a common bus
5. When twisted-pair cable is used as the network media it can:
   a. be no longer than 100 meters
   b. only have one pair of wires per cable
   c. use BNC type connectors
   d. be unshielded or shielded
6. A single optical fiber strand consists of two parts. What are those two parts?
7. What are the three most commonly used network access-control protocols?
8. The Open Systems Interconnect (OSI) model is:
   a. used to connect two networks
   b. a reference model of network architecture and a suite of protocols
   c. a reference model of the physical topologies of a network
   d. a network media connector
9. A "frame" is a data link layer "packet" that contains information required by the physical media. What are the three basic sections that make up a frame?
10. In what type of industrial network would you typically find the master/slave protocol and bus type topology used?
    a. Information Networks
    b. Control Networks
    c. Device Networks
    d. All of the above
11. The network address assigned to a device is:
    a. a letter
    b. a number
    c. a word
    d. a ten-digit code

CHAPTER 21 Start-Up and Troubleshooting

Objectives

After completing this chapter, you should have the knowledge to:

- Understand start-up procedures listed in the manufacturers' literature.
- Explain how input devices are tested.
- Explain how to test output devices using a push button or another input device.
- Explain safety considerations when testing output devices.
- Describe how voltage readings are taken to check input and output modules.

START UP

Careful start-up procedures are necessary to prevent damage to the driven equipment and the programmable controller system, or, more importantly, injury to personnel.

Prior to beginning a system start-up procedure, it is important to check and verify that the system has been installed according to the manufacturer's specifications, and that the installation meets local, state, and national codes. Special attention should be given to system grounding.

Before applying power to the controller, complete the following steps:

Step 1. Verify that the incoming power matches the jumper-selected voltage setting of the power supply. Figure 21–1 shows a typical power supply with the jumper position indicated.

Note: *Almost all power supplies are shipped from the manufacturer with the voltage setting in the HIGH voltage (240 V) position.*

Step 2. Verify that a hardwired safety circuit or other redundant EMERGENCY STOP device (described in chapters 2 and 10) has been installed and is in the open position.

Step 3. Check all power and communication cables to ensure that connector pins are straight, and not bent or pulled out.

Step 4. Connect all cables, making sure that connectors are fully inserted into their sockets. Secure connectors as applicable.

Step 5. Ensure that all modules are securely held in the I/O rack, and that field wiring arms (if applicable) are fully seated and locked.

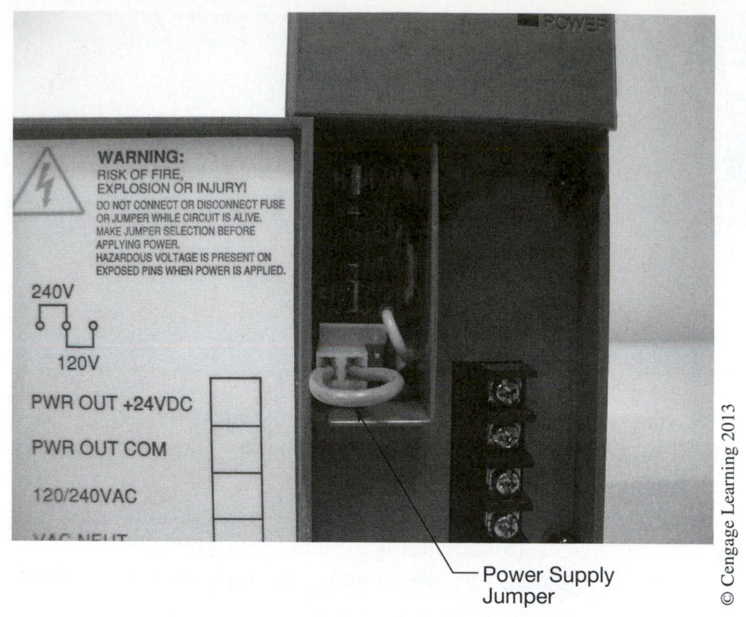

© Cengage Learning 2013

Figure 21–1 Power Supply with Jumper

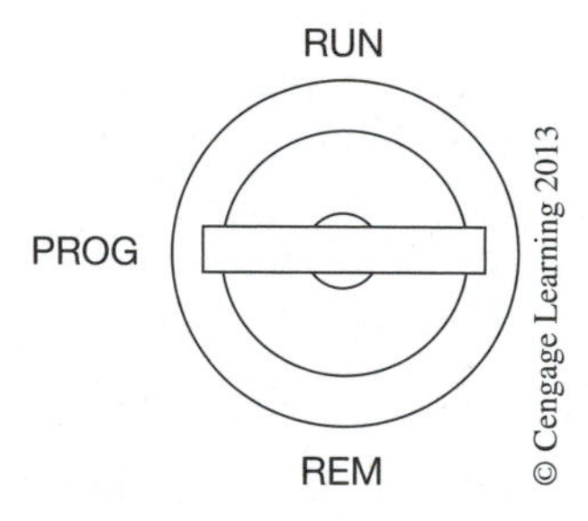

© Cengage Learning 2013

Note: *"REM" means the PLC mode (Run, Program, etc.) can be controlled remotely by a programming device*

Figure 21–2 Key Switch in Program Position

Step 6. Place the PLC processor key switch to a safe position, as indicated in Figure 21–2.
Step 7. Double-check the setting(s) of all DIP switches.

Caution: Before proceeding further, make sure that the safety circuit or other EMERGENCY STOP devices are *OFF,* or open, and that power is removed from all **output** devices.

Apply power and observe processor indicator light(s) for proper indications.

When power is applied and the safety switch is closed, the power supply should provide the necessary DC voltage for the processor and I/O rack. If the proper voltage is present, the input indicator LEDs of the input modules will be functioning. Any input device that is closed, or *ON,* will have an illuminated LED (Figure 21–3).

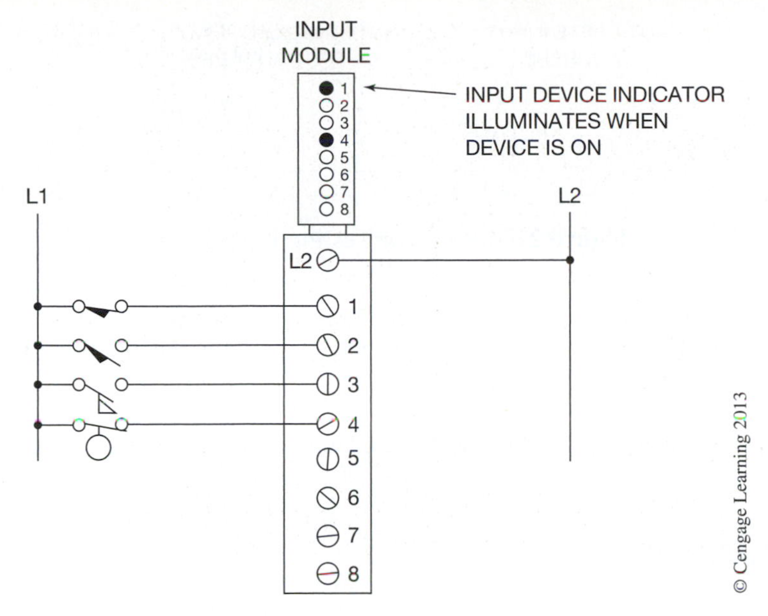

Figure 21–3 Input Module Indicators

TESTING INPUTS

Each input device can be manipulated to obtain open and closed contact conditions.

Caution: *Do not* activate the input devices mounted on equipment by hand, because unexpected machine motion could cause injury. Use a wooden stick or other nonconducting material to activate input devices mounted on equipment.

Each time an input device is closed, the corresponding LED on the input module should illuminate. Failure of an LED to illuminate indicates:

1. Improper input device operation.
2. Incomplete or incorrect wiring; check to be sure that the input device is wired to the correct input module and proper terminal.
3. Loss of power to the input device.
4. Defective LED and/or input module.

To further check the system, a program can be developed and entered into the processor that uses each input device address. With the PLC in the test or disable output state, the status of the input devices may now be monitored by using the computer programmer, or by LED indicators on a hand-held programmer. On a monitor, the input contact becomes intensified or goes to reverse video, depending on the model of PLC, when the instruction is true. An EXAMINE ON instruction intensifies or goes to reverse video when the input device is *ON,* or closed. An EXAMINE OFF instruction intensifies or shows reverse video when the input is *OFF,* or open. Figure 21–4 shows a rung for testing input devices.

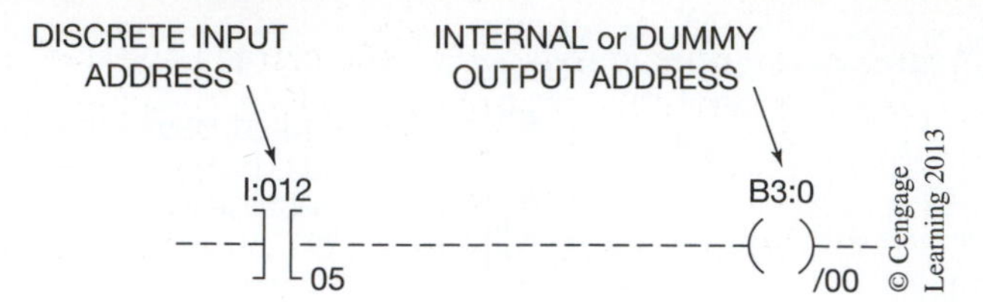

Figure 21–4 Rung for Testing Input Devices

A second method for testing the status of inputs in the processor (for PLCs with the option) is by monitoring the status (1 or 0) of the desired input address through the tag monitor or INPUT image file using the monitor of a computer programmer.

Once all input devices have been tested and checked out as operational and properly terminated, the output devices can be tested.

TESTING OUTPUTS

Before testing output devices, it must be determined which devices can safely be activated and which devices should be disconnected from the power source. Figure 21–5 shows a motor starter with the motor disconnected for safety. In this configuration, the motor starter coil (the output device) is activated for checkout without energizing the motor. This prevents unwanted machine motion that might cause injury to personnel or damage the machine.

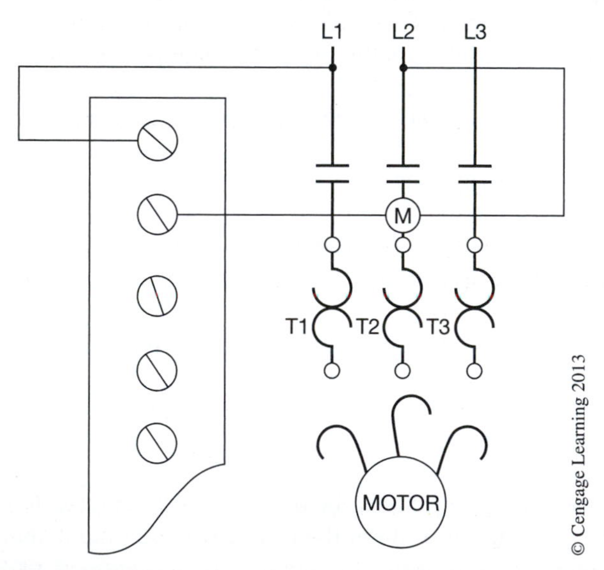

Figure 21–5 Disconnecting Motor Leads for Safety

For outputs that can be safely started, be sure equipment is in the start-up position, is properly lubricated, and is ready to run.

There are two methods used to test output devices. The first method uses a push button or another convenient input device that is part of the control panel. The push button is programmed to energize each output, one at a time.

The second method uses the FORCE function of the PLC to energize outputs, one at a time.

When using a push button (or another input device), the address of the push button is programmed in series with the output device to be tested (shown in Figure 21–6).

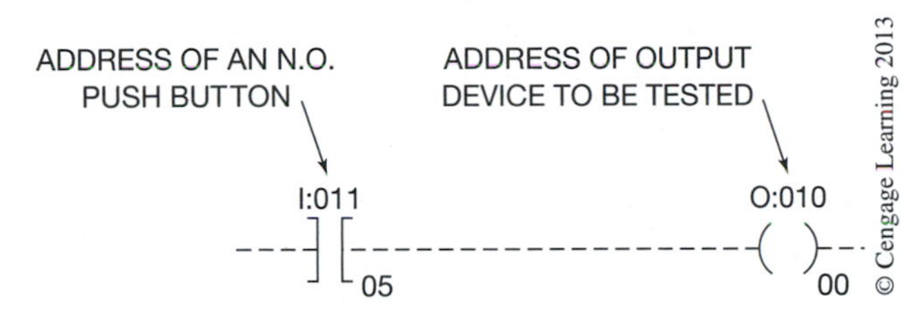

Figure 21–6 Rung for Testing Output Devices

Once the rung has been programmed and entered into processor memory, the processor is placed in the *run* mode. Pressing the push button (I:011/05) illuminates the output indicator on the output module for address O:010/00. If there is an output connected, verify that the output device is energized. If the output indicator does not illuminate, using the monitor, verify that the input instruction I:011/05 is intensified or showing reverse video to indicate an *ON* condition. Double-check the output address, and verify that the output instruction indicates an *ON* condition. If both instructions indicate an *ON* condition and the output address is correct, a defective module is likely the problem.

If the output module indicating LED is illuminated, but a connected output device does not energize, check the following:

1. *Wiring* to the output device.
2. *Operation* of the output device.
3. Proper *potential* to the output device.
4. Output device wired to correct output module and proper terminal.

A second method for testing output devices (for PLCs with the option) is the FORCE feature. The FORCE feature allows the user to turn an output device *ON* and *OFF* without using a push button or adding logic. Figure 21–7 shows an Allen-Bradley FORCE table that allows any output address to be forced *ON* and/or *OFF*.

By moving the cursor to the desired output address, the FORCE ON function can be initiated, by placing a 1 in the address location of the desired output (see Figure 21–7, address O:000/03) and enabling Forces, which should illuminate the output module indicating LED and turn *ON* the output device, if one is connected. If the LED does not illuminate, and Forces are shown as being enabled, one should suspect the module. If the LED lights, but the output device does not energize, proceed as previously described. To FORCE OFF the output, place a 0 in the address location (see Figure 21–7, address O:000/06) and to remove the FORCE simply replace the 1 or 0 with a period (".").

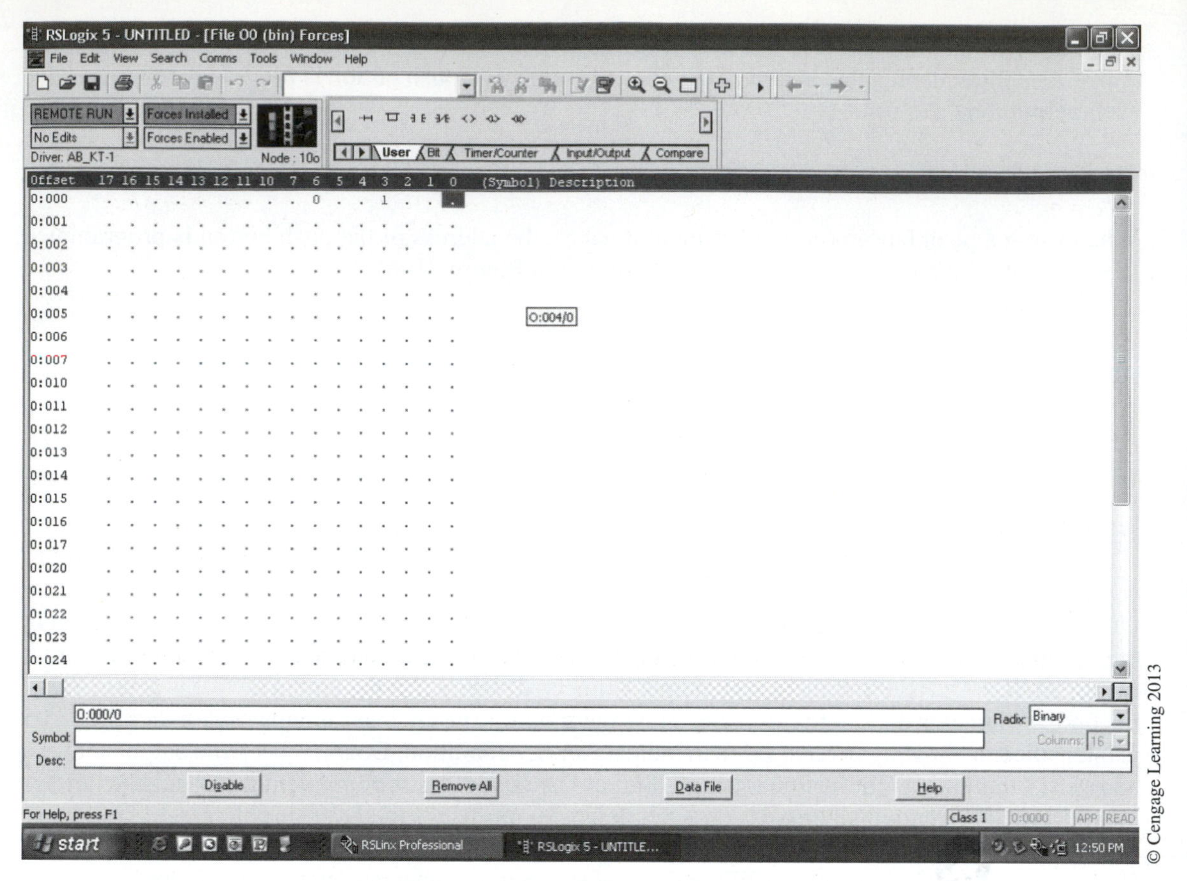

Figure 21–7 Allen-Bradley Force Table

FINAL SYSTEM CHECKOUT

After all input and output circuits have been tested and verified, the electrician or technician is ready for the final system checkout.

Reconnect any output loads (motors, solenoids, etc.) that were previously disconnected. In the case of motors, correct rotation needs to be established before the complete machine or process can be tested. Using a momentary push button that is part of the control panel (or using one installed specifically for this purpose), load a rung of logic into the processor as previously discussed for testing outputs (Figure 21–6).

Caution: Because this part of the test causes machine motion, make sure the machine is operational and all personnel are in the clear. Station someone at the EMERGENCY STOP or disconnect location to de-energize the system, if necessary.

Close the push button and immediately release or open it. This momentary operation of the push button is called jogging or bumping, and allows the output (motor starter) to energize only momentarily. The motor starter is only energized long enough to determine the direction of rotation of the motor. If the rotation is wrong, reverse any two motor leads (assuming 3-phase power) and repeat the test for verification. Continue testing all output loads previously disconnected until all of them function correctly. Once all machine components are tested and correct rotations are established, total machine operation testing can be accomplished.

For final system checkout, the following steps should apply:

Step 1. Place the processor in the *program* mode.

Step 2. Clear the memory of any previous rungs used for testing.

Step 3. Using a programming device, enter the program (ladder diagram) into memory.

Step 4. Place the processor in the *test* or *disable output* mode, depending on the PLC, and verify correctness of the program.

Note: *In the test, or disable output mode, the outputs cannot be energized. All logic of the circuit is verified, input devices function, but no outputs come on. This step must not be skipped if injury to personnel or damage to equipment is to be avoided.*

Step 5. Once the circuit operation has been verified in the *test* or *disable output* mode, the processor can be placed in the *run* mode for final verification.

Step 6. Make changes to the program as required (timer settings, counter presets, and the like).

Step 7. Once the circuit is in final form, and the machine or process is running correctly, it is recommended that a copy of the program be made.

TROUBLESHOOTING

The key word to effective troubleshooting is *systematic*. To be a successful troubleshooter, the technician must use a *systematic* approach.

A systematic approach should consist of the following steps:

Step 1. Symptom recognition.

Step 2. Problem isolation.

Step 3. Corrective action.

The electrician and/or technician should be aware of how the system normally functions if he or she expects to successfully troubleshoot the system. When prior knowledge of system operation is not possible, the next best source of information, if applicable, is the operator. Don't hesitate to ask the operator what the symptoms are and what he or she thinks the problem might be. If no operator is available, the next best source of information is the PLC system itself. Although the PLC can't talk, it can communicate in various ways to show what the problem is. There are status lights on the processor, power supply, and I/O rack that indicate proper operation, as well as status lights that alert the troubleshooter to the problem. The status lights of a typical processor with built-in power supply indicate:

1. DC POWER ON—If this LED is not lit, there is a fault in the DC power supply. Check the power supply fuse and/or incoming power.
2. MODE—Indicates which operating mode the processor is in (*run, halt, test, program,* etc.). The fault may simply be that the key switch is in the wrong position.
3. PROCESSOR FAULT—When this status light is on, it indicates a fault within the processor. This is a major fault, and requires changing the processor module.
4. MEMORY FAULT—This status light illuminates when a parity error exists in the transmission of data between the processor module and the memory module. Replace only one module at a time. If the first module does not correct the problem, reinstall the original module and then replace the second module. If replacing the second module doesn't clear the problem, replace both modules.
5. I/O FAULT—This light indicates a communication error between the processor and the I/O rack. Check that the communication cables are fully inserted into their sockets. If available, connect a programming device with a monitor to the processor, and look for error codes and/or fault messages for further diagnostic assistance.

Note: *Refer to manufacturer's literature for explanation of error codes.*

6. STANDBY BATTERY LOW—When this LED is illuminated, the RAM backup batteries are low and need to be replaced. Although this is not a fault condition, failure to replace batteries results in losing the program when the system is shut down or a power failure occurs.

Status lights on the I/O modules also assist with troubleshooting problems that involve input and output devices.

If the operator confirms that the solenoid that activates a brake isn't working, the first step is to determine the address of the solenoid. Once the address is known, the programming device can be used to ensure that the output circuit to the solenoid has been turned *ON*. Are all input devices closed that should be closed? Has the rest of the rung logic been completed? If the answers are yes, then it is necessary to determine the hardware location. (Some PLCs use the address to specify the hardware location, whereas others do not.)

Output modules have LED indicators that illuminate when each of the output circuits is turned *ON*. If the LED is lit for the location of the solenoid, it indicates that the problem is not with the output module, but with the circuit from the module to the solenoid, or with the solenoid itself.

A voltage check from L2 to the terminal, as shown in Figure 21–8, verifies the conclusion that the module is working properly, but that the problem is either in the wiring to the solenoid, or in the solenoid itself. Further voltage checks will locate the problem.

Note: *When testing AC output modules, the high internal resistance of most analog or digital meters acts like a series voltage divider when measuring across an open load (Figure 21–9). The result is a reading of nearly full voltage even after the triac has been turned* OFF. *For accurate readings, a 10 K ohm resistor can be placed in parallel with the meter leads as shown in Figure 21–10, or a solenoid-type tester (Wiggins) with low internal resistance can be used.*

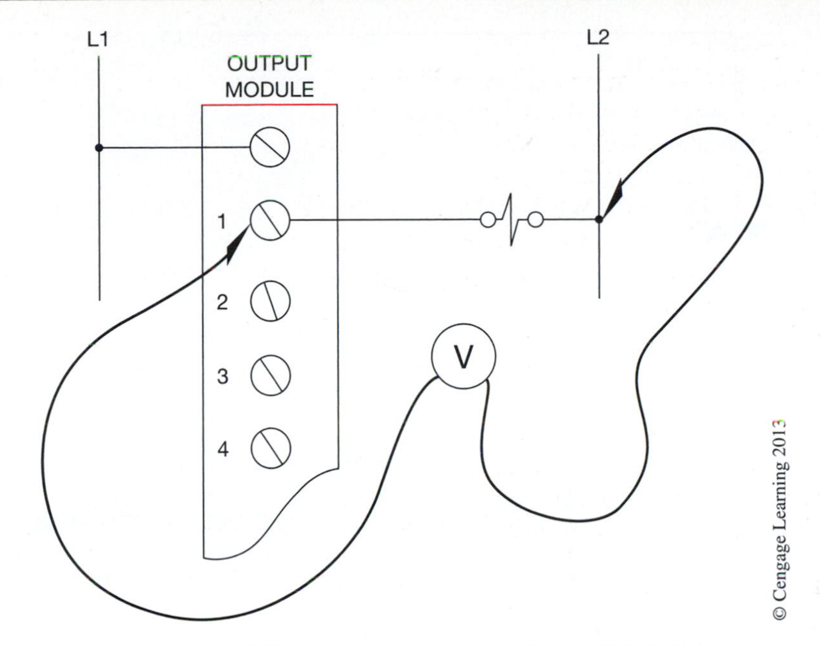

Figure 21–8 Testing Voltage on an Output Module

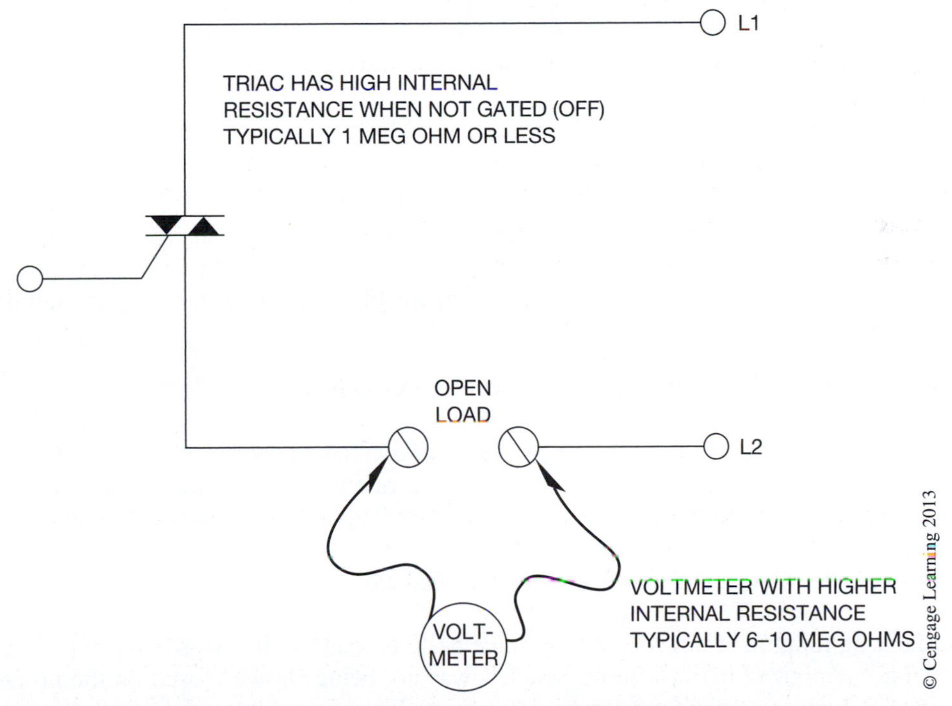

Figure 21–9 Series Voltage Divider Effect When Reading Across an Open Load

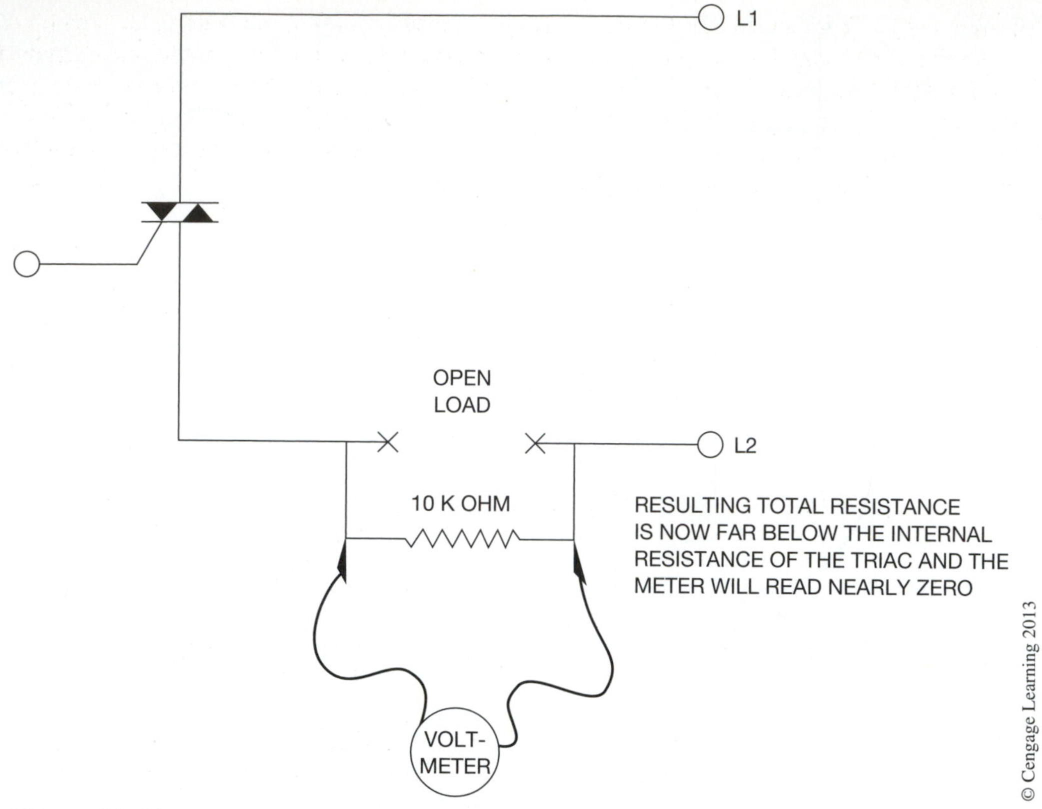

Figure 21–10 Adding a Resistor in Parallel with Meter to Reduce Voltage Divider Effect

If the indicator LED has not been lit, the first reaction might be to change the output module. Instead, look first for a blown-fuse indicator LED. A fuse might be all that's needed to make the solenoid operational. If, however, there was no blown fuse, replacing the output module should correct the problem.

Note: *Remove all power from the I/O rack before changing modules.*

Some PLC manufacturers offer deluxe output modules that have two indicator LEDs. One indicates that the logic from the processor has been received to turn on the output; the second LED comes on when the triac, or power transistor, has been turned *ON*. These two LEDs should come on simultaneously, unless the PLC is in the *test* mode. In the *test* mode, only the logic LED is lit because the output circuits are isolated, and kept from being turned *ON*.

Troubleshooting input modules follows the same basic procedure. If it was determined that the solenoid had not energized because limit switch 1 was not being shown closed on the programming device, the LS-1 address would need to be determined. From the address, determine the terminal on the input module that LS-1 is connected to.

An illuminated LED indicates that the limit switch is closed, but that the state of the switch (*ON*) is not being communicated to the processor. Exchanging the input module should make the system operational. However, had the indicator LED for LS-1 not been lit, there could be several other possible problems, such as a bad limit switch, faulty wiring from LS-1 to the input module, or a bad input module. Closing the limit switch and taking a voltage check, as shown in Figure 21–11, determines if the limit switch and associated wiring is operational.

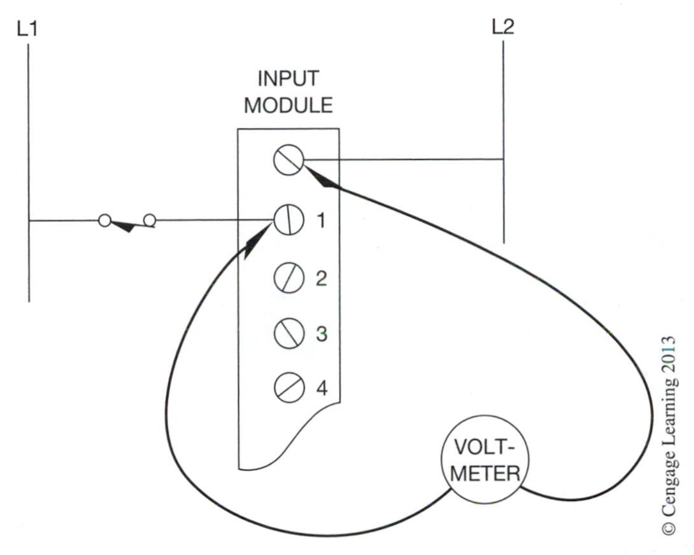

Figure 21–11 Testing Voltage on an Input Module

A voltage reading equal to the applied voltage indicates that the limit switch and wiring are operating, but that there is a faulty input module. No voltage reading indicates a problem with either the limit switch or its wiring. Further voltage checks will isolate the problem.

Caution: Use a high-impedance voltage meter when taking voltage readings on PLC input circuits. Failure to use a high-impedance voltage meter could cause unexpected machine operation by unintentionally activating inputs such as Start and Jog buttons, limit switches, etc.

Similar to the deluxe output modules, there are also input modules that have two indicating LEDs. The first LED indicates that the input device has closed, and a voltage signal has been received by the input module; the second LED indicates that the status of the input device (*ON*) has been communicated to the processor.

Note: *Once the PLC program has been proven, or checked out, for complete and accurate operation, any future problems normally will be bad field wiring and/or bad field devices.*

Chapter Summary

Start-up procedures check each part of the PLC system for proper installation and operation. Safety must *always* be the overriding factor when testing or operating a PLC system. Care must be taken to prevent unexpected or incorrect machine motion if injury to personnel and/or damage to equipment is to be avoided.

Once the system start-up is complete and the system is operational, problems or faults can occur. To successfully troubleshoot the system, a systematic approach must be used. This systematic approach includes recognizing the symptoms, isolating the problem, and taking corrective action. A variety of indicator and status lights, as well as error messages and fault codes, assist the electrician or technician in troubleshooting a given system.

The information covered in this chapter is intended to be general in nature and is not specific to any particular PLC. For more specific information on start-up and troubleshooting procedures, refer to the manufacturer's operating manual that accompanies the PLC.

Review Questions

1. A programmable controller system should be installed according to:
   a. manufacturer's specifications
   b. local electrical codes
   c. state electrical codes
   d. national electrical codes
   e. all of the above
2. Explain why a safety circuit or other EMERGENCY STOP device is important.
3. Describe briefly how input devices are tested.
4. List two methods for testing output devices.
5. Why should output devices be disconnected before testing?
6. Draw an input module complete with input devices and power connected (L1–L2), and show how a voltage reading is taken to verify the operation of an input device.
7. On a deluxe input module, what do the two LED indicators represent?
8. On a deluxe output module, what do the two LED indicators, other than the blown-fuse indicator, represent?
9. Define the term *jogging*.
10. List the three steps of systematic troubleshooting.

CHAPTER 22

PLC Programming Examples

Objectives

After completing this chapter, you should have the knowledge to:

- Better understand PLC instructions.
- Apply various PLC instructions together.
- Apply simple PLC control logic solutions.

This chapter will provide the reader with examples of PLC programming code/logic utilizing many of the instructions covered in previous chapters. The examples presented here are intended to help the reader gain a better understanding of the various PLC instructions and how they can be combined to provide simple control logic solutions. The examples are not intended to take the place of application-specific requirements. In most cases there is more than one solution to a problem; the right solution and subsequent logic depends on having a thorough understanding of the machine or process to be controlled.

To help in understanding the following PLC programming examples, PLC memory addresses will most often not be shown. Instead a description will be used followed by the type of memory address in bold type. As an example, shown in Figure 22–1 is an N.O. contact (EXAMINE ON) with the description "Start Push Button" followed in bold by **"Digital Input";** this memory address would be a digital input from a real-world button labeled "Start."

Figure 22–1 Digital Input Description Example

The following are examples of other memory types used in this chapter:

Digital Output—A digital address in the output memory area of the PLC that a real-world device is being controlled from, such as a pilot light, motor starter, solenoid, etc.

Digital Memory—A digital address in a user memory area of the PLC that is used primarily for internal or dummy relays, such as "Control Relay CR-1."

Analog Input—A word address in PLC memory containing a decimal value from a real-world analog input device such as a temperature sensor, pressure transmitter, etc.

Analog Output—A word address in PLC memory containing a decimal value being sent to a real-world analog output device such as a valve, variable speed drive, panel meter, etc.
Analog Memory—A word address in a user memory area of the PLC containing a decimal value from calculations or simply used to store values for in comparison type instructions.

Timer and counter address will only have descriptions such as "½ Second Flasher Timer," "½ Second Flasher Timer Done," etc.

EXAMPLE 1—PUSH ON/PUSH OFF CIRCUIT

In some applications it is desirable to have a single, N.O., momentary push button used to both turn *ON* and *OFF* a digital output(s) such as lights, motors, solenoids, etc. When one begins to think about the solution and subsequent logic it becomes clear that there is certainly more than one solution possible, as is most often the case. Which is the best solution is a question that challenges all PLC programmers. Only after carefully studying what the output controls and taking into account any safety or fail-safe considerations can the best solution be chosen.

In our example of the "Push ON/Push OFF" circuit, we have chosen to control a light and always have it be in the *OFF* state when the PLC begins executing the program on a restart of the processor (fail-safe position on restart). Figure 22–2 shows the logic we have selected. In our example

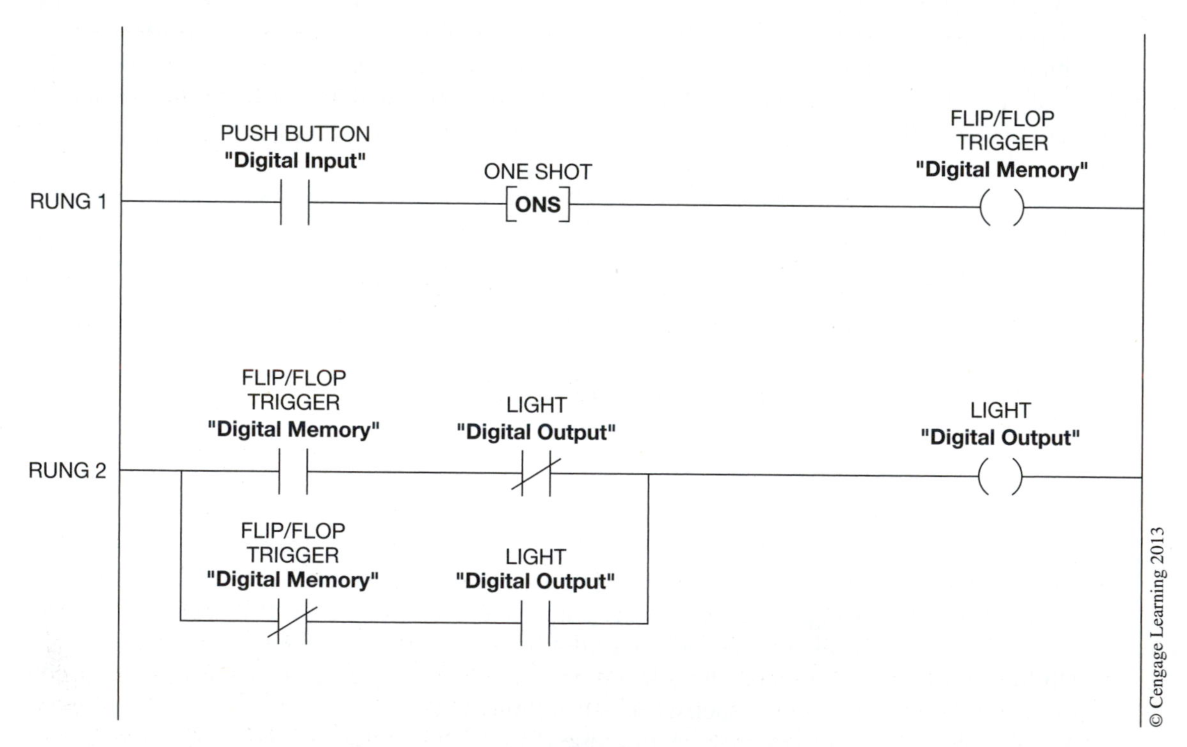

Figure 22–2 Push ON/Push OFF Circuit

we are using a common logic circuit called the *flip/flop* circuit to provide the means of alternating between the two states, *ON* and *OFF*. The basic *flip/flop* circuit is shown in Rung 2 of Figure 22–2. It gets the name *flip/flop* because any time the input memory address labeled "Flip/Flop Trigger" is true or *ON* and the PLC processor scans the rung, the output, the light to be controlled in this case, will change state or alternate, hence the name *flip/flop*. In Rung 1 of Figure 22–2 the real-world push button is used to turn on the "Flip/Flop Trigger" output that is used to change the state of the flip/flop circuit. The "ONS" instruction in Rung 1 is a One-Shot instruction used to keep the "Flip/Flop Trigger" output *ON* for only one scan, which is required for proper operation of the flip/flop circuit. Refer back to Chapter 11 for a review of the "ONS" instruction.

The basic flip/flop circuit shown here is a very useful and universal circuit that can be used for many different applications, as you will soon see.

If the PLC you are using does not have a One-Shot instruction, one can be made by using the logic shown in Figure 22–3 in place of Rung 1 in Figure 22–2.

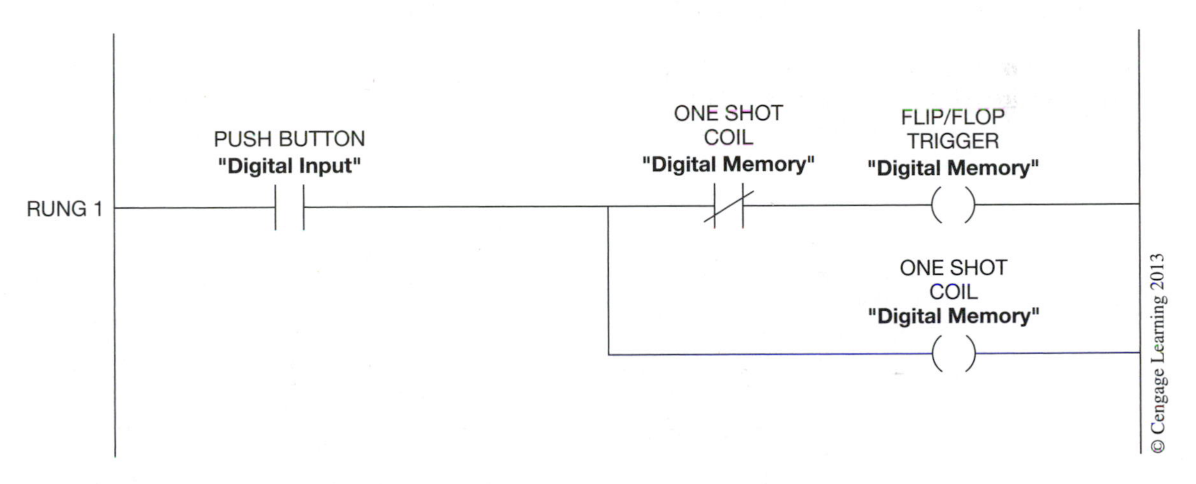

Figure 22–3 Push Button with One-Shot Coil

EXAMPLE 2—½ SECOND PILOT LIGHT FLASHER CIRCUIT

Many times when designing control circuits, it is desirable to flash pilot lights at various times to alert operators to abnormal or pending conditions. The power and flexibility of PLCs makes this a very easy addition to any control circuit design. In the example presented here, we will design a universal flasher circuit that will provide for a digital memory bit labeled "½ Second Flasher" that can be used throughout your PLC program to flash pilot lights as desired. The one-half (½) second flash rate used in this example is a common flash rate for pilot lights, but any flash rate can be chosen based on your application or needs. The same basic *flip/flop* circuit used in Example 1 will also be used here to provide the basis for our flasher. As you will recall from Example 1, any time the "Flip/Flop Trigger" memory bit is *ON*, the output of the *flip/flop* circuit will change state. In that example, the push button and one-shot were used to turn on the "Flip/Flop Trigger" memory bit. Now if we could have someone push the button every ½ second we would have ourselves a ½ second flasher. Of course having someone push the button every ½ second is not practical, but if we replace the rung containing the push button, one-shot, and "Flip/Flop Trigger" output with an On-Delay Timer set for ½ second that automatically resets itself, we will have our ½ second flasher. Figure 22–4 shows the flasher circuit using the On-Delay Timer and *flip/flop* circuit.

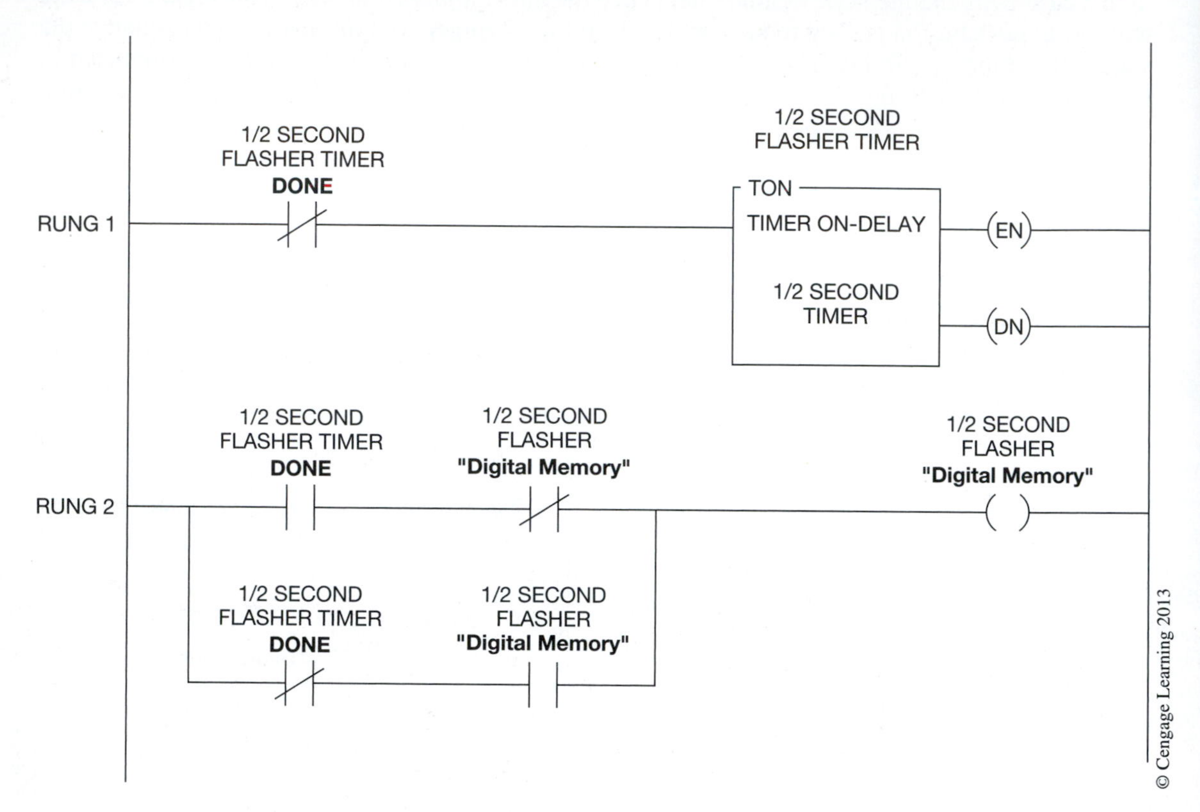

Figure 22–4 ½ Second Flasher Circuit

After reviewing the flasher circuit in Figure 22–4, did you notice that we did not use a one-shot instruction as was required in Example 1? That is because the On-Delay timer in Rung 1 is self-resetting, meaning that the done bit of the On-Delay timer is only *ON* for one scan before being used to reset the timer.

EXAMPLE 3—MOTOR STARTER FAULT-MONITORING LOGIC

When controlling industrial motors with PLCs, it is often desirable to monitor and know when motor starters fail to operate as directed. If a motor starter should fail to operate, then that information can be used in your PLC program to provide for the safe shutdown of other equipment, as well as alert an operator or maintenance person to the problem. Motor starters can fail to operate for many reasons, but some of the more common reasons are: motor overload trips, low or no control voltage, open starter coils, welded contacts, etc. The logic presented in this example is designed to monitor and detect a failure of a motor starter to pull in and/or drop out when directed. In order to detect a motor starter failure, an N.O. auxiliary (aux) contact mounted on the starter must be wired to an input module of the PLC (refer to Chapter 10, "Overload Contacts," for additional information on this input and the circuit presented here). The status of this input, as well as the PLC output to the starter, becomes the basis for the monitoring and fault detection circuit shown in Figure 22–5.

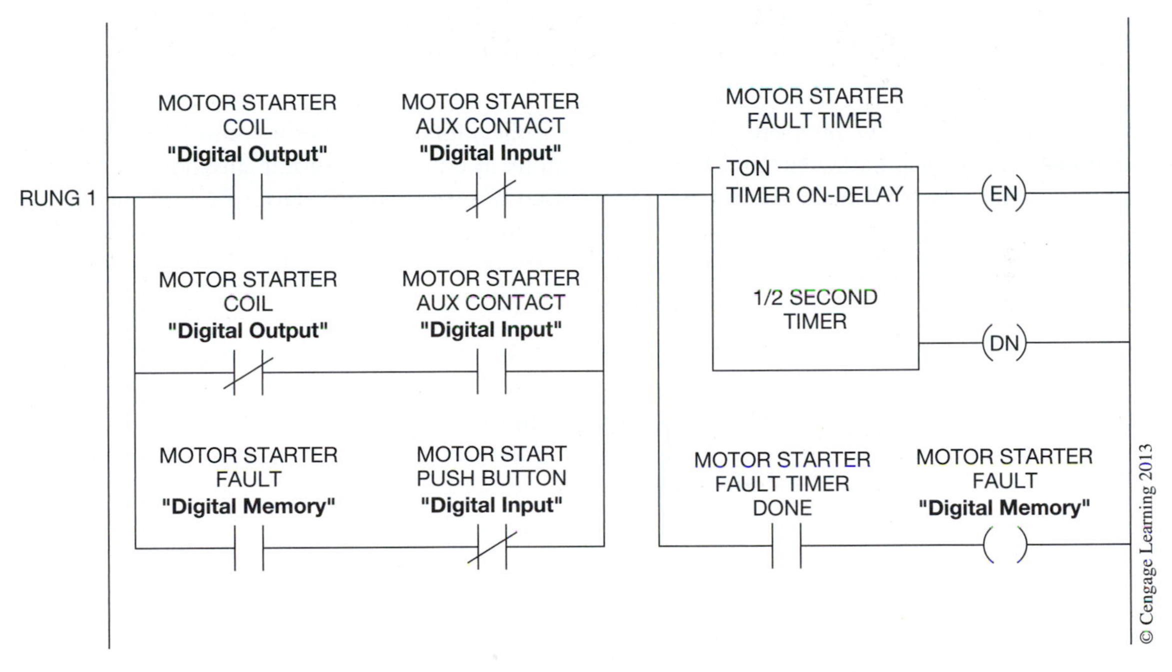

Figure 22–5 Motor Starter Fault-Monitoring Logic

In this circuit, a single motor starter can be monitored for a failure both to *pull in* and/or *drop out* by evaluating the state of the output with that of the input. If the output to the starter is *ON* and the input from the starter auxiliary contact is *OFF,* then this would be considered a failure to pull in, and conversely, if the output to the starter is *OFF* and the input is *ON* then this would be considered a failure to drop out. This logic is shown in the first two branches of Figure 22–5. If a motor starter failure is detected, then the third branch of Figure 22–5 acts as a holding circuit for the fault condition until it can be acknowledged and reset by an operator or maintenance person. In our example, we have chosen to use the "Motor Start" push button to also act as the fault reset button, but you could just as well use a separate reset button, or for that matter, any other condition that fits your application or needs.

The On-Delay timer in Figure 22–5 is required to allow for the physical movement (time) of the motor starter when energizing and de-energizing the starter. The ½ second time delay chosen for the On-Delay timer is typically more than adequate for most applications, but can be adjusted as necessary. If the On-Delay timer should time out, then the "Motor Starter Fault" memory bit will turn *ON* and remain *ON* until reset. This "Motor Starter Fault" memory bit can be used in your PLC program as necessary to disable the motor starter output rung, flash pilot lights, shut down other equipment, etc.

EXAMPLE 4—THREE-WIRE MOTOR CONTROL LOGIC WITH FAULT MONITORING, PILOT LIGHT, AND FLASHER CIRCUIT

In this example, we will show how the ½ Second Pilot Light Flasher and Motor Starter Fault-Monitoring Logic shown previously can be used in a basic three-wire motor control circuit. In our example we will use both Start and Stop push buttons to control the motor. In addition to starting the motor, the Start push button will also be used to reset a motor starter fault if one should occur. Along with the Start and Stop push buttons, a pilot light will also be used to indicate when the motor is running, as well as when a motor fault occurs. Shown in Figure 22–6 is the logic for Example 4.

Rung 1 of Figure 22–6 is your basic three-wire motor control circuit for controlling the motor output. In addition to the Start and Stop push buttons in Rung 1, you will also notice an N.C. contact labeled "Motor Starter Fault" that is in series with the motor output coil. This contact is used to turn *OFF* the output to the motor starter during a motor fault condition. Rung 2 of Figure 22–6 is the Motor Starter Fault-Monitoring Logic that was described in Example 3, always monitoring the motor starter for abnormal conditions. Rungs 3 and 4 set up a ½ second flasher circuit to be used in Rung 5 for flashing the motor pilot light when a motor fault should occur. Keep in mind that this flasher circuit could be located at the beginning of your program and used in any of the pilot light circuits in your program as needed. Rung 5 is the Motor Pilot Light circuit with Motor Fault Flasher included.

EXAMPLE 5—TIME-BASED EVENTS

Often it is desirable to trigger an event at some predetermined time. If the time is in seconds or minutes, then a basic On-Delay or Off-Delay timer may be all that is needed, but if the time is in hours or days then additional PLC logic is required. In the examples presented here, our goal is to show two methods that can be used to trigger an event at a predetermined time that a timer instruction alone is unable to do. In the following two examples, we shall assume we are operating a ball mill that requires routine service by the plant maintenance personnel every 168 hours or 7 days of ball mill operation. The event to be triggered at the end of the 168 hours is a maintenance warning light that is used to alert the maintenance personnel that the ball mill requires service. In our first example, we will use a counter and timer instruction to keep track of the number of hours the ball mill has operated and trigger our maintenance warning light. In the second example we shall accomplish the same thing, but instead of a counter and timer, we shall use math and compare instructions along with a timer.

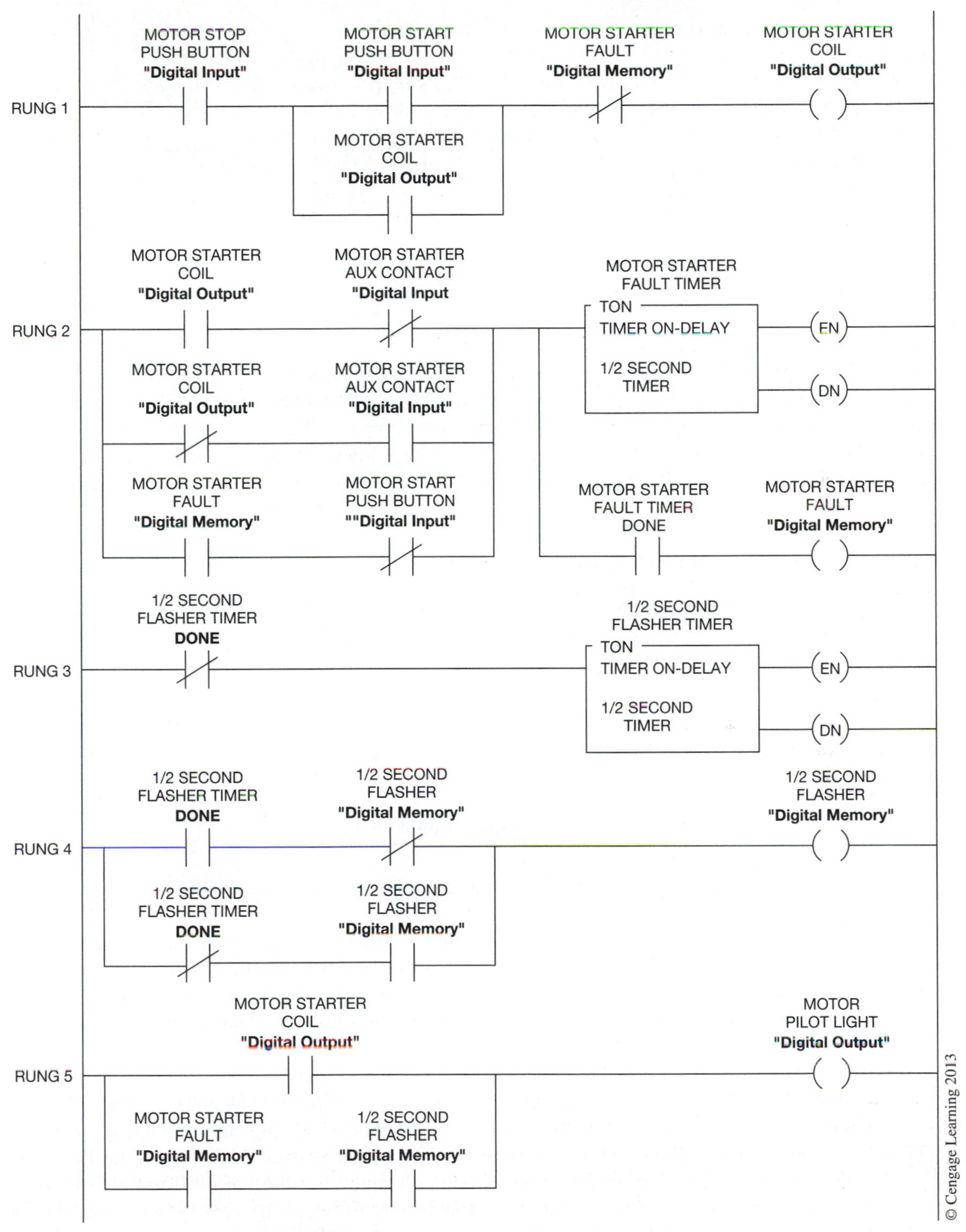

Figure 22–6 Motor Control and Monitoring Logic

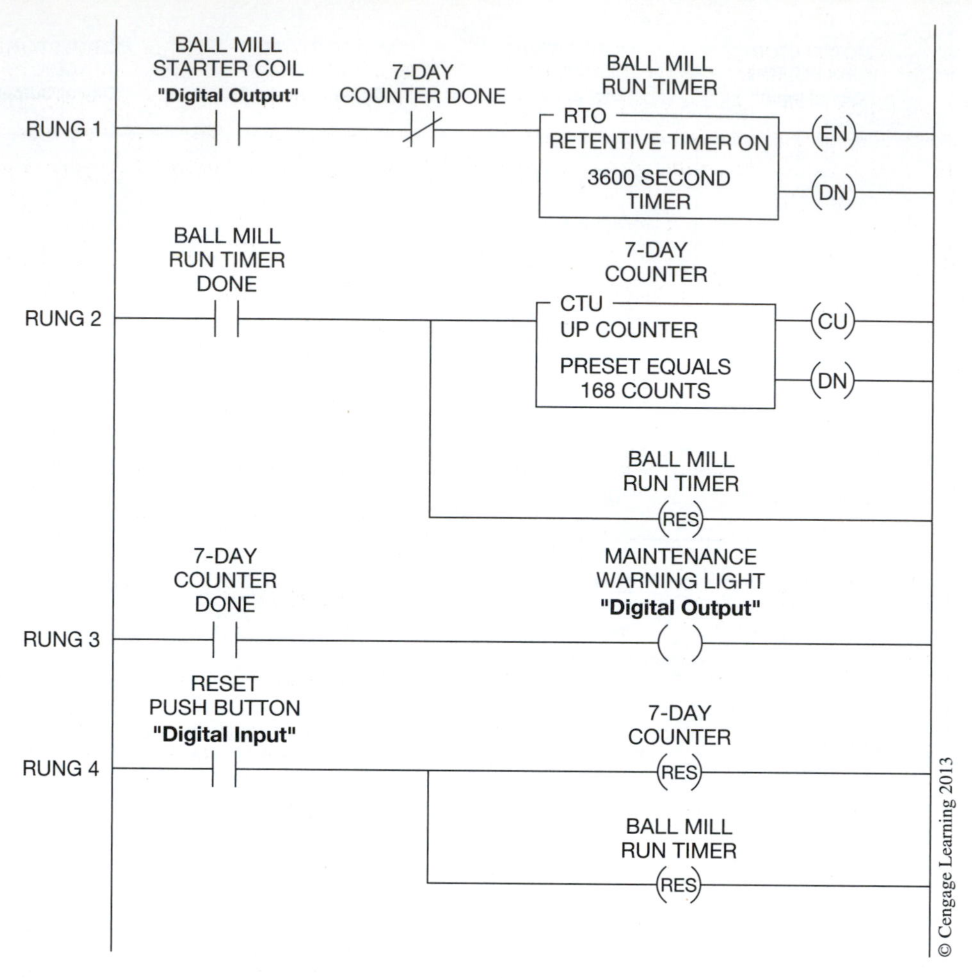

Figure 22–7 Time-Based Logic Using Counter

Shown in Figure 22–7 is the logic for our time-based event using a counter and timer.

In Rung 1 of Figure 22–7 the On-Delay Retentive Timer is enabled whenever the ball mill is running. A retentive On-Delay timer was chosen so that any accumulated time would not be lost in the event the mill was stopped and then restarted. The On-Delay timer is set to time in seconds with a preset time of 3600 seconds or one hour. When the On-Delay timer has timed out at the end of one hour, the done bit of the timer is used to increment the counter (CTU) in Rung 2. At the same time the counter is incremented, the retentive On-Delay timer is also reset, allowing it to begin another timing cycle. The counter instruction in Rung 2 is programmed with a preset count of 168 counts, which is equal to 168 hours and the time required to trigger our maintenance warning light. When the counter accumulated value equals the preset, the done bit of the counter is used in Rung 3 to turn *ON* our maintenance warning light, alerting the maintenance personnel that the ball mill requires service. A "Reset" push button is used to reset the counter and timer to begin the operation over as shown in Rung 4. The "Reset" push button could be an integral part of the maintenance warning light.

The done bit of the counter is also used to keep the timer and counter from counting above the preset time of 168 hours. If you wish to track the hours beyond 168 for display or HMI purposes, then remove the counter done bit from Rung 1, providing the counter instruction you are using will increment beyond the preset value of the counter. Remember that the accumulated value of a counter has a maximum value.

Shown in Figure 22–8 is the logic for our time-based event using a Timer, Math, and Compare Instructions. The same PLC logic used in Rung 1 of Figure 22–7 will also be used in this example.

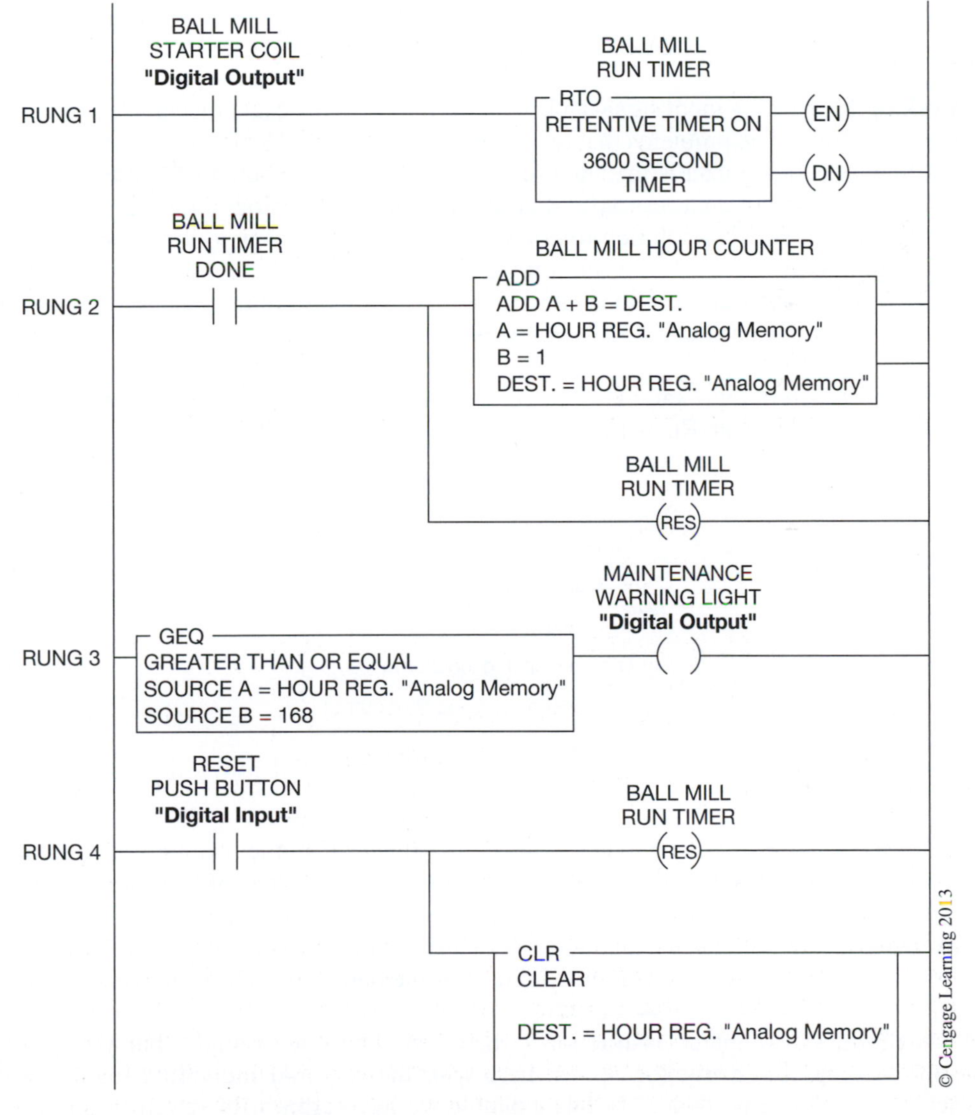

Figure 22–8 Time-Based Logic Using Math and Compare

In fact, the primary difference between the two examples is that the counter instruction in Figure 22–7 is replaced with a math "Add" instruction, and the counter done bit in Rung 3 is replaced with a compare "Greater Than or Equal" instruction. To reset the accumulated hours in the analog memory word, a math "Clear" instruction is used in place of the counter reset instruction in Rung 4 of Figure 22–7, as is shown in Figure 22–8. In this example we have chosen to allow the accumulated hours to increment above 168 if not reset. If you do not wish to have the accumulated hours increment above 168, then what would you do?

Note: *Remember that the length of the program affects scan time, which in turn affects timer accuracy and total time. The actual time it takes to count to 168 hours may be 168 hours plus.*

EXAMPLE 6—ANALOG SIGNAL FILTER ALGORITHM

When working with analog input signals it is not uncommon to have analog signals that may contain some degree of noise. The problems that this noise can cause vary, depending on the severity of the noise and how the analog data is intended to be used. Common problems include digital displays that become hard to read, unnecessary and erratic operation of analog output devices, process errors or disturbances, etc. The solution to noisy analog input signals is to implement some kind of signal conditioning, either hardware- or software-based. In this PLC programming example, you will see one possible software solution that requires very little programming to implement and is an improvement over other methods, such as running or weighted averages.

The PLC logic shown in this example is based on a math algorithm that uses **recursion** to provide a filtering effect similar to an RC network. The formula is shown in Figure 22–9.

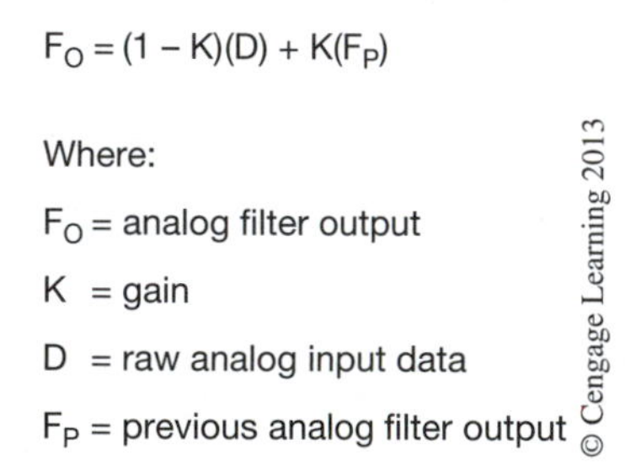

Figure 22–9 Analog Filter Equation

To adjust the amount of filtering needed, change the value of K in the formula. Maximum filtering occurs when K approaches 1.0, and if K is zero, the filter is off. Values of K between 0.80 and 0.90 provide the best results. When you are using this filter, the equation shown in Figure 22–9 should not be executed any faster than the analog data acquisition update time. Keep in mind that this filter will introduce some process *lag*. The PLC logic to implement this filter is shown in Figure 22–10.

The Allen-Bradley "Compute" instruction is being used in our example, but you could easily replace the "Compute" instruction shown here with the required individual math instructions, "Subtract," Multiply," and "Add." In this example we have arbitrarily set our filter update time

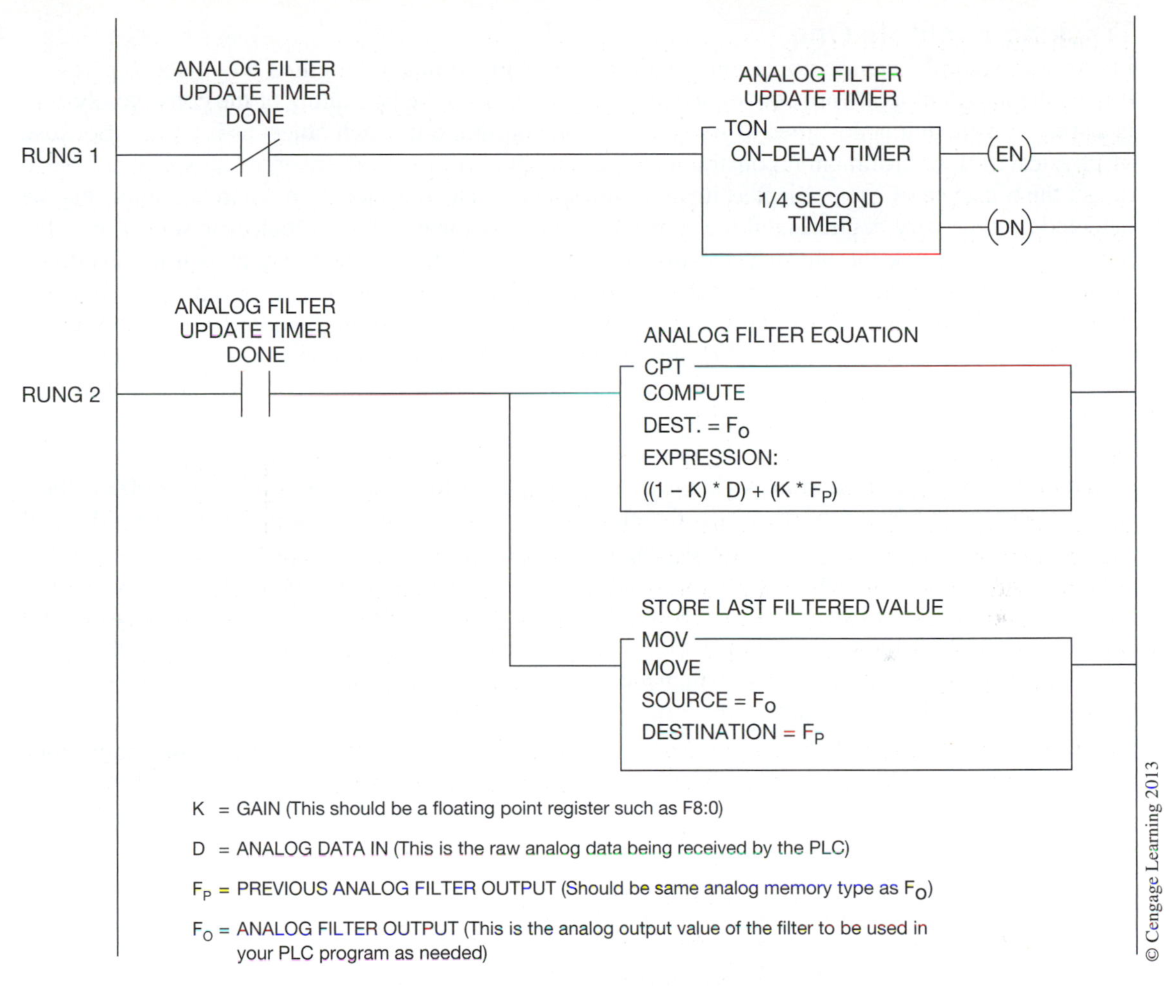

Figure 22–10 Analog Filter Logic

to one-quarter of a second, as indicated in our On-Delay timer in Rung 1. Make sure you adjust the On-Delay timer in Rung 1 to match your analog data acquisition update time. The "Move" instruction in Rung 2 stores the current filtered analog value for use in the next calculation as the previous value.

EXAMPLE 7—PARTS CONVEYOR TRACKING LOGIC

In many industrial plants it is often desirable to track objects as they are transported from one end of a conveyor system to the other. There can be many reasons for tracking an object, including tracking the physical location of an object; tracking the information about an object such as part number, length, destination, etc.; or just simply keeping track of the number of objects. Every tracking requirement is unique, both in the hardware and PLC logic needed. For this reason, it is only possible to show several basic examples of tracking logic.

Tracking Example One

In our first example, we are going to track the physical location and length of an object or part as it travels the length of a 50-foot conveyor. Located 25 feet from the beginning of the conveyor system is a spray nozzle that is designed to spray a cleaning solution onto each object as it passes. Because of physical and environmental constraints we are unable to mount a sensor at the spray nozzle to detect the presence of each object as it passes the sprayer. The only location where an object can be detected is at the very beginning of the conveyor system by means of a photoelectric sensor. For this reason, we must track the physical location along with the length of each object as it travels down the conveyor system in order to turn *ON* and *OFF* the spray nozzle at the appropriate time. In addition to the photoelectric sensor, an incremental encoder is mounted on the conveyor system and is designed to produce a pulse (*ON/OFF* signal) for every 1 inch that the conveyor travels. These two digital PLC inputs, Photoelectric Sensor and Incremental Encoder, will be used to track the position and length of each object as it travels down the conveyor system.

Based on the above information we know that the length of the conveyor is 50 feet or 600 inches, which also equals 600 encoder pulses; therefore, the spray nozzle located at 25 feet would equal 300 encoder pulses. By taking the encoder input and using it to operate a "Bit Shift Left" instruction with a length of 600 bits, we are able to have a bit location that corresponds to each inch of conveyor length. Figure 22–11 shows the conveyor, photoelectric sensor, spray nozzle, and 600-bit array laid out on the length of the conveyor system. As you study Figure 22–11 you will note that the spray nozzle is located at bit location 300, which we know equals 25 feet. For each inch that the conveyor system travels, all bits in the array are shifted one position to the left toward bit location 600.

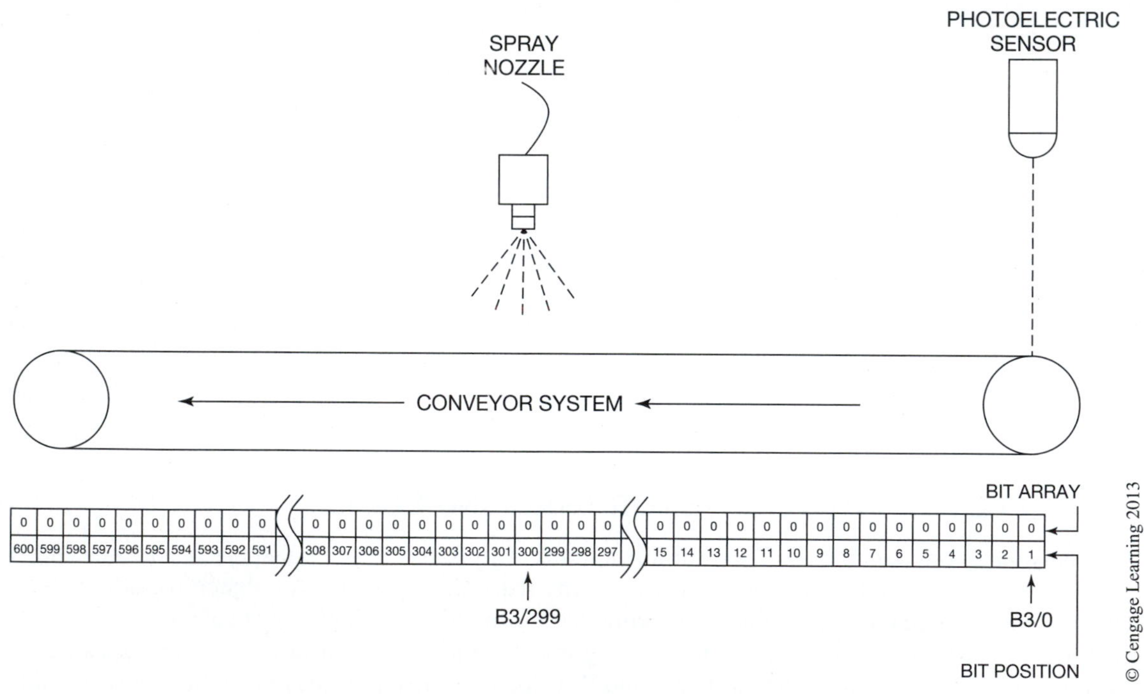

Figure 22–11 Conveyor Tracking System Layout

If we use the photoelectric sensor input as the input bit address to the "Bit Shift Left" instruction, we will in effect be setting the bit in position 1 of the array to *ON* or 1 any time that the photoelectric sensor is blocked by an object. Each time the conveyor moves one inch, the status of the photoelectric sensor, *ON* or *OFF*, is loaded into the bit array and shifted one position to the left. If an object on the conveyor system is 12 inches in length and blocks the photoelectric sensor as it passes, then our bit array will have 12 consecutive bits that will be *ON*, matching the length of the object (see Figures 22–12a and b). As the object travels down the conveyor system towards the spray nozzle, the 12 bits matching the object's length are also being shifted through the bit array, in effect tracking the object's position on the conveyor system. When the first bit of the string of 12 bits reaches the 300th bit location in the array, the spray nozzle is energized, spraying cleaning solution on the object, and it remains *ON* until all 12 bits have passed the 300th bit location (see Figure 22–12c).

Shown in Figure 22–13 is the PLC logic for our bit array tracking system described above. We have shown the "Bit Shift Left" instruction with a file or array address of #B3:0. We have also shown the file or array address for the Photoelectric Sensor and Spray Nozzle in Figures 22–11 and 22–12 for clarity. The "Conveyor System Running" digital input in Rung 2 is there to ensure that the "Spray Nozzle" output operates only when the conveyor is moving.

If the conveyor system can be operated in both the forward and reverse directions, then the addition of a "Bit Shift Right" instruction with the same file or array address as that of the "Bit Shift Left"

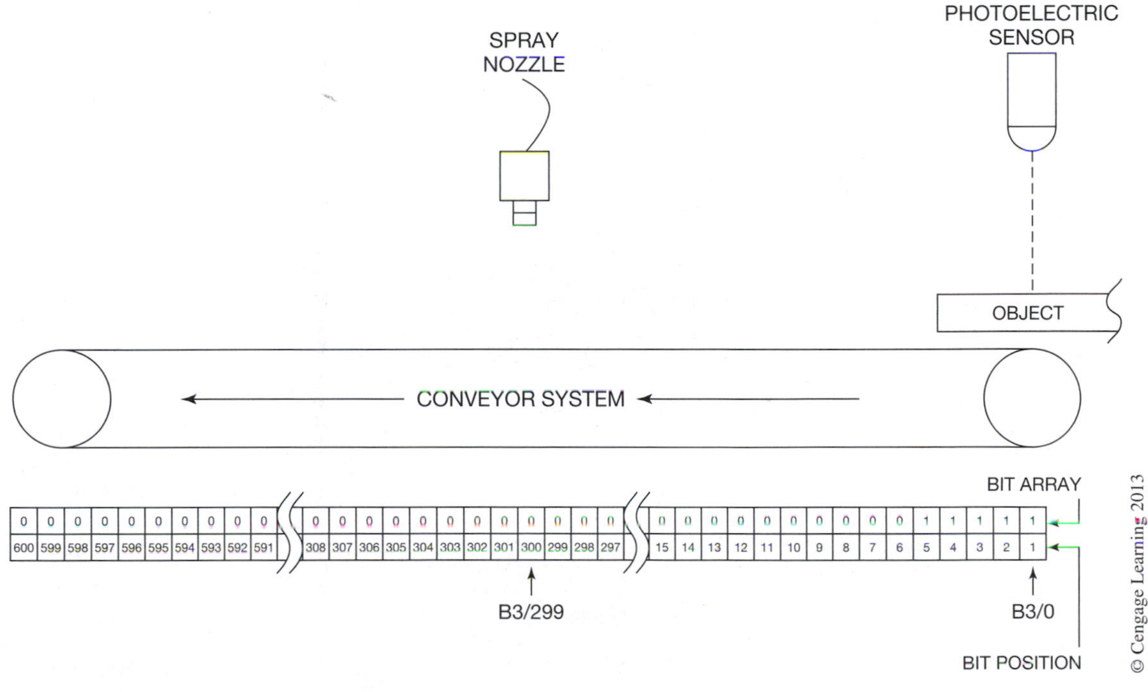

Figure 22–12a Object at Photoelectric Sensor

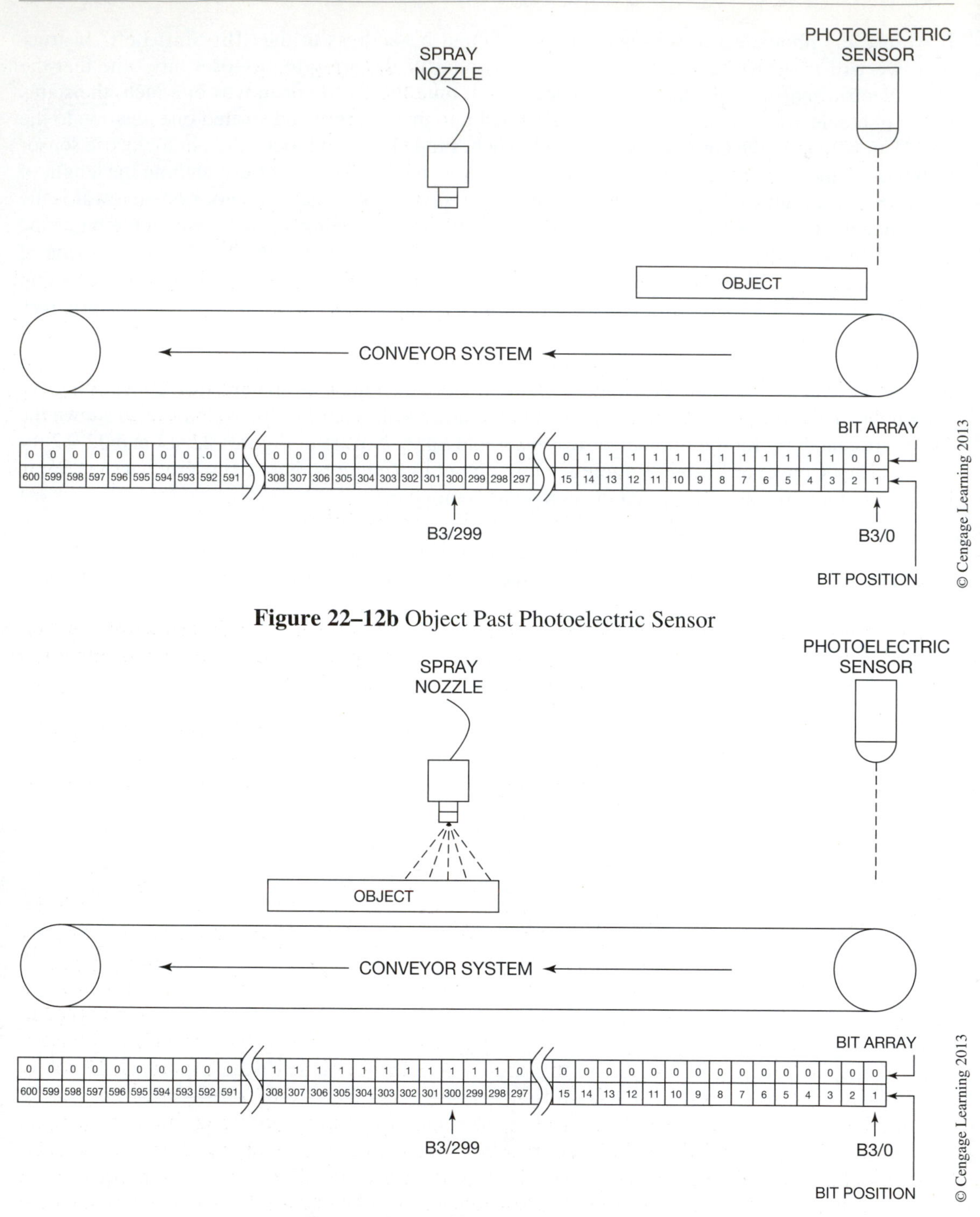

Figure 22–12b Object Past Photoelectric Sensor

Figure 22–12c Object at Spray Nozzle

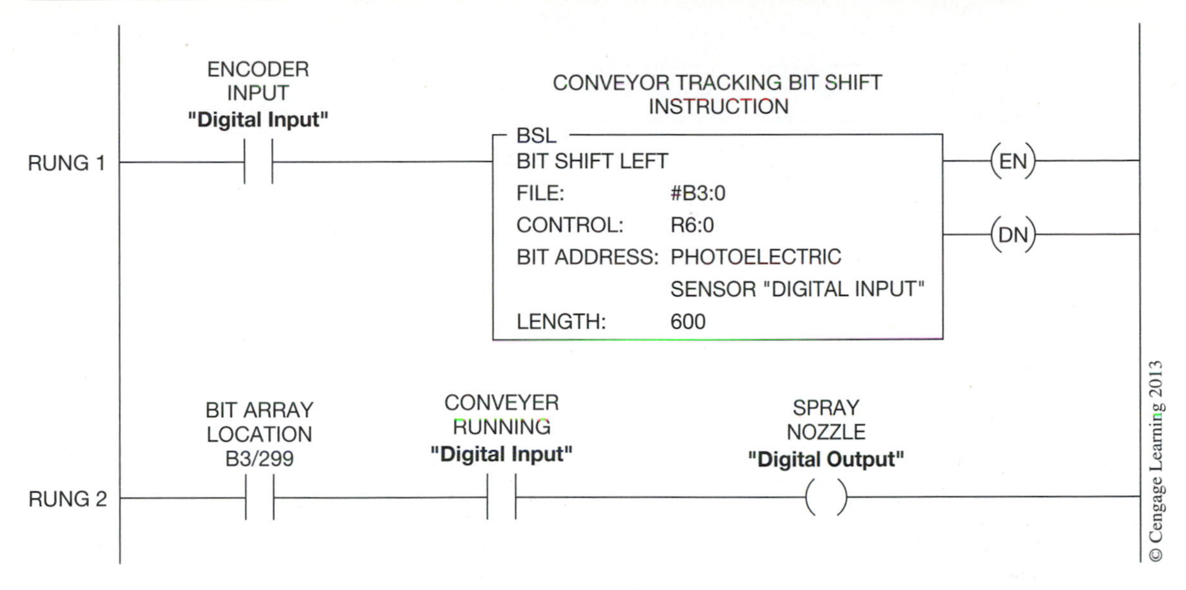

Figure 22–13 Bit Array Tracking Logic

should be used. In addition to the "Bit Shift Right" instruction, each bit shift instruction must be enabled *only* when the conveyor system is operating in the appropriate direction, using either the status of the motor starters or a **quadrature** encoder (see Figure 22–14).

Note: *When using an encoder with standard digital PLC input modules, make sure that the encoder pulses are not changing faster than can be read by the PLC, including scan time, or missed pulses could occur, causing inaccurate operation. Many PLC manufacturers sell special encoder or counter modules for high-speed operation.*

Tracking Example Two

In our second tracking example, we are going to track the part number of an object as it travels the length of the same 50-foot conveyor. We will use the same encoder and digital inputs as in the first example. The only difference between this example and the first is that we will replace the spray nozzle with a Diverter Gate and add a Bar Code Reader to the system. Instead of loading 1s and 0s into a bit array, we will be loading part numbers into a word array. The Diverter Gate will be designed to divert an object off the conveyor system if its part number matches the reject part number stored in PLC memory. The diverter must also be activated for the length of the object. Figure 22–15 shows the conveyor, bar code reader, photoelectric sensor, diverter gate, and 600-word array laid out on the length of the conveyor. The Diverter Gate, like that of the spray nozzle in the first example, is also located 25 feet or 300 words from the start of the conveyor system. The Bar Code Reader, located just ahead of the photoelectric sensor, reads the bar code number or part number of each object and stores the part number in PLC analog memory word address (N7:601) to be used by the word array tracking logic.

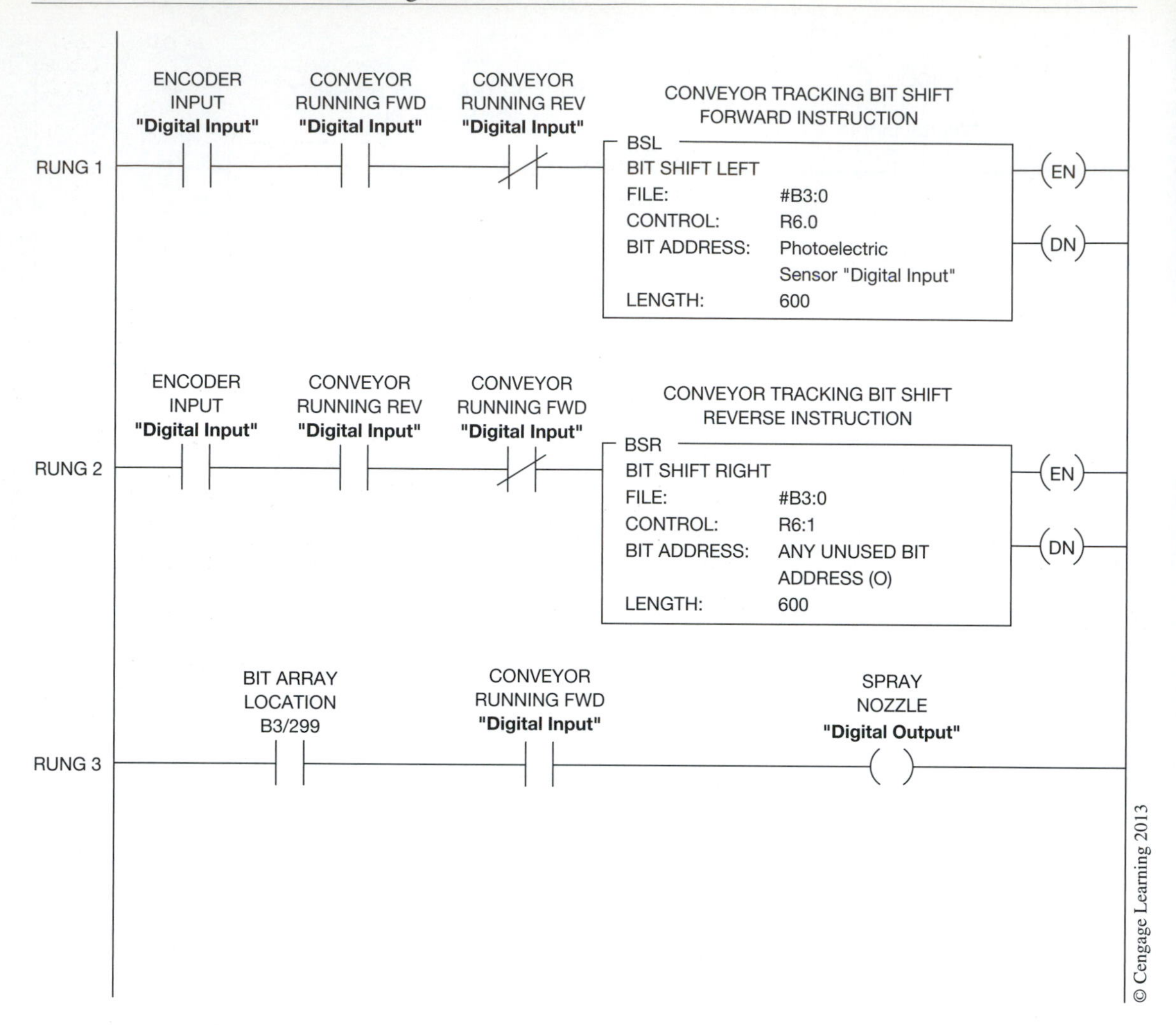

Figure 22–14 Bit Array Tracking Logic Forward/Reverse

In this example, we will need to use some type of word shift instruction that will allow us to shift part numbers through a word array, as was done in the first example with the "Bit Shift Left" instruction. As you will recall from Chapter 16, most PLCs have some type of asynchronous shift register (FIFO) instructions, such as Allen-Bradley's "FFL" and "FFU" instructions, that can be used to shift word values into a stack or array. We could easily use the "FFL" and "FFU" instructions to shift our part numbers through the array, but we have chosen to use the Allen-Bradley file copy instruction (COP) because it only requires one instruction, not two; is faster to execute than the FIFO instructions; and is simpler to implement for our application. Figure 22–16 shows the PLC logic for our part number tracking system.

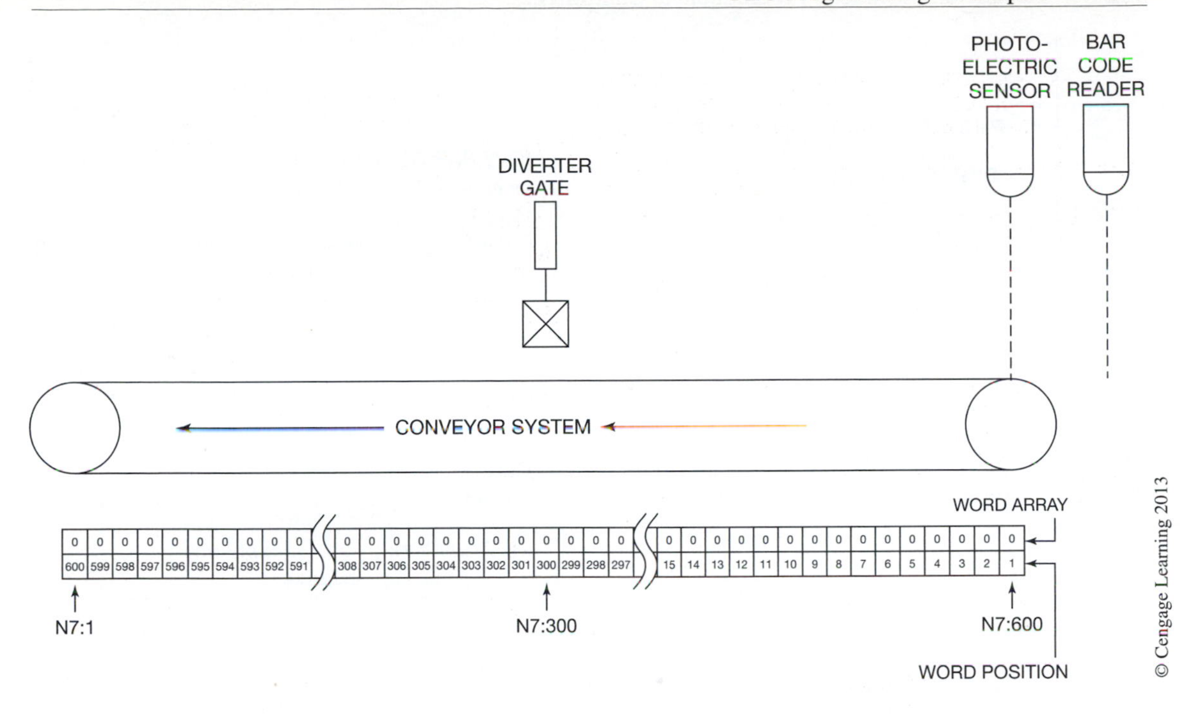

Figure 22–15 Conveyor Bar Code Tracking System

The move “MOV” instruction in Rung 1 is used to move the part number, read by the Bar Code Reader, into the first word of the array when the photoelectric sensor is detecting an object. Rung 2 clears the first word of the array if no part is detected by the photoelectric sensor. Just as in the first example, in which we were loading 1s or 0s into a bit array, here we are loading part numbers and/or 0s into the word array based on the photoelectric sensor input.

Rung 3 is the copy “COP” instruction that is used to shift the part numbers through the word array when the encoder input is true. When the “COP” instruction is used in this manner, the values in the word array are shifted from the highest word location (N7:600) to the lowest word location (N7:0); see Figure 22–15. The one-shot “ONS” in Rung 3 is required to ensure the Copy Instruction only moves the data in the word array one position whenever the encoder input is *True*.

By studying Figure 22–15 we know that the Diverter Gate is located at word position 300 in the word array, just as the Spray Nozzle was located at bit position 300 in the bit array. Rung 4 in Figure 22–16 uses an equal-to “EQU” instruction to see if the part number at word position 300 in the word array is equal to the reject part number stored in PLC memory, and if true, energizes the Diverter Gate, diverting the object off the conveyor system. As with the spray nozzle, the diverter gate is *ON* for the full length of the object.

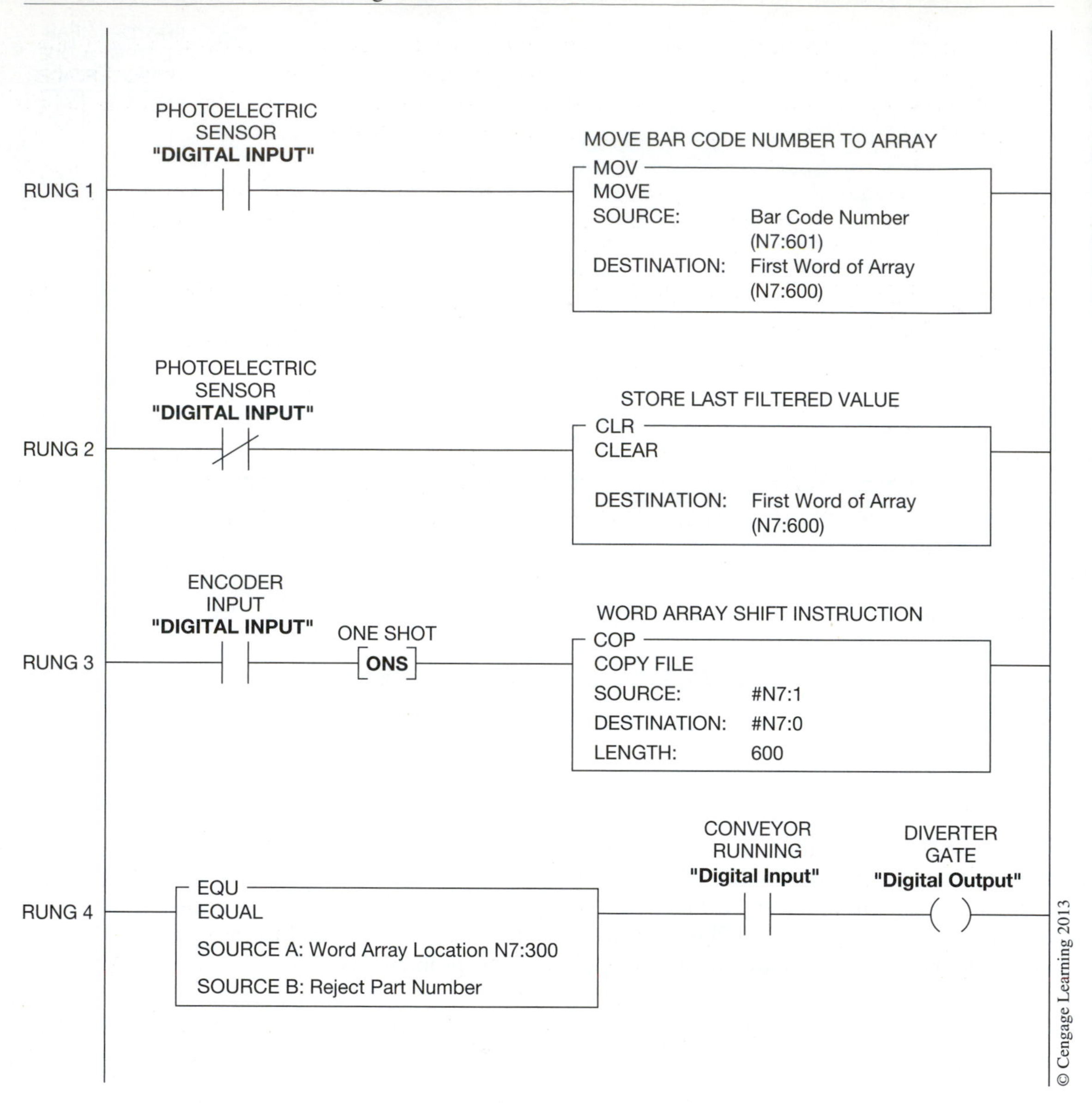

Figure 22–16 Word Array Tracking Logic

Chapter Summary

After working through the examples presented in this chapter, you should begin to see how the various PLC instructions can be combined to provide control logic solutions. As was stated at the beginning of this chapter, there is typically more than one solution to a problem and the right solution depends on having a thorough understanding of the machine or process to be controlled. Along with understanding what you are controlling, you must also understand how the various PLC instructions work and how the PLC executes them. Think of the various PLC instructions as tools in your tool bag: They are only as good as the person who knows how to use them.

When developing PLC logic always think about safety to personnel and preventing damage to equipment. Never create an unsafe condition; always look to prevent it. SAFETY FIRST!

Review Questions

No time for questions! It is time to go out and start programming.

GLOSSARY

ADDRESS: A location in processor memory.

ANALOG INPUT MODULE: A module that converts an analog input signal to a binary or BCD number for use by the processor.

ANALOG OUTPUT MODULE: A module that provides an output proportional to a binary or a BCD number provided to the module by the processor.

ANALOG SIGNAL: A continuous signal that depends directly on magnitude (voltage or current) to represent some condition. For example, a voltage might represent the speed of a motor (5 V corresponding to 200 rpm, 10 V corresponding to 400 rpm, etc.).

AND LOGIC: Logic that has a series relationship. Two devices that are in series would both have to be true to pass the logic. Device one AND two need to be true for the logic to pass.

ARITHMETIC CAPABILITY: The ability of a PLC to perform addition, subtraction, multiplication, division, and other math functions.

ARRAY: An array is similar to a data file and lets you group data of the same data type using a common name. An array tag occupies a contiguous block of memory.

ASCII: Acronym for **A**merican **S**tandard **C**ode for **I**nformation **I**nterchange. It is a seven- or eight-bit code for representing alphanumerics, punctuation marks, and certain special characters for control purposes.

ASSIGNMENT: Use an assignment statement to assign values to tags when using structured text programming.

ASYNCHRONOUS SHIFT REGISTER: An instruction that shifts data one word at a time into a file or register.

BANDWIDTH: The amount of data that can be transmitted in a fixed amount of time. For digital devices, the bandwidth is usually expressed in bits per second (bps) or bytes per second. For analog devices, the bandwidth is expressed in cycles per second, or Hertz (Hz).

BAUD: The seven or eight bits that make up a character. A character can be a letter, number, symbol, etc.

BAUD RATE: A unit of data transmission speed equal to the number of code elements (characters) per second. For example, 300 baud is thirty characters—letters, numbers, symbols—per second. 1200 baud is 120 characters per second.

BINARY: A numbering system that uses a base of 2. There are two digits (1 and 0) in the binary system.

BINARY CODED DECIMAL (BCD): One of several numbering systems used with PLCs. This unique numbering system uses four binary digits to represent each decimal digit from 0 to 9. Groups of four binary digits are grouped together to display decimal numbers. Twelve bits can represent a three-digit number. Sixteen bits are needed to represent a four-digit number.

BIT: An acronym for **B**inary dig**IT**. A bit can be only one of two possible states: *ON* or *OFF*, high or low, logic 1 or logic 0, etc.

BOOL EXPRESSION: A BOOL expression uses BOOL tags, relational operators, and logical operators to compare values or check if conditions are true or false. A BOOL expression can be found in sequential function chart programming.

BOOLEAN ALGEBRA: Shorthand notation for expressing logic functions.

BOOLEAN EQUATION: Expression of relations between logic functions and/or elements.

BOOT: A term used in the computer world to indicate that a computer has been turned on, and the software has been loaded.

BRANCH: A parallel logic path within a user program rung.

BREAKDOWN VOLTAGE: The voltage at which a disruptive discharge takes place, either through or over the surface of insulation.

BUFFER: A temporary storage area where information is held while the printer or other device catches up with the transmission speed of the data.

BUS: A central cable that connects all devices on a local-area network (LAN). It is also called the backbone.

BUS TOPOLOGY: The physical configuration of a communications network in which all devices are connected to a central cable, called the bus or backbone. Bus networks are relatively inexpensive and easy to install for small networks. PLC systems use a bus topology.

BYTE: A sequence of binary digits usually operated upon as a unit (normally eight bits).

CASCADING: A programming technique that extends the ranges of timer and/or counter instructions beyond the maximum values that normally may be accumulated.

CENTRAL PROCESSING UNIT (CPU): Another term for **PROCESSOR.**

CHARACTER: One symbol of a set of elementary symbols, such as a letter of the alphabet, a decimal numeral, a punctuation mark, etc.

CLOCK: A device (usually a pulse generator) that generates periodic signals for synchronization or timing.

CMOS: An acronym for Complementary **M**etal **O**xide **S**emiconductor. A family of very low-power, high-speed integrated circuits.

COAXIAL CABLE: A type of cable that consists of a center wire surrounded by insulation and then a grounded shield of braided wire. The shield minimizes electrical and radio frequency interference. Coaxial is the primary type of cabling used by the cable television industry and is also used for many PLC networks, such as ControlNet.

CODE: A system of symbols (bits) for representing data (characters).

COMPARE FUNCTION: A program instruction that compares numerical values for "equal," "less than," "greater than," etc.

COMPATIBILITY: The ability of various specified units to replace one another, with little or no reduction in capability.

COMPUTER INTERFACE: A device designed for data communication between a central computer and another unit, such as a PLC processor.

CONSTRUCT: A conditional statement used to trigger structured text code when using structured text programming.

CONTACT SYMBOLOGY DIAGRAM: Commonly referred to as a ladder diagram, it expresses the user-programmed logic of the controller in relay-equivalent symbols.

CONTROL VARIABLE (CV): An electrical or pneumatic signal that is used to control the position, speed, etc., of devices such as valves, motors, and pumps that have a direct influence on the process in which they are placed.

COUNTER: A device that can count up or down in response to transitions (*OFF* or *ON*) of an input signal and opens and/or closes contacts when a predetermined count is reached. Counters are internal to the processor and are not real-world devices.

CPU: An abbreviation for **C**entral **P**rocessing **U**nit. It is used interchangeably with **PROCESSOR.**

CRC: Short for cyclic redundancy check, a common technique for detecting data transmission errors. Transmitted messages are divided into predetermined lengths that are divided by a fixed divisor. According to the calculation, the remainder number is appended onto and sent with the message. When the message is received, the computer or PLC recalculates the remainder and compares it to the transmitted remainder. If the numbers do not match, an error is detected.

CRT: The **C**athode **R**ay **T**ube, which is an electronic display tube similar to the familiar TV picture tube. It is more commonly called a monitor.

CURSOR: A means for indicating on a monitor the point at which data entry or editing is to occur.

DATA MANIPULATION: The process of altering and/or exchanging data between storage words.

DATA TRANSFER: The process of exchanging data between PLC memory words and/or areas.

DECREMENT: A term used with counters to indicate that the value of the counter has decreased. When a counter value goes from 4 to 3, it is said to have decremented by 1.

DEFAULT: The initial setting of a value, or the initial assignment of a file by the software. Default values may or may not be changeable, depending on the software and the PLC manufacturer.

DEFAULT DRIVE: The drive that will be used if no other drive has been specified.

DERIVATIVE: Something derived from another. In the case of a process controller, the change in the output by an amount proportional to the rate of change of the process variable.

DIGITAL: The representation of numerical quantities by means of discrete numbers. It is possible to express in binary digital form all information stored, transferred, or processed by dual-state conditions; for example, *ON*/*OFF*, open/closed, etc.

DINT: A data type that stores a 32-bit signed integer value.

DIP SWITCH: Dual In-Line Package Switch; multiple switches installed in a small device or package that are used to set PLC parameters.

DISCRETE INPUT: An input that is either *ON* or *OFF*. Examples of discrete inputs are limit switches, push buttons, float switches, etc.

DISCRETE INPUT MODULE: A module that converts signals from real-world input devices to logic level signals for use by the processor.

DISCRETE OUTPUT: An output that is either *ON* or *OFF*. Examples of discrete outputs are solenoids, motor starter coils, pilot lights, etc.

DISCRETE OUTPUT MODULE: A module that converts the logic levels of the processor to an output signal to control a real-world output.

DISKETTE: A magnetic medium for storing information that can later be read by the computer or PLC.

DOUBLE PRECISION ARITHMETIC: Two words are used to store the results of arithmetic operations, thereby increasing the size of the value that can be stored. (See **SINGLE PRECISION ARITHMETIC.**)

DUMP: A term used when information stored in memory is copied or recorded onto magnetic tape or disk.

DUPLEX: A means of two-way data communication. Also see **FULL DUPLEX** and **HALF DUPLEX.**

EAROM: A type of programmable memory that can be erased or altered electrically. The term stands for **E**lectrically **A**lterable **R**ead **O**nly **M**emory.

EEPROM OR E2PROM: A memory chip that can be programmed using a standard programming device, and can be erased when the proper signal is applied to the erase pin. The initials stand for **E**lectrically **E**rasable **P**rogrammable **R**ead **O**nly **M**emory.

ELECTRICAL NOISE: Noise, or voltage spikes, that is generated whenever inductive loads such as relays, solenoids, motor starters, and motors are operated by "hard contacts" such as push buttons, selector switches, and relay contacts.

ELECTRICAL-OPTICAL ISOLATOR: A device that couples different voltage levels using a light source and detector in the same package. It is used to provide electrical isolation between line voltage input and output circuitry and the processor.

ELEMENT: A program instruction (N.O. contact, timer, counter, etc.) displayed on a monitor. An element is also an addressable unit of data that is a sub-unit of a larger unit.

EMI: Electromagnetic induction. The term used to describe electrical noise.

ENABLED: A term used to indicate that a function or operation has been activated.

EVEN PARITY: The condition that occurs when the sum of a string of binary digits, 1s and 0s, is an even number. Parity is used for error checking.

EXAMINE OFF: An EXAMINE OFF PLC instruction is a true precondition if its addressed bit is *OFF* (0). It is false if the bit is *ON,* or 1.

EXAMINE ON: An EXAMINE ON instruction is a true precondition if its addressed bit is *ON* (1). It is false if the bit is *OFF,* or 0.

EXPRESSION: An expression is part of a complete assignment or construct statement and evaluates to a number or to a true or false state when using structured text programming.

FALSE: When relating to PLC instructions, an *OFF* state or condition.

FAULT: Any malfunction that interferes with normal operation.

FIFO: First-In First-Out. A reference to the way that information is stored and removed from a file or register.

FILE: A group of words, usually consecutive, that is used to store information.

FORCE: A mode of operation or instruction that allows the operator (as opposed to the processor) to control the state of an input or output device.

FORCE OFF FUNCTION: A feature that allows the user to de-energize any input or output by means of the programmer, independent of the PLC program.

FORCE ON FUNCTION: A feature that allows the user to energize any input or output by means of the programmer, independent of the PLC program.

FORTRAN: An acronym for **FOR**mula **TRAN**slation, a scientific programming language.

FRAME: A packet of transmitted information.

FULL DUPLEX (FDX): A mode of communications in which data may be simultaneously transmitted and received by both ends (sender/receiver).

GROUND: A conducting connection, intentional or accidental, between an electric circuit or equipment chassis and the earth ground.

GROUND POTENTIAL: Zero voltage potential with respect to earth ground.

HALF DUPLEX (HDX): A mode of data transmission capable of communicating in two directions, but in only one direction at a time.

HARD CONTACTS: Any type of physical switch contacts contrasted with electronic switching devices, such as triacs and transistors.

HARD COPY: Any form of printed document such as a ladder diagram, program listing, data table configuration, etc.

HARD DRIVE: A storage system that consists of an inflexible (hard) disk, as opposed to a floppy disk, that is used to store files, directories, software programs, etc.

HARDWARE: The mechanical, electrical, and electronic devices that constitute a programmable logic controller and its application.

HARDWARE KEY: A piece of hardware that is required for a program to run. It may be in the form of a plug or connector plugged into the printer port that allows the software to run. Without the hardware key installed, the software does not work, or limits the access to certain portions of the program.

HARDWIRED: Devices interconnected through physical wiring.

HEAT SINK: Heat sinks work on convection and are used to dissipate the heat generated by electronic devices.

HEXADECIMAL: The numbering system that represents all possible statuses of four bits with sixteen unique digits (0–9 then A–F).

HIGH = TRUE: A signal type wherein the higher of two voltages indicates a logic state of 1, or *ON*. (See **LOW = TRUE.**)

HOLDING REGISTER: A register or file that holds a value or values for comparison or for use in a user program.

HUMAN MACHINE INTERFACE (HMI): A computer type device that provides the ability of a human to monitor and/or control a machine or process. Most HMIs are standard computers running special graphical interface software designed to communicate to one or more PLC controllers and/or other intelligent devices.

IEEE: An acronym for **I**nstitute of **E**lectrical and **E**lectronics **E**ngineers.

IMAGE TABLE: An area in PLC memory dedicated to I/O data. Ones and zeros (1s and 0s) represent *ON* and *OFF* conditions, respectively. During every I/O scan, each input controls a bit in the *input image table,* and each output is controlled by a bit in the *output image table.*

INCREMENT: A term used with counters to indicate that the value has increased. When a counter has counted up from 3 to 4, it is said to have incremented by 1.

INPUT DEVICES: Devices such as limit switches, pressure switches, push buttons, etc., that supply data to a programmable logic controller. These real-world inputs are of two types: those with common returns, and those with individual returns (referred to as isolated inputs). Other inputs include analog devices and digital encoders.

INSTRUCTION: A command or order that causes a PLC to perform a single prescribed operation.

INT: A data type that stores a 16-bit integer value.

INTEGRAL: Belonging to or forming a necessary part of a whole; to make up or complete as a whole. In the case of a process controller, the change in the output by an amount proportional to the error and the duration of the error.

INTERFACING: Interconnection of a PLC with its input and output devices and data terminals through various modules and cables. Interface modules convert PLC logic levels into external signal levels, and vice versa.

INTERPOSING RELAY: A relay that is added to a PLC circuit to handle current values larger than can be handled by one terminal of an output module.

INTERRUPTIBLE: Interruptible refers to a timer that can be interrupted but still retain its accumulated time.

I/O: An abbreviation for **I**nput/**O**utput. For example, a group of input modules and output modules would be referred to as I/O modules.

I/O ELECTRICAL ISOLATION: Separation of the field-wiring circuits from the logic level circuits of the PLC. This is typically achieved using electrical-optical isolators mounted in the I/O module.

I/O MODULE: The printed circuit assembly that interfaces between the user devices and the PLC.

I/O RACK: A chassis that contains I/O modules.

I/O SCAN TIME: The time required for the PLC to monitor all inputs, read the user program, and control all outputs. The I/O scan repeats continuously.

I/O SECTION: Interfaces the different signals from real-world devices and sensors to signals the CPU can use.

JOGGING: The momentary operation of a motor to test rotation or to cause small movements of the driven equipment.

JUMPER: A short length of conductor used to make a connection between terminals, around a break in a circuit, or around a device.

KEYED: A term used with PLCs to indicate that a keying device has been installed to prevent I/O modules from being installed in the wrong slot.

KILO-: A prefix used with units of measurement to designate quantities 1000 times as great (as in kilowatt). The exception to K having a value of 1000 is when referring to computer memory. Computer memory is counted using the binary numbering system and one kilo (or K) of memory is 1024, not 1000 ($2^{10} = 1024$).

LADDER DIAGRAM: A complete control scheme normally drawn as a series of contacts and coils arranged between two vertical supply lines so that the horizontal lines of contacts appear similar to rungs of a ladder. A ladder diagram is normally the reference document used by the operator when entering the control program. (See **CONTACT SYMBOLOGY DIAGRAM.**)

LADDER DIAGRAM PROGRAMMING: A method of writing a user's PLC program in a format similar to a relay ladder diagram.

LANGUAGE: A set of symbols and rules for representing and communicating information (data) among people, or between people and machines.

LATCH: A device that continues to store the state of the input signal after the signal is removed. The input state is stored until the latch is reset.

LATCH INSTRUCTION: A PLC instruction that causes an output to stay *ON,* regardless of how briefly the instruction is enabled. (It can only be turned *OFF* by an **UNLATCH INSTRUCTION** in a separate rung.)

LATCHING RELAY: A relay constructed so that it maintains a given position by mechanical means until released mechanically or electrically.

LEAST SIGNIFICANT DIGIT (LSD): The digit that represents the smallest value. In the number 102, the 2 is the least significant number, or digit.

LED: Acronym for **L**ight **E**mitting **D**iode.

LIFO: Last-In First-Out. A reference to the way information is stored or removed from a file or register.

LIMIT SWITCH: A switch that is actuated by some part or motion of a machine or equipment to alter the electrical circuit associated with it.

LIQUID CRYSTAL DISPLAY (LCD): A reflective visual readout. Because its segments are displayed only by reflected light, it has extremely low power consumption, as contrasted with an LED display, which emits light.

LOAD: 1. The power delivered to a machine or apparatus. 2. A device intentionally placed in a circuit or connected to a machine or apparatus to absorb power and convert it into the desired useful form. 3. To place data (e.g., a ladder diagram) into the processor's memory.

LOCAL AREA NETWORK (LAN): A computer or PLC network that spans a relatively small area. Most LANs are confined to a single building or group of buildings. However, one LAN can be connected to other LANs over any distance via telephone lines and radio waves. A system of LANs connected in this way is called a wide-area network (WAN).

LOGIC LEVEL: The voltage magnitude associated with signal pulses representing ones and zeros (1s and 0s) in binary computation.

LOW = TRUE: A signal type wherein the lower of two voltages indicates a logic state of 1, or *ON.* (See **HIGH = TRUE.**)

MAIN ROUTINE: The first routine to execute and be used to call (execute) other routines.

MALFUNCTION: Any incorrect functioning within electronic, electrical, or mechanical hardware. (See **FAULT.**)

MANIPULATION: The process of controlling bits or words within a program to obtain the required program outcomes.

MASK: Bits in a word that are used to prevent other bits in a different word from being used. If there is a 1 in the bit location of the mask, the corresponding bit in the output word is enabled and can be turned

ON and *OFF*. If the mask bit is set to 0, the corresponding bit in the output word is disabled, and does not allow the output to be turned *ON*, even if the program called for the bit to be turned *ON*.

MASK WORD: A word used to mask or selectively screen out data or bits of a word of memory.

MASTER/SLAVE: A type of network access method in which a master device controls communication traffic on a network. The master of a network typically polls every slave device to check if it has a message to transmit. In a master/slave configuration only the master can initiate communication. The slave can only reply if it receives a special message token that explicitly enables the slave to reply.

MCR: The Master Control Relay is used to disconnect all power to a PLC system.

MECHANICAL DRUM CONTROLLER: A type of **SEQUENCER** that operates switches by means of pins or cams placed on a rotating drum. The switch sequence may be altered by changing the pin or cam pattern.

MEMORY: The section of the programmable logic controller that stores the user program and other data. The storage may be either temporary or semipermanent.

MEMORY PROTECT: The hardware capability to prevent a portion of the memory from being altered by an external device. This hardware feature can be under actual key and lock control, or may use passwords that are referred to as software keys.

MENU: A display on the computer or PLC monitor that offers options or gives the operator choices to select from.

METROPOLITAN AREA NETWORK (MAN): A data network designed for a town or city. In terms of geographic area, MANs are larger than local-area networks (LANs), but smaller than wide-area networks (WANs). MANs are usually characterized by very high speed connections using fiber optic cable or other digital media.

MIDDLE DIGIT (MD): The middle digit of a three-digit number.

MILLIAMPERE (mA): One thousandth of an ampere: 10^{-3} or 0.001 ampere.

MILLISECOND (ms): One thousandth of a second; 10^{-3} or 0.001 second.

MINI-PLC: A scaled-down version of a standard PLC with small I/O capability.

MNEMONIC: A shorthand notation used with PLC instructions, such as OTE, the mnemonic for Output Energized; BST, for Branch Start; and so forth.

MODE: A selected method of operation (for example, *run, test,* or *program).*

MODEM: Acronym for **MO**dulator/**DEM**odulator. A device used to transmit and receive data by frequency-shift-keying (FSK). It converts FSK tones into their digital equivalents, and vice versa.

MODULE: An interchangeable "plug-in" item containing electronic components that may be combined with other interchangeable items to form a complete unit.

MOST SIGNIFICANT DIGIT (MSD): The digit representing the greatest value. In the number 102, the 1 is the most significant number, or digit.

MOTOR CONTROLLER: A device, or group of devices, that serves to govern, in a predetermined manner, the electrical power delivered to a motor.

NAND LOGIC: NAND logic is a combination of a NOT gate and an AND gate. By placing the invert, or NOT, symbol at the output of the AND gate the output can only be true when one or both of the inputs are false, or set to 0.

NEMA STANDARDS: Consensus standards for electrical equipment approved by the majority of the members of the **N**ational **E**lectrical **M**anufacturers **A**ssociation.

NESTING: A programming technique that has a "branch-within-a-branch." Depending on the PLC manufacturer, nesting may or may not be allowed.

NETWORK: A group of connected logic elements used to perform a specific function. A network can range from one element to a complete matrix of elements, plus coil(s) as desired by the user. The size and configuration of the matrix or rungs varies with PLC manufacturers.

NETWORK ACCESS CONTROL: The means used to control access to a data network. Common network access control methods include Master/Slave, Token Passing, and CSMA/CD.

NETWORK MEDIA: In computer and PLC networks, media refers to the cables linking processors together. There are many different types of transmission media, the most popular being twisted-pair wire, coaxial cable, and fiber optic cable.

NODE: A common connection point between two or more contacts or elements in a circuit. In communication networks, a processing location. A node can be a PLC, a computer, or some other device, such as a printer. Every node has a unique network address, sometimes called a Node address, Data Link Control (DLC) address, or Media Access Control (MAC) address.

NOISE: Extraneous signals; any disturbance that causes interference with the desired signal or operation.

NOISE SPIKE: Voltage or current surge produced in the industrial operating environment.

NONVOLATILE MEMORY: A memory that is designed to retain its information even though its power supply is turned off.

NOR LOGIC: NOR logic is the combination of a NOT gate and an OR gate. The NOR gate will only be logically true when both inputs, NOT and OR, are false, or set to 0.

NOT LOGIC: The NOT gate, often referred to as the inverter, will have only one input lead and one output lead. If the input is *OFF,* or set to 0, then the output will be *ON,* or set to 1. If the input is *ON,* or set to 1, then the output will be *OFF,* or set to 0.

OCTAL NUMBERING SYSTEM: A numbering system that uses a base of 8. Only the digits 0 through 7 are used.

ODD PARITY: The condition that occurs when the sum of 1s and 0s in a binary word is an odd number. Parity is used for error-checking.

OFF-DELAY TIMER: 1. In relay panel application, a device in which the timing period is initiated upon de-energization of its coil. 2. In a PLC, an instruction that starts the delay whenever the timer rung goes false.

OFF-LINE PROGRAMMING: A method of programming that is done while the processor is not communicating with the outputs.

OFFSET: To offset is to equalize or compensate. In the case of linear interpolation formulas, the amount to be added +/− to the final calculated value.

ON-DELAY TIMER: 1. In relay panel applications, a device in which the timing period is initiated upon energization of its coil. 2. In a PLC, an instruction that starts the delay whenever the timer rung goes true.

ON-LINE OPERATION: Operations in which the programmable logic controller is directly controlling the machine or process.

ON-LINE PROGRAMMING: A method of programming by which rungs in the program may be inserted, changed, or deleted while the processor is running and controlling the process equipment.

OPERAND: 1. Either of the two numbers used in a basic computation to produce an answer. For example, in the computation $2 \times 3 = 6$, 2 and 3 are the operands. 2. Data required for the operation of a special function.

OPTICAL FIBER: A technology that uses glass (or plastic) threads (fibers) to transmit data. A fiber optic cable consists of a bundle of glass threads, each of which is capable of transmitting messages modulated onto light waves.

OPTICALLY COUPLED/OPTICAL ISOLATION: The use of a light emitting diode and a photo transistor to communicate a signal or state to the processor. Optical coupling is used in input and output modules to isolate the logic level signal for line voltage sources.

OR LOGIC: Logic that has a "parallel" relationship. When two devices are in parallel, if either device one OR device two is true, the logic passes.

OUTPUT CIRCUIT: An output module point, real-world device (e.g., motor starter, digital readout, solenoid, etc.), and its associated wiring. The output module's function is to convert processor signal levels to field voltage levels necessary to control the real-world devices.

OUTPUT DEVICES: Devices such as solenoids, motor starters, etc., that receive data (control) from the programmable logic controller.

OUTPUT SIGNAL: A signal provided by the processor to the real-world output devices that controls their status (*ON* or *OFF*).

OVERLOAD: A load greater than that which a device is designed to handle.

PACKET: A piece of a message transmitted over a packet type network. One of the key features of a packet is that it contains the destination address in addition to the data.

PARALLEL COMMUNICATIONS: A type of communication or information transfer whereby a group of digits (bytes) is transmitted simultaneously. This is different from serial communications in which the data bits are transmitted one at a time—sequentially—in a string.

PARENT: A name given to a directory that has subdirectories.

PARITY: A method of testing the accuracy of binary numbers used in recorded, transmitted, or received data.

PARITY BIT: An additional bit added to a binary word to make the sum of the number of 1s in a word always even or odd.

PARITY CHECK: A check that tests whether the number of 1s in an array of binary digits is odd or even.

PASSWORD: A word used to gain access to a program or process. A password serves the same function as a hardware key, except that it is not a piece of hardware, but rather a word that when entered at the keyboard, gains access for the user to the program.

PC: Abbreviation for **P**ersonal **C**omputer (also used as an abbreviation for **P**rogrammable **C**ontroller). To avoid the confusion caused by using the same abbreviations for two different types of systems, the programmable logic controller is now most often referred to as a **PLC.**

PEER-to-PEER: A type of network in which each station has equivalent capabilities and responsibilities. This differs from client/server or master/slave architectures, in which some stations are dedicated to serving the others. Peer-to-peer networks are often called token passing networks. Peer-to-peer networks are generally simpler, often found in PLC networks, but lack performance under heavy loads.

PILOT DEVICE: A device used in a circuit that performs a control function only. Pilot devices are limited to 10 amps of current-carrying capacity, and are to be used in control circuits only. They are not designed to control the power and current required by the operating equipment.

PLC: See **PROGRAMMABLE LOGIC CONTROLLER.**

POWER SUPPLY: Supplies the DC power for the CPU and for the I/O section. The voltage is typically +5 volts. The power supply can be internal with the processor, rack-mounted, or externally mounted as a separate unit. A separate power supply is required if DC voltage is required for the actual input and/or output devices.

PRIORITY: Order of importance.

PROCESS VARIABLE (PV): Refers to the physical measurement of a process component such as temperature, level, pressure, flow, speed, etc. The physical measurement is converted into an electrical or pneumatic signal that varies proportionally with a change in the measured quantity. Devices such as transducers and transmitters are used to convert the physical measurement into a corresponding electrical or pneumatic signal.

PROCESSOR/PROCESSOR UNIT: The part of the programmable logic controller that performs logic-solving, program storage, and special functions within a PLC system. It scans all the inputs and outputs in a predetermined order. The processor monitors the status of the inputs and outputs in response to the user-programmed instructions, and energizes or de-energizes outputs as a result of the logical comparisons made through these instructions.

PROGRAM: A sequence of instructions executed by the processor to control a machine or process. Also, a set of related routines and tags found in the Logix5000 controllers.

PROGRAMMABLE LOGIC CONTROLLER (PLC): A solid-state control system that has a user-programmable memory for storage of instructions to implement specific functions such as I/O control logic, timing, counting, arithmetic, and data manipulation. A PLC consists of the processor, input/output interface, memory, and programming device that typically uses relay-equivalent symbols. The PLC is purposely designed as an industrial control system to perform functions equivalent to a relay panel or a wired solid-state logic control system.

PROGRAMMER: A device that is needed to enter, modify, and troubleshoot the PLC program, and check the condition of the processor. The programmer may be hand-held, dedicated desktop-type, or a personal computer.

PROGRAM PANEL (PROGRAMMER): A device for inserting, monitoring, and editing a program in a programmable logic controller.

PROGRAM SCAN TIME: The time required for the processor to execute all instructions in the program one time.

PROM: Acronym for **P**rogrammable **R**ead **O**nly **M**emory. A type of read only memory that requires an electrical operation to generate the desired bit or word pattern. In use, bits or words can be accessed on demand, but cannot be changed.

PROPORTIONAL: A comparative relation between magnitudes as to size, quantity, number, etc. A proportional amount is a large or small proportion of a total amount or part. In the case of a process controller, it is the change in the output by an amount proportional to the error.

PROTECTED MEMORY: Storage (memory) locations reserved for special purposes or use by the processor into which data cannot be entered directly by the user.

PROTOCOL: A defined means of establishing criteria for receiving and transmitting data through communication channels.

RACK: A PLC chassis that contains modules. Some PLC manufacturers, like Allen-Bradley, use the term "rack" to indicate a given number of I/O points rather than to identify a specific piece of hardware. In the Allen-Bradley scheme, a chassis could contain a number of racks. With most other manufacturers, however, rack and chassis are used interchangeably, and mean the hardware that holds the various modules, power supplies, etc.

RADIO AREA NETWORK (RAN): A communication network that uses radio waves as the medium that interconnects the computers or PLCs to form a network. Most RANs are used in the control field to interconnect PLCs where no other means is possible or practical. RANs can quite often be found in the water and wastewater industries.

RAM: Acronym for **R**andom **A**ccess **M**emory. Random access memory is a type that can be read from (accessed) or written into by the user.

RANDOM ACCESS: See **RAM.**

RATED VOLTAGE: The maximum voltage at which an electrical component can operate for extended periods without damage or undue degradation.

READ/WRITE MEMORY: A memory into which data can be placed (*write* mode) or accessed (*read* mode). The *write* mode destroys previous data; the *read* mode does not alter stored data.

REAL: A data type that stores a 32-bit IEEE floating-point value.

REGISTER: A word or group of words used to store numerical values.

REPORT: A display of data, or a printout, containing data and/or information that is useful to the user or operator. Reports can include operator messages, part records, production lists, etc. Reports are normally stored in a memory area separate from the user's program.

REPORT GENERATION: The printing or displaying of user-formatted application data by means of a programming device. Report generation can be initiated by means of either the user's program or a programming device keyboard.

RETENTIVE: To retain a value or time.

RETENTIVE OUTPUT: An output that remains in its last state (*ON* or *OFF*), depending on which of its two program rungs (one containing a **LATCH INSTRUCTION,** the other an **UNLATCH INSTRUCTION**) was the last to be true. The retentive output remains in its last state when both rungs are false. It also remains in its last state if power is removed from, then restored to, the PLC.

RETENTIVE TIMER: A PLC instruction that accumulates the amount of time, whether continuous or not, when the preconditions of its rung are true, and which controls one or more outputs after the total accumulated time is equal to the preset time. When the rung is false, the accumulated time is retained. Moreover, if the outputs have been energized, they remain *ON*. Additionally, the accumulated time and energized outputs are retained if power is removed from, then restored to, the PLC.

RING TOPOLOGY: All devices are connected in the shape of a closed loop, so that each device is connected directly to two other devices, one on either side of it. Ring topologies are relatively expensive and difficult to install, but they offer high bandwidth and can span large distances.

ROM: Acronym for **R**ead **O**nly **M**emory. A read only memory is a solid-state digital storage memory whose contents cannot be altered by the user.

ROUTINE: A set of logic instructions in a single programming language in a Logix5000 controller and similar to a program file.

RS-232C: An Electronic Industries Association (EIA) standard for data transfer and communication.

RTD: Resistance Temperature Detector, or temperature probe that can be connected to a special PLC input module to indicate temperature reading.

RUNG: A grouping of PLC instructions that control one output or storage bit. Some PLCs can have multiple outputs on the same rung. A rung is also referred to as a network.

SCAN: The time required for a program to make one complete scan through memory and update the status of all inputs and outputs.

SCHEMATIC: A diagram of a circuit in which symbols illustrate circuit components.

SCR: An acronym for **S**ilicon **C**ontrolled **R**ectifier. The SCR is used to convert AC current to DC current.

SEQUENCER: A controller that operates an application through a fixed sequence of events.

SERIAL COMMUNICATION: A type of communication or information transfer within a programmable logic controller whereby the bits are handled sequentially rather than simultaneously as they are in parallel communications. Serial operation is slower than parallel operation for equivalent clock rate. However, only one channel is required for serial operation.

SETPOINT (SP): The desired operating point that the process system is to operate at. The setpoint value typically has the same range as the process variable being measured and is determined by the operator of the system.

SHIELDING: The practice of confining the electrical field around a conductor to the primary insulation of the cable by putting a conducting layer around the cable insulation.

SIGNED BIT: A way to indicate that a number is negative or positive.

SINGLE PRECISION ARITHMETIC: Only one word is used to store the results of math or arithmetic operations. Single precision arithmetic is limited to a maximum value of 999. (See **DOUBLE PRECISION ARITHMETIC.**)

SINT: A data type that stores an 8-bit signed integer value.

SLOPE (RATE): An inclined or slanting direction downward or upward from the horizontal, as in the case of an *xy* graph. It is the degree of the incline or amount of change.

SOFTWARE: The manufacturer's program that controls the operation of a programmable logic controller.

SOLID-STATE: Circuitry designed using only integrated circuits, transistor, diodes, etc.; no electromechanical devices such as relays are utilized. High reliability is obtained with solid-state logic.

SOLID-STATE DEVICES (SEMICONDUCTORS): Electronic components that control electron flow through solid materials (e.g., transistors, diodes, integrated circuits, etc.).

STAR TOPOLOGY: All devices are connected to a central hub. Star networks are relatively easy to install and manage, but bottlenecks can occur because all data must pass through the hub. Today switches are used in place of hubs, all but eliminating bottlenecks.

STATE: The logic condition, 1 or 0, in PLC memory, or at a circuit's input or output.

STEP: A step represents a major function of your process and contains actions. Steps can be found in sequential function chart programming.

STORAGE: Synonymous with **MEMORY.**

STORAGE MEMORY: That part of the memory that stores the status of the input and output devices, numeric values for timers and counters, numeric values for arithmetic functions, status of internal relays, and information stored in holding and storage registers.

SURGE: A transient variation in the current and/or voltage at a point in the circuit.

SWITCHING: The action of turning a device *ON* and *OFF.*

SYMBOLIC NAME: A user designation for an application I/O device (e.g., S-1, LS-4, or SOL-7).

SYNCHRONOUS SHIFT REGISTER: An instruction that shifts information one bit at a time within a word or from one word to another.

SYSTEM PROMPT: The system prompt indicates the current drive for the computer. If the prompt is C:\, then the current drive is the C drive.

TAG: A named area of memory where data is stored in the controller. Tags are the basic mechanism for allocating memory in the Logic5000 controllers.

TASK: A scheduling mechanism for executing a program in the Logix5000 controllers.

THUMBWHEEL SWITCH: A rotating numeric switch used to input numeric information to a controller.

TIMER: In relay-panel hardware, an electromechanical device that can be wired and preset to control the operating interval of other devices. In a PLC, a timer is internal to the **PROCESSOR,** meaning that it does not exist in the real world, but can be controlled by a user-programmed instruction. A timer instruction has greater accuracy and timing range than a hardware timer.

TOGGLE SWITCH: A panel-mounted switch normally used for *ON* or *OFF* switching.

TOPOLOGY: The shape of a local area network (LAN) or other communications system. Topologies are either physical or logical. There are four principal topologies used in LANs: bus, star, ring, and tree.

TRANSFORMER COUPLING: One method of isolating I/O devices from the controller.

TRANSITION: A transition is the physical conditions that must occur or change in order to go to the next step in a sequential function chart programming.

TREE: A command that is used to display the organization of all the directories, subdirectories, and files on a given disk or hard drive.

TREE TOPOLOGY: A tree topology combines characteristics of linear bus and star topologies. It consists of groups of star-configured workstations connected to a linear bus backbone cable.

TRIAC: A solid-state component capable of switching alternating current.

TRUE: As related to a PLC instruction, an enabling logic state or *ON* condition. (See **FALSE.**)

TRUTH TABLE: A matrix that shows all the possible states (*ON* or *OFF*) of a single input device or combination of input devices, and the corresponding state (*ON* or *OFF*) of the output device(s).

TTL: Abbreviation for **T**ransistor–**T**ransistor **L**ogic, a family of integrated circuit logic. (Usually 5 volts is high, or 1, and 0 volts is low, or 0.)

TUTORIALS: Text included in software that provides the user with helpful information on how the software works or how to use functions of the software. Typically can be accessed by using the "Help" option.

TWISTED-PAIR CABLE: A type of cable that consists of two independently insulated wires twisted around one another. The use of two wires twisted together helps to reduce crosstalk and electromagnetic induction. Twisted-pair cable is the least expensive type of local area network (LAN) cable. Most computer and some PLC networks contain twisted-pair cabling at some point along the network.

TWOS COMPLEMENT: A convention for binary representation of negative and positive decimal numbers.

UNCONDITIONAL: A term applied to an output (or other instruction) that is always true.

UNLATCH INSTRUCTION: A PLC instruction that causes an output to unlatch, or turn *OFF*, regardless of how briefly the instruction is enabled. (It can only be turned back *ON* by a **LATCH INSTRUCTION** in a separate rung.)

UPDATING: A term used to indicate that the processor has scanned and checked the status of all input and output devices. After the status of the inputs and outputs is known, the data table is updated to reflect the current status.

UPWARD COMPATIBILITY: The ability of a new version of software to support previous editions of the same software. If version 6.6 of a particular software allows documents and files from previous versions (5.2 and 6.0) to be read, the software is said to have upward compatibility.

USER MEMORY: The portion of memory that is set aside for the storage of the user program (i.e., ladder diagrams, program messages, etc.).

UV ERASABLE PROM: An erasable programmable read only memory that can be erased or cleared (set to 0) by exposure to intense ultraviolet light. After being cleared, it may be reprogrammed.

VALUE: 1. A number that represents a computed or assigned quantity. 2. A number contained in a register or file word.

VOLATILE MEMORY: A memory that loses its information if the power is removed.

WATCHDOG TIMER: A timer that is used within the PLC processor to verify that the program scan has been completed correctly in the allotted amount of time. If there is a program error, the scan is not completed in the prescribed amount of time, and the watchdog timer times out and indicates that there is a problem with the circuit.

WIDE AREA NETWORK (WAN): A computer network that spans a relatively large geographical area. Typically, a WAN consists of two or more local area networks (LANs). Computers connected to a wide area network are often connected through public networks, such as the telephone system. They can also be connected through leased lines or satellites. The largest WAN in existence is the Internet.

WORD: A grouping, or a number of bits, in a sequence that is treated as a unit.

WRITING OVER: A term used to indicate that information will be replaced (or the existing information will be written over) by new information.

XOR LOGIC: The XOR logic gate will only turn the output *ON* when either input A or B is *ON*—but not both *ON*.

INDEX

D

R

S